Study Guide

to accompany

Life

Sixth Edition

The Science of Biology

Study Guide

to accompany

Life

Sixth Edition

The Science of Biology

Purves, Sadava, Orians, and Heller

Prepared by

V. Christine Minor
Clemson University

Edward M. Dzialowski and **Warren W. Burggren**
University of North Texas

Lindsay Goodloe
Cornell University

Nancy Guild
University of Colorado at Boulder

Sinauer Associates, Inc.

W. H. Freeman and Company

The Cover
Giraffes *(Giraffa camelopardalis)* near Samburu, Kenya.
Photograph © BIOS/Peter Arnold, Inc.

Study Guide to accompany *Life: The Science of Biology,* **Sixth Edition**

Address editorial correspondence to:
Sinauer Associates, Inc., P.O. Box 407
23 Plumtree Road, Sunderland, MA, 01375 U.S.A.
Fax: 413-549-1118
Internet: www.sinauer.com; publish@sinauer.com

Address orders to:
W. H. Freeman and Company
VHPS/W. H. Freeman & Co. Order Department
16365 James Madison Highway,
U. S. Route 15, Gordonsville, VA 22942 U.S.A.
www.whfreeman.com

ISBN 0-7167-3951-8
Printed in U.S.A.

4 3 2

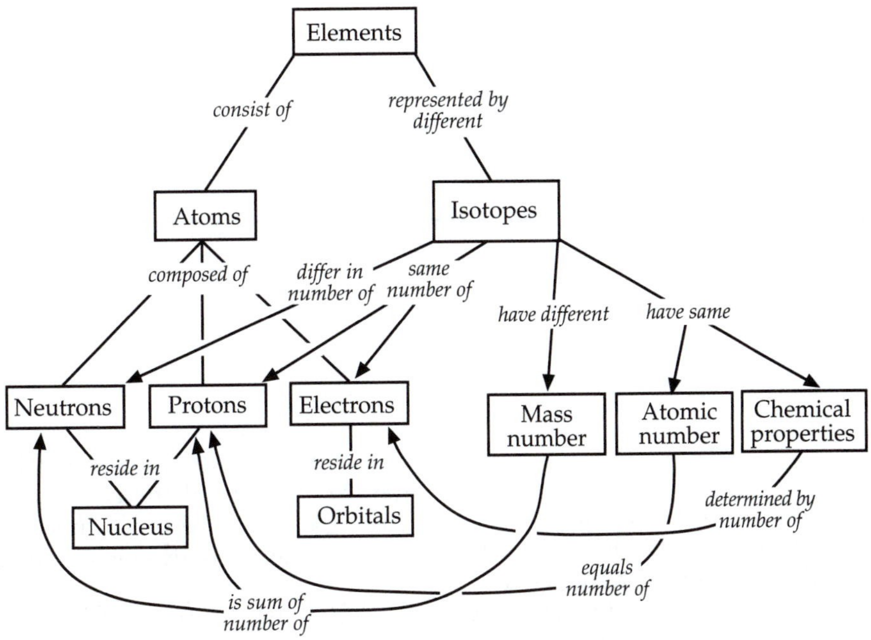

not provided in the text, try to draw them yourself. This is especially important when attempting to understand structures and pathways. You will find that the study guide also refers you to specific tutorials and activities on the website. Each of these visual tools will aid in your understanding of the material.

Once you have read the material, summarized it in your own words, and reviewed all of the figures and tables, it is time to do another quick review. Scan the summaries in your text again and read over the key concepts in this study guide. We recommend further that you construct a "mind map" of the main points of the chapter. This will assure you that you see how concepts are interrelated and that you understand the organization of the material. An example of a "mind map" is above. Main concepts are in boxes, and supporting concepts are listed below the illustration. Arrows are used to connect concepts and show relationships.

Finally, when you are comfortable with chapter content, move to the questions section of this guide. We recommend working through all of the questions before checking your answers. Any questions that you do not understand or that you answer incorrectly should be flagged as material you need to review. Remember that your goal is to understand the material completely, not just to answer these particular questions.

Experiments have suggested that the average attention span of a reader is approximately 12 minutes. You cannot expect to sit down and master an entire chapter in one sitting. Break up your study into short segments of approximately 30 minutes each. This will give you time to get down to business, maximize your attention, and learn without becoming frustrated or drained. At the end of this time move to some other activity or area of study. When you return to biology you will find your mind is ready to absorb more.

This entire process should take place well in advance of your exams. You cannot learn even one chapter adequately the day before a test. Mastering biology requires daily study and review. If you keep up with learning the concepts as your course proceeds, you will find that reviewing the text, study guide summaries and questions is adequate preparation. The lesson for learning biology is slow and diligent, daily work.

Doing Well on Biology Exams

Your instructors are your best resource for doing well on exams. They are there to assist you in learning the material that is presented in your textbook. They are experts in their fields and understand how the concepts presented are vital to your biology education. Follow their lead in preparing for tests.

Preface

To the Student

Biology is an incredibly exciting field of study, but in order to appreciate new discoveries and discussions, it is necessary to have a firm grasp of underlying concepts and ideas. Your textbook is designed to give you a comprehensive overview of important biological phenomena. It will also serve as a resource to you in future studies. Together with your instructor, your textbook will provide you with important information for beginning your study of biology.

This study guide is designed to supplement, not to replace, your text and your instructor. It was written for you, the student, in language that you can understand, but it does emphasize proper usage of biological terminology. Important concepts and ideas have been synthesized into short, easy-to-read segments that provide an overview of the biological phenomena discussed in your text. This summary should serve as a preview tool, an understanding check, and a review tool. Each chapter summary is followed by a series of questions. These questions have been grouped into two categories. *Knowledge and Synthesis Questions* are designed to determine if you have retained information from the chapter, and if you can put together various concepts in order to answer questions. *Application Questions* ask you to apply the knowledge you gleaned from the chapter to answer questions that are more open ended. These questions require you to have assimilated several concepts and to think a little beyond what you have just read. Your instructor may ask questions similar to these on your tests or may use an entirely different approach to questioning your knowledge, but these questions will be a good check of how well you understand the material. Brief answers to the questions are provided at the end of each chapter. These answers are not exhaustive but are designed, rather, to point to the correct concepts in your book and in the summaries. Because of the nature of many of the *Application Questions*, your answers should be much more expansive than the short explanations given in the study guide.

Strategies for Studying Biology

Each individual has his or her own unique study pattern. However, the authors believe that there are some successful strategies that are universal. We recommend that you first preview a chapter in your text. In this initial preview it is important to note the organization of the chapter and the main points, and to go over the chapter summary at the end. By referring to the study guide at this point you will further understand the organization of the material. If you follow these steps, you might have some specific questions in your mind that you will expect your reading to answer.

Reading a text is an active process. We recommend that you always have a pencil and paper available. Jotting notes in margins also serves as an excellent mental trigger when it comes time to review. As you read each section of your text, see if you can summarize it in your own words. Compare your summaries to those provided at the end of the chapter and in your study guide. You want to assure yourself that the main points you are noting match those that the author has selected. Refer back to those questions that came up as you previewed the chapter. You should be able to answer your own questions by the time you have finished your reading.

As you are reading, take time to review all figures and tables. Your text makes use of figures to illustrate points, describe pathways, and give visual representation to complex topics. Often these figures are more helpful than the paragraphs of written explanation. In places where figures might be beneficial to you but are

Test performance is directly linked to classroom attendance and daily study. Sit front and center in your lecture room and pay close attention to what your instructor writes or provides in overhead displays or in other presentations. Your instructor will give you guidance about the most important points to study. You should take careful notes and ask for clarification when necessary.

You should always compare your classroom and lecture notes to your text and the study guide. Mark in the books those points your instructor emphasizes. Were there specific questions that the instructor brought up in class? Do these correspond to the discussion questions at the end of your chapter or in the study guide? If so, chances are you will see them again on the test. This practice should be part of your daily study regimen.

Approach the test itself with confidence. Read the directions carefully and read the questions completely. One of the biggest mistakes students make on exams has to do with not following instructions. We recommend that you read carefully through the entire test even before you begin to answer the first question. You may find that early questions are answered partially in later ones, and that an overview of the whole exam helps to trigger your memory of the material.

If your test contains multiple choice questions, treat each of the answer options as if it were a separate true/false question. In other words, mentally fill in the question with each answer and ask yourself if the resulting statement is true or false. Be sure to read each of the options, even if the first one strikes you as the correct one. You may be dealing with a question that has multiple correct answers or asks you to select "all of the above" or "none of the above." If you are unsure about how to answer a question, begin by ruling out answers that you know are incorrect. By eliminating some possibilities, your chance of selecting the right one is greatly improved! If you narrow the choices down to two possible answers and just can't decide between them, go with your gut response. You just may be right. And when you go over your answers, change your response only if you know you answered a question incorrectly. If you aren't sure, it may be best to leave the answer as is. Your first instincts are often correct.

Essay questions require you not only to think through the concepts, but also to decide how you will present them. Organized, concise essay answers are always preferred to rambling answers with little direction. Take a look at the point value of each question. This is often an indication of how many "points" you need to make in your discussion. A 10-point question rarely can be answered with a single sentence. Look carefully at what your instructor is asking. If you are required to "discuss"a concept or a problem, do not present a list or some scattered phrases. Write in complete sentences and paragraphs unless you are asked specifically to list or itemize the material. Be sure that you address all points in the question. Essay questions frequently have multiple parts to test your understanding of the links between various concepts.

Always be aware of the time you have to complete your test. Answer the questions you know swiftly, and allow yourself time to go back and really think about those you are struggling with. Be sure to concentrate on the questions with the highest point value. Those are the questions testing the most critical concepts and they are the ones that will have the most influence on your grade.

Once you think you have completed the test, take a few minutes to go back over it. If there are any questions about which you are still unclear, be sure to ask your instructor for clarification.

Learn from Your Mistakes

After your instructor returns your exam or posts the answers, review the mistakes you have made. Take this opportunity to clarify misunderstandings and relearn material when necessary. You will find this helpful not only for the final exam, but also as you approach upper level courses. Material reviewed, corrected, and relearned is more likely to be retained in the long run.

Get Help before It Is Too Late

Many students come to college without adequate study habits and/or are ill prepared for the rigor of biology at the college level. But this is a problem that can be solved fairly easily. Many colleges and universities have academic centers that assist with locating tutors, teaching study habits, and acquainting you with a variety of other resources. Make good use of these facilities. They are there for you.

Acknowledgments

We would like to thank our colleagues and our home institutions for support, assistance, and the use of equipment necessary to produce this study guide. We also would like to thank our families for enduring the pitter-patter of keyboards well into the night and for allowing us the time to produce this work. The questions found in this book come from our own teaching experiences, those of our colleagues, and from the authors of previous editions of this text. We thank the authors of *Life: The Science of Biology* for a well-written and wonderfully organized text on which to base this study guide. We have done our very best to be factual and accurate in our summaries and in the questions we have written. Any mistakes that might be found here should be brought to our attention as soon as possible for correction. We would like to thank the staff at Sinauer Associates for their assistance during writing, editing, and production. Our heartfelt thanks go out especially to Kathaleen Emerson and Chelsea Holabird for their editorial assistance and for their ability to pull this project together in spite of ornery authors with lots of other obligations!

Please let us know what you think of the study guide, and feel free to contact the author and publisher with corrections, questions, and clarifications.

V. Christine Minor
Clemson University

Edward M. Dzialowski and Warren W. Burggren
University of North Texas

Lindsay Goodloe
Cornell University

Nancy Guild
University of Colorado at Boulder

Contents

PART FOUR The Evolution of Diversity

PART FIVE The Biology of Flowering Plants

PART SIX The Biology of Animals

PART SEVEN Ecology and Biogeography

1 An Evolutionary Framework for Biology

Important Concepts

The fact of evolution has changed how biology is approached.

- An understanding of the age of the earth through fossil evidence and radioactive dating allows scientists to conceptualize change over time.
- Buffon was one of the first to claim that descent with modification takes place. This early statement of evolution attempted to explain similarities in mammal limbs that remained even when the resulting structure was no longer functional.
- Lamark, a student of Buffon, postulated a theory of acquired characteristics. Though his theory eventually was discredited, it was a shift in thinking to the possibility of change through time.
- Darwin and Wallace independently proposed the theory of evolution through natural selection. This theory is based on two facts and a resulting inference:
  Fact: Reproductive rates must be balanced by mortality rates.
  Fact: Organisms are variable, but offspring inherit features from their parents.
  Inference: Differences among individuals influence how well they survive and reproduce. Those traits that help an individual to survive and reproduce are maintained. This process is called natural selection.
- Evolutionary change takes place over long periods of time and without any goals.

Particular evolutionary results had to occur before others.

- Life arose approximately 4 billion years ago.
- The aggregation of complex chemicals led to the existence of life.
- The first known cells are approximately 3.8 billion years old. A membrane-enclosed cell allows maintenance of an internal environment. Cells can capture and use energy (metabolize) and replicate. Early cells were simple prokaryotes. Over time, increasingly complex cells with diverse patterns of metabolism developed.
- Approximately 2.5 billion years ago photosynthetic organisms arose. These cells could take light energy from the sun and convert it to chemical energy with oxygen as a byproduct. Increasing numbers of photosynthetic organisms increased atmospheric oxygen levels, allowing for the evolution of aerobic metabolism. Photosynthetic organisms also contributed to the insulating ozone layer that shields the earth from radiation and modifies temperature. This shield eventually allowed for life on land.
- Sexual reproduction brings about genetic recombination and variation. Variation allows adaptation to environmental change. Organisms inherit characteristics that increase survival and reproductive success in specific environments.
- Eukaryotic cells appeared approximately 1.5 billion years ago. These cells arose when larger prokaryotic cells engulfed and assimilated smaller cells, resulting in membrane-enclosed compartments (organelles) that carry out specific cellular functions.
- Multicellularity allows cells to specialize and organism size to increase. The development of resting spores and the ability of cells to clump was the first step toward multicellularity. Specialized sex cells led to new patterns of cell division and increased possibilities for genetic variation.
- With increased complexity of organisms, the maintenance of a constant internal environment (homeostasis) became increasingly demanding, leading to the evolution of new homeostatic mechanisms.
- Multicellular organisms require controlled growth and development. This development is controlled genetically, and small alterations can produce very large effects.

Speciation leads to diversity.

- Species are genetically independent groups of organisms.
- Anything that results in the isolation of two groups of organisms such that they can no longer mate and produce viable offspring may ultimately cause speciation.
- Ancestral organisms are neither "primitive" nor lower than "derived" organisms. All living organisms are successfully adapted to their environment.

We may study biology as a size hierarchy or an evolutionary hierarchy.

- A size hierarchy often goes from smallest to largest: atom, molecule, cell, tissue, organ, organism, population, community, ecosystem, biosphere.
- Each level brings emergent properties not seen at preceding levels. These properties result from the interaction of

constituent parts of the system and from collective properties of the aggregation that the individual units lack.

- Evolutionary hierarchies result from groupings intended to show evolutionary relationships, including evolutionary "distance" of an organism from a common ancestor. This classification scheme is based on fossil evidence, physical structure, and DNA.
- Figure 1.10 in your book shows the three domains and six resulting kingdoms.
- Organisms are referred to by their genus and species names (e.g., *Homo sapiens*).

Scientists ask specific types of questions and reason in a methodical way.

- All scientific inquiry begins with particular observations that lead the observer to question why things are as they appear. Questioning leads to the forming of hypotheses.
- What sets science apart from other methods of inquiry is that those tentative explanations are testable and potentially falsifiable. From a hypothesis, predictions can be made and tested and conclusions can be drawn. This process is called the hypothetico-deductive (H-D) approach. (Review the amphipod example in your book as it illustrates each step in the H-D approach.)
- A hypothesis that withstands the test of time without being falsified is called a theory.
- Scientific experiments are controlled. This means that the variables of interest are isolated and compared, but other variables that may influence the outcome are held as steady as possible.
- For scientific results to be accepted they have to hold up to many rounds of testing. Scientific "fact" is based on many forms of tested evidence.

Biological science has an impact on every human being and is the basis for many public policy decisions.

- The environmental stresses of population growth, the emergence of agricultural and medical technologies, and other modern challenges have increased the need for everyone to understand biological information.

The Big Picture

- Evolution provides us with a context for understanding relationships among organisms. Dating techniques, fossils, and DNA evidence point to an origin of life some 4 billion years ago. Heritability of characteristics from generation to generation, mortality rates that counter reproductive rates, and natural selection have led to the current host of organisms inhabiting the planet. There are specific characteristics of life, and the evolutionary roots of these characteristics can be traced through time.
- Biology is studied either as a hierarchy of size, starting with atoms and their interactions and proceeding up to the biosphere as a whole, or as an evolutionary hierarchy. When studying organisms from an evolutionary perspective, scientists are concerned with evolutionary relatedness and common ancestors.
- Science follows a specific hypothetico-deductive approach. The necessity of a testable and falsifiable hypothesis sets science apart from other methods of inquiry. Control of variables and repeated testing lends credibility to the method.

Common Problem Areas

- Many students have trouble remaining open-minded enough to begin to understand evolutionary biology. Neither your instructors nor this book are asking you to set aside your beliefs. We are asking you to learn science. We encourage you to set aside prejudices and begin to understand the evidence supporting evolution.
- Refrain from learning the hypothetico-deductive approach merely as a series of steps. Think about how each step in the process follows from the preceding step. There are two characteristics of science that set it apart from other modes of inquiry. One is the setting of hypotheses. The other is the continual process by which it is carried out. With each conclusion leading to a new set of hypotheses, science has the capacity to alter our overall base of knowledge.
- You most likely have studied a 5-kingdom system of classification to this point. The domain structure and 6 kingdoms will be new to you. Be careful not to assume you know this!

Study Strategies

- Many students find that understanding the hypothetico-deductive approach makes understanding some of the evidence behind evolution easier. If you find you are having trouble, work on that section first.
- Drawing a time line or examining the calendar model of Figure 1.1 will help you understand the chronology of the major evolutionary events discussed in your book.
- Review the following activity on the website:
  Activity 1.1 The Major Groups of Organisms

Test Yourself

Knowledge and Synthesis Questions

1. Life arose on earth approximately how many years ago?
   a. 3.8 billion
   b. 3.8 million
   c. 4000
   d. 1.5 billion
2. Which of the following molecules are components of all living organisms?
   a. Carbohydrates
   b. Lipids
   c. Nucleic Acids
   d. All of the above
3. Which of the following statement(s) regarding evolutionary biology are not true?
   a. Darwin and Wallace postulated that mortality rates match reproductive rates, that characteristics are heritable, and that those characteristics most suited to the environment are selected for.

b. Darwin postulated that characteristics could be acquired in a single generation.
c. Darwin was the first to come up with an evolutionary theory.
d. b and c

4. Photosynthesis was a major evolutionary milestone because
a. photosynthetic organisms contributed oxygen to the environment, which led to the evolution of aerobic organisms.
b. photosynthesis led to conditions that allowed life to arise on land.
c. photosynthesis is the only metabolic process that can convert light energy to chemical energy.
d. All of the above

5. Sexual reproduction was evolutionarily important because
a. previously all "daughter" organisms were clones of the parent organism.
b. reproduction with variation is a major characteristic of life.
c. variation is necessary for evolution.
d. All of the above

6. The evolution of multicellularity led to an increased need for maintaining a constant internal environment. This is referred to as
a. homeopathy.
b. homeostasis.
c. environmental maintenance.
d. adaptation.

7. Which of the following are necessary for speciation to occur?
a. Reproductive isolation of two groups of organisms.
b. A large amount of mutation within a population.
c. A reduction in the number of individuals in a population.
d. None of the above

8. Which of the following represents a size hierarchy from smallest to largest?
a. Atoms, cells, molecules, communities, organs
b. Atoms, molecules, cells, organs, communities
c. Molecules, atoms, cells, organs, communities
d. Atoms, cells, communities, organs, molecules

9. The deepest division of the evolutionary history of life is the
a. kingdom.
b. genus.
c. domain.
d. species.

10. Organisms are named using first their __________ and then their __________.
a. species, genus
b. genus, species
c. domain, genus
d. genus, domain

11. In the amphipod example in your book, researchers isolated amphipods and placed them in containers in the laboratory where all conditions between test groups were identical except for the variable of interest. This is an example of a __________ experiment.
a. repeated
b. controlled
c. laboratory
d. None of the above

12. For a hypothesis to be scientifically valid it must be __________ and it must be possible to __________it.
a. testable, prove
b. testable, reject
c. controlled, prove
d. controlled, reject

Application Questions

1. Discuss how the process of scientific inquiry is different from other forms of inquiry. Include in your discussion a description of the hypothetico-deductive approach.
2. Your book states that evolution proceeds without a "goal." With this idea in mind, would you say that organisms that are alive currently are more "advanced" than those that have become extinct?
3. Define adaptation in your own terms. If organisms do not acquire characteristics in a single generation, how do they "adapt" to an environment?
4. Look over a recent newspaper to see how many articles are directly related to biology. Select one article and discuss how the researchers followed or did not follow the hypothetico-deductive approach.

Answers

Knowledge and Synthesis Answers

1. **a.** There is evidence of the beginnings of life some 3.8 billion years ago.
2. **d.** All living organisms contain carbohydrates, lipids, nucleic acids, and proteins.
3. **d.** Evolutionary ideas began nearly 100 years before Darwin and Wallace published their ideas on natural selection. For these ideas to emerge it was necessary to accept that the earth was ancient, because evolution occurs slowly over extended periods of time.
4. **d.** Photosynthesis allowed the accumulation of oxygen in the atmosphere and contributed to the formation of an ozone layer. The presence of oxygen allowed the evolution of aerobic organisms, and the ozone layer shielded the earth from harmful radiation. These two phenomena contributed to the evolution of terrestrial life.
5. **d.** Variation among organisms allows for natural selection. For variation to occur, sexual reproduction, which allows for variation between parent and offspring, is necessary.
6. **b.** Homeostasis refers to consistency in an internal environment.
7. **a.** For speciation to occur, reproductive isolation, such that two groups can no longer reproduce and produce viable offspring, is necessary.
8. **b.** See Figure 1.9 in your text.

9. **c.** The knowledge of domains has come primarily from DNA evidence.
10. **b.** An example of this is *Homo sapiens.*
11. **b.** For experiments to be scientifically valid they must be controlled.
12. **b.** Scientific hypotheses are set apart from mere conjecture by being testable and falsifiable.

Application Answers

1. The process of scientific inquiry is different because a hypothesis must be testable and it must be possible to reject it. The hypothetico-deductive approach begins with observations that lead to questions. From the questions, hypotheses are formed that are probable explanations for the phenomena being observed. Predictions are formed from the hypothesis and tested. Conclusions are drawn from the tests. These conclusions may, in turn, lead to additional hypotheses.
2. Extinct organisms were adapted to the environment during their own time. Current organisms are best adapted to the environment at this time. It cannot be said that one is more advanced than the other.
3. Adaptation refers to the way a group of organisms (a population) interacts with its environment. The presence of genetic variation allows for the population to change over time, with those organisms best suited to the environmental conditions surviving and reproducing. Adaptation comes from natural selection.
4. Often an article reporting on science news does not provide enough information for the reader to know whether a particular study followed the hypothetico-deductive approach. It pays to be a smart reader and consumer in this respect.

2 Small Molecules: Structure and Behavior

Important Concepts

All matter is composed of the same chemical elements.

- Atoms, containing a defined number of positively charged protons and neutral neutrons in their nucleus, and negatively charged, moving electrons in their outer shells, combine to form all matter. The elements comprise atoms of one type.
- Six elements compose most of every living organism: carbon, hydrogen, oxygen, nitrogen, phosphorus, and sulfur.
- The number of protons in an element determines its atomic number.
- Isotopes of the same element have the same number of protons but different numbers of neutrons. Radioisotopes are unstable isotopes that give off energy. This decay eventually causes the atom to change.
- Protons and neutrons have mass and contribute to atomic weight. Atomic weights on a periodic chart are averages of all isotopes of an element.
- The behavior of electrons determines how chemical bonds will form. Electrons orbit the nucleus in a defined way. These patterns of orbit are called electron shells. A specific number of electrons can occupy a shell. The outermost shell is the chemically active shell. Stable atoms have eight electrons in the outermost shell. This phenomenon is referred to as the octet rule.
- Two or more linked atoms compose a molecule. Linking stabilizes the outer electron shell.

Chemical bonds hold molecules together.

- Atoms share pairs of electrons to stabilize their outer shells in covalent bonding. This type of bond is very stable and can be broken only with a great deal of energy. All molecules have a three-dimensional shape. The shape of a molecule effects how it behaves.
- Multiple covalent bonds may exist. Double bonds involve two pairs of shared electrons. Triple bonds share three pairs, but are rare.
- Covalent bonding between atoms of the same element involves equal sharing of electrons. Bonds between different elements generally do not share electrons equally. Unequal sharing results in partial (δ) charges in molecules because one nucleus attracts the electrons more strongly. One atom of a molecule may be partially negative while the other is partially positive. This balance of partial charges results in a "polar" molecule with a δ^- pole and δ^+ pole.
- Partial charges allow hydrogen bonding. The δ^+ of H is attracted to the δ^- of another atom. This is a weak bond that is easily broken because no electrons are shared, but this type of bond is important in stabilizing the 3-D shape of large molecules like proteins and DNA.
- Ions occur when atoms lose electrons to other atoms to stabilize their outer shells. When an atom gains or loses an electron it becomes a charged ion.
  - Cations are positively charged and have one fewer electron than protons.
  - Anions are negatively charged and have one more electron than protons.
- More than one electron can be gained or lost from an atom or molecule. A group of covalently bonded atoms may also gain or lose electrons to form complex ions.
- Ionic bonds form between ions of opposite charge. Ionic bonds are not as strong as covalent bonds. Ions can also attract the partial charges of polar molecules.
- Polar molecules are hydrophilic, or "water loving," because of the partial charges. Nonpolar molecules are hydrophobic, or "water hating." Attractions of nonpolar molecules are enhanced by van der Waals forces. Polar molecules tend to aggregate with other polar molecules and nonpolar molecules aggregate with other nonpolar molecules.

Chemical reactions occur when atoms combine and form new bonds.

- Reactions involve reactants that are altered to produce products. Matter is neither created nor destroyed, but changed. Some reactions produce energy and others require energy.
- Chemical bonds hold potential energy. When these bonds are broken, release of this energy is important to living systems.
- Oxidation–reduction reactions are important as they release energy used by living cells.

Water is an essential molecule composing 95 percent of a living organism and providing many biologically important functions.

- Water is a polar molecule and the solvent in which most biological reactions take place. Its ability to form hydrogen bonds and 3-D shape give it these properties.
- Ice floats because of the 3-D shape that results when water molecules are frozen. Ice insulates ponds and lakes in the winter, allowing life to continue beneath the frozen layer.
- Much energy is required to melt molecular ice and much energy is lost to freeze water. Thus, bodies of water help moderate temperature.
- Water has a high specific heat in the liquid state. This property also contributes to temperature moderation.
- Water has a high heat of vaporization which means much energy is required for it to go from a liquid to a gaseous state. This makes evaporating water an effective coolant.
- The polar nature of water contributes to the cohesive strength of water and its surface tension, both of which are biologically important.
- Solutions are formed when a substance is dissolved in water or another liquid.

Quantitative terms are important to understand for both biology and chemistry.

- Molecular formula = the number and types of atoms in a molecule.
- Molecular weight = sum of the atomic weights of all atoms in the molecule.
- Mole = amount of an ion or compound in grams whose weight is numerically equal to its molecular weight.
- Avogadro's number = 6.02×10^{23} molecules per mole. This number relates the number of molecules of any substance to its weight.

Acids and bases are biologically important.

- Acids dissolve in water and release H^+ ions.
- Bases dissolve in water and attach to H^+ ions.
- Acids and bases do not all ionize in the same way. Some ionize easily and completely and are called strong acids or bases. Others only partially or reversibly ionize and are called weak acids or bases.
- Water is a very weak acid because two molecules are required for ionization. Though ionization of water is uncommon, it is biologically important.
- The pH of a solution refers to the concentration of H^+ ions. Specifically, pH = $-\log10\ [H^+]$. Lower pH indicates a higher H^+ concentration or a more acidic solution. A higher pH indicates a lower H^+ concentration or a more basic solution.
- Buffers—mixtures of weak acids and their corresponding base—are used to prevent fluctuations in the pH of a solution.

Molecules have specific properties that enhance their biological activity.

- Organic molecules always contain carbon and may contain oxygen, hydrogen, nitrogen, or phosphorus.
- All molecules have a specific 3-D shape.
- Isomers are molecules that have the same atoms but are differently arranged. Thus, isomers have their own unique 3-D shape.
- Functional groups are parts of larger molecules with specific properties.

The Big Picture

- Understanding the chemical building blocks of all matter is essential to understanding the biology of organisms. Atoms, containing protons, neutrons, and electrons, form elements or combine to form molecules. Molecules may be held together with covalent or ionic bonds and stabilized in three-dimensional conformations by hydrogen bonds. Chemical reactions change reactants to products without creating or destroying matter.
- Water has unique properties that make it biologically important. An essential property is its ability to act as a solvent. How a substance ionizes in water determines if it is acidic or basic. Acids have a high concentration of H^+, whereas bases have a low concentration of H^+. The pH scale allows us to compare the concentrations of H^+ in a variety of solutions.
- The properties of molecules determine their function and how they will react with other molecules.

Common Problem Areas

- pH tends to confuse many people. Remember that it is a concentration of H^+ ions. Those that have a higher concentration are acidic and those that have less are basic. Also remember that pH is a log scale, and that a low pH value is an acid and a high pH value is a base.
- Learning quantitative terminology and how solutions are made is often difficult. Completing practice problems is the best way to learn this material.
- People are sometimes overwhelmed by the periodic table. Consult Figure 2.1 in your book to make sure you understand the components. Do not try to memorize the table.
- Use Table 2.1 to understand the differences between the various types of bonds and interactions.

Study Strategies

- The best way to learn the quantitative terms and pH material is to work on many practice problems. Also take advantage of any laboratory experience making solutions. Every good biologist makes many solutions over time.
- Use the figures in your book to help you visualize what atoms and molecules look like.
- Be careful not to get lost in terminology. Think about what each term means and focus on understanding the concept rather than memorizing the term.

- Review the following tutorial and activities on the website:
  Tutorial 2.1 Chemical Bonds
  Activity 2.1 Chemical Functional Groups
  Activity 2.2 Features of Amino Acids
  Activity 2.3 Electron Orbitals
  Activity 2.4 Chemical Bond Identification

Test Yourself

Knowledge and Synthesis Questions

1. The atomic number of an element refers to the number of
   a. protons and neutrons in an atom.
   b. protons in an atom.
   c. electrons in an atom.
   d. neutrons in an atom.

2. Which of the following statements concerning electrons is *not* correct?
   a. Electrons orbit the nucleus of an atom in defined orbitals.
   b. The outer shell of all atoms must contain eight electrons.
   c. An atom may have more than one valence shell.
   d. Electrons are negatively charged particles.

3. The element with which of the following atomic number would be most stable?
   a. 1
   b. 3
   c. 12
   d. 18

4. How can you differentiate between an element and a molecule?
   a. Molecules may be composed of different types of atoms whereas elements are always composed of only one type of atom.
   b. Molecules are composed of only one type of atom whereas elements are composed of different types of atoms.
   c. Molecules are elements.
   d. Molecules always have larger atomic weights than elements.

5. The strongest chemical bonds occur when
   a. two atoms share electrons in a covalent bond.
   b. two atoms share electrons in an ionic bond.
   c. hydrogen bonds are formed.
   d. van der Waals forces are in effect.

6. You have discovered that a molecule is hydrophilic. What else do you know about this molecule?
   a. It cannot form hydrogen bonds.
   b. It is a polar molecule.
   c. It has a partial positive region and a partial negative region.
   d. b and c

7. The stability of the three-dimensional shape of many large molecules is dependent on
   a. covalent bonds.
   b. ionic bonds.
   c. hydrogen bonds.
   d. van der Waals attractions.

8. The molecular weight of glucose is 180. If you added 180 grams of glucose to a 0.5 liter of water, what would be the molarity of the resulting solution? (See Figure 2.1 for a periodic table.)
   a. 18
   b. 1
   c. 9
   d. 2

9. Which of the following properties makes water an excellent solvent?
   a. The pH of water is 7.
   b. Water is a polar molecule.
   c. Water has a high surface tension.
   d. Water has a high heat capacity.

10. Cola has a pH of 3; blood plasma has a pH of 7. The hydrogen ion concentration of cola is __________ than the hydrogen ion concentration of blood plasma.
    a. 4 times greater
    b. 4 times lesser
    c. 400 times greater
    d. 10,000 times greater

11. If solution A has a pH of 2 and solution B has a pH of 8, which of the following statements is true?
    a. A is basic and B is acidic.
    b. A is acidic and B is basic.
    c. A is a base and B is an acid.
    d. A has a greater [OH-] than B.

12. One mole of glucose ($C_6H_{12}O_6$) weighs
    a. 180 grams.
    b. 42 atomic mass units.
    c. 96 grams.
    d. 342 grams.

13. Which of the following is not an organic molecule?
    a. $C_6H_{12}O_6$
    b. CH_3OH
    c. H_2SO_4
    d. CH_3COOH

Application Questions

1. In the following figures, place electrons in the appropriate shells based on the atomic numbers of the elements carbon ($_6C$), nitrogen ($_7N$), sodium ($_{11}Na$), chlorine ($_{17}Cl$), argon ($_{18}Ar$), and oxygen ($_8O$).

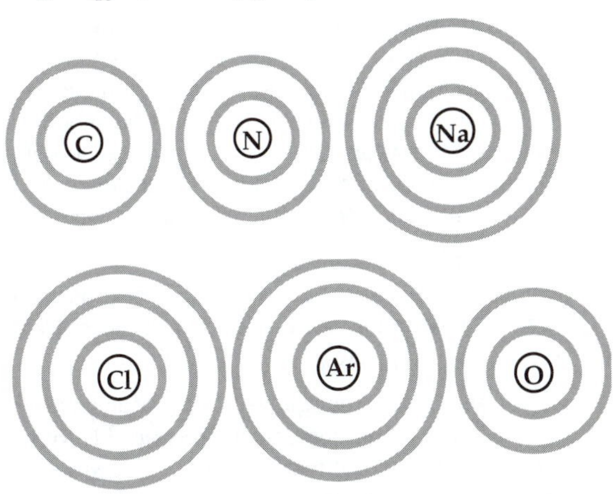

2. Calcium has an atomic number of 20. Draw structures for Ca and Ca^{2+}. What is different about these structures? Why is the Ca ion Ca^{2+} but the Li ion is Li^{+}?
3. N atoms can form triple bonds with each other. What is a triple bond? How many electrons are shared between two N atoms?
4. Water is a polar molecule. This property contributes to cohesion and surface tension. Draw six water molecules. In your drawing, indicate how hydrogen bonding between molecules contributes to cohesion and surface tension. (Be sure to include appropriate covalent bonds in each molecule.)
5. Rank the following solutions in order from the most acidic to the most basic.
   Lemon juice pH=2
   Mylanta pH=10
   Sprite pH=3
   Drain cleaner pH=15
   Seawater pH=8
   Of the above, which has the highest concentration of H^+ ions? Which has the lowest concentration of H^+ ions?
6. If you have 12 moles of a substance, how many molecules do you have of that substance? Suppose the substance has a molecular weight of 342. How many grams of that substance would you have to dissolve in a liter of water to make a 12 M solution?

Answers

Knowledge and Synthesis Answers

1. **b.** The atomic number refers to the number of protons in an atom. The atomic weight can vary with the isotope and is the total of the number of protons and neutrons in an atom.
2. **b.** The first energy shell requires only two electrons to be full; subsequent shells may hold eight or more electrons.
3. **d.** The element with atomic number 18 (Argon) is the most stable because its outermost electron shell is full.
4. **a.** An element is always composed of only one type of atom. Molecules contain many types of atoms that are bonded together.
5. **a.** Covalent bonds, resulting from the sharing of a pair (or more) of electrons, form the strongest types of bonds. Ionic bonds, where electrons are donated or received, are the second strongest. Hydrogen bonds are the weakest bonds and result from the attraction of partial charges to one another. van der Waals forces are not bonds but attractions between nonpolar molecules.
6. **d.** Hydrophilic molecules are polar and have a partial positive and a partial negative pole. These molecules are said to be hydrophilic because they easily form hydrogen bonds with water.
7. **c.** Hydrogen bonds, though weak individually, are quite effective in large numbers and are responsible for maintaining the structural integrity of many large molecules such as proteins and DNA.
8. **d.** The molarity of the solution is 2. A 1 M solution is made by adding 180 grams to 1 liter of water. Adding the same amount to half a liter would result in a solution with twice the molarity.
9. **b.** The polar nature of water makes it an excellent solvent because it can bind ions and other polar molecules. This prevents solutes from coming out of solution. The other listed properties also result from water's polar nature.
10. **d.** pH is a logarithmic scale. Each increase or decrease in pH value is by a factor of 10. Therefore, a difference between pH 3 and pH 7 would be a difference of $10 \times 10 \times 10 \times 10$ or 10,000.
11. **b.** A solution with a pH of 2 is acidic and one with a pH of 8 is basic. The lower the pH value, the higher the concentration of H^+ ions.
12. **a.** Each carbon atom weighs 12 atomic mass units, each hydrogen atom is 1 atomic mass unit, and each oxygen is 16 atomic mass units. Adding these yields a molecular weight of 180 atomic mass units. Since one mole is equal to the atomic weight in grams, one mole of glucose equals 180 grams.
13. **c.** Organic molecules must contain carbon.

Application Answers

1.

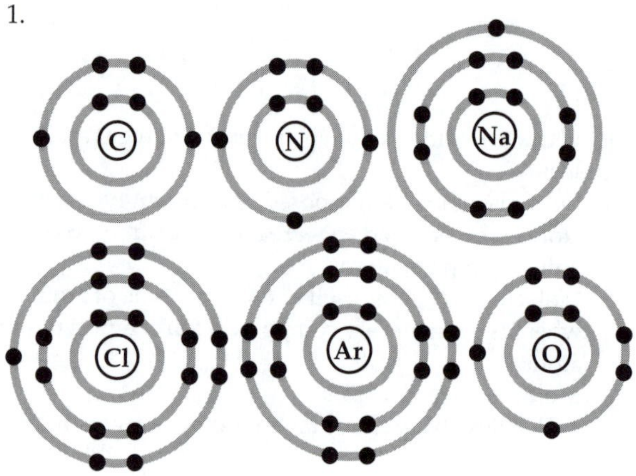

2.

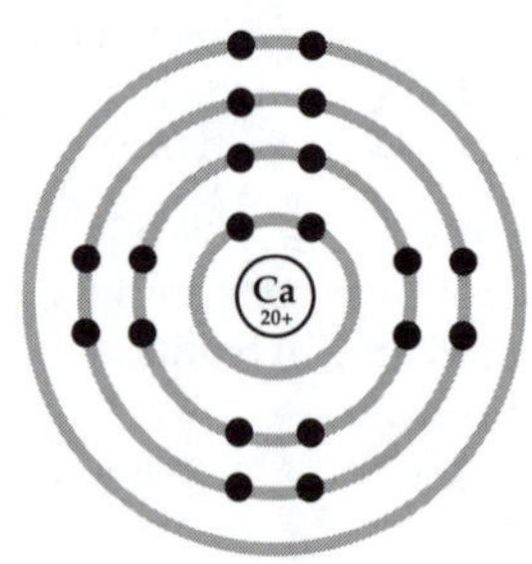

Calcium (Ca) has two electrons in its outer shell.

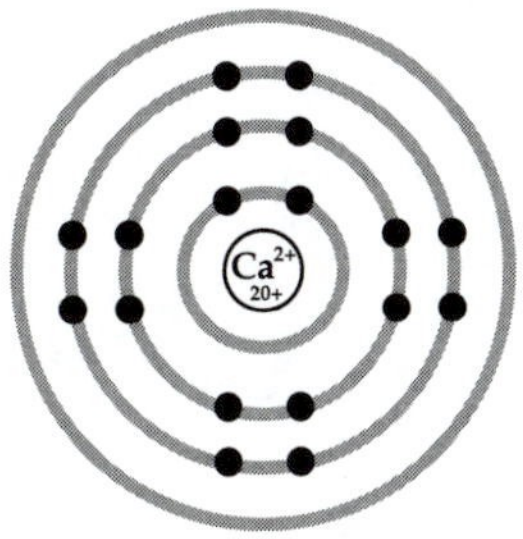

Calcium ion (Ca^{2+}) has lost its two outer electrons and therefore has a positive charge of 2^+ because it has two more protons than electrons. Lithium has only one electron in its outer shell so when lost, it gains a positive charge of only 1^+.

3. Three pairs of electrons are shared between two N atoms in a triple bond.
4. The partially positive hydrogens of one water molecule are attracted to the partially negative oxygens of another molecule of water. This attraction tends to lead to water molecules "sticking" together and creating surface tension.

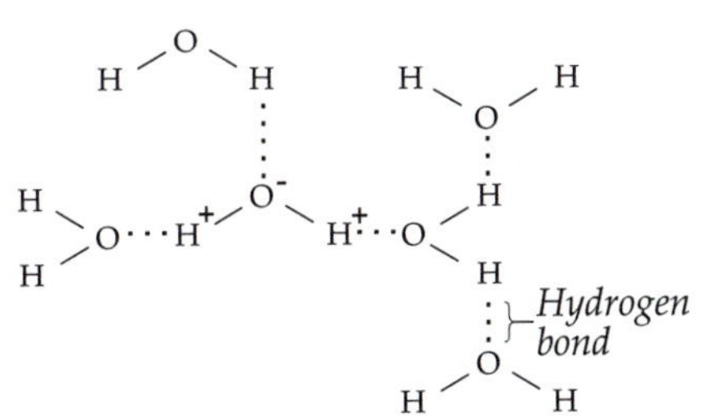

5. Lemon juice pH=2 (highest concentration of H^+ ions)
   Sprite pH=3
   Seawater pH=8
   Mylanta pH=10
   Drain cleaner pH=15 (lowest concentration of H^+ ions)
6. You would multiply Avogadro's number (6.02×10^{23}) times 12 to determine the number of molecules. You have 7.224×10^{24} molecules of this substance. A 12 M solution would be made by multiplying the formula weight of the molecule (342 for this molecule) times 12. This gives you the number of grams required to make a 12 M solution, 4104 grams.

3 Macromolecules: Their Chemistry and Biology

Important Concepts

Macromolecules are very large polymers with a wide range of biological functions.

- Polymers are made from the covalent bonding of monomers in a process called condensation. See Figure 3.2*a* in your text.
- Polymers may be broken down into their constituent monomers through hydrolysis. See Figure 3.2*b* in your text.
- Both condensation and hydrolysis reactions require an input of energy.
- The size and structure of each polymer is related to its biological function.

Polymers of amino acids are called proteins and function in structural support, catalysis, defense, and movement.

- Enzymes are specialized proteins that catalyze (speed up) chemical reactions in cells. The shape of proteins is integral to this function.
- Proteins may be made of single chains of amino acids or multiple interacting chains plus nonprotein prosthetic groups.
- Twenty amino acids are found in proteins. All amino acids have the following composition:

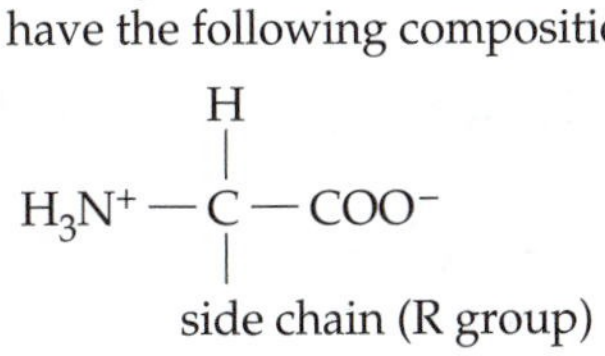

- Only the side chains vary between amino acids.
- The shape and structure of proteins are influenced by the properties of the side chains of the constituent amino acids. Five amino acids have charged side chains, four have polar side chains, and three have special functions.
- Cysteine's terminal –SH can form disulfide bridges with other cysteines and helps to determine how a protein chain folds.
- Glycine is small and fits into folds of proteins.
- Proline has limited bonding ability and often contributes to looping in proteins.
- Amino acids are chained together during condensation. The resulting bonds are called peptide linkages. See Figure 3.4 in your text.
- Peptide chains always begin with an N terminus and end with a C terminus.
- The nature of the peptide bond contributes to the form and function of proteins.

Proteins have levels of structural organization.

- Primary structure refers to the sequence of amino acids.
- Secondary structure refers to regular repeated patterns of the amino acid chain and is stabilized by hydrogen bonds.
- The a helix consists of a coiling of a single chain.
- The β pleated sheet occurs when two chains form H bonds between themselves.
- Tertiary structure refers to bending and folding of an amino acid chain. This structure is stabilized by disulfide bridges, hydrophobic interactions, van der Waals forces, and ionic interactions. R groups are responsible for this level of folding.
- Quaternary structure refers to the three-dimensional interactions of multiple protein subunits.

Protein shape influences function.

- The ability of proteins to bond noncovalently to other molecules makes them biologically important. This property allows them to function as carrier molecules, receptors, and structural elements.
- Binding sites in proteins are of a specific shape, allowing for selective and specific interactions with other molecules. The particular shape and the ability to interact are a result of the protein's 3-D shape and the properties of its constituent amino acids.
- Because a protein's 3-D structure is stabilized by relatively "weak" bonds, environmental conditions such as temperature, pH changes, or altered salt concentrations can change the shape of the protein into an inactive form. This is called denaturation.
- Chaperonins are a class of proteins that prevent denatured proteins from bonding inappropriately.

Sugar polymers called carbohydrates serve the roles of primary energy source and carbon skeleton in biological systems.

- Monosaccharides are simple sugar monomers that are used in the synthesis of complex carbohydrates. More complex arrangements include disaccharides (2 monosaccharides), oligosaccharides (3–20 monosaccharides), and polysaccharides (large, complex carbohydrates).
- Simple sugars (monomers) are biologically important as an energy source (glucose) and as a structural backbone for RNA and DNA (ribose and deoxyribose).
- Sugars may have variant amounts of carbon atoms and may exist as different isomers.
- Dehydration synthesis between monomers results in glycosidic linkages. See Figure 3.13.
- Polysaccharides are very large chains of monomers. The structure of polysaccharides contributes to their function. For instance, starch and cellulose are both chains of glucose, but the glycosidic bonds are in different orientations and yield very different chemical properties. See Figure 3.14.
- Added functional groups give carbohydrates altered functions. Sugar phosphates and amino sugars have significant biological activity.

The nucleic acids (DNA and RNA) store, transmit, and use information.

- The primary role of nucleic acids is to store and transmit hereditary information.
- Nucleic acids are made up of nucleotide monomers. Each nucleotide monomer consists of a pentose (5-C) sugar, a phosphate group, and a nitrogenous base. The difference between RNA and DNA is one oxygen atom. DNA has one fewer oxygen atom in the ribose sugar and is thus called deoxyribonucleic acid.
- They also differ in that DNA has the nitrogenous base thymine but in RNA thymine is replaced by uracil.
- The backbone of DNA consists of alternating sugars and phosphates and the bases project outwards. The sugars bond to the phosphates in phosphodiester linkages.
- RNA is a single strand of nucleotides, whereas DNA is a double strand. The two strands in DNA are stabilized by hydrogen bonds.
- Only four types of nucleotides are found in DNA: 2 purines (adenine and guanine) and 2 pyrimidines (thymine and cytosine).
- Nucleotide bases are capable of complementary base pairing, which allows for double stranded DNA, replication and production of RNA from a DNA template. Complementary base pairing is facilitated by the size and shape of the molecules and by hydrogen bonding.
- The uniformity of the DNA molecules makes it suited for carrying information. Base sequence provides the code and the regularity of the molecule allows it to be "read" in any cell.
- Similarities in base sequences of DNA can be used to help determine evolutionary relationships among organisms because it is a conserved and regular molecule.
- Nucleotides have additional biological functions. ATP and GTP serve as energy sources. CAMP serves as a messenger in many cellular processes.

Lipids are chemically diverse and provide many biological functions. They are unique in that they are water insoluble.

- Insolubility is due to many nonpolar covalent bonds.
- Lipids aggregate due to their nonpolar nature and van der Waals forces.
- Fats and oils, composed of fatty acids and glycerol, are also known as triglycerides and function primarily in the storage of energy.
- Each triglyceride contains 3 fatty acids where the carbon atoms may be single-bonded (saturated fatty acids) or are double-bonded (unsaturated fatty acids). The nature of the bonds affects the shape of the molecule and determines the melting point.
- Phospholipids compose cellular membranes. The replacement of a hydrophobic fatty acid with hydrophillic phosphate group allows phospholipids to have a "water loving" head and "water hating" tails. The hydrophobic tails tend to aggregate with the phosphate heads facing an aqueous environment. This bilayer effect allows the establishment of an inside and an outside aqueous environment separated by a membrane.
- Carotenoids and steroids are synthesized from a series of isoprene units. Carotenoids trap light energy during photosynthesis. Steroids serve as signaling molecules and are synthesized from cholesterol.
- Some vitamins are lipids that are not synthesized in the human body. Vitamins A, E, D, and K are all lipid-derived.
- The nonpolar nature of lipids makes them an excellent water repellant. Wax coatings on hair, feathers, and leaves are all lipid-derived.

Macromolecules interact and may covalently bond with one another.

- It is very common to encounter glycoproteins, lipoproteins, or glycolipids with specific functions.

The Big Picture

- Macromolecules provide for specific biological activity and provide energy storage, information storage, structural components, catalytic activity, and other functions within the cell. Biologically important macromolecules fall within four basic classes: proteins, carbohydrates, nucleic acids, and lipids. All macromolecules are synthesized from smaller molecules through condensation and can be broken down through hydrolysis.
- Macromolecules all have regular primary structures held by covalent bonds, but many, especially proteins, can be folded into complex shapes held by ionic or hydrogen bonds or attractive forces.

- Chemically active side chains or functional groups contribute to the specific functions of macromolecules.

Study Strategies

- Make a table of macromolecule characteristics. This will help you see the patterns of similarity and the differences.
- To avoid being overwhelmed, remember that each macromolecule is synthesized from smaller components that give the molecules its specific function.
- Avoid the temptation to memorize structures. If you understand condensation and hydrolysis and know the basic components of the macromolecules, memorizing the structures of the large macromolecules is unnecessary.
- Review the following tutorial and activities on the website:
  Tutorial 3.1 Macromolecules
  Activity 3.1 Forms of Glucose
  Activity 3.2 Nucleic Acid Building Blocks
  Activity 3.3 DNA Structure

Test Yourself

Knowledge and Synthesis Questions

1. Which of the following statements concerning polymers is *not* true?
   a. Polymers are synthesized from monomers during condensation.
   b. Polymers are synthesized from monomers during dehydration.
   c. Polymers consist of at least two types of monomers.
   d. Both b and c

2. You are a biochemist and have recently discovered a new macromolecule. Studies of the bond types found in this macromolecule reveal many hydrogen bonds and peptide linkages. You most likely have found what type of macromolecule?
   a. Carbohydrate
   b. Lipid
   c. Protein
   d. Nucleic Acid

3. An α helix is an example of which level of protein structure?
   a. Primary
   b. Secondary
   c. Tertiary
   d. Quaternary

4. You have isolated a monomer with the following components: a phosphate group, a sugar, and a nitrogen containing base. Polymers synthesized from this monomer belong to what class of macromolecule?
   a. Carbohydrate
   b. Lipid
   c. Protein
   d. Nucleic Acid

5. Cellulose and starch are composed of the same monomers. Which of the following result in their being structurally and functionally different?
   a. They have different types of glycosidic linkages.
   b. They have different numbers of glucose monomers.
   c. They are held together by different bond types.
   d. None of the above

6. DNA utilizes the bases guanine, cytosine, thymine, and adenine. In RNA, _________ is replaced by _________.
   a. adenine, arginine.
   b. thymine, uracil.
   c. cytosine, uracil.
   d. cytosine, arginine.

7. The pairing of purines with pyrimidines to create a double-stranded DNA molecule is called
   a. complementary base pairing.
   b. phosphodiester bonding.
   c. antiparallel synthesis.
   d. dehydration.

8. Lipids, as a complex class of macromolecules, are similar in that they are synthesized from
   a. glycerol.
   b. fatty acids.
   c. steriod precursors.
   d. cholesterol.

9. Amino acids are linked together into proteins by which of the following bond types?
   a. Covalent bonds
   b. Peptide bonds
   c. Phosphodiester bonds
   d. Both a and b

10. Which of the following characteristics differentiate carbohydrates from other macromolecule types?
    a. Carbohydrates are constructed of monomers that always have a ring structure.
    b. Carbohydrates never contain nitrogen.
    c. Carbohydrates consist of a carbon bonded to a hydrogen and a hydroxyl group.
    d. None of the above

11. Which of the following statements about carbohydrates is *not* true.
    a. Monomers of carbohydrates have 6 carbon atoms.
    b. Monomers of carbohydrates are linked together during dehydration.
    c. Carbohydrates are energy storage molecules.
    d. None of the above

12. What would you expect to be true of the R groups of amino acids located on the surface of protein molecules found within the interior of biological membranes?
    a. The R groups would be hydrophobic.
    b. The R groups would be hydrophilic.
    c. The R groups would be polar.
    d. The R groups would be able to form disulfide.

Application Questions

1. Differentiate between primary, secondary, tertiary, and quaternary structure as it relates to proteins. If a protein is immersed in a pH solution far from its optimum, which structures are most likely to disassociate first and why?
2. Dietary guidelines encourage people to stay away from saturated fats. What is meant by the term "saturated fat"? Why is this fat type of more concern than unsaturated fats in the diet? How are the two fats structurally different?
3. Draw a phospholipid. What characteristics of phospholipids make them perfectly suited for membranes? What do you think might happen if phospholipids did not form a bilayer? How might they arrange themselves in an aqueous environment?
4. Complex carbohydrates should be a mainstay of your diet. What properties make them excellent food sources?
5. Discuss how a protein's three-dimensional structure makes it perfect for acting as a carrier and receptor molecule. Why are proteins uniquely suited for this function, whereas other macromolecules are not?
6. What properties of DNA structure make it well suited for its function as an informational molecule?
7. Amino acid R groups have a variety of chemical properties. How do these different properties contribute to the final three-dimensional shape of the molecule?
8. You have isolated a protein with the following amino acid sequence: RSCFLA. Using Table 3.2 in your book, draw this protein. In your drawing, label the N terminus, the C terminus, and show all peptide linkages. How many water molecules were generated in the synthesis of this protein?
9. Explain how the pattern of charges in a macromolecule can affect its final form and function. You may use any of the four macromolecule classes to illustrate your point.
10. Consider the following triglycerides (*A* and *B*) in answering a – d.

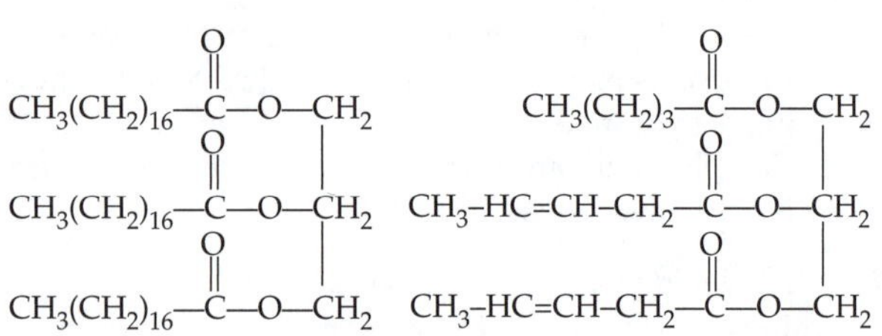

A *B*

a. In *B*, circle the remnant of the glycerol portion of the triglyceride.
b. Which triglyceride (*A* or *B*) is probably a solid at room temperature? __________ Explain your answer.
c. Which triglyceride (*A* or *B*) is probably derived from a plant? __________ Explain your answer.
d. How many water molecules result from the formation of triglyceride *B* from glycerol and three fatty acids?

11. Use the base-pairing rules for DNA and RNA to label the complementary strand of RNA (right) to the single strand of DNA (left) shown below, where C = cytosine, G = guanine, A = adenine, T = thymine, and U = uracil.

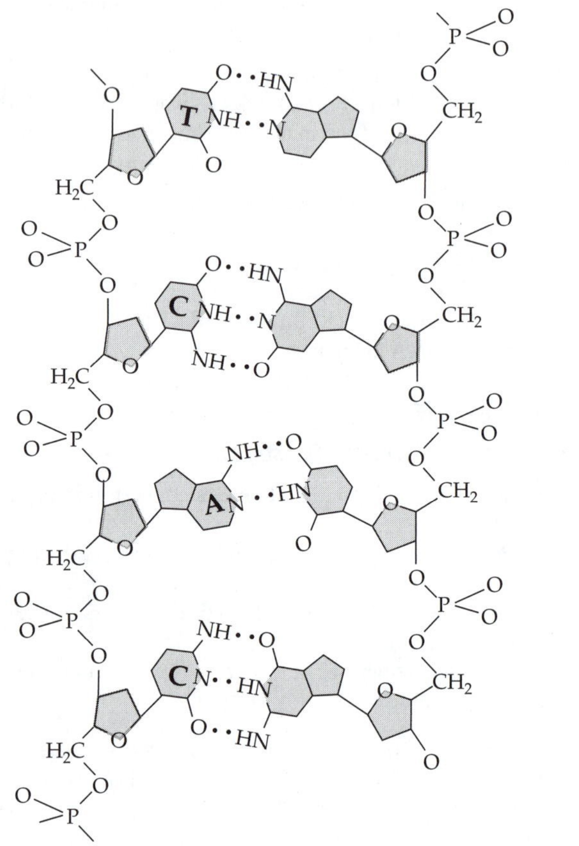

Circle and label an example of a nucleotide and a nucleoside in the figure showing the double-stranded DNA and RNA hybrid. Based on the orientation of the sugar molecules, label the four ends of the molecule as 3′ or 5′ in the figure showing the double-stranded DNA and RNA hybrid.

Answers

Knowledge and Synthesis Answers

1. **d.** Polymers are synthesized via condensation and broken down via hydrolysis. Polymers may be made of a single type of monomer or from multiple types of monomers.
2. **c.** Proteins consist of amino acids joined by peptide linkages. Higher levels of structure are stabilized by hydrogen bonds.
3. **b.** α helices are examples of secondary structure and are maintained by hydrogen bonds between the amino acid residues.

4. **d**. These are all constituents of a nucleotide, the monomer that makes up nucleic acids.
5. **a**. Starch has β-glycosidic linkages, whereas cellulose has α-glycosidic linkages. This leads to structural and functional differences between the two macromolecules.
6. **b**. RNA and DNA differ by one oxygen molecule in their ribose sugar and in the substitution of uracil (RNA) for thymine (DNA).
7. **a**. Complementary base pairing results from the attraction of charges and the ability to pair purines with pyrimidines through hydrogen bond formation.
8. **b**. All lipids are synthesized from fatty acid precursors.
9. **b**. The peptide bond that links amino acids to form proteins is a type of covalent bond.
10. **c**. Carbohydrates always have carbons bonded to hydrogens and hydroxyl groups. They may have a variety of other associate molecules in addition to these.
11. **a**. Carbohydrate monomers may have different numbers of carbon atoms. Five carbon monomers (pentoses) and 6 carbon monomers (hexoses) are both common.
12. **a**. The interior of the plasma membrane is hydrophobic; therefore, an imbedded protein would have hydrophobic residues.

Application Answers

1. See Figure 3.5 in your text for diagrams of primary, secondary, tertiary, and quaternary structures. The quaternary structure is the least stable and breaks down first in nonoptimum conditions. Protein structures continue to denature by unfolding the tertiary structures, then secondary structures. Primary structure is maintained by covalent bonds and is the last to break down. Disruptions in tertiary and quaternary structure are often reversible. Disruptions in primary or secondary structure are generally irreversible.
2. Saturated fats contain only single bonds and are "saturated" with hydrogen. This allows saturated fat molecules to pack together densely. This characteristic is easily seen in that most saturated fats are solid at room temperature. Unsaturated fats contain double bonds that affect the shape of the molecule. They are not "saturated" with hydrogens. This characteristic keeps them from packing together tightly and they tend to be liquid at room temperature. Saturated fats are dangerous because of this "packing" ability.
3. See Figures 3.21 and 3.22 in your book for drawings of phospholipids. The hydrophilic head and the hydrophobic tail of phospholipids allow them to have an "inside" that resists an aqueous environment and an "outside" that can reside in such an environment. When they exist as a bilayer, the hydrophobic tails aggregate together. If they did not exist in two layers the tails would still try to aggregate; This would result in a spherical aggregation of phospholipids called a micelle in which the tails are arranged toward the center of the sphere, away from the aqueous environment, and the heads are immersed in the aqueous environment.
4. Complex carbohydrates are easily broken down into glucose monomers which provide nearly all cellular energy. By storing glucose monomers in large carbohydrates, the osmotic strain on any give cell is reduced without sacrificing availability of energy.
5. The three-dimensional nature of proteins allows them to form binding sites. These binding sites are uniquely shaped to interact with other molecules.
6. DNA structure is highly regular and conserved which makes it usable by all cells, but the different bases allow it to encode specific genetic information.
7. The size of the R group, the charge of the R group, and any special binding properties all contribute to the final orientation of a protein molecule.

8.

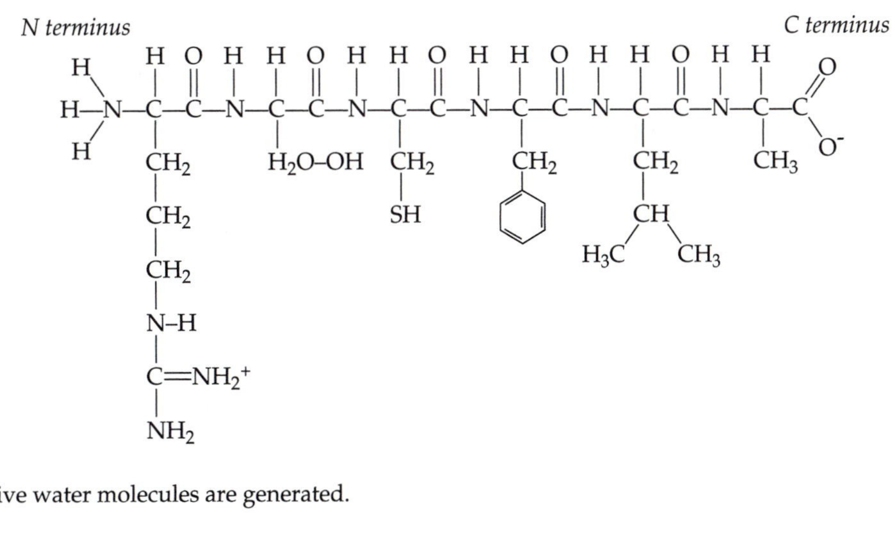

Five water molecules are generated.

9. One example is charge distribution in proteins affects folding, with hydrophobic regions turning inward and hydrophilic regions turning outward. Another excellent example is phospholipids. See the answer to Question 3.

10. a.

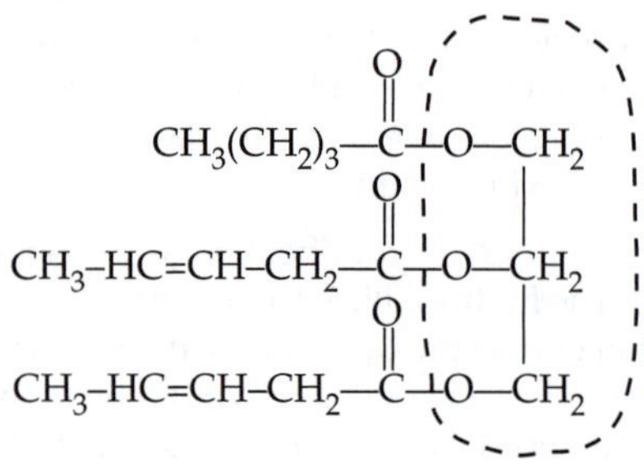

b. Triglyceride *A* is probably solid at room temperature; its fatty acid chains are saturated (no double bonds) and relatively long, both characteristics of solid, animal-derived triglycerides.

c. Triglyceride *B* is probably derived from a plant; its fatty acid chains are unsaturated (double bonds) and relatively short, both characteristics of liquid, plant-derived triglycerides.

d. ***Three*** water molecules will result. A water molecule results for each of the three fatty acids added to glycerol by a condensation reaction.

11.

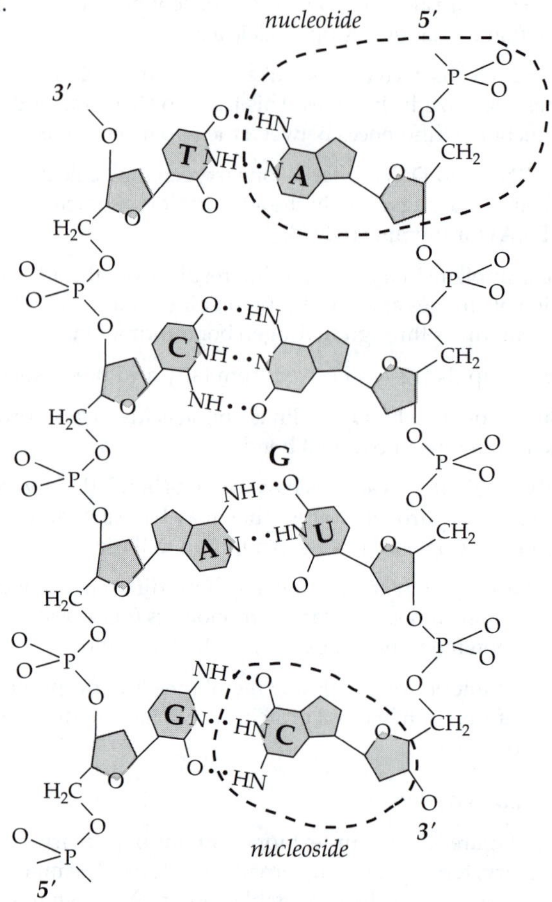

4 The Organization of Cells

Important Concepts

The cell is the basic unit of life.

- The cell theory states that all organisms are composed of cells and that all cells come from preexisting cells.
- Cell size is limited by surface area-to-volume ratio. Very large cells are not feasible because as volume increases surface area does not increase at the same rate. Cells with large volumes do not take in enough material across their surface to support the metabolic activity of the volume. Some cells circumvent this situation by means of long cellular extensions (e.g., nerve cells).
- Because cells are small, either light microscopes or electron microscopes must be used to visualize them. Light microscopes, using lenses and light, resolve objects as small as 0.2 μm. Electron microscopes, using electron beams, can resolve objects as small as 0.5 nm. Staining techniques are used to assist in visualizing cellular components.
- All cells are surrounded by a plasma membrane composed of a phospholipid bilayer with many embedded and protruding proteins. These proteins make the membrane selectively permeable, permitting some small molecules to pass through but preventing others from doing so. The membrane allows for maintenance of the internal cellular environment.

Cells are classified as either prokaryotic or eukaryotic.

- Prokaryotic cells do not have internal membrane compartments and are generally smaller than eukaryotic cells.
- Prokaryotic cells are found only in domains Archaea and Bacteria. Prokaryotes are one-celled organisms.
- Prokaryotic cells have a cell membrane and a nucleoid containing the genetic material which is not membrane-bound. The cell consists of the liquid cytosol and insoluble particles that act as subcellular machinery (i.e., ribosomes for protein synthesis).
- Some prokaryotic cells have special features
  - Cell walls: This rigid structure is outside the cell membrane and is composed of peptidoglycans. Type of cell wall is frequently used as a characteristic to identify bacteria.
  - Outer membrane: This membrane is outside the cell wall. It is polysaccharide rich, and some of the polysaccharides are disease-causing toxins.
  - Capsule: This slime layer is outside of the wall and outer membrane. Though it provides protection, it is not necessary for the cell's survival.
  - Membrane structures: Photosynthetic prokaryotes have stacks of folded membranes on which photosynthesis is carried out. Other prokaryotes have membrane folds called mesosomes which function in energy reactions and cell division.
  - Flagella: These tiny protein machines cause motion of the cell.
  - Pili: These are protein extensions that assist in adherence.
- Eukaryotic cells are larger than prokaryotic cells and are distinguished by having membrane-bound internal components.
- Each organelle (membrane-bound compartment) has its own internal environment uniquely suited to its function. The membrane assists in regulating that environment.

Some organelles function to store and process information.

- The nucleus stores DNA and is the site of DNA duplication, DNA regulation of cellular activity, ribosome assembly, and RNA synthesis.
- The nucleus is bounded by a double membrane called the nuclear envelope. Small openings in the nuclear envelope called nuclear pores allow passage of RNA and ribosomes to the cell cytoplasm. The outer membrane of the nuclear envelope is continuous with the endoplasmic reticulum.
- Inside the nucleus DNA molecules and proteins make up chromatin. At cell division, chromatin condenses to form chromosomes. Chromatin is organized on the nuclear matrix and the nuclear lamina. The lamina helps maintain nuclear shape.
- Ribosomes found in the cytoplasm, the endoplasmic reticulum, and the mitochondria act as information transcription centers and guide the synthesis of proteins from nucleic acid blueprints.

The endomembrane system functions in manufacturing and packaging cellular products.

- The endoplasmic reticulum (ER) is a complex of membrane sacs throughout the cell and is continuous with the nuclear membrane.
- The endoplasmic reticulum is classified into two types based on the presence or absence of attached ribosomes.
  - Rough endoplasmic reticulum (RER) is studded with active ribosomes. Since ribosomes actively aid in the synthesis of new proteins, the RER is important for storage, transport, and modification of the proteins.
  - Smooth endoplasmic reticulum also functions in protein modification and transport but, more importantly, functions in the modification of chemicals taken in by the cell and as the site of hydrolysis of glycogen and synthesis of steroids.
- The Golgi apparatus, composed of flattened membrane stacks called cisternae and vesicles, is important in protein modification, packaging, and polysaccharide synthesis in plants.
- Different regions of the Golgi have different enzymes and functions. The *cis* region is closest to the nucleus or ER. The *trans* region is closest to the cell surface. Proteins are released from the ER in a vesicle and are transported to the *cis* region, where the vesicle membrane fuses with Golgi membrane to release the contents. Vesicles in turn pinch off of the *cis* Golgi for transport between cisternae or away from the Golgi.
- Lysosomes are "digestion centers" within a cell. They contain a host of powerful enzymes in a slightly acidic environment. They break down engulfed molecules and cellular waste.

Mitochondria and plastids function in energy transformation.

- Mitochondria convert potential chemical energy from glucose to ATP, a form usable by the cell.
- Mitochondria have two membranes, an outer smooth membrane and a highly folded internal membrane. The folds are called cristae, and the remaining internal space is called the matrix. The matrix holds ribosomes and DNA. Enzymes used during cellular respiration are embedded in the cristae.
- Plastids are not produced in animal cells. Plant cells have several different types of plastids, each with unique functions.
- Chloroplasts are the site of photosynthesis where light energy is converted to chemical energy.
- Chloroplasts have two membranes. The innermost membrane is folded into stacks of thylakoids, called grana, and hold chlorophyll and enzymes for photosynthesis. The liquid content of the chloroplast is called stroma, and contains ribosomes and DNA.
- Other plastid types include chromoplasts (which contain pigments), and leucoplasts (storage for starch).

The endosymbiotic theory explains the origin of mitochondria and chloroplasts.

- Mitochondria are unique organelles in that they have their own DNA and are double membrane-bound. They also reproduce within the cell. In other words, they act very much like cells within cells, but they are still under nuclear control.
- The endosymbiotic theory suggests that a small photosynthetic prokaryote was engulfed and retained by a larger cell, with each getting mutual benefit from the other (endosymbiosis). Mitochondria may have origins as an engulfed respiring prokaryote. See Figure 4.18.
- There is evidence to support this theory, but it is not as yet conclusive.

Other specialized organelles are found in some but not all cells.

- Peroxisomes function to break down harmful peroxide by-products.
- Vacuoles, found in plants and protists, function in storage of waste products, as a food store, or in water regulation, depending on the organism.

The cytoskeleton provides support, shape, and movement for the cell.

- Microfilaments, composed of a double helix of actin, stabilize cell shape and assist with contraction of the cell. These filaments are involved in muscle cell contraction, cell division, cytoplasmic streaming, and cell shape.
- Intermediate filaments are found only in multicellular organisms and function in stabilizing structure and resisting tension. These make up the lamina, form desmosomes, and anchor the nucleus.
- Microtubules are long hollow tubes of a protein called tubulin which allow rigidity of the cell and act as a framework of movement for motor proteins. They have a very specific structure that can be quickly added to or reduced.
- Cilia and flagella are powered by microtubules. Though cilia and flagella differ in size and function, they have the same basic structure: a "9 + 2" arrangement of microtubules. See Figure 4.24.
- Movement of cilia and flagella occurs when the microtubules slide past one another. The sliding is caused by molecules of dynein changing shape while bound to the microtubules. See Figure 4.26.
- Centrioles are also made up of microtubules and are involved in cell division.

Extracellular structures are outside of the plasma membrane and provide support, protection, and anchoring for the cell.

- Plants have semirigid cell walls composed of cellulose, which functions to support and protect the cell. Plasmodesmata (small holes in the cell walls) allow connection between plant cells.
- The extracellular matrices of some animal cells function to

hold cells together, filter materials, orient cell movement, and assist with chemical signaling. Some cells, such as bone cells, secrete an elaborate and rigid matrix.

The Big Picture

- All living things are composed of cells, and all cells come from preexisting cells. There are two basic cell forms: the small prokaryotic cell with no organelles and the eukaryotic cell with many membrane-bound organelles.
- The organelles of eukaryotic cells have a variety of functions, including information storage and transmission, modification and packaging of cellular products, and energy transformation. See figure 4.7 of your text to understand the structure and function of each component.
- The cytoskeleton of the cell provides support, structure, and protection. Cytoskeletal components also provide movement of the cell and within the cell.

Study Strategies

- Terminology is the most difficult part of this material. It is easier to learn if you understand the concepts before you memorize the terms.
- Use Figure 4.7 while learning the organelles and their components. Visualizing the jobs each must do and how their functions relate to the structure makes learning easier.
- Review the following tutorial and activities on the website:
  Tutorial 4.1 Eukaryotic Cell Tour
  Activity 4.1 Lysosomal Digestion
  Activity 4.2 Know Your Techniques
  Activity 4.3 The Scale of Life

Test Yourself

Knowledge and Synthesis Questions

1. You have found a mass of cells in the sediment surrounding a thermal vent in the ocean floor. The salinity in the area is quite high. Upon microscopic examination of the cells you find no evidence of membrane-bound organelles. How would you classify this cell?
   a. As a eukaryotic cell
   b. As a prokaryotic cell
   c. As a member of domain Archea or Bacteria
   d. Both b and c

2. Centrifugation of a cell results in the rupture of the cell membrane and the contents compacting into a pellet in the bottom of the centrifuge tube. Bathing this pellet with a glucose solution yields metabolic activity including the production of ATP. One of the contents of this pellet is most likely which of the following?
   a. Cytosol
   b. Mitochondria
   c. Lysosomes
   d. Golgi bodies

3. Eukaryotic cells are thought to be derived from prokaryotic cells that underwent phagocytosis without digestion of the phagocytized cell. This mutualistic relationship is explained by the
   a. endosymbiotic theory.
   b. cell theory.
   c. evolutionary theory.
   d. parasite theory.

4. Though science fiction has produced stories like "The Blob," we don't see very many large single-celled organisms. Which of the following tends to limit cell size?
   a. Ability to maintain a continuous large membrane
   b. Ability to reproduce a large cell
   c. Surface area-to-volume ratios
   d. All of the above

5. Microscopes are used to resolve images that cannot be seen with the unaided eye. Electron microscopes use _________ to resolve images, whereas light microscopes use _________ to resolve images .
   a. light and lenses, diffraction of electron beams
   b. diffraction of electron beams, light and lenses
   c. lasers, light and lenses
   d. None of the above

6. Match the following cellular functions to the organelle in which they occur:

____ Lysozome	a. DNA synthesis
____ Nucleus	b. Protein folding
____ RER	c. Packaging of materials for export
____ SER	d. Protein synthesis
____ Golgi apparatus	e. Breakdown of phagocytized material
____ Chromoplast	f. Photosynthesis
____ Mitochondrion	g. Cellular respiration
____ Chloroplast	h. Pigment production

7. Which of the following organelles are double membrane-bound?
   a. Nucleus
   b. Chloroplast
   c. Mitochondrion
   d. All of the above

8. Movement of cells is accomplished in both prokaryotes and eukaryotes with which of the following structures?
   a. Cilia
   b. Pili
   c. Dynein
   d. Flagella

9. Which of the following statements is true regarding mitochondria and chloroplasts?
   a. Some cells produce chloroplasts.
   b. Mitochondria and chloroplasts may be found in the same cell.
   c. Mitochondria and chloroplasts are not found in the same cell.
   d. Chloroplasts can revert to mitochondria in certain conditions.

10. Which of the following best describes ribosomes?
    a. Ribosomes guide protein synthesis.
    b. Ribosomes are found only in the nucleus or on the RER.
    c. There are no ribosomes in the mitochondria.
    d. All of the above

11. Nuclear DNA exists as a complex of proteins called _________ that condenses into _________ during cellular division.
    a. chromosomes, chromatin
    b. chromatids, chromosomes
    c. chromophors, chromatin
    d. chromatin, chromosomes

12. Rough endoplasmic reticulum and smooth endoplasmic reticulum differ
    a. only by the presence or absence of ribosomes.
    b. both in the presence or absence of ribosomes and in their function.
    c. only in microscopic appearance.
    d. None of the above

Application Questions

1. A cell is responsible for making a protein that is secreted from the cell. Trace the production of that protein from the nucleus through all necessary organelles to the point of release from the cell.
2. Explain how microtubules and dynein function to make cilia and flagella move.
3. Compare and contrast prokaryotic and eukaryotic cells.
4. Explain the significance of organelles. What are the costs and benefits of having large compartmentalized cells?
5. What is the primary function of a cell membrane? What characteristics of membranes allow them to contribute to metabolic activity?
6. The organelles that contain their own DNA are all double membrane-bound. Relate this observation to the endosymbiotic theory.
7. Vacuoles in plants hold waste material. Why do plants tend to retain cellular wastes whereas animal cells do not?
8. Compare and contrast the prokaryotic cell wall with the plant cell wall.
9. There are structural similarities between mitochondria and chloroplasts. Using the idea that form follows function, why are there similarities in these two organelles?
10. The extent of the presence of a cytoskeleton has become known relatively recently. Taking the advances in microscopy as a reference point, discuss how it is possible to better understand the cytoskeleton.

Answers

Knowledge and Synthesis Answers

1. **b**. There are several characteristics that suggest this is a prokaryote. It survives in high salinity and high heat, but the sure indication is that it contains no membrane-bound organelles.
2. **b**. The pellet is undergoing cellular respiration, a function that occurs in the mitochondria. You can also assume that if the single membrane of the cell itself is ruptured, other single membrane-bound organelles would be ruptured as well.
3. **a**. The endosymbiotic theory is used to explain the presence of DNA as well as the two membranes of the mitochondrion and the chloroplast.
4. **c**. As volume increases, the surface area available for exchange does not increase proportionally. There comes a point at which the surface is not large enough for maintenance of the metabolic activity of the cell.
5. **b**. A concentrated beam of electrons is focused on an object and allows resolution of structures as small as 5 nm. Light microscopy, using lights and lenses, can only resolve down to approximately 0.2 μm.
6. **a**. Nucleus
   b. SER
   c. Golgi apparatus
   d. RER
   e. Lysozome
   f. Chloroplast
   g. Mitochondrion
   h. Chromoplast
7. **d**. The nucleus, the mitochondria, and the chloroplasts are the only double membrane-bound organelles.
8. **d**. Though the flagella are of different structure, they serve the same role in prokaryotes and eukaryotes.
9. **b**. Mitochondria and chloroplasts may be found in the same cell. All eukaryotic cells must contain mitochondria.
10. **a**. Ribosomes, found in the nucleus, the cytosol, and in organelles such as the mitochondria, RER, and chloroplasts, complex with RNA to guide protein transcription.
11. **d**. The complex of proteins and DNA is called chromatin. Chromosomes are only present during cell reproduction.
12. **b**. Both the structure and the function of RER and SER differ.

Application Answers

1. The protein could be synthesized on ribosomes in the nucleus, or synthesis could take place in the cytosol or on the RER. From there it most likely would be transported along the RER for further modification before being packaged in a bit of ER membrane and transported to the *cis* region of the Golgi. Further modification might take place in the cisternae of the Golgi before the protein is encased in a final vesicle that fuses with the plasma membrane of the cell and the contents are released to the outside.
2. Dynein molecules bind to pairs of microtubules in the flagella or cilia. With the addition of cellular energy, the dynein molecules undergo a conformational change that results in microtubules sliding past one another, resulting in a whiplike motion of the flagella.

3.

Prokaryotic Cells	Eukaryotic Cells
Small in size	10 or more times greater in size
No membrane-bound organelles	Membrane-bound organelles
Found only in domains Archea and Bacteria	Found in all domains other than Archea and Bacteria
Survive and metabolize in chemical environments with no oxygen	Metabolize mainly in oxygen-rich environments
DNA is in nucleoid	DNA is in nucleus

4. Organelles allow different metabolic environments to exist in the same cell. This partitioning of jobs allows for greater specialization but comes at an energy cost. Eukaryotic cells are more energy expensive.
5. A cell membrane exists to form an inside and an outside of a cell. The presence of an inside and an outside allows the establishment of different environments. In addition, membranes hold integral proteins with a variety of chemical properties and activities. This allows for the enzymatic activity associated with membranes. Stacks of membranes, such as those in mitochondria and chloroplasts, increase the amount of chemical activity in an area.
6. See Figure 4.18 for a description of the origin of double membranes from endosymbiosis.
7. The presence of cell walls in plants limits what can be expelled outside of the plant cell. Holding wastes in an internal compartment circumvents this problem.
8.

Prokaryotic Cell Walls	Plant Cell Walls
Functions in protection and support	Functions in protection and support
Composed of peptidoglycans	Composed of cellulose
Located outside the cell membrane	Located outside the cell membrane
Influences cell shape	Influences cell shape

9. Both mitochondria and chloroplasts are involved in energy transformation activities that require many enzymes. The stacking or folding of membranes provide enzymatic activity centers for these reactions.
10. To visualize the components of the cytoskeleton, electron and transmission microscopy is necessary. With increasing powers of resolution it is possible to see more and more cellular components and to understand them.

5 Cellular Membranes

Important Concepts

Biological membranes have a specific content and structure.

- Biological membranes are composed of lipids, proteins, and carbohydrates. Lipids provide the structure and barrier functions; proteins are involved in creating channels and transporting materials across the lipid barrier; and carbohydrates are involved in signaling and adhesion.
- Lipids composing membranes are phospholipids with hydrophilic heads and hydrophobic tails. They arrange themselves into a bilayer with tails touching and heads extending into the aqueous environment inside and outside the cell. This arrangement allows fluid movement of the two layers on top of one another and sealing of any disruptions in the membrane.
- Though the basic structure of a bilayer is always the same, the inner and outer halves differ in lipid composition and thus have slightly different properties.
- Proteins are distributed in membranes "as needed," and the numbers of proteins and their placement varies greatly based on cell type. Proteins may be embedded in or extend across membranes. Hydrophobic protein regions are found in the hydrophobic environment of the membrane. Hydrophilic regions extend out from the membrane. Those proteins that penetrate the phospholipid bilayer are called integral proteins. Those not embedded are referred to as peripheral proteins.
- The fluid mosaic model describes how the phospholipids and proteins are independent of one another and may move rapidly throughout the membrane. This happens for the most part because no covalent bonding occurs among the constituents. Some proteins may be anchored to cytoskeletal components or bound to certain lipids. Such proteins have very specific functions.
- The "inside" and "outside" of a membrane often have different properties due to the different characteristics of the transmembrane proteins on the two sides.
- Carbohydrates in membranes serve as recognition sites for other cells and molecules. Carbohydrates are frequently bound to lipids or proteins, forming glycolipids or glycoproteins respectively. Because of the nature of carbohydrates, they contribute to cell adhesion.

Cells must be able to recognize and adhere to other cells to form tissues, organs, and organisms.

- Cells have recognition proteins. Often the proteins on two cells that adhere to one another are the same; these molecules and the bonds they form are referred to as homotypic. Occasionally the two proteins are different, but have complementary binding sites. In this case the molecules and the bonding are called heterotypic .
- Cell adhesions are specific to the organism and to the cell type. This specificity is due to recognition proteins.

Cells must make connections (cell junctions) to other cells in order to communicate and stabilize tissues.

- Tight junctions are found specifically in epithelial cells and function to prevent substances from moving through the intercellular space. They also restrict migration of phospholipids and membrane proteins. Tight junctions help tissues to move materials in specific directions.
- Desmosomes are structural connections between cells that "reinforce" a tissue. Because the desmosome involves the cytoskeleton and crosses the cell membrane, the connections have integrity and are not easily broken.
- Gap junctions facilitate communication between cells. They are made of protein connexons that span adjacent plasma membranes.

Membranes are selectively permeable and allow some substances, but not others, to pass through.

- Passive transport, either through the lipid bilayer, via channel proteins or by carrier molecules, does not require an input of energy.
- Diffusion is a process of random movement toward equilibrium and is a net movement from greater concentration to lesser concentration. The rate of diffusion depends on the size of diffusing substance, electrical charge of diffusing substance, temperature of solution, and concentration gradient.
- Diffusion within small areas such as cells may occur rapidly, but it occurs more slowly with increasing distance.
- Membranes alter the rate of diffusion. A membrane is said to be permeable to those substances that can pass through it and impermeable to those that cannot. If a substance can

pass through a membrane, it will diffuse until concentrations on either side of the membrane are equal. If a substance cannot permeate the membrane it cannot diffuse across it.

- In simple diffusion, small molecules pass through a membrane. The more lipid-soluble it is, the faster it diffuses. Charged and polar molecules do not readily pass through a membrane due to the membrane's internal hydrophobic nature. The exception to this rule is water.
- Osmosis is the diffusion of water across a membrane. Water will move across a membrane to equalize solute concentrations on either side if the solute cannot move. Water moves from areas of high solute concentration in an attempt to equalize solute concentrations on both sides of the membrane.
  - Isotonic solutions have the same solute concentrations.
  - Hypertonic solutions have a higher solute concentration than the solutions with which they are being compared.
  - Hypotonic solutions have a lower solute concentration than the solutions with which they are being compared.
- Turgor pressure refers to pressure within cells and is directly related to osmosis.Turgor pressure can build up only if there is a cell wall to limit cell expansion.
- Facilitated diffusion may be aided by protein-lined channels in the plasma membrane. The proteins allow polar molecules and ions to pass based on their chemical composition. These channels may be "gated." This occurs when a protein changes shape in response to some signal and either opens or closes off a channel. Passage through channels is still governed by factors affecting rates of diffusion.
- Facilitated diffusion may be aided by carrier proteins. Carrier molecules transport molecules across a membrane. Diffusion in this case is not only limited by concentration gradient, but also by the number of available carrier molecules. Such substances can diffuse only if a carrier molecules is available to take it.

Active transport is directional, works against a concentration gradient, and requires an input of energy.

- Three types of protein systems are involved in active transport.
  - Uniport transporters move a single solute in one direction.
  - Symport transporters move two solutes in the same direction.
  - Antiport transporters move two solutes in opposite directions.
- Primary active transport requires the direct participation of ATP and transport cations.
- Secondary active transport uses ATP indirectly by coupling solute transport with the ion concentration gradient established by primary transport.

Endocytosis and exocytosis move large molecules into and out of cells.

- Endocytosis brings large substances into the cell. This is accomplished by the cell membrane, which folds around the substance to form a vesicle. See Figure 5.14.
  - In phagocytosis fairly large substances (even entire cells) are engulfed. Once inside the cell the vesicle fuses with a lysosome for digestion.
  - In pinocytosis, small fluid-filled vesicles form.
  - Receptor-mediated endocytosis is found in animal cells, and captures specific macromolecules from the environment. Receptors for the macromolecules cluster together in coated pits reinforced by clathrin. The resulting vesicle becomes clathrin-coated until it is well inside the cell, where it loses its coat and fuses with a lysosome.
- Exocytosis moves materials out of the cell. Binding proteins on cell-produced vesicles bind with receptor proteins of the cytoplasmic side of the cell membrane. The vesicle membrane fuses with the cell membrane and the contents of the vesicle are released outside the cell membrane.

Membranes have multiple biological functions.

- Though the primary membrane functions are to act as barriers and binding sites, membranes also have roles in information processing, energy transformation, and the organization of chemical reactions.
- In order to carry out these functions, membranes are necessarily dynamic. They are continually being formed, modified, and degraded.

The Big Picture

- Cellular membranes are a dynamic composition of a phospholipid bilayer, integral and peripheral proteins, and carbohydrates. The nature of the constituent phospholipids allows the formation of a barrier that is semipermeable. Small hydrophilic molecules can traverse the membrane via simple diffusion, water may cross the membrane by osmosis, small charged ions may pass through via protein channels, and some other molecules may be shepherded through via carrier molecules. Transport of other molecules must be active and requires energy input; either directly from ATP or as coupled to ATP-driven transport. Larger substances depend on endocytosis and exocytosis for transport into and out of the cell.
- In multicellular organisms, adhesion and cell-to-cell connections are important for structure and communication. Tight junctions allow multiple cells to form a sheet that directs movement of substances in a single direction. Desmosomes enhance structural support of sheets of cells that are prone to abrasion. Gap junctions allow communication by connecting adjacent cells.

Common Problem Areas

- The biggest challenge when studying membranes is understanding diffusion and osmosis. It is very easy to get the terminology confused, especially the terms "hypertonic" and "hypotonic" when discussing osmosis. Looking at the Latin roots of the words is helpful. "Hyper-" generally means excess, and "hypo-" generally means "less than." "Tonic" refers to solute concentration. Therefore, "hypertonic" means excess solutes and "hypotonic" means fewer solutes.

- Secondary active transport also tends to be confusing. Remember that secondary active transport does not use ATP directly, but is tightly coupled to the transport of ions that does require ATP.

Study Strategies

- When studying the fluid mosaic model, think about the properties of the constituent molecules. This will make understanding the membrane's structure much easier.
- Draw diagrams of the movement of water and solutes when working with diffusion and osmosis. Diagrams help you to visualize what is happening across a membrane.
- Review the following tutorials and activities on the website:
  Tutorial 5.1 Membrane Transport
  Tutorial 5.2 Receptor-Mediated Endocytosis
  Activity 5.1 The Fluid-Mosaic Model
  Activity 5.2 Animal Cell Junctions

Test Yourself

Knowledge and Synthesis Questions

1. Which of the following statements regarding cellular membranes is *not* true?
   a. The hydrophobic nature of the phospholipid tails limits the migration of polar molecules across the membrane.
   b. Integral proteins and phospholipids move rapidly and fluidly throughout the membrane.
   c. Phospholipids flip back and forth from one side of the bilayer to the other.
   d. Glycolipids and glycoproteins serve as recognition sites on the cell membrane.

2. Which of the following contributes to differences in the two sides of the cell membrane?
   a. Differences in peripheral proteins
   b. Different domains expressed on the ends of integral proteins
   c. Differences in phospholipid types
   d. All of the above

3. Which of the following cell membrane components serve as recognition signals for interactions between cells?
   a. Recognition proteins
   b. Glycolipids or glycoproteins
   c. Phospholipids
   d. Integral proteins

4. Which of the following types of junctions are responsible for communication between cells?
   a. Tight junctions
   b. Desmosomes
   c. Gap junctions
   d. None of the above

5. You are monitoring the diffusion of a colored molecule across a membrane. Which of the following will result in the fastest rate of diffusion?
   a. An internal concentration of 5 percent and an external concentration of 60 percent.
   b. An internal concentration of 60 percent and an external concentration of 5 percent.
   c. An internal concentration of 35 percent and an external concentration of 40 percent.
   d. a and b

6. If a red blood cell with an internal salt concentration of about 0.85 percent is placed in a saline solution (salt solution) that is 4 percent, which of the following will most likely happen?
   a. The red blood cell will loose water and shrivel.
   b. The red blood cell will gain water and burst.
   c. The turgor pressure in the cell will greatly increase.
   d. The cell will remain the same.

7. Which of the following represents a hypotonic solution?
   a. The solution has a greater solute concentration than the one it is being compared to.
   b. The solution has a lesser solute concentration than the one it is being compared to.
   c. The solution has an equal solute concentration to the one it is being compared to.
   d. None of the above

8. Which of the following statements regarding osmosis is *not* true?
   a. Osmosis refers to the movement of water along a concentration gradient.
   b. In osmosis, water moves to equalize solute concentrations on either side of the membrane.
   c. If osmosis occurs across a membrane, then diffusion is not occurring.
   d. The movement of water across a membrane can affect the turgor pressure of some cells.

9. Channel proteins allow ions that would not normally pass through the cell membrane to go through the channel. What properties of the proteins are responsible for this?
   a. The channels are often composed of charged or polar R groups.
   b. The channels are often composed of hydrophobic R groups.
   c. a and b
   d. None of the above

10. Which of the following limits the movement of molecules when carrier-mediated facilitated diffusion is involved?
    a. Concentration gradient
    b. Availability of carrier molecules
    c. Temperature
    d. All of the above

11. Active transport differs from passive transport in that active transport
    a. requires energy.
    b. always requires direct input of ATP.
    c. moves molecules against a concentration gradient.
    d. a and c

12. Single-celled animals like amoeba engulf entire cells for food. Which of the following represents the manner in which amoeba "eat"?
    a. The amoeba binds only what it has receptors for, the amoeba's cell membrane surrounds the cell to be digested, a vesicle forms, and the vesicle fuses with a lysosome for digestion.
    b. The amoeba's cell membrane surrounds the cell to be digested, a vesicle forms, and the vesicle fuses with a lysosome for digestion.
    c. The cell is taken into the amoeba's vacuole, a vesicle is formed, and the vesicle fuses with a lysosome for digestion.
    d. None of the above

13. Sodium and potassium pumps are common in many cells. Which of the following are necessary for the pumps to work?
    a. ATP-driven pumping proteins
    b. A signal to activate the pumps
    c. A concentration gradient to work against
    d. All of the above

14. Bacterial cells are often found in very hypotonic environments. Which of the following characteristics keeps them from continuing to take on water from their environment?
    a. The presence of a cell wall allows a buildup of turgor pressure that prevents any more water from entering the cell.
    b. The presence of a cell wall allows a buildup of tonic pressure that prevents any more water from entering the cell.
    c. The cell expels water as fast as it takes it up.
    d. None of the above

15. Which of the following may affect the rate of diffusion?
    a. Temperature
    b. Molecule size
    c. Concentration gradient
    d. All of the above

Application Questions

1. Draw a diagram of a cell membrane and label the phospholipids, integral proteins, peripheral proteins, and carbohydrates. Using your diagram, describe the fluid mosaic model.
2. One characteristic of cell membranes is their ability to "self seal." Describe what properties of cell membranes prevent them from rupturing easily.
3. When scientists want to rupture cells to study the contents, they use a detergent and physical force. Why are both force and a detergent necessary?
4. Proteins for export are frequently modified in the endoplasmic reticulum, passed to the Golgi apparatus, and then exported from the cell. Explain the role of membranes in a protein's journey from the ER to the extracellular environment.
5. A marathon runner has just come into your emergency room with severe dehydration. You must decide what type of solution to pump into his veins so that he survives. Your choices are: pure water, 0.9 percent saline, and 1.5 percent saline. You cannot remember which solution is correct, so you treat blood samples that you have placed on microscope slides with each solution and observe what happens (Hint: blood cells are approximately 0.9 percent saline). Describe what you think will happen to the blood cells when exposed to each solution. Which solution you would choose to rehydrate the runner.
6. Compare and contrast active and passive transport.
7. You are interested in the epithelial tissue that lines the intestine. This single sheet of cells must absorb nutrients but prevent many other substances from crossing it. Describe which cell-to-cell connections exist to allow movement of substances into this tissue, resistance from abrasion, and communication between cells.
8. Barrier formation is only one function of the cell membrane. Describe some other functions of the membrane and discuss how the membrane is suited for that function.

Answers

Knowledge and Synthesis Answers

1. **c.** Because of the hydrophobic tails and hydrophilic head of the phospholipids, it is impossible for them to flip back and forth from side to side of the membrane.
2. **d.** The cell membrane is asymmetric and has different properties and functions on the cytoplasmic side versus the extracellular side. These properties arise from differences in the constituents of the membrane.
3. **b.** Both glycolipids and glycoproteins serve as recognition signals.
4. **c.** Gap junctions are involved in chemical and electrical signaling between cells.
5. **d.** Diffusion may take place in either direction across a membrane and always follows a concentration gradient. The larger the gradient, the faster diffusion will occur.
6. **a.** The cell will lose water in an attempt to equalize solute concentrations on both sides of the membrane.
7. **b.** Hypotonic solutions have lower solute concentrations than the solution they are being compared to.
8. **c.** Diffusion and osmosis are not mutually exclusive and may take place at the same time.
9. **a.** The charged or polar lining of the channel proteins allows passage of polar and charged molecules.

10 **d**. Anything that affects the rate of diffusion will affect carrier-mediated facilitated diffusion. In addition to those effects, carrier-mediated facilitated diffusion also relies on the availability of carrier molecules.

11. **d**. Active transport works against a concentration gradient and requires energy to do so. That energy does not always have to be directly supplied in the form of ATP.

12. **b**. See Figure 5.14.

13. **d**. Sodium and potassium pumps are forms of primary active transport.

14. **a**. Turgor pressure limits osmosis, and once a cell is turgid, no more water may be taken on.

15. **d**. Temperature, molecule size, molecule charge, and concentration gradients all affect the rate at which diffusion takes place.

Application Answers

1. Refer to Figure 5.1 in your text.
2. The nonpolar regions of the fatty acids tend to aggregate together to the exclusion of water. This property leads to "sealing" of ruptures in the membrane.
3. Detergents disrupt cell membranes because of their interactions with the fatty acids in the membranes. Physical forces allow the detergents to reach the phospholipid tails and break up their aggregations.
4. The protein leaves the ER encased in a membrane vesicle. Signal molecules on the surface of the membrane bind to receptor molecules on the membrane of the Golgi. The membranes fuse and the protein contents of the vesicle enter the Golgi for modification. Upon leaving the Golgi, the protein is encased in membrane, which again, with the appropriate signals and receptors, binds to the cytoplasmic surface of the cell membrane. The vesicle membrane and the cell membrane fuse and the contents of the vesicle are released to the extracellular side of the membrane.
5. In pure water, the blood cells will take on water through osmosis, swell, and eventually rupture. In 0.9 percent saline, the cells should neither gain nor lose a significant amount of water and should remain the same. In a 1.5 percent saline solution, the cells should lose water and shrivel. You might jump to the conclusion that for quick rehydration the patient needs a hypotonic solution. But remember that you will be adding this substance directly to the bloodstream—the patient would end up with ruptured cells. You want to use a solution isotonic to the patient's blood cells, 0.9 percent, to avoid rupture of cells.
6. The main difference between active and passive transport is that active transport goes against a concentration gradient and requires energy, while passive transport diffuses passively and does not require energy.
7. Tight junctions allow multiple cells to form a sheet that directs movement of substances in a single direction. Desmosomes enhance structural support of sheets of cells that are prone to abrasions. Gap junctions allow communication by connecting adjacent cells.
8. Membranes function in processing energy transformation and in the organization of chemical reactions. Integral and peripheral proteins contribute to these functions. The membrane serves as a holding site for the catalytic enzymes associated with these processes.

6 Energy, Enzymes, and Metabolism

Important Concepts

All cells need energy.

- No cell manufactures energy; therefore, energy must be obtained from an energy source and converted (transformed) to a usable form.
- Energy transformations are linked to chemical transformations in cells. These chemical transformations constitute metabolism.
- Kinetic energy is energy in motion and does work.
- Potential energy is stored energy. It may be converted to kinetic energy.
- Metabolic processes either store or release energy in chemical bonds.
  - Anabolic reactions link simple molecules to create complex molecules and store energy in the resulting bonds.
  - Catabolic reactions break down complex molecules and release stored energy.
- First law of thermodynamics: Energy is neither created nor destroyed. Energy can be converted from one form to another.
- Second law of thermodynamics: Not all energy can be used. When energy is converted from one form to another, some becomes unusable.
  - Total energy (enthalpy [H]) = usable energy (free energy [G]) + unusable energy (entropy [S]) × absolute temperature (T).
- Because we cannot measure absolute energy, we can only measure change (Δ) in energy. We determine the change in usable energy as follows:
  $\Delta G = \Delta H - T\Delta S$
  - If ΔG is negative, free energy is released and the reaction is exergonic.
  - If ΔG is positive, free energy is required and the reaction is endergonic.
- With each conversion, entropy (S) tends to increase; therefore, processes tend toward disorder and randomness.
- All biochemical reactions must either release (–ΔG) or take up (+ΔG) usable energy. Reactions that proceed over halfway to completion without an input of energy are called spontaneous. Nonspontaneous reactions require an input of energy.
- Reactions that are spontaneous in one direction are nonspontaneous in the other. The direction of the reaction depends on concentrations of reactants versus products. The point at which the forward reaction occurs at the same rate as the reverse reaction is termed chemical equilibrium.
- The further toward completion a reaction's equilibrium lies, the more free energy is given off (larger –ΔG).
- A larger +ΔG means equilibrium favors the reverse reaction (equilibrium falls toward the reactant). A ΔG near zero indicates that both products and reactants have free energy.

The role of ATP in cells

- ATP is needed to capture, transfer, and store free energy (usable energy) in cells to do work.
- Hydrolysis of ATP to ADP and P_i releases free energy (ΔG = –12kcal/mol). The equilibrium falls toward ADP production.
- The formation of ATP is endergonic. It requires as much energy to form ATP as ATP hydrolysis releases. This allows ATP to couple endergonic and exergonic reactions.
- ATP is formed when coupled with the exergonic reactions of cellular respiration. ATP is hydrolyzed to fuel endergonic reactions like protein synthesis. See Figure 6.9.

Enzymes catalyze biochemical reactions by lowering activation energy.

- Catalysts speed up reactions without being consumed in the process. Most biological catalysts are protein enzymes.
- Energy barriers exist between reactants and products. Energy must be added to get past this barrier. You may think of it as a little shove to get the reaction going. This is called activation energy (E_a). See Figures 6.11 and 6.12.
- For reactions to take place, the reactants must be slightly destabilized and turned into "state species" with higher free energy than that of either the reactants or the products. This is accomplished with the help of a catalyst.
- Enzymes are highly specific biological catalysts. They bind specific areas on their surfaces called active sites. The enzyme's specificity comes from the shape of its active

site. Only specific substrates can fit in and bind with an enzyme's active site. Once bound, the site is referred to as an enzyme-substrate complex.

- The enzyme alters the conformation of the substrate, thus lowering the activation energy for the reaction. See Figure 6.14. From this, product is formed and the enzyme is retained unchanged.
  $E + S \rightarrow ES \rightarrow E + P$
- Enzymes do not alter the equilibrium of a reaction; therefore, both the forward and reverse reactions are sped up.
- Enzymes speed up reactions by orienting substrates for maximum chemical interactions, adding changes to substrates, or by inducing strain on the substrate.
- Three main factors influence the rate of enzyme catalyzed reactions.
  - Temperature: Increases in temperature may aid in reduction of activation energy by adding kinetic energy; however, this may also result in denaturing of enzymes.
  - pH: Changes in pH influence the shape of enzymes and may drastically alter their catalytic ability.
  - Substrate concentration: Rate of reaction increases as substrate increases until all enzyme active sites are occupied. At that point, no amount of additional substrate will increase the reaction rate because no more enzyme is available for catalysis. Enzyme efficiency is measured in turnover number, or how fast an enzyme can convert substrate to product and free up its active site.

The structure of an enzyme determines its function.

- Enzymes are very large macromolecules with small active sites for binding small metabolites.
- Binding of small molecules to the active sites depends on H-bonds, charge interactions, and van der Waals forces.
- Binding of substrate by an enzyme changes the enzyme's shape. The enzyme's shape can be altered by the substrate. This is called "induced fit."

Cofactors, coenzymes, and prosthetic groups all contribute to an enzyme's activity.

- Cofactors are inorganic ions.
- Coenzymes are organic molecules that bind in the active site along with the substrate.
- Prosthetic groups are part of the protein itself.
- All three have roles in altering the shape of an enzyme for maximum activity.

Metabolism is a series of enzyme-catalyzed reactions organized into pathways in which the products of one reaction become the reactants for the next in the series.

- Enzymes are regulators of metabolism.
- Enzymes are subject to regulation as well.
- Inhibitors bind to enzymes and slow the rate of reaction.
  - Irreversible inhibition: Inhibitors covalently bond to an enzyme and destroy the active site
  - Reversible inhibition: Inhibitors bind to and alter the active site but the inhibition is reversible depending on the concentration of the inhibitor and substrate. Some reversible inhibitors are called competitive inhibitors because they compete with the substrate for the active site. Others are called noncompetitive inhibitors because they bind elsewhere but alter the active site so that the substrate cannot bind or the rate of binding is reduced.

Allosteric enzymes consist of multiple subunits that bind to and influence each other.

- Alteration of one subunit has an impact upon other subunits in the complex.
- Effector molecules regulate allosteric enzymes by binding to an allosteric site and changing the form and function of the entire enzyme. Effectors may function as activators or inhibitors.
- Allosteric enzymes have multiple "types" of subunits with different jobs
  - Catalytic subunits bind substrate in the active site.
  - Regulatory subunits bind effectors in one or more allosteric sites.
- Binding in either type of subunit affects the other.
- Allosteric enzymes may have multiple active sites, this allows coupling of activity.
- Allosteric enzymes may also have multiple sites for multiple levels of regulation.
- Allosteric interactions help control metabolism. The first step in an enzyme-mediated pathway is referred to as the commitment step. Once initiated, the pathway is followed to completion. Frequently, the final product allosterically inhibits the enzyme of the commitment step. This prevents overproduction of the final product. This process is called end-product inhibition.

The Big Picture

- Cells need energy in order to carry out their functions. Metabolism is a series of energy-transferring reactions that fuel the processes of the cell. All involved reactions either give off energy (exergonic) or require energy to proceed (endergonic). Energy may be stored in chemical bonds (potential energy) and released when the bonds are broken to do work (kinetic energy). ATP serves as the energy shuttle of many metabolic processes.
- Enzymes aid biological reactions by lowering the activation energy required to start the reaction. Each enzyme has a specific three-dimensional conformation that interacts specifically with the substrate. Interaction between the enzyme and substrate results in optimum orientation for a reaction to take place. Enzymes are proteins and are affected by temperature, pH, concentrations, or reactants and inhibitors.
- Metabolism is regulated by enzymes. Most metabolic pathways are under allosteric control. Enzyme complexes allow for interactions and regulation of adjacent active sites. Often the final product of a specific pathway regulates the commitment step of the pathway itself.

Common Problem Areas

- The chemistry and physics involved in this material are often overwhelming and you may question why you are learning this in biology. Remember that the biological processes are based on physical and chemical properties. To understand the function of organisms it is necessary to understand the basics of chemistry and physics as they pertain to energy transfer.
- Allosteric regulation can be confusing because the terms "allosteric regulation" and "allosteric enzyme" are related but slightly different. Remember that an allosteric enzyme is a protein complex in which each subunit interacts with the others. The fact that it is a complex leads to allosteric regulation, in which something that affects one subunit affects the entire complex.

Study Strategies

- With this chapter, the best strategy is to take "small bites." You may find the concepts to be very unfamiliar; therefore, make sure you understand each section before proceeding. This chapter contains a large amount of terminology. Making a terminology/vocabulary list may be helpful.
- When studying the energy equations, be sure that you understand how any change in either free energy, unusable energy, or absolute temperature will affect the total energy of the system. Also focus on what a larger or smaller ΔG value means in terms of how reactions proceed and whether they require or release energy.
- When thinking of enzyme mediated reactions, be sure to consider that altering anything on the left side of the equation is going to lead to changes in the amount of product formed. Always remember that enzymes are conserved across the equation.
- Review the following tutorial and activities on the website:
  Tutorial 6.1 Enzyne Catalysis
  Activity 6.1 ATP and Coupled Reactions
  Activity 6.2 Free Energy Changes

Test Yourself

Knowledge and Synthesis Questions

1. ATP is necessary for the conversion of glucose to glucose 6-phosphate. The energy released by the splitting of ATP into ADP and P_i is what type of energy?
   a. Potential
   b. Kinetic
   c. Entropic
   d. Enthalpic

2. Before ATP is split into ADP and P_i it holds what type of energy?
   a. Potential
   b. Kinetic
   c. Entropic
   d. Enthalpic

3. Which of the following statements concerning energy transformations are true?
   a. Increases in entropy reduce usable energy.
   b. Energy may be created during transformation.
   c. Potential energy increases with each transformation.
   d. Increases in temperature decrease total amount of energy available.

4. A reaction has a ΔG of –20kcal/mole. This reaction is
   a. endergonic and equilibrium is far toward completion.
   b. exergonic and equilibrium is far toward completion.
   c. endergonic and the forward reaction occurs at the same rate as the reverse reaction.
   d. exergonic and the forward reaction occurs at the same rate as the reverse reaction.

5. ATP hydrolysis releases energy to fuel cellular functions. ATP hydrolysis is
   a. endergonic.
   b. exergonic.
   c. chemoautotrophic.
   d. None of the above

6. Enzymes are biological catalysts and function by
   a. increasing free energy in a system.
   b. lowering activation energy of a reaction.
   c. lowering entropy in a system.
   d. increasing temperature near a reaction.

7. Which of the following contribute to the specificity of enzymes.
   a. Each enzyme has a narrow range of temperature and pH optima.
   b. Each enzyme has a specific active site that interacts specifically with a particular substrate.
   c. Substrates themselves may alter the active site slightly for optimum catalysis.
   d. All of the above

8. Coenzymes and cofactors as well as prosthetic groups assist enzyme function by
   a. stabilizing three-dimensional shape and maintaining active sites.
   b. assisting with the binding of enzyme and substrate.
   c. a and b
   d. None of the above

9. Which of the following are unique features of enzymes?
   a. They are not consumed by the enzyme-mediated reaction.
   b. They are not altered by the enzyme-mediated reaction.
   c. They lower activation energy.
   d. All of the above

10. Ascorbic acid, found in citrus fruits, acts as an inhibitor to catecholase, the enzyme responsible for the browning reaction in fruits such as apples, peaches, and pears. One possibility for its function could be that ascorbic acid is very similar in size and shape to catechol, the substrate of the browning reaction. If this is true, then this inhibition is most likely an example of __________ inhibition.
    a. indirect
    b. competitive
    c. direct
    d. noncompetitive

11. Refer to question 10. Suppose further studies indicate that ascorbic acid is not similar to catechol in size and shape but that the pH of the ascorbic acid solution is denaturing catecholase. If this is true, then this inhibition is most likely an example of _________ inhibition.
    a. competitive
    b. irreversible
    c. direct
    d. noncompetitive

12. Metabolism is organized into pathways. The pathway is linked in which of the following manners?
    a. All cellular functions feed into a central pathway.
    b. All steps in the pathway are catalyzed by the same enzyme.
    c. The product of one step in the pathway functions as the substrate in the next step.
    d. Products of the pathway accumulate and are secreted from the cell.

13. Which of the following represents an enzyme-catalyzed reaction?
    a. $E + P \rightarrow E + S$
    b. $E + S \rightarrow E + P$
    c. $E + S \rightarrow P$
    d. $E + S \rightarrow E$

14. Which of the following graphs of enzyme mediated reactions represents an allosteric enzyme?
    a. A graph similar to 6.22*a*
    b. A graph similar to 6.22*b*
    c. A graph similar to 6.26 (change axis labels to those in 6.22)
    d. None of the above

15. Allosteric enzymes differ in that
    a. they have multiple subunits.
    b. changes in one subunit affect the functions of the other subunits.
    c. some subunits are catalytic and others are regulatory.
    d. All of the above

Application Questions

1. It is expected that approximately 10 percent of energy that passes between levels in a food web is retained as usable energy (free energy). This appears to contradict the first law of thermodynamics. Explain the first law of thermodynamics and discuss why this apparent loss of usable energy does not contradict the law.

2. Amylase is a digestive enzyme that breaks down starch and is secreted in the mouth of humans. Amylase functions well in the mouth but ceases to function once it hits the acidic stomach environment. Explain why amylase does not function in the stomach.

3. The ultimate goal of metabolism is to drive ATP synthesis. ATP is considered the energy currency of the cell. Discuss how ATP couples endergonic and exergonic reactions and why it is so important in cellular functions.

4. Inhibition of enzymes can come from competition for the active site or from alterations in the enzyme's three-dimensional shape that alter the active site. Explain how inhibition in general allows for allosteric control of a pathway.

5. Explain how substrate concentration affects the rate of an enzyme-mediated reaction.

6. You are studying a new species never before studied. It lives in acidic pools in volcanic craters where temperatures reach 100°C. You determine that it has a surface enzyme that catalyzes a reaction leading to its protective coating. You decide to study this enzyme in the laboratory. Under what conditions would you most likely find optimum activity of this enzyme?

7. Explain how free energy, total energy, temperature, and entropy are related.

8. Use a graph to explain how temperature affects enzyme activity.

Answers

Knowledge and Synthesis Answers

1. **b**. The released energy is available to do work, therefore, it is kinetic energy.

2. **a**. Potential energy is energy held within chemical bonds that may be converted to working kinetic energy.

3. **a**. Total energy = Free energy + entropy × temperature. Any increase in entropy is necessarily going to reduce free energy.

4. **b**. A negative ΔG indicates an exergonic reaction (energy is liberated). A large ΔG indicates that equilibrium lies toward completion.

5. **b**. ATP hydrolysis results in a ΔG of –12kcal/mol.

6. **b**. Enzymes reduce activation energy and speed up reactions.

7. **d**. Enzymes are specific to particular substrates that may actually "adjust" the fit of the active site. They also function in specific, narrow optimum ranges of pH and temperature.

8. **c**. Cofactors, coenzymes, and prosthetic groups assist with the maintenance of an enzyme's three-dimensional shape, and the conformation of the active site.

9. **d**. Enzymes are not consumed or altered in any way during an enzyme-mediated reaction and they function to lower the activation energy of a reaction.

10. **b**. Competitive inhibitors compete for the active site with the substrate.

11. **b**. A low pH may actually denature the enzyme, which is an irreversible alteration of the three-dimensional shape.

12. **c**. Within a given pathway, the products of the preceding step act as substrates for subsequent steps.

13. **b**. Substrate is converted to product and the enzyme is unchanged.

14. **b**. Allosteric enzymes produce sigmoidal curves on reaction rate graphs. See Figure 6.22*b*.

15. **d**. Multiple subunits allow for multiple sites of activity and control.

Application Answers

1. The first law of thermodynamics states that energy cannot be created nor destroyed and that it may be converted from one form to another. In a food web, food consumed by one level provides the energy to be converted during metabolism. The second law of thermodynamics states that not all energy can be used. In transfer between levels, approximately 90 percent of the energy is converted to unusable heat energy. Therefore, there is a net loss of usable energy during each conversion.
2. The pH optimum of amylase is approximately 7. At that pH the protein has the three-dimensional shape to allow starch to bind to its active site and catalyze its hydrolysis. At stomach pH (approximately 2), the protein is denatured and its three-dimensional shape and active site are lost; therefore, it can no longer catalyze the reaction.
3. The conversion of ATP to ADP and P_i releases approximately 12kcal/mol of energy. This energy release fuels (endergonic) reactions in the cell. Equilibrium of the reaction is far to the right and favors the formation of ADP. In the converse, the formation of ATP from ADP and P_i is energy intensive and can be coupled to highly exergonic reactions within the cell. Thus ATP functions as an energy shuttle between endergonic and exergonic reactions. The small size of the molecule and its ubiquitousness allows it to be available and move freely within the cell.
4. In an allosterically controlled pathway, the end-product frequently acts as a control for the commitment step of the pathway. Once adequate end-product is available, it inhibits the commitment step either by competing for the active site or by indirectly altering the active site. This control may occur on an allosteric subunit of the enzyme which then alters the active site on the functional subunit.
5. Increasing substrate concentration will result in an increased rate of reaction until all available active sites are occupied. At that point, no amount of substrate increase will increase the rate of reaction. See Figure 6.16.
6. Optimum enzyme activity is most likely going to be under acidic conditions at high temperatures. Optimum activity is found when the enzyme is its most functional shape. This occurs under most enzymes' native conditions.
7. Total energy = free energy + unusable energy × absolute temperature
8. See Figure 6.26.

7 Cellular Pathways That Harvest Chemical Energy

Important Concepts

All metabolic activities are similar in their properties.

- Metabolic processes occur in pathways.
- Each step of a pathway is catalyzed by a specific enzyme.
- Organelles serve to compartmentalize pathways.
- Enzymes serve as regulators for pathways.

Glucose is the energy source most often used by living organisms.

- Most organisms, with the exception of some bacteria, use glucose as a metabolic fuel source
- Cells oxidize glucose ($C_6H_{12}O_6$) for energy:

  $C_6H_{12}O_6 + 6O_2 \rightarrow 6CO_2 + 6\ H_2O$ + energy
- The metabolism of glucose requires one or more of the following pathways:
  - Glycolysis: Glucose is converted to pyruvate with a small net energy release.
  - Cellular respiration: In the presence of O_2, pyruvate from glycolysis is converted to CO_2 and energy in the form of ATP.
  - Fermentation: In the absence of O_2, pyruvate is converted to lactic acid or ethanol with a small net energy release.
- Complete conversion of glucose to CO_2 in the presence of O_2 releases –686kcal/mol of energy. Incomplete breakdown through fermentation produces substantially less.

Redox reactions are the energy shuttling mechanism of metabolism.

- Energy is transferred from compound to compound through transfer of electrons in redox reactions.
- The gain of electrons or hydrogen atoms is called reduction.
- The loss of electrons or hydrogen atoms is called oxidation.
- Oxidation and reduction are always coupled. In order for a material to lose an electron or a hydrogen atom, another material must accept it.
- Oxidizing agents accept electrons and become reduced. In metabolism, O_2 is the oxidizing agent.
- Reducing agents donate electrons and become oxidized. In metabolism, glucose is the reducing agent.
- The net ΔG of a redox reaction is negative; therefore, energy is liberated.

Just as ATP is an energy shuttle in metabolism, NAD is an electron shuttle in the redox reactions of metabolism.

- NAD may exist as either NAD^+ or $NADH + H^+$.
- Two electrons are transferred in the conversion of NAD^+ to $NADH + H^+$ and the reaction liberates ~50 kcal/mol.
- Though glycolysis results in a net gain of two ATP, some steps are endergonic and require energy input.
- Refer to Figure 7.7 in your text to follow each step of the conversion of glucose to two molecules of pyruvate. The first five steps are energy requiring steps. Subsequent steps are energy liberating.
- The following are key points to remember about glycolysis:
  - Glycolysis begins with a 6-C sugar that is ultimately converted to two 3-C pyruvate molecules.
  - The energy intensive steps of glycolysis involve phosphorylating and converting the 6-C sugar into a molecule that can be split into two 3-C phosphorylate molecules. ATP is used as a source for phosphate groups and energy.
  - The energy producing steps occur when phosphate groups are transferred to ADP to make ATP.
  - The conversion of glyceraldehyde 3-phosphate (G3P) to 1,3-bisphosphoglycerate (BPG) is significant because it is an oxidation process and energy is held in 2 $NADH + H^+$ after this conversion. How this energy is used depends on whether the cell is respiring aerobically or going through fermentation.

    $NAD^+ + 2H \rightarrow NADH + H^+$

    This energy is reclaimed in the oxidation of $NADH + H^+$ by O_2 and is coupled to ATP synthesis in metabolism.

    $NADH + H^+ + \frac{1}{2}O_2 \rightarrow NAD^+ + H_2O$

Energy is released from glucose in different ways depending on whether oxygen is present.

- Metabolism begins with the conversion of glucose to pyruvate whether O_2 is present or not.
- If O_2 is available, cellular respiration consisting of pyruvate oxidation, the citric acid cycle, and the respiratory chain takes place.
- If O_2 is unavailable, pyruvate is converted to lactate or alcohol through fermentation reactions only.
- See Figure 7.5 for starting and end-products of each of the above processes.
- See Table 7.1 for the cellular locations of each of the above processes.

Glycolysis is the process of converting glucose to pyruvate.

- Glycolysis is a fundamental pathway occurring in the cytoplasm of the cell. It occurs during both aerobic or anaerobic cell respiration.
- Glucose is converted to two molecules of pyruvate with a net production of two ATP and two NADH + H^+.
  - Two ATPs are used in glycoysis and four ATPs are produced, resulting in a net ATP gain of two.
  - Additional energy is stored in NADH + H^+, but its liberation is dependent on further metabolic processes.
  - Substrate-level phosphorylation is driven by energy liberated through oxidation, and results in ATP formation.

Fermentation in anaerobic conditions is necessary to recycle NAD.

- Fermentation produces no additional ATP yield.
- NADH + H^+ is recycled to NAD by the reduction of pyruvate.
- If NAD is not recycled, glycolysis cannot proceed.
- Fermentation results in the conversion of pyruvate to lactate or alcohol + CO_2.

Aerobic conditions allow the oxidation of pyruvate to acetate, which is combined with CoA to form acetyl CoA.

- Pyruvate oxidation occurs on the inner mitochondrial membrane.
- CO_2 is liberated and NADH + H^+ is produced.

The citric acid cycle oxidizes acetate to CO_2 and forms $FADH_2$, NADH + H^+, and ATP.

- Figure 7.9 details the reactions of the citric acid cycle
- The concentrations of molecules involved in the citric acid cycle are maintained at constant levels and have various other roles in the cell.
- CoA is recycled, two CO_2 molecules are formed as are three NADH + H^+, one FAD, and one ATP per turn of the cycle. Remember that each glucose molecule yields two acetyl CoA entering the cycle, so yields-per-glucose are doubled.
- The majority of the citric acid cycle takes place in the mitochondrial matrix.

Energy is liberated from reduced NAD^+ and FAD through electron transport in the respiratory chain.

- Electrons are passed through a series of membrane-associated electron carriers in the mitochondria.
- Movement of electrons results in a proton gradient being established across the membrane.
- Diffusion of protons through a proton channel drives ATP synthesis.
- Figure 7.12 details the respiratory chain. Note that there are three enzyme complexes as well as cytochrome *c* and ubiquinone that shuttle electrons from NADH + H^+ and $FADH_2$. O_2 is the ultimate acceptor, resulting in H_2O as a by-product.
- The purpose of the electron shuttling is to pump protons across the mitochondrial membrane against the concentration gradient. This also polarizes the membrane (creates a charge differential). These coupled effects are called the proton-motive force.

ATP is produced as protons move back across the mitochondrial membrane.

- ATP synthase couples proton movement with the synthesis of ATP (chemiosmotic mechanism).
- Because the ATP synthase reaction may go in either direction, ATP is removed from the mitochondrial matrix immediately to favor ATP formation. Also, the proton gradient is maintained to favor ATP synthesis.
- If the proton flow is uncoupled from ATP synthase, the resulting energy is lost as heat. This phenomenon is used in thermoregulation.

If O_2 is not available, ATP is generated via fermentation.

- Fermentation occurs in the cell cytoplasm.
- Glycolysis occurs at a much higher rate under anaerobic conditions to maintain ATP levels.
- There are two types of fermentation: lactic acid fermentation and alcoholic fermentation.
- The purpose of fermentation is to reduce pyruvate (or a metabolite) and return NAD for use in glycolysis.

Fermentation and cellular respiration have different energy yields.

- Fermentation yields a net total of two ATP regardless of whether it is lactic acid fermentation or alcoholic fermentation (see Figures 7.15 and 7.16).
- Aerobic respiration yields a total of 36 ATP (see Figure 7.17).
- The general formula for aerobic cellular respiration is:

$$C_6H_{12}O_6 + 6O_2 \rightarrow 6O_2 + 6\ H_2O + 36\ ATP$$

Glycolysis and cellular respiration occur while other metabolic process are occurring.

- Components of these pathways are also components of other metabolic processes.
- Figure 7.19 shows how other metabolic pathways relate to glycolysis and cellular respiration.
- Just as intermediates can be broken down (catabolyzed) to release energy, they can also be synthesized (anabolyzed) for storage or use by the cell.
- Metabolism, both anabolism and catabolism, is regulated by enzymes to maintain stable concentrations of intermediates.

Metabolic enzymes are allosterically controlled.

- Metabolism responds to both positive and negative feedback provided by intermediates and end-products.
- Each process has a specific control point for regulation.
- This control point for glycolysis is phosphofructokinase and is inhibited by ATP. This allows glycolysis to speed up during fermentation and slow down during cellular respiration.
- The control point for the citric acid cycle is isocitrate dehydrogenase. NADH + H^+ and ATP inhibit the enzyme, and NAD^+ and ADP act as activators.
- If the citric acid cycle slows due to abundant ATP, glycolysis slows as well.
- If excess acetyl CoA is produced and ATP is abundant, acetyl CoA can be shuttled to fatty acid synthesis for storage.

Evolutionary refinements are evident in metabolic pathways.

- Regulatory systems and pathways in fine balance are a result of natural selection.

The Big Picture

- Metabolic processes occur in pathways and are regulated by enzymes. These pathways may be further controlled by compartmentalization into organelles in eukaryotic cells. Because metabolic processes are energy transferring reactions, ATP and NAD are necessary for shuttling energy and electrons between steps of the pathways. Redox reactions are the basis of metabolism. As one compound is oxidized, another is reduced.
- Glucose provides cellular energy. It may be metabolized in the absence of oxygen through glycolysis and fermentation or in the presence of oxygen through glycolysis and cellular respiration in the form of the citric acid cycle, electron transport, and oxidative phosphorylation. Anaerobic respiration produces significantly less cellular energy than aerobic respiration in the form of ATP.
- Each step of the catabolic pathways is regulated by the products of subsequent pathways and depends on oxygen being present as the final electron acceptor. This level of control is achieved through allosteric regulation.

Common Problem Areas

- The biggest mistake you can make in learning glucose metabolism is to focus on memorizing the pathways without understanding what is happening. When you begin your study, do not focus on the pathways themselves. Focus on understanding the beginning and end-products of each pathway and why the pathway does what it does. Once you conceptually understand why the pathway is present, move on to learning the pathway itself.
- Students often fail to see how regulation of the pathways occurs. By understanding how the presence or absence of intermediates affects the entire pathway, you can better understand the pathways themselves.

Study Strategies

- Remember that ATP and NAD are energy currencies. They move energy from pathway to pathway. Think about what roles ATP and NAD have in redox reactions to completely understand the pathways.
- Do not memorize pathways. Make sure you *understand* the pathways. Also understand how affecting one pathway can alter another pathway.
- This chapter is very visual in nature. You should spend a significant amount of time with the diagrams.
- Review the following tutorial and activities on the website:
  Tutorial 7.1 Electron Transport and ATP Synthesis
  Activity 7.1 Glycolysis
  Activity 7.2 Pyruvate Oxidation
  Activity 7.3 The Citric Acid Cycle
  Activity 7.4 The Respiratory Chain

Test Yourself

Knowledge and Synthesis Questions

1. Which of the following cellular metabolic processes can occur in the presence *or* the absence of oxygen?
   a. The citric acid cycle
   b. Electron transport
   c. Glycolysis
   d. Fermentation
2. Which of the following statements regarding glycolysis is/are true?
   a. A 6-C sugar is broken down to a 3-C product.
   b. Two ATP molecules are consumed.
   c. A net sum of two ATP molecules is generated.
   d. All of the above
3. During which process is most ATP generated in the cell?
   a. Glycolysis
   b. The citric acid cycle
   c. Electron transport coupled with chemiosmosis
   d. Fermentation

4. The main purpose of the electron transport chain is to
   a. cycle NADH + H^+ back to NAD^+.
   b. use the intermediates from the citric acid cycle.
   c. break down pyruvate.
   d. All of the above

5. Cellular respiration is allosterically controlled. Which of the following act as inhibitors at the various control points?
   a. ATP
   b. NADH + H^+
   c. a and b
   d. None of the above

6. Which of the following best describes the role of the mitochondrial membrane?
   a. The membrane acts as an anchor for the membrane-associated enzymes of cellular respiration.
   b. The membrane allows for the establishment of a proton-motive force.
   c. a and b
   d. None of the above

7. In a redox reaction between G3P and NAD^+ yielding BPG and NADH + H^+, __________ is oxidized and __________ is reduced.
   a. G3P, NAD^+
   b. BPG, NADH + H^+
   c. G3P, NADH + H^+
   d. NAD^+, NADH + H^+

8. Which of the following are true regarding redox reactions?
   a. Oxidizing agents accept electrons.
   b. A molecule that accepts electrons is said to be reduced.
   c. Redox reactions involve electron transfers.
   d. All of the above

9. Cyanide poisoning inhibits aerobic respiration at cytochrome *c* oxidase. Which of the following are results of cyanide poisoning at the cellular level?
   a. Oxygen is not reduced to water.
   b. ATP cannot be synthesized because electron transport is never completed.
   c. Cells (with the exception of brain cells) must switch to anaerobic respiration.
   d. All of the above

10. Not all cells can switch between aerobic and anaerobic respiration. If a cell cannot switch to aerobic respiration in the absence of oxygen, what happens?
    a. The cell concentrates oxygen to continue aerobic respiration.
    b. The cell dies.
    c. The cell uses fermentation.
    d. None of the above

11. The main purpose of cellular respiration is to
    a. convert energy stored in the chemical bonds of glucose to an energy form that the cell can use.
    b. convert potential to kinetic energy.
    c. convert kinetic to potential energy.
    d. create energy in the cell.

12. Which of the following statements concerning the synthesis of ATP is true?
    a. ATP synthesis cannot occur without the presence of ATP synthase.
    b. The proton-motive force is the establishment of a charge and concentration gradient across the mitochondrial membrane.
    c. The proton-motive force is necessary to drive protons back across the membrane through channels established by the ATP synthase channel protein.
    d. All of the above

Application Questions

1. Why is oxygen necessary for aerobic respiration?
2. Glycolysis yields two molecules of pyruvate, two ATP, and two NADH + H^+ regardless of whether oxygen is present or not. What are the fates of these molecules in the absence of oxygen? What would happen if NADH + H^+ was not recycled?
3. Explain how the proton-motive force drives chemiosmosis.
4. The fate of acetyl CoA differs according to how much ATP is present in the cell. Explain what happens to aceytl CoA when ATP is limited, and compare that to what happens when acetyl CoA is abundant. How do these processes help regulate metabolism?
5. Cellular respiration occurs simultaneously along with many other cellular processes. Describe generally how cellular respiration interacts with other cellular metabolic events.
6. Compare and contrast energy yields from aerobic respiration and fermentation.
7. One of the by-products of aerobic cellular respiration is carbon dioxide. Assuming you begin with labeled glucose, trace the fate of that molecule until carbon dioxide is released.
8. Identify the commitment steps of glycolysis, the citric acid cycle, and electron transport. What regulators affect each of these commitment steps?

Answers

Knowledge and Synthesis Answers

1. **c.** Glycolysis proceeds during both fermentation and cellular respiration. Only in cellular respiration is oxygen needed as the terminal electron acceptor of the pathway.
2. **d.** During glycolysis, 6-C glucose is broken down into two 3-C pyruvate molecules. In the process, four total ATP are produced but two are consumed, leaving a net production of two ATP molecules.
3. **c.** Most of the ATP produced during cellular respiration is produced as electron transport and chemiosmosis are coupled with oxidative phosphorylation.
4. **a.** The electron transport chain is responsible for oxidizing NADH + H^+ back to NAD^+.

5. **c**. Both ATP and NADH + H^+ allosterically control metabolism. ATP controls both phosphofructokinase and isocitrate dehydrogenase, which are commitment steps for glycolysis and the citric acid cycle respectively. NADH + H^+ controls isocitrate dehydrogenase.
6. **c**. The mitochondrial membrane is necessary for the anchoring of proteins as well as the establishment of a barrier across which a gradient can be established.
7. **a**. A molecule is oxidized when it loses electrons or protons and is reduced when it gains electrons or protons.
8. **d**. Oxidizing agents accept electrons and cause oxidation of another molecule. Reducing agents donate electrons and cause the reduction of another molecule.
9. **d**. Cyanide stops aerobic cellular respiration because cytochrome oxidase loses the ability to reduce oxygen to water. Those cells that can switch to anaerobic respiration (fermentation) do so. Brain cells cannot make that switch and so are killed.
10. **b**. In the absence of oxygen, cells that cannot go through fermentation die.
11. **a**. Cellular respiration is the cell's way of converting potential energy in the chemical bonds of glucose to potential energy that the cell ultimately can use.
12. **d**. The proton-motive force results in a concentration and charge gradient across the mitochondrial membrane. In order for that gradient to equalize, the protons must flow through a channel protein. If this channel protein has an associated ATP synthase, ATP is generated as protons flow through.

Application Answers

1. Oxygen acts as the terminal electron acceptor in the electron transport pathway. Without it, NADH + H^+ cannot be cycled back to NAD^+. The accumulated NADH + H^+ acts as an inhibitor to the citric acid cycle and effectively shuts it down. Therefore, in the absence of oxygen a cell only can undergo glycolysis.
2. In the presence of oxygen, pyruvate would be converted to acetyl CoA and cycle through the citric acid cycle to produce additional ATP and NADH + H^+. The NADH + H^+ from both glycolysis and the citric acid cycle would be oxidized during electron transport and ATP would ultimately be synthesized via chemiosmosis and oxidative phosphorylation. The two ATP would be used as cellular energy.
3. The proton-motive force results in a concentration and charge gradient across the mitochondrial membrane. In order for that gradient to equalize, the protons must flow through a channel protein. If this channel protein has an associated ATP synthase, ATP is generated as protons flow through.
4. If ATP is limited, acetyl CoA enters the citric acid cycle and cellular respiration utilizes it to ultimately produce ATP. If ATP is abundant, acetyl CoA is shuttled to fatty acid synthesis, thus storing the energy in chemical bonds.
5. Consult Figure 7.19 to see where different metabolic pathways in the cell interact.
6. Fermentation yields only two ATP. Cellular respiration yields 38 ATP. Though this difference is great, organisms can survive quite well relying on fermentation because the rate at which glycolysis occurs is increased nearly tenfold.
7. Follow Figures 7.6 and 7.10. Carbon dioxide is liberated as α-ketoglutarate is oxidized to succinyl CoA.
8. This control point for glycolysis is phosphofructokinase and is inhibited by ATP. This allows glycolysis to speed up during fermentation and slow down during cellular respiration. The control point for the citric acid cycle is isocitrate dehydrogenase. NADH + H^+ and ATP inhibit the enzyme and NAD^+ and ADP act as activators. Electron transport is controlled by the amount of NADH + H^+ fed in and by NADH-Q reductase.

8 Photosynthesis: Energy from the Sun

Important Concepts

The photosynthetic production of oxygen by green plants is necessary for most organisms to obtain the energy for life.

- Photosynthesis uses CO_2, water, and light to produce carbohydrate and O_2.

  $6CO_2 + 12H_2O \rightarrow C_6H_{12}O_6 + 6O_2 + 6H_2O$
- In plants, water for photosynthesis must be acquired by the roots and transported to leaves where photosynthesis takes place.
- In plants, CO_2, O_2, and water vapor are exchanged through pores in the leaf's surface.
- Light is required for photosynthesis to occur.
- There are photosynthetic prokaryotes and protists. Their water- and gas-exchange mechanisms are different, but the reactions are similar.
- Photosynthesis in eukaryotes occurs in chloroplasts.
- Photosynthesis occurs in a two-pathway process. The light reactions produce ATP from light energy via photophosphorylation. The Calvin–Benson cycle uses ATP and NADPH + H^+ from the light reactions to trap CO_2 and produce sugar. *Neither* pathway operates in the absence of light.

The interaction of light and pigment is fundamental to photosynthesis.

- Light comes in discrete packets called photons and has wavelike properties.
- Light has different properties depending on its wavelength. People perceive different wavelengths of visible light as different colors. See Figure 8.5.
- Photosynthesis utilizes the energy in light. Light energy is proportional to a photon's frequency. Frequency depends on wavelength and speed. Because of this, shorter wavelengths of light have greater energy.
- Light intensity refers to the amount of energy falling on a defined area.
- Pigments absorb photons and move into an excited state. The energy increase is equal to the energy of the photon.
- Pigment color is due to the wavelengths of light that are reflected. Chlorophyll appears green because blue and red light are absorbed and green light is reflected.
- Compounds have unique absorption spectra depending on which wavelengths of light are absorbed (Figure 8.7).
- An action spectrum can be measured. This spectrum indicates the effectiveness of light at particular wavelengths, allowing differentiation between pigments.
- Photosynthesis relies on chlorophyll *a*, chlorophyll *b*, and accessory pigments. Chlorophylls are more abundant. Figure 8.9 illustrates the structure of chlorophyll. Carotenoids and phycobilins allow utilization of light across the visible spectrum.

In photosynthesis, a pigment molecule in the excited state passes the absorbed energy along.

- Once light is absorbed by a pigment, two things can happen: The energy can be released as fluorescence, or it can be passed to another pigment.
- Photosynthetic antennae function to pass the energy from light from one pigment to another until the reaction center is reached and light energy is converted to chemical energy. In plants, the reaction center is chlorophyll *a*.
- Excited chlorophyll acts as a reducing agent and begins a cascade that ultimately drives ATP synthesis.
- Noncyclic electron flow produces ATP and NADPH + H^+. Water is oxidized to form O_2, H^+, and electrons. Electrons move from water to chlorophyll, through electron carriers, and ultimately to $NADP^+$. Release of free energy drives ATP synthesis through chemiosmosis.
- Noncyclic electron flow requires two photosystems to continuously absorb light. Photosystem II takes slightly higher energy (680nm) than photosystem I (700nm).
- Figure 8.12 shows the flow of electrons from the splitting of water to the reduction of $NADP^+$. ATP is formed as electrons are transported.
- Cyclic electron flow produces ATP but no NADPH + H^+. This process is necessary because the Calvin–Benson cycle requires more ATP than NADPH + H^+. Noncyclic flow produces equal amounts of ATP and NADPH; therefore, cyclic flow is necessary to generate the extra ATP.

- Cyclic electron flow involves only photosystem I and cycles back to chlorophyll (Figure 8.13). The amount of NADPH + H^+ regulates whether cyclic or noncyclic electron flow occurs. If NADPH + H^+ is abundant, electrons are accepted by plastioquinone rather than NADP.
- ATP synthesis in either cyclic or noncyclic electron flow is produced via chemiosmosis. Just as in the mitochondrion, a proton motive force coupled to ATP synthase is established in the chloroplast as electrons are passed along the transport chain.

The Calvin–Benson cycle incorporates carbon dioxide into sugars.

- Energy in the form of ATP and NADPH + H^+ from the light reactions are used to convert CO_2 to sugars. Because ATP and NADPH + H^+ are required, these reactions occur only in the light.
- The Calvin–Benson cycle occurs in the stroma of the chloroplasts.
- The cycle was revealed using radiolabeled CO_2 and observing where it was incorporated. Following this process, researchers revealed the pathway by which CO_2 is converted to sugar. See Figure 8.15 to understand how this was done.
- CO_2 is "fixed" in ribulose biophosphate (RuBP), which is split into two molecules of 3-phosphoglycerate by an enzyme called rubisco. Rubisco is the world's most abundant protein and regulates the Calvin–Benson cycle.
- ATP and NADPH + H^+ from the light reactions is then used to convert the fixed CO_2 to carbohydrate (G3P).
- Finally, ATP is used to regenerate RUBP so that additional CO_2 may be fixed.
- The resulting carbohydrate (G3P) is converted to starch or sucrose, either for storage or for conversion to glucose and fructose for cellular fuel.

Rubisco also mediates the process of photorespiration.

- Rubisco can also fix O_2 at the expense of CO_2 to produce glycolate. Glycolate is ultimately converted to glycine in the peroxisome and then to serine and CO_2 in the mitochondrion.
- ATP and NADPH from the light reaction are used in photorespiration, but CO_2 is released instead of used to make carbohydrate.
- Whether CO_2 or O_2 is fixed depends on relative concentrations of CO_2 and O_2. Excess O_2 (abundant, for example, on hot, dry days) forces photorespiration.

Photorespiration is metabolically expensive; therefore, plants have evolved mechanisms to avoid it.

- To avoid photorespiration, the level of CO_2 around rubisco must be high. This is difficult to achieve, since O_2 levels in air are much higher than CO_2 levels. If hot conditions force the closing of stomata, CO_2 levels are quickly depleted.
- Plants that fix CO_2 to 3PG are called C_3 plants and are very sensitive to CO_2 / O_2 levels Photorespiration occurs frequently and limits the range of conditions under which C_3 plants can grow.
- Many tropical plants are C_4 plants. They fix carbon through phosphoenol carboxylase and oxaloacetate using the enzyme PEP carboxylase, which has no oxygenase activity and fixes CO_2 at very low levels. The resulting 4-C compound diffuses to photosynthetic cells and their chloroplasts, releases CO_2, and is recycled. The released CO_2 goes through the Calvin–Benson cycle including rubisco as normal. The early fixation process is a CO_2 concentrating mechanism.
- CAM plants open their stomata only at night and store CO_2 for use during light. CO_2 is fixed as oxaloacetate and converted to malic acid. CO_2 is stored in malic acid until it is transferred to the chloroplasts or photosynthetic cells when light is available.

Metabolic pathways in plants involve photosynthesis and respiration.

- Because plants are autotrophic, they can synthesize all needed molecules from CO_2, H_2O, and NH_4.
- All cells, whether photosynthetic or not, go through respiration and they do so in the light and the dark.
- The Calvin–Benson cycle and the respiratory pathway are closely linked. 3PG can be converted to pyruvate and enter respiration. G3P can be converted to enter glycolysis.
- Figure 8.22 illustrates how plant metabolism is interconnected.

The Big Picture

- Photosynthesis allows organisms with the appropriate pigments and metabolic processes to convert light energy from the sun to chemical energy that can be used in the cell or stored. This conversion process forms the basis of all food webs and is vital to life on earth. Photosynthetic organisms fix carbon dioxide into sugars. This process requires water and light energy and the by-products are oxygen and water.
- The process of photosynthesis occurs in two steps. The first step utilizes light energy to start chemical energy generating the electron transport system. The movement of electrons through the electron transport system generates cellular energy in the form of ATP that can be used in the fixing of carbon dioxide. The fixing of carbon dioxide is mediated by an enzyme called rubisco and is concentration dependent.
- Rubisco can also act as an oxygenase and allow a plant to go through a metabolically expensive process called photorespiration. Plants that grow in conditions favoring photorespiration have evolved a wide variety of systems to avoid photorespiration.

Common Problem Areas

- There is the temptation to learn pathways without understanding why they occur. Be sure to focus on why the plant has a particular pathway before attempting to learn the pathway itself.

- It is very common to think that the light reactions happen in the light and the Calvin–Benson cycle occurs in the dark. This is not the case! The light reactions must occur simultaneously with the Calvin–Benson cycle in order to supply the needed energy in the form of ATP.
- Remember that plants respire as well as photosynthesize and that photorespiration and respiration are different processes. All plant cells go through cellular respiration. Cellular respiration provides the energy necessary to drive other cellular processes. Photorespiration is energy-expensive and is of little benefit to the plant. Photosynthesis occurs in specialized photosynthetic cells and happens at the same time as respiration.

Study Strategies

- The process of photosynthesis is easier to understand when you visualize what is happening. Use the figures in your book to understand where energy is flowing. Remember that what the plant is doing is taking light energy and converting it to chemical energy that can be stored. Watch for where energy is flowing and how.
- The properties of chlorophyll and light itself greatly affect how well a plant photosynthesizes. Take some time to understand the properties of light and how pigments interact with light.
- Rubisco is the most abundant protein on the planet. Focus on how the relative concentrations of oxygen and carbon dioxide dictate how the enzyme functions. Also understand that C_4 and CAM plants have modifications to avoid photorespiration.
- Review the following tutorial and activities on the website:
  Tutorial 8.1 Photophosphorylation
  Activity 8.1 Calvin-Benson Cycle
  Activity 8.2 C_3 - C_4—Leaf Anatomy

Test Yourself

Knowledge and Synthesis Questions

1. The main purpose of photosynthesis is to
   a. consume carbon dioxide.
   b. produce ATP.
   c. convert light energy to chemical energy.
   d. produce starch.

2. Which of the following best represent the components that are necessary for photosynthesis to take place?
   a. Mitochondria, accessory pigments, visible light, water, and carbon dioxide
   b. Chloroplasts, accessory pigments, visible light, water, and carbon dioxide
   c. Mitochondria, chlorophyll, visible light, water, and oxygen
   d. Chloroplasts, chlorophyll, visible light, water, and carbon dioxide

3. Chlorophyll is suited for the capture of light energy because
   a. certain wavelengths of light raise it to an excited state.
   b. in its excited state chlorophyll gives off electrons.
   c. chlorophyll's structure allows it to attach to thylakoid membranes.
   d. All of the above

4. Plants give off oxygen because
   a. oxygen results from the incorporation of carbon dioxide into sugars.
   b. plants do not respire since they photosynthesize.
   c. water is the initial proton donor, leaving oxygen as a photosynthetic by-product.
   d. All of the above

5. Cyclic and noncyclic electron flow is used in plants to
   a. meet the ATP demands of the Calvin–Benson cycle.
   b. avoid producing excess NADPH + H^+.
   c. balance ATP and NADPH + H^+ ratios in the chloroplast.
   d. All of the above

6. Which of the following statements concerning the light reactions of photosynthesis are true?
   a. Photosystem I can operate independently of Photosystem II.
   b. Photosystems I and II are activated by different wavelengths of light.
   c. Photosystems I and II transfer electrons and create proton gradients across the thylakoid membrane.
   d. All of the above

7. ATP is produced during the light reactions via
   a. carbon dioxide fixation.
   b. chemiosmosis.
   c. reduction of water.
   d. All of the above

8. Because of the properties of chlorophyll, plants need adequate __________ light to grow properly.
   a. green
   b. blue and red
   c. infrared
   d. ultraviolet

9. Which of the statements concerning the Calvin–Benson cycle are *false*.
   a. Light energy is not required for the cycle to proceed.
   b. Carbon dioxide is assimilated into sugars.
   c. RuBP is regenerated.
   d. None of the above

10. Which of the following statements concerning rubisco are *true*.
    a. Rubisco is an enzyme.
    b. Rubisco catalyzes both the beginning steps of photorespiration and the Calvin–Benson cycle.
    c. Rubisco is the most abundant protein on earth.
    d. All of the above

11. Which of the following begins the Calvin–Benson cycle and is the commitment step that results in the entire pathway being carried out?
    a. 3PG is reduced to G3P using ATP and NADPH + H^+.
    b. The regeneration of RuBP.
    c. Carbon dioxide and RuBP join forming 3PG.
    d. As a cycle, it can start at any point.

12. The Calvin–Benson Cycle results in the production of
    a. glucose.
    b. starch.
    c. rubisco.
    d. G3P.

13. Which of the following statements regarding photorespiration are true?
    a. Photorespiration is a metabolically expensive pathway.
    b. Photorespiration is avoided when carbon dioxide is abundant.
    c. Photorespiration results in a loss of usable carbon dioxide.
    d. All of the above

14. The fixation of carbon dioxide by PEP carboxylase functions to
    a. concentrate carbon dioxide for use in photosynthetic cells.
    b. allow plants to close stomata without having photorespiration occur.
    c. allow plants to photosynthesize in the dark.
    d. a and b

15. CAM plants differ from C_4 plants in that
    a. carbon dioxide is stored as malic acid.
    b. photosynthesis can occur at night in these plants.
    c. their stomata close during periods that favor photorespiration.
    d. they use PEP carboxylase to fix carbon dioxide.

16. Which of the following statements are true regarding the relationship between photosynthesis and cellular respiration in plants?
    a. Photosynthesis occurs in specialized photosynthetic cells.
    b. Cellular respiration occurs in specialized respiratory cells.
    c. Cellular respiration and photosynthesis can occur in the same cell.
    d. a and c

Application Questions

1. Plants consume carbon dioxide and give off oxygen; how is this so if plants must also undergo cellular respiration?
2. Why do plants undergo both the light reactions of photosynthesis and the Calvin–Benson Cycle? Why don't they simply use the ATP produced in the light reactions of photosynthesis to drive cellular processes?
3. Why do plants undergo both photosynthesis and cellular respiration? Why don't they simply use the ATP produced in the light reactions of photosynthesis to drive cellular processes?
4. Rubisco has both carboxylase and oxygenase activities. These processes compete with one another. What determines which function the enzyme has? What conditions favor photosynthesis? What conditions favor photorespiration?
5. How did the evolution of photosynthetic organisms lead to the evolution of life on land?
6. Compare and contrast C_3 and C_4 plants.
7. The Calvin–Benson cycle was once referred to as the "dark" reactions of photosynthesis. Why is this a misnomer?
8. Explain the differences between cyclic and noncyclic electron flow. Why are both processes necessary?
9. How do accessory pigments enhance photosynthetic activity in plants?
10. Why are plants green?

Answers

Knowledge and Synthesis Questions

1. **c.** Photosynthetic organisms are the only life forms capable of trapping light energy and converting it to chemical energy. Because of this they form the basis of food chains.
2. **d.** Chloroplasts are the site of the photosynthetic reactions; chlorophyll is excited by photons of light and serve as reaction centers for the photosystems; visible light is necessary to excite chlorophyll and accessory pigments; water is the initial electron donor for the pathway; and carbon dioxide is necessary to make energy storage precursor molecules.
3. **d.** The "tails" of chlorophyll molecules are associated with the thylakoid membranes of the chloroplasts. This close membrane association assists with establishing the proton motive force that will allow ATP synthesis. When excited by light, the chlorophyll moves into an exited state and passes electrons to acceptor molecules. This begins to set up the proton gradient across the membrane that will drive ATP synthesis.
4. **c.** Water is split at Photosystem II to donate electrons to the reaction center. The resulting protons are moved across the membrane to establish the proton motive force and oxygen is given off as a by-product.
5. **d.** ATP is required at higher levels in the Calvin–Benson cycle than NADPH + H^+ is; therefore, there must be a mechanism for producing additional ATP. Cyclic electron flow provides that mechanism. If noncyclic electron flow were to be sped up to meet ATP needs, an excess of NADPH + H^+ would result. Shifting between cyclic and noncyclic flow balances ATP/NADPH + H^+ ratios.
6. **d.** Photosystems I and II operate depending on whether electron flow is cyclic or noncylic. Activity is controlled by the ATP levels in the chloroplast. Photosystem II is activated by light of a higher energy level than photosystem I. Both photosystems transfer electrons and create proton gradients across the thylakoid membranes.

7. **b.** ATP synthesis occurs when protons flow through a channel protein with an associated ATPase. This is a classically chemiosmotically driven process.
8. **b.** Chlorophyll and accessory pigments absorb light in the blue and orange-red wavelengths of visible light. Green light is reflected; therefore, plants appear green.
9. **a.** Light energy is required for the Calvin–Benson cycle to proceed. ATP synthesis is dependent on light energy and the Calvin–Benson cycle is dependent on ATP.
10. **d.** Rubisco, the most abundant enzyme on earth, has both oxygenase and carboxylase activities.
11. **c.** The first step of the Calvin–Benson cycle is the fixation of carbon dioxide into 3PG. This is the regulatory step and requires ATP and NADPH + H^+.
12. **d.** The Calvin–Benson cycle produces only G3P. G3P can then be metabolized into storage products like sugars and starch.
13. **d.** Photorespiration uses as much ATP as photosynthesis, but results in no energy gains for the plant and has carbon dioxide as a lost by-product. If carbon dioxide is abundantly available, rubisco acts as a carboxylase rather than an oxygenase.
14. **d.** Plants do not photosynthesize in the dark. PEP carboxylase allows the fixation of carbon dioxide and the concentration of it at rubisco.
15. **a.** CAM plants functionally store carbon dioxide as malic acid.
16. **d.** Cellular respiration occurs in *every* living cell.

Application Questions

1. Plant cells undergo cellular respiration in all cells. Therefore, all cells consume oxygen. Photosynthesis occurs in specialized cells that consume both carbon dioxide and oxygen. Because atmospheric oxygen levels are high, there is excess oxygen available for the plant to utilize; therefore, oxygen continues to be evolved from the plant.
2. The light reactions of photosynthesis produce ATP. ATP cannot be stored for use later (such as when light is not available); therefore, there has to be a mechanism for that energy to be stored. The Calvin–Benson cycle allows the energy to be stored in the chemical bonds of G3P and moved to longer term storage in the form of carbohydrates.
3. Though photosynthesis produces all the necessary energy for a plant, a plant cannot be continuously photosynthetically active. Therefore, a plant stores energy in carbohydrates. Cellular respiration is necessary to break down stored carbohydrates.
4. Whether rubisco acts as a carboxylase or an oxygenase depends on the relative ratio of oxygen to carbon dioxide. At higher carbon dioxide levels it acts as a carboxylase. At low carbon dioxide levels it acts as an oxygenase. The oxygenase activity is enhanced under conditions favoring photorespiration (hot, dry weather that forces the closing of stomata). Oxygenase activity is favored when stomata can remain open and light intensity is optimal.
5. Increases in photosynthesis lead to an oxygen rich atmosphere. Abundance of oxygen led organisms to evolve aerobic metabolism and the ability to live a terrestrial existence once atmospheric oxygen levels reached a high enough point.
6. Refer to Table 8.1 in your book.
7. Light is required for both the light reactions of photosynthesis and the Calvin–Benson cycle. The Calvin–Benson cycle depends on the ATP generated during the light-dependent reactions.
8. See Figures 8.13 and 8.14. See also the answer to 5 under "Knowledge and Synthesis Questions."
9. Accessory pigments allow utilization of light in all wavelengths of the visible spectrum. The energy absorbed is channeled to the reaction centers of the photosystems.
10. The primary pigments in plants are chlorophylls. Chlorophylls absorb blue and orange-red wavelengths of light and reflect green light, thus making plants appear green. See the absorption spectra and action spectra of chlorophyll in Figures. 8.7 and 8.8.

9 Chromosomes, the Cell Cycle, and Cell Division

Important Concepts

Cell division

- Cell division requires an environmental signal, subsequent DNA replication, segregation of the newly replicated chromosomes to opposite poles of the cell, and the division of the cytoplasm to form two daughter cells.
- Cells divide asexually to produce two daughter cells with identical genetic information.
- In eukaryotic cells, asexually dividing cells segregate their chromosomes during mitosis. For organisms that reproduce sexually, haploid gametes are formed during meiosis.

Prokaryotes divide by fission.

- Prokaryotic cells divide by increasing in size, replicating their DNA (a single circular chromosome), and dividing into two new cells through the process of fission.
- DNA replication in prokaryotic cells requires a number of different proteins and special sites on the chromosome where replication begins (the *ori* site) and terminates (the *ter* site).

Eukaryotic cells divide by mitosis or meiosis.

- Eukaryotic cells have more than one chromosome and these chromosomes are linear.
- Eukaryotic cells replicate their chromosomes, segregate those chromosomes during mitosis, and divide their cytoplasm during cytokinesis.
- DNA replication, mitosis, and cell division occur in a special sequence during the reproductive cycle of the cell, also known as the cell cycle.
- There are two major phases of the cell cycle, interphase and mitosis. Interphase is divided into three distinct parts, G1 (or Gap1), S (or DNA Synthesis), and G2 (or Gap2).
  - G1 is the phase of the cell cycle in which the decision and subsequent preparation for DNA synthesis occurs.
  - S is the phase of the cell cycle in which all of the chromosomes are replicated.
  - During G2, the cell prepares for mitosis.
- During M (mitosis), the cell segregates the newly replicated chromosomes to opposite poles of the cell. At the end of M, cytokinesis occurs and cell division is complete.

Both kinases and cyclins play a key role in cell cycle transitions.

- Cdk, a kinase that phosphorylates other proteins in the cell, is catalytically active when it is bound to another protein, cyclin.
- Different Cdk-cyclin protein complexes signal the transition of G1 to S, the initiation of DNA replication, the condensation of the chromosomes in G2, and the entry into the cell cycle.
- Other proteins, in response to environmental signals, regulate Cdk-cyclin complexes.

Eukaryotic chromosome structure

- When eukaryotic chromosomes are replicated, the daughter chromosomes, termed sister chromatids, are still joined at the centromere.
- In a eukaryotic nucleus, DNA molecules are wrapped around beadlike particles of protein called nucleosomes. Nucleosomes are composed of eight histone molecules, two from each class. Histones are very basic proteins.
- The nucleosomes are further condensed around a central axis into a 30 nm fiber, with the assistance of another histone, histone 1. These fibers of DNA and nucleosomes are further compacted into loops. During mitosis more compaction occurs, and the sister chromatids become visible. See Figure 9.7.

Preparation for separation of the chromosomes

- Separation of the sister chromatids into the daughter cells is facilitated by microtubules.
- Microtubules grow from a microtubule-organizing center, the centrosome, and form a spindle along which the chromosomes will move. Later in the cell cycle, some of these microtubules attach to the chromosomes and help separate them to opposite poles of the cell.
- Centrosome duplication (controlled by a Cdk-cyclin complex) occurs during S phase.
- Duplicated centrosomes migrate to opposite poles of the cell during the G2-to-M transition.

Mitosis

- Mitosis can be divided into five phases: prophase, prometaphase, metaphase, anaphase, and telophase.

- During prophase, the newly replicated chromosomes condense so that the sister chromatids (which are held together by cohesion proteins localized at the centromere) are visible. Kinetochores assemble at the centromere of each newly replicated chromosome.
- Prometaphase signals the disintegration of the nuclear lamina, the breakup of the nuclear envelope into small vesicles, and the attachment of microtubules to the kinetochores of each newly replicated chromosome.
- Metaphase signals the positioning of all of the kinetochores at the equatorial (metaphase) plate. This positioning of the chromosomes is facilitated by repulsive forces from each pole and by the cohesion proteins at each centromere. Kinetochores of each sister chromatid are selectively attached to microtubules emanating from opposite poles of the cell.
- The end of metaphase occurs when the cohesion proteins break down. DNA topoisomerase II unravels the interconnected DNA at the centromere and the sister chromatids separate.
- During anaphase, the chromosomes move to opposite poles of the spindle assisted by dynein motors acting at the kinetochores, by the shortening of the microtubules at the poles, and by the sliding of polar microtubules which pushes the poles further apart.
- At telophase, the poles break down, the chromosomes decondense, and the nuclear envelope begins to form around the newly segregated chromosomes, resulting in two identical nuclei.

Cytokinesis

- During cytokinesis, the cell membrane contracts due to the interaction of actin and myosin microfilaments at the cell membrane. For a summary of these mitotic events, see Figure 9.8.
- Upon completion of cytokinesis, there are two distinct cells, each with a full complement of chromosomes.
- In plant cell cytokinesis, a cell plate is formed between the newly segregated chromosomes. This cell plate is derived from Golgi vesicles, which fuse to form a plasma membrane. The plasma membrane secretes plant cell wall materials to complete cell division.

Cell reproduction

- There are two kinds of reproduction: asexual, in which the replicated chromosomes are separated by mitosis, and sexual, in which two different cells (gametes) fuse to form a zygote. This fusion is called fertilization. Gametes each contain a single set of chromosomes resulting from meiosis, and the zygote contains two sets of chromosomes, one set from each parent.
- Diploid organisms contain pairs of homologous chromosomes, one set of chromosomes coming from each parent. Haploid cells contain only one set of chromosomes.
- Mature organisms can exist primarily in the diploid or haploid state or can alternate between haploid and diploid states during their life cycles.
- Haplontic organisms are haploid as mature organisms. They produce spores that fuse to form a zygote, which is diploid. The zygote then undergoes meiosis to become haploid again. Those cells divide mitotically to become the mature organism.
- Diplontic organisms are diploid in their mature state; only their gametes are haploid.
- Some plants have mature forms that alternate between diploid and haploid stages (see Figure 9.12).
- Fusion of gametes to form a zygote increases the diversity and perhaps overall survival of that organism, because its genetic makeup is derived from two different parents, each contributing one set of chromosomes.
- Mitotic chromosomes can be characterized by their shape, number, and size, and each organism has a distinctive set, called a karyotype.

Meiosis

- Meiosis is distinguished from mitosis by two nuclear divisions (meiosis I and II), but only one round of DNA replication, resulting in a reduction in the number of chromosomes from diploid to haploid.
- Each haploid cell has one complete set of chromosomes, and the combinations of alleles on the chromosomes of each gamete are different. The advantage of creating gametes with different combinations of alleles is genetic diversity in the future zygote, potentially increasing its ability to survive.
- There are five phases in each meiotic division: prophase, prometaphase, metaphase, anaphase, and telophase.
- In prophase of meiosis I, the newly replicated chromosomes pair with their homologs. Since there are four chromatids (two from each homolog), these paired chromosomes are called tetrads or bivalents.
- The synaptonemal complex is the aggregate of proteins that promotes the synapsing of the chromatids during prophase. Crossing over between sister chromatids occurs during prophase, at special sites called chiasmata (see Figure 9.15).
- Prometaphase of meiosis I marks the disappearance of the nuclear envelope, the formation of the spindle, and the attachment of the microtubules to the kinetochore. Unlike mitosis, in which there are two kinetochores per replicated chromosome, in meiosis I there is only one kinetochore per chromosome. As a result, when the chromosomes are separated to opposite poles by microtubules attached to the kinetochore, the entire homolog is pulled to the pole.
- In some organisms the completion of meiosis I is followed by an interphase termed interkinesis, while other organisms enter meiosis II immediately after meiosis I.
- No DNA replication occurs before meiosis II begins. The number of chromosomes on the equatorial plate is haploid, even though each chromosome still consists of two sister chromatids. Unlike in mitosis, each chromatid is different from its sister chromatid, due to crossing over in meiosis I.

- Segregation of each chromatid occurs as in mitosis, with microtubules forming a spindle and attaching to the kinetochores (of which there are now two) on each chromosome.
- Each of the four haploid nuclei resulting from meiosis is genetically different due to crossing over in prophase I, and due to the random segregation of each chromatid to the daughter cells. In humans there are 23 homologs and thus 2^{23} ways these chromosomes can segregate into the daughter nuclei.

Abnormal numbers and kinds of chromosomes

- Aneuploidy, an abnormal number of chromosomes, occurs in cells whose chromosomes fail to segregate in meiosis I (both homologs go to the same pole) or meiosis II (both chromatids go to one pole).See Figure 9.18. One of the causes of nondisjunction in meiosis I is the failure to form chiasmata.
- Translocations, another chromosome abnormality, occur when part of one chromosome becomes attached to another chromosome.
- Polyploids are cells having three ($3n$) or four ($4n$) or more sets of chromosomes and result from the lack of cytokinesis after mitosis.

Cell death

- Cell death can occur in cells in two ways: by necrosis (cells poisoned or starved of essential nutrients) or by programmed cell death, known as apoptosis. Apoptosis is a normal process during development.
- Signals for cell death include a lack of mitotic signals and recognition of DNA damage. See Table 9.2.

The Big Picture

- In the presence of an environmental signal, cells undergo a specific set of steps to replicate their DNA; segregate those newly replicated chromosomes, and divide their cytoplasm into two daughter cells. These steps ensure that the daughter cells receive a complete set of genetic instructions.
- Diploid organisms produce haploid gametes, which fuse during fertilization to form new diploid organisms.
- During the generation of these gametes, crossing over between sister chromatids and random segregation of those chromatids create genetically diverse gametes, leading to increased species diversity and survival.

Common Problem Areas

- Be sure you can distinguish the differences between mitosis and meiosis. In mitosis, the replicated chromosomes are segregated to the daughter cells, resulting in each daughter cell's having two sets of chromosomes (diploid) like its parent cell. After every DNA replication, there is one cell division. In meiosis, the replicated chromosomes undergo two cell divisions (for every one round of DNA replication), resulting in cells that have only one set of chromosomes (haploid). Furthermore, the homologous chromosomes pair in prophase of meiosis I (which is not the case in mitosis) and crossing over occurs.

Study Strategies

- Make several sequential drawings that show a cell with four chromosomes and what those chromosomes look like after DNA replication and during each phase of mitosis. Repeat these sequential drawings for meiosis I and II, using different colors for each homologous chromosome. In your drawings be sure to include crossing over in prophase I of meiosis.
- Draw a picture of the cell cycle and indicate the points at which the cyclin-Cdk complexes act during that cell cycle. Refer to Figure 9.5 as well as the information in your text.
- List the five phases of mitosis and describe what happens in each phase. Repeat this process for meiosis.
- Review the following tutorials and activities on the website:
  Tutorial 9.1 Mitosis
  Tutorial 9.2 Meiosis
  Activity 9.1 Images of Meiosis
  Activity 9.2 Images of Mitosis
  Activity 9.3 Sexual Life Cycle
  Activity 9.4 The Mitotic Spindle

Test yourself

Knowledge and Synthesis Questions

1.Which of the following is true of mitosis?
   a. The chromosome number in the resulting cells is halved.
   b. DNA replication is completed prior to the beginning of this phase.
   c. The chromosome number of the resulting cells is the same as that of the parent cell.
   d. b and c

2. Which of the following is true of meiosis?
   a. The chromosome number in the resulting cells is halved.
   b. DNA replication occurs after the beginning of this phase.
   c. The homologs do not pair during prophase I.
   d. The chromosome number of the resulting cells is the same as that of the parent cell.

3. Which of the following is true of kinetochores?
   a. They are localized at the centromere of each chromosome.
   b. They are the sites where microtubules attach to separate the chromosomes.
   c. They are organized so that there is one per sister chromatid in meiosis.
   d. All of the above

4. Which of the following is true of the mitotic spindle?
   a. It is composed of actin and myosin microfilaments.
   b. It is composed of kinetochores at the metaphase plate.
   c. It is composed of microtubules, which help separate the chromosomes to opposite poles of the cell.
   d. It originates only at the centrioles in the centrosomes.

5. Imagine that there is a mutation in the Cdk2 gene such that its gene product is nonfunctional. What kind of effect would this mutation have on a nondividing differentiated cell?
   a. The cell would be unable to replicate its DNA.
   b. The cell would not be able to enter G1.
   c. The cell would be unable to reproduce itself.
   d. There would be no effect, because differentiated cells do not enter the cell cycle.

6. Imagine that there is a mutation in the Cdk2 gene such that its gene product is nonfunctional. What kind of effect would this mutation have on a dividing cell?
   a. The cell would be unable to replicate its DNA.
   b. The cell would be unable to enter G1.
   c. The cell would be unable to reproduce itself.
   d. a and c

7. The cyclin D-Cdk4 complex is only active during G1 of the cell cycle in humans. Why?
   a. The environmental signals to stimulate the cell cycle are active only in G2.
   b. Cyclin D is destroyed at the end of G1 and thus cannot form a catalytically active cyclin D-Cdk4 complex.
   c. Cdk4 is destroyed at the end of G1 and cannot form a catalytically active protein cyclin D-Cdk4 complex.
   d. Cyclin D is only needed to activate DNA synthesis.

8. Which of the following is true of chromatids?
   a. They are replicated chromosomes still joined together at the centromere.
   b. They are identical in mitotic chromosomes.
   c. They are identical in meiotic chromosomes.
   d. a and b

9. Histones are positively charged because
   a. the majority of the ions in the nucleus of the cell are negatively charged.
   b. histones interact with acidic residues of proteins found in the nucleus.
   c. the basic side chains of histone proteins interact with the negatively charged DNA.
   d. histones have a majority of acidic residues in their protein sequence.

10. Chromosome movement during anaphase is the result of
    a. the molecular motors at the kinetochores that move the chromosomes toward the poles.
    b. molecular motors at the centrosome that pull the microtubules toward the poles.
    c. shortening of the microtubules at the centrosome that pull the chromosomes toward the poles.
    d. a and c

11. Programmed cell death (apoptosis)
    a. occurs in cells that have been deprived of essential nutrients.
    b. occurs only in cells that have damaged DNA.
    c. is a natural process during development.
    d. is signaled by the initiation of mitosis.

Application Questions

1. How is cell division different in prokaryotic cells and eukaryotic cells?
2. By administering a drug that inhibits cytokinesis, you have created peaches that are tetraploid. How many sets of chromosomes do these peaches have? (What is the ploidy of these chromosomes?) Will these peaches produce gametes that are fertile? What if the peaches were triploid?
3. How does cytokinesis differ in animal and plant cells?
4. Describe how two meters of DNA in a typical human cell is localized into the nucleus, which is 5 μm in diameter.
5. Describe two ways that the genetic diversity of organisms is increased during meiosis.
6. What is the fate of cells that exit the cell cycle?

Answers to Questions

Knowledge and Synthesis Answers

1. **d.** Mitosis occurs after DNA replication and results in cells having the same number of chromosomes as the parent cell.
2. **a.** Meiosis occurs after one round of DNA replication, homologous chromosomes pair during prophase I of meiosis, and the resulting cells have half the number of chromosomes as the parent cell.
3. **d.** Kinetochores are assembled at the centromere of each chromosome and are the sites where microtubules attach to segregate the chromosomes. In meiosis there is only one fused kinetochore per chromosome; in mitosis there are two kinetochores per chromosome.
4. **c.** The mitotic spindle is composed of microtubules, not actin and myosin filaments. The spindle originates from the centrosome, which may or may not have centrioles.
5. **d.** Only dividing cells go through the cell cycle; nondividing differentiated cells do not enter the cell cycle.
6. **d.** Dividing cells do enter the cell cycle and cyclin-Cdk complexes signal transitions in the cell cycle. The cyclin E-Cdk2 complex and the cyclin A-Cdk2 complex are signals for DNA replication. Without functional Cdk2, no catalytically active cyclin-Cdk complexes can form. The cell will be unable to replicate its DNA and will not progress through the cell cycle to reproduce itself.
7. **c.** Cyclin is destroyed at the end of G1 and therefore cannot interact with Cdk4 to form a catalytically active complex.
8. **d.** Chromatids are highly condensed, newly replicated chromosomes, which will be segregated to the daughter cells. After DNA replication, chromatids are still attached to one another at the centromere. Meiotic sister chromatids are different from one another due to crossing over in prophase of meiosis I. Mitotic sister chromatids are identical.

9. **c.** The positive charges on histone proteins are due to the large number of basic amino acid residues found in these proteins. These positive charges interact with the negatively charged phosphate sugar backbone of DNA during assembly of the DNA on the nucleosome.
10. **d.** Chromosomes are attached to the microtubules at their kinetochores. There are molecular motors at the kinetochores which help move the chromosomes to opposite poles. Chromosomes are also pulled toward the poles by the shortening of the kinetochore microtubules.
11. **c.** Programmed cell death occurs during the development of many organisms (for instance, tadpoles lose their tails to become adult frogs). One of the stimuli for programmed cell death is DNA damage, but it is not the only cause of death. Necrosis (cell death that is not programmed) occurs when cells have been deprived of essential nutrients. The initiation of mitosis is part of the cell cycle, in which cells reproduce, and is not a step in programmed cell death.

Application Answers

1. In prokaryotic cells there is only one circular chromosome which is attached to the membrane. As the cell enlarges to prepare for division, the newly replicated daughter chromosomes, which are still attached to the membrane, are separated at opposite sides of the cell. During fission, the cell membrane pinches in and cell wall components are synthesized between the daughter cells. In eukaryotic cells, there are more chromosomes and they are linear. The cell undergoes a sequential set of steps called the cell cycle, in which the chromosomes are replicated and then separated to opposite poles of the cell. Microtubules are used to segregate the chromosomes equally into the daughter cells, and actin and myosin filaments cause the cell membrane to contract to form two daughter cells.
2. Peaches that are tetraploid have four sets of chromosomes. Since there is an even number of chromosomes ($4n$), each replicated homologous chromosome will be able to find a replicated homolog to pair with at meiosis and will produce fertile gametes. These gametes will be diploid. Triploid cells will not be fertile because one of the three homologs will not find its pair during prophase of meiosis I and the single homologs will be segregated randomly into the daughter cells.
3. In animal cells cytokinesis results from the interaction of myosin and actin filaments which causes the cell membrane to pinch in and divide the cytoplasm into two cells. In plant cells, a cell plate forms between the newly segregated chromosomes and Golgi vesicles fuse at that site to form the new cell membranes. Cell wall components are then secreted between the plasma membranes to complete cytokinesis.
4. See Figure 9.7.
5. Genetic diversity is increased during crossing over of prophase I of meiosis so that each gamete has chromosomes with different combinations of alleles. During meiosis, each homologous chromosome is randomly segregated to one of the two poles, leading to 2^{23} different combinations of homologous chromosomes per gamete.
6. Cells that exit the cell cycle no longer divide. In multicellular organisms, cells that have differentiated into particular structures have exited the cell cycle.

10 Genetics: Mendel and Beyond

Important Concepts

Genes are units of inheritance.

- Mendel studied the inheritance of characteristics (such as flower color) and traits (a particular feature, such as a red, purple or white color) as they were passed from one generation of pea plants to another.
- He used plants that were true-breeding (which meant that the observed trait must be the only form for many generations). To test if plants were true-breeding, Mendel allowed these plants to self-fertilize for many successive generations to determine if the trait was consistently expressed in every generation.
- In genetic crosses, the two individuals in the initial cross are the parents (the P generation) and their offspring are the first filial generation (the F_1 generation). The F_2 (second filial generation) are the offspring from a cross of the F_1 generation.
- Monohybrid crosses are matings between two true-breeding parents, each expressing one different trait, i.e., a cross between one true-breeding parent with a smooth seed and one true-breeding parent with a wrinkled seed. Mendel found that in these crosses, the F_1 generation expressed only one trait (smooth), and in the F_2 generation both the smooth and the wrinkled traits were expressed in ratios of 3:1 (smooth to wrinkled). He termed the F_1 trait (the smooth seed) dominant and the F_2 trait (the wrinkled seed) recessive.
- Mendel's theory stated that particles of inherited traits existed in pea plants and that these particles were assorted independently during gamete formation. Each of the gametes received one of these particles, while the zygote received two particles for a trait, one from each parent.
- Alleles are different forms of the same gene. The gene for seed shape in peas has a smooth allele and a wrinkled allele.
- True-breeding parents are homozygous for that trait and have two copies of the same allele, one on each homologous chromosome.
- Heterozygous individuals have two different alleles for a gene, one on each homologous chromosome. For example, the F_1 generation of a monohybrid cross between pea plants with smooth seeds and pea plants with wrinkled seeds is heterozygous for seed shape, having one smooth allele and one wrinkled allele.
- The phenotype of an organism is the physical appearance of that organism (i.e., seeds that are smooth or wrinkled).
- Mendel's theories have been restated as laws. Mendel's first law, the law of segregation, states that when any individual produces gametes, alleles separate so that each gamete receives only one allele for that trait.
- Punnett squares can be utilized to analyze the potential outcome of a genetic cross. All of the possible gametes from one parent are lined up on one side of the Punnett square and all of the gametes of the other parent are lined up on the other side. Genotypes of the progeny are predicted by placing both alleles of the gametes in the square (see Figures 10.4, 10.6, and 10.7).
- Test crosses are used to determine if a given individual is heterozygous or homozygous for a dominant allele. That individual is crossed with an individual homozygous for the recessive trait (the test cross). If the test individual is heterozygous, then one-half of the offspring from this cross will express the dominant trait and one-half will express the recessive trait. If the individual from this test cross is homozygous for that trait, then all of the offspring will express the dominant trait (see Figure 10.6).
- Mendel's second law, the law of independent assortment, states that all of the alleles of different genes assort independently during gamete formation. When parents that are heterozygous for two traits—for example, seed shape and seed color—are crossed, all of the alleles are expressed in the F_2 generation. In the F_2 progeny from this dihybrid cross, both yellow and green seed color are observed as well as wrinkled and smooth seeds. Review the Punnett square in Figure 10.7.

Probability calculations can be used to predict the outcome of genetic crosses.

- The product rule states that when predicting the outcome of two independent events, that outcome is equal to the product of the probability of each of those individual events (see Figure 10.9).
- The sum rule is used to predict the probability of an event that can occur in two or more different ways, and is equal to the sum of the individual probabilities (see Figure 10.9).

Pedigrees can be used to predict how traits are inherited from generation to generation.

- If a trait is due to a rare dominant allele:

- the affected person has an affected parent;
- on average, one-half of the offspring of an affected person will be affected;
- the phenotype is seen in both sexes.

- If a trait is due to a rare recessive allele:
  - the unaffected parents have affected children;
  - on average, one-fourth of the children from unaffected parents express the trait;
  - the phenotype occurs in both sexes.

Mutations are changes in the DNA sequence that are stable and passed on to the progeny (heritable).

- "Wild type" refers to traits that occur in most individuals in nature, while "mutant" refers to traits that are different from wild type.
- A polymorphic trait is a trait that has many different forms, with each individual phenotype present in less than 99 percent of the population.

There are several extensions of Mendel's laws.

- Multiple alleles: There can be more than one allele for a particular trait (for example, rabbit fur color; see Figure 10.12).
- Incomplete dominance: Individuals that are heterozygotic for a particular trait appear as intermediate forms of the parents. For example, a cross between a red-flowered plant and a white-flowered plant yields F_1 plants with pink flowers.
- Codominance: Both alleles are expressed in the individual (for example, blood types in humans).
- Pleiotrophic: A single allele can have more than one effect in an individual. For example, in the Siamese cat, a single allele can affect pigmentation and crossed eyes.
- Epistasis: The phenotypic expression of one gene is masked by another gene. For example, an allele that prevents pigmentation in fur can mask other alleles that produce a specific pigment in that individual, producing an individual lacking any color, an albino.
- Hybrid vigor: Individuals heterozygous for a particular trait are superior to the homozygote. For example, hybrid varieties of corn produce much better yields than their homozygous parents.
- Polygenes: More than one gene can affect a particular phenotype, with each allele intensifying or diminishing the phenotype (for example, human skin pigmentation).
- Environmental effects: Environmental conditions such as light, temperature, or nutrition can affect the expression of a particular trait. Both penetrance and expressivity are related to environmental effects.
  - Penetrance is the proportion of individuals within a group with a particular genotype that actually express that genotype.
  - Expressivity is the degree a particular genotype is expressed in an individual.
- Linkage: Different alleles on the same chromosome do not assort independently, as first noted by Morgan in his studies on *Drosophila*.
  - Linkage groups are the full set of loci on a given chromosome.
  - Recombination frequencies between alleles that are linked (due to crossing over during prophase of meiosis I) can be used to map positions of genes on a chromosome. Recombination frequencies are expressed in map units. A map unit corresponds to a frequency of 0.01. Recombination frequencies are also expressed in Centimorgans, which is equal to the recombination frequency multiplied by 100.

Sex is determined in different ways in different species.

- Monoecious organisms, (earthworms, pea plants, corn, etc.) produce both female and male gametes.
- Dioecious organisms, (vertebrates, some insects) produce only male or only female gametes resulting in two separate sexes.
- In bees, sex determination is decided by number of chromosomes and environmental conditions. Diploid bees are female workers, and diploid bees that are fed royal jelly can become queens. Haploid bees are males.
- In female grasshoppers, sex is determined by the presence of a sex chromosome and its ratio to the other chromosomes (called autosomes). In female grasshoppers, there are two X chromosomes; in male grasshoppers; there is only one X chromosome.
- In mammals, there are two X chromosomes in females, and one X chromosome and one Y chromosome in males. The *SRY* (sex determining region on the arm of the Y chromosome) determines maleness in males. *SRY* is responsible for the primary sex determination in males and causes the embryo to develop sperm-producing tissue, the testes. Males without the *SRY* gene develop as females. The *DAX 1* gene, which encodes the anti-testes factor, is found on the X chromosome and inhibits the development of testis tissue. In XY individuals (males), the *SRY* gene product inhibits the effect of *DAX 1* and testes develop.
- Secondary sex determination, the outward manifestation of male and female characteristics, is determined by levels of testosterone or estrogen. Secondary sex characteristics in humans include body type, breast development, and body hair.
- In *Drosophila*, sex is determined by the ratio of X chromosomes to autosomes, XX is female (as is XXY), and XY is male (as is XO). The Y chromosome is not necessary for sex determination but is required for male fertility.
- In birds, moths, and butterflies, ZZ (XX) are male and ZW (XY) are female. In these species the egg determines the sex of the offspring.

Sex-linked inheritance is governed by loci on the sex chromosome.

- In mammals and other organisms, including some insects, females have two copies of each X-linked gene while males only have one copy of the X-linked genes.
- Males are hemizygous for X-linked traits. Recessive X-linked traits, such as hemophilia, red-green color blind-

ness, and muscular dystrophy (all in humans), are expressed more often in males than in females.

- If an allele is X-linked, reciprocal crosses do not have the same result. For instance a cross between a red-eyed female *Drosophila* and a white-eyed male *Drosophila* will generate all red-eyed flies. The reciprocal cross between a white-eyed female and a red-eyed male will produce an F_1 generation in which all female flies have red eyes and all male flies have white eyes.
- Males with the X-linked traits can only pass those traits to their daughters.
- Females who are carriers of an X-linked trait pass that trait on average to one-half of their sons and one-half their daughters. Since sons always receive the X chromosome from their mother, one-half of those sons, on average, will express the X-linked trait. Daughters who receive the X-linked trait will be carriers, since their other X chromosome is donated by their father.
- X-linked traits can skip a generation; a trait that is expressed in the grandfather may be expressed in his grandson.
  - There are approximately 20 known genes in humans that are Y-linked. These traits are always passed from father to son.

Cytoplasmic inheritance differs from the Mendelian pattern.

- Mitochondria and plastids contain genomes that are passed to the zygote from the gamete that makes the largest cytoplasmic contribution (the egg in humans).
- These organelles are generally so numerous (and each contains multiple numbers of chromosomes) that genes for cytoplasmic traits are polyploid.
- Genes in organelles mutate faster, leading to multiple alleles for a particular non-nuclear–encoded trait.
- Mutations in mitochondrial genes affect the electron transport chain and the synthesis of ATP. These mutations have a strong effect on tissues that are sensitive to ATP concentrations, particularly the nervous and muscular systems.

The Big Picture

- Genes are units of inheritance that are passed down to the progeny by the fusion of gametes to form a new organism.
- Meiosis generates gametes that contain only one allele for each gene and gametes whose alleles are segregated independently (unless those alleles are linked).
- By studying the expression of particular phenotypes in successive generations, one can begin to understand the alleles that gave rise to those phenotypes (i.e., rare recessive, rare dominant, sex-linked).
- Exceptions to Mendel's rules also help explain the interactions of gene products of particular alleles that lead to the observed phenotype.

Common Problem Areas

- It is easy to be overwhelmed by all of the exceptions to Mendel's rules. Initially, make sure you understand Mendel's laws of segregation and independent assortment. Examine the exceptions to these rules by thinking about the molecular role of these alleles. For example, what is the explanation for a codominant phenotype at the molecular level, such as the AB blood type in humans? Both gene products are synthesized by the cell and modify the cell surface glycoproteins, producing both A and B antigens. Or what is the molecular explanation for epistasis? The enzyme required to produce an initial substrate necessary for pigmentation is missing, even though the other gene products (further down the biochemical pathway) for producing a particular color are present in the cell.

Study Strategies

- Review the figures of meiosis in Chapter 9 to see how Mendel's laws of independent assortment and segregation apply. Draw the chromosomes with different alleles (*S* on one chromosome, *s* on the other) and follow how those alleles are segregated to the gametes during meiosis. Repeat this process, but draw two separate pairs of chromosome with each bearing a different allele (for instance *Ss* and *Yy*). Show how each of these alleles sorts independently in the gametes (review Figure 10.8).
- Use the Punnett square to predict the outcome of different monohybrid and dihybrid crosses. Use the probability rules to predict the genotypes and phenotypes of these same crosses and compare the two methods.
- After reviewing Mendel's laws of independent assortment and allele segregation, make a list of the exceptions to those laws with specific examples for each exception.
- Review the linkage problem on page 193 to familiarize yourself with allele loci on a chromosome and how recombination frequencies affect the genotypes of the gametes.
- Review the following tutorial and activities on the website:
  Tutorial 10.1 Independent Assortment
  Activity 10.1 Concept Matching I
  Activity 10.2 Concept Matching II
  Activity 10.3 Homozygous or Heterozygous?

Test Yourself

Knowledge and Synthesis Questions

1. In the beginning of Chapter 10, hemophilia is mentioned as a trait carried by the mother and passed to her sons. What is the pattern of inheritance for this trait?
   a. Hemophilia is an allele carried on one of the mother's autosomal chromosomes.
   b. Hemophilia is an allele carried on the Y chromosome because more males have this genetic disorder than females.
   c. Hemophilia is an allele carried on the X chromosome and can be directly inherited by the son from the father or the mother.
   d. Hemophilia is carried on the X chromosome and can only be inherited by the son if the mother is a carrier.

2. Originally, genetic inheritance was thought to be a function of the blending of traits from the two parents. Which exception to Mendel's rules is an example of blending?
   a. Polygenic inheritance
   b. Incomplete dominance
   c. Codominance
   d. Pleiotropism

3. True-breeding plants
   a. produce the same offspring when crossed for many generations.
   b. result from a monohybrid cross.
   c. result from a dihybrid cross.
   d. result from crossing over during prophase I of meiosis.

4. What is the probability that a cross between a true-breeding pea plant with smooth seeds and a true-breeding pea plant with wrinkled seeds will produce F_1 progeny with smooth seeds?
   a. 1/2
   b. 1/4
   c. 0
   d. 1

5. What is the pattern of inheritance for a rare recessive allele?
   a. Every affected person has an affected parent.
   b. Unaffected parents can produce children who are affected.
   c. Unaffected mothers have affected sons and daughters who are carriers.
   d. None of the above

6. What is the pattern of inheritance for a rare dominant allele?
   a. Every affected person has an affected parent.
   b. Unaffected parents can produce children who are affected.
   c. Unaffected mothers have affected sons and daughters who are carriers.
   d. None of the above

7. What is the pattern of inheritance for a sex-linked allele?
   a. Every affected person has an affected parent.
   b. Unaffected parents can produce children who are affected.
   c. Unaffected mothers have affected sons and daughters who are carriers.
   d. None of the above

8. Penetrance and expressivity are related to
   a. the increased expression of a particular trait when a hybrid species is formed.
   b. quantitative traits that diminish or intensify a particular phenotype.
   c. the organism's environment affecting the expression of a particular genotype.
   d. the expression of one gene masking the effects of another gene.

9. Sex determination in grasshoppers, humans, and *Drosophila* is similar because
   a. females are hemizygous.
   b. males have one X chromosome and females have two X chromosomes.
   c. all males always have one Y chromosome in all three species.
   d. the ratio of autosomes to sex chromosomes is the same in all three organisms.

10. Linked genes are genes that
    a. assort independently.
    b. segregate equally in the gametes during meiosis.
    c. always contribute the same trait to the zygote.
    d. are found on the same chromosome.
    e. recombine during mitosis.

11. Cytoplasmic inheritance
    a. results from polygenic nuclear traits.
    b. is the result of gametes contributing equal amounts of cytoplasm to the zygote.
    c. is determined by genes on DNA molecules in mitochondria and chloroplasts.
    d. follows Mendel's law of segregation.

Application and Synthesis Questions

1. Draw a pedigree for three generations in which the grandfather has red–green color blindness and his daughter is a carrier. This daughter has four sons. Predict how many of the sons will be color-blind.
2. Draw a sample pedigree with three generations in which the paternal grandfather has a rare dominant autosomal trait.
3. Draw a sample pedigree with three generations in which the maternal grandmother and maternal grandfather are carriers of a rare recessive autosomal trait.
4. Cytoplasmic traits in certain species of trees are passed from the male plant to all of its progeny. Compare this observation to cytoplasmic inheritance in humans.
5. You are a genetics counselor who is working with a 21-year-old pregnant woman who has just discovered that her father has Huntington's chorea, a rare dominant autosomal trait. This disease usually develops in middle-aged individuals, so people carrying this trait do not find out they have this genetic disorder until midlife. What are the chances that the child she is carrying will develop Huntington's chorea? (Assume that her husband's family has no history of Huntington's chorea.) What is the chance that she has Huntington's chorea?

Answers to Questions

Knowledge and Synthesis Answers

1. **d.** Hemophilia is an X-linked trait, and can only be inherited by the son from his mother's X chromosome. The father contributes the Y chromosome to his son (not his X chromosome) and thus cannot pass any of his X-linked alleles to his son.
2. **b.** Incomplete dominance results in the progeny's expressing an intermediate form of the two parental alleles. (In a cross between red-flowered plants and

white-flowered plants, the expression of pink-flowered plants would be a "blend" of the parental traits). Codominance is not an example of blending because both alleles are fully expressed in the individual.

3. **a.** Monohybrid and dihybrid crosses produce heterozygous individuals; true-breeding individuals are always homozygous.
4. **d.** This is an example of a monohybrid cross. All of the F_1 progeny would have smooth seeds. (The F_1 generation would all have the genotype *Ss*, producing the phenotype of smooth seeds because the smooth allele, *S*, is dominant to the wrinkled allele, *s*).
5. **b.** Rare recessive alleles can be carried by both parents but not expressed in those parents. The parents will be heterozygous for this allele (*Aa*). Their children will have a one-fourth probability of expressing that recessive allele (*aa*).
6. **a.** Rare dominant alleles are heterozygous (*Aa*) and every affected individual has one allele that is *A* (dominant). The individual with the *A* allele had to receive that allele from one of his parents.
7. **c.** The most common sex-linked alleles are X-linked and are passed from a mother to her son (since the mother always donates one of her X chromosomes to her son and the father always donates his Y chromosome to his son). Daughters can also receive the X-linked allele from their mothers, but the father donates the other X chromosome, so daughters can be carriers.
8. **c.** Penetrance and expressivity are related to the effects the environment has on a particular phenotype. Answer **a** refers to hybrid vigor, answer **b** refers to quantitative traits, and answer **d** refers to epistasis.
9. **b.** In these three species, females have two X chromosomes and males have one X chromosome. The ratio of X chromosomes to autosomes is important (and different in each organism) in *Drosophila* and grasshoppers, but not in humans. In all three species, males have one Y chromosome, but the Y chromosome is required for male fertility, not for *Drosophila* to be male (in *Drosophila*, male flies can be XO).
10. **d.** Linked genes by definition are on the same chromosome and thus do not assort independently, do not contribute the same trait to the zygote, and do not recombine during mitosis or segregate equally to the gametes during meiosis.
11. **c.** The genes on the mitochondria and chloroplast chromosomes are passed on to all of the progeny from the gamete that contributes the majority of the cytoplasm.

Application Answers

1. See Figure 10.26, generation II, III, and IV. One-half of her sons could be color-blind.
2. See Figure 10.11, generations II, III, and IV.
3. See Figure 10.10.
4. In humans, the gamete with the largest cytoplasmic contribution is the egg, so cytoplasmic inheritance is passed from the female parent to all of her children. In certain tree species, the male gamete contributes the majority of the cytoplasm to the zygote, so that all of the mitochondria and chloroplasts in the zygote are inherited from the male parent.
5. She has a 50 percent chance that she will get Huntington's chorea. Since the trait is an autosomal dominant allele, one-half of her father's gametes will contain the homologous chromosome carrying that allele and one-half of his gametes will contain the homologous chromosome that carries the wild type allele. If she received the Huntington's allele, her child has a 50 percent chance of receiving this allele from her. The product rule is used to predict the probability that her child will inherit the Huntington's allele: 1/2 (the probability that she has the Huntington's allele) x 1/2 (the probability her child will inherit this allele from her) = 1/4 (the probability her child has the allele). Her child has a 25 percent chance of carrying the Huntington's chorea allele, and thus of developing the disease.

STUDY GUIDE

11 DNA and Its Role in Heredity

Important Concepts

Specific discoveries revealed that DNA is the genetic material.

- Staining techniques showed that DNA is present in the nucleus.
- Proportional amounts of DNA were found in somatic and germ cells.
- Different organisms showed varying amounts of DNA in their nuclei.
- Griffith's transformation experiments indicated that some factors could be passed from cell to cell.
- Avery, MacLeod, and McCarty's work supported DNA as being the transforming factor.
- Hershey and Chase demonstrated definitively that DNA was the factor responsible for infection of bacteriophage.

A series of related yet independent experiments elucidated the structure of DNA.

- Franklin and Wilkins prepared X-ray crystallographic images which were critical to understanding the structure of DNA.
- Chargaff provided rules for relative ratios of pyrimidines and purines, the building blocks of DNA.
- Watson and Crick coupled results from previous experimentation and model building and revealed the three-dimensional structure of DNA.

The form of DNA is tied to its function.

- The double helix is antiparallel. The 5′ end of one strand pairs with the 3′ end of its complementary strand.
- Base pairs are complementary; they consist of a pyrimidine pairing with a purine via hydrogen bonding. This, along with hydrophobic interactions, stabilizes the molecule and maintains a constant diameter. The complementary base pairing also allows for a mechanism of replication.
- DNA forms a right-handed twist.
- Multiple ways of ordering bases allows for many genetic variations. Any alteration of the sequence of bases may result in mutation.

DNA must be able to be replicated in order to be heritable.

- For replication to take place, a template strand, all four deoxyribonucleoside triphosphates (dATP, dCTP, dGTP, and dTTP), and DNA polymerase are all required.
- Meselson and Stahl established that DNA replication is semiconservative, with each new molecule composed of a parent and daughter strand.
- During replication, DNA is denatured and each resulting strand serves as a template for a new strand. A new strand is built by complementary pairing of each nucleotide with the parent strand via hydrogen bonds. New nucleotides are linked at the 3′ end of the elongating strand via covalent bonds.
- Replication occurs at replication complexes. DNA binds to replication complexes at replication origins. Prokaryotes have one replication origin whereas eukaryotes may have many. Replication proceeds in both directions from the replication origin.
- Replication complexes are composed of many proteins: helicases unwind parent DNA, binding proteins prevent parent strands from reannealing, DNA polymerases elongate daughter strands, and DNA ligases anneal breaks in the sugar–phosphate backbone.
- New daughter strands must begin with a short RNA primer, formed using the parent strand as a template. Once a primer is formed, DNA polymerase may then be added to the strand.

Replication proceeds in a specified direction.

- Nucleotides can only be added to the 3′ end of a strand. Thus, the parent strand is read from the 3′ to 5′ direction, and the daughter strand elongates from the 5′ to 3′ direction.
- The leading daughter strand elongates from the 5′ to 3′ direction as the template 3′ end is revealed.
- The lagging strand template is exposed from the 5′ end, but nucleotides can only be added at the 3′ end. This is accomplished via short Okazaki fragments. This process requires many primers, and the Okazaki fragments are formed in short backward-directed strands that are ligated (connected) after synthesis.

Many mistakes are made during initial synthesis of DNA; therefore, DNA proofreading and repair mechanisms are essential.

- DNA polymerase proofreads and repairs base mismatches.
- Mismatch repair systems scan for possible errors in newly synthesized DNA.

- Excision repair mechanisms replace abnormal bases with functional bases.

Advanced techniques in molecular biology have allowed us to determine the sequence of bases in DNA and to make multiple copies of DNA from a single strand.

- DNA sequencing is now a highly automated process in which modified bases are used as markers.
- The polymerase chain reaction (PCR) allows the rapid amplification of a single short DNA segment into many copies.

The Big Picture

- Knowing that DNA is the genetic material has allowed scientists to understand how hereditary information is passed on at the molecular level. It has also allowed us to understand how mutations that change that hereditary information occur.
- Advances in our knowledge of how DNA is replicated provides new technologies for understanding the genetic code and studying it in a variety of organisms.

Common Problem Areas

- Students frequently get lost in the history of discovery. Do not focus on *who* and *when*, but instead on *how* the experiments provide evidence to support DNA as the genetic material.
- Understanding DNA replication is a highly visual process. Use the figures in your text to "see" what is happening. Make your own diagrams of these processes.
- 5′ and 3′ ends are often confused on template versus daughter strands. On both strands, the 3′ end corresponds to a site where a hydroxyl group is (or was), and the 5′ end corresponds to a site where a phosphate tail is (or was).
- Take advantage of laboratory activities that simulate or actually allow you to experience sequencing of PCR. Nearly all molecular research labs utilize one or both of these techniques.

Study Strategies

- Focus on how scientific evidence has led to the base of knowledge we have.
- Review the following tutorial on the website:
  Tutorial 11.1 DNA Replication

Test Yourself

Knowledge and Synthesis Questions

1. Griffith's experiments showing the transformation of R strain pneumococcus bacteria to S strain pneumococcus bacteria in the presence of heat-killed S strain bacteria gave evidence that
   a. an external factor was affecting the R strain bacteria.
   b. DNA was definitely the transforming factor.
   c. S strain bacteria could be reactivated after heat killing.
   d. All of the above

2. Experiments by Avery, MacLeod, and McCarty supported DNA as the genetic material by showing that
   a. both protein and DNA samples provided the transforming factor.
   b. DNA was not complex enough to be the genetic material.
   c. only samples with DNA provided transforming activity.
   d. even though DNA was molecularly simple, it provided adequate variation to act as the genetic material.

3. Hershey and Chase used radioactive ^{35}S and ^{32}P in experiments to provide evidence that DNA was the genetic material. These experiments pointed to DNA because
   a. progeny viruses retained ^{32}P but not ^{35}S.
   b. retention of ^{32}P in progeny viruses indicated that DNA was passed on.
   c. loss of ^{35}S in progeny viruses indicated that proteins were not passed on.
   d. All of the above

4. X-ray crystallography provides information about the ______________ of DNA but is limited because of the ______________ of DNA. The technique is based on the pattern of ______________ off the atoms in the molecule.
   a. structure; difficulty of purification; light absorption
   b. dimensions; molecular weight; diffraction
   c. molecular weight; shape; diffraction
   d. dimensions; linearity; light absorption

5. Chargaff's rules of base pairing states that
   a. the ratio of purines to pyrimidines is roughly equal in all tested organisms.
   b. the ratio of A to T is roughly equal in all tested organisms.
   c. the ratio of A + T and G + C is roughly equal in all tested organisms.
   d. a and b

6. Watson and Crick's model allowed them to visualize
   a. the molecular bonds of DNA.
   b. how the purines and pyrimidines fit together in a double helix.
   c. that the two strands of the DNA double helix were antiparallel.
   d. All of the above

7. A fundamental requirement for the function of genetic material is that it must be
   a. conserved among all organisms with very little variation.
   b. passed intact from organism to organism.
   c. replicable.
   d. found outside of the nucleus.

8. Evidence indicating that DNA replication was semiconservative came from
   a. DNA staining techniques.
   b. DNA sequencing.
   c. density gradient studies using "heavy" nucleotides.
   d. None of the above

9. Current evidence indicates that replication complexes are attached to stationary nuclear components and that DNA is threaded through these complexes. Which of the following best describes the role of the replication complex?
   a. The complex acts as an enzymatic center for DNA replication.
   b. The complex binds specifically to replication origins, then controls the rate at which replication occurs.
   c. The complex is the initiating site of replication forks.
   d. All of the above

10. The primary function of DNA polymerase is to
    a. add nucleotides to the growing daughter strand.
    b. seal nicks along the sugar–phosphate backbone of the daughter strand.
    c. unwind the parent DNA double helix.
    d. prevent reassociation of the denatured parent DNA strands.

11. The lagging daughter strand of DNA is synthesized in what appears to be the "wrong" direction. This synthesis is accomplished by
    a. ligating (connecting) short Okazaki fragments that are synthesized in short spurts in the "right" direction.
    b. primase.
    c. using multiple primers and DNA polymerase I.
    d. a and b

12. RNA primers are necessary in DNA synthesis because
    a. DNA polymerase can only add to an existing strand of nucleotides.
    b. DNA polymerase can only add to an existing RNA strand.
    c. DNA primase is the first enzyme in the replication complex.
    d. All of the above

13. Proofreading and repair occur
    a. at anytime during or after synthesis of DNA.
    b. only before DNA methylation occurs.
    c. only in the presence of DNA polymerase.
    d. only in the presence of an excision repair mechanism.

14. DNA replication is an ____________ process and ____________ energy.
    a. exergonic; does not require
    b. endothermic; does require
    c. endergonic; does require
    d. endodontic; does not require

15. *T. aquaticus* DNA polymerase is not denatured during the heat cycling required to denature DNA. This property allowed advances in what technique?
    a RFLP analysis
    b. PCR
    c. Sequencing
    d. EPA

16. Thirty percent of the bases in a sample of DNA extracted from eukaryotic cells is adenine. What percentage of cytosine is present in this DNA?
    a. 10%
    b. 20%
    c. 30%
    d. 40%

17. Which of the following represents a bond between a purine and a pyrimidine (in that order)?
    a. C–T
    b. G–A
    c. G–C
    d. T–A

18. Which of the following statements about DNA replication is false?
    a. Okazaki fragments are the initiators of continuous DNA synthesis along the leading strand.
    b. Replication forks represent areas of active DNA synthesis on the chromosomes.
    c. Error rates for DNA replication are often less than one in every billion base pairings.
    d. Ligases and polymerases function in the vicinity of replication forks.

19. Which of the following would not be found in a DNA molecule?
    a. Purines
    b. Ribose sugars
    c. Phosphates
    d. Sulfur

20. If a nucleotide lacking a hydroxyl group at the 3′ end is added to a PCR, what would be the outcome?
    a. No additional nucleotides would be added to a growing strand containing that nucleotide.
    b. Strand elongation would proceed as normal.
    c. Nucleotides would only be added at the 5′ end.
    d. *T. aquaticus* DNA polymerase would be denatured.

Application Questions

1. Given the following parent strand sequence, what would the daughter strand sequence look like?
   5′- G C T A A C T G T G A T C G T A T A A G C T G A -3′
2. Diagram the double helix. Be sure to label those properties that make it most suited as the genetic material.
3. Diagram a replication fork as it would be seen in a replicating segment of DNA. In your diagram label the 5′ and 3′ ends of each parent strand and daughter strand. Indicate which new strand is the leading and which is the lagging strand of the daughter DNA.
4. Based on your diagram in Question 3, construct a flow chart that represents what occurs during DNA replication. In your chart, indicate the roles of helicase, DNA polymerase, binding proteins, nucleotides, parent DNA strands, DNA ligase, RNA primase, RNA primers, and primosomes.
5. Explain the difference between conservative and semiconservative models of DNA replication. What results supported the semiconservative model? What would the results have looked like had the conservative model of DNA replication been accurate? Are there any other potential hypotheses?
6. Explain the role of Okazaki fragments in the synthesis of the lagging daughter strand.

7. Differentiate between proofreading, mismatch repair, and excision repair. Which of these repair mechanisms is responsible for repairing a mutation that occurs in an adult cell from overexposure to the sun? Justify your answer.
8. Explain how modified nucleotides are used in DNA sequencing.

Answers to Questions

Knowledge and Synthesis Answers

1. **a.** The experiments showed only that something was causing the R strain pneumococcus to develop the S strain virulence. Additional experiments were necessary to provide evidence that the transforming factor was indeed DNA.
2. **c.** These researchers were able to isolate nearly pure DNA samples. It was only these samples that provided transformation activity.
3. **d.** Only DNA incorporates radioactive P. Radioactive S is incorporated into proteins. The fact that progeny viruses retained radioactive P and not radioactive S indicates that DNA is heritable and protein is not.
4. **b.** X-ray crystallography allows researchers to measure the distances between atoms and determine the physical dimension of a molecule. The technique becomes quite difficult with high-molecular-weight molecules. Wilkins's contributions were significant because he was able to isolate nearly perfect strands of the molecule. The resulting images are patterns of how the X-rays are diffracted off the constituent atoms of the molecule.
5. **d.** Chargaff found that the relative ratios of purines to pyrimidines were equal. Adenine and guanine are purines and cytosine and thymine are pyrimidines; therefore, ratios of adenine and thymine should be equal. Chargaff also found that there is no conserved ratio between specific pairs (e.g., A + T and G + C).
6. **d.** Model building by Watson and Crick allowed a three-dimensional visualization of the size, bond angles, base pairings, and overall structure of the DNA molecule. The information contributing to their model came from a variety of experiments performed by other scientists.
7. **c.** Replication of DNA is a fundamental requirement for its function as the genetic material. DNA must be correctly replicated in each cell of an organism, and must be passed from parent to offspring.
8. **c.** Density gradient labeling allowed Meselson and Stahl to track parent versus daughter strands of DNA. Their research showed that DNA replication was a semiconservative process with each "generation" containing one parent strand and one daughter strand.
9. **d.** Replication complexes bind to replication origins. The enzymes contained in the complex are responsible for unwinding the parent DNA and setting up replication forks. Additional enzymes within the complex are responsible for the actual synthesis of the new strands of daughter DNA.
10. **a.** DNA polymerase adds nucleotides to an existing nucleotide strand.
11. **d.** Okazaki fragments are short pieces of newly synthesized DNA. The production of each of these fragments is dependent on a beginning RNA primer. The small fragments are ultimately ligated (connected) together to form the lagging strand.
12. **a.** DNA polymerase cannot initiate the building of a nucleotide strand; it can only add to an existing strand. Thus, RNA primers are necessary to begin DNA synthesis.
13. **a.** For the integrity of DNA to be maintained, repair mechanisms must be active during synthesis, modification, and utilization of DNA.
14. **c.** DNA replication is an energy-consuming process that must have an input of energy to proceed. Energy is provided in the breaking of the triphosphate tails of each nucleotide.
15. **b.** PCR depends on continuous cycles of heating, in which DNA is denatured, followed by cooling which then allows synthesis to occur. The original limitation of the technique was that DNA polymerases were denatured along with the DNA itself. Because *T. aquaticus* polymerase is active at high temperatures, it is not denatured at the same temperature as double-stranded DNA. This allows many heat cycles without the addition of extra polymerase enzyme.
16. **b.** If 30 percent of DNA is adenine, then by Chargaff's rule 30 percent will be thymine. The remaining 40 percent of the DNA is cytosine and guanine. Since the ratio of cytosine to guanine must be equal, then each accounts for 20 percent of the bases.
17. **c.** Guanine is a purine and its paired pyrimidine is cytosine.
18. **a.** Okazaki fragments are found only along the lagging strand.
19. **d.** Sulfur is a constituent of many protein molecules, but is not found in DNA.
20. **a.** A hydroxyl group at the 3′ position of a nucleotide is necessary for the binding of any additional nucleotides. If this hydroxyl group were absent, no other nucleotides could be added to a growing strand.

Application Answers

1. 3′–CGATTGACACTAGCATATTCGACT–5′
2. See Figure 11.6 in the text

3.–4. See Figure 11.16 in the text

5. In the conservative model of DNA replication, the parent DNA remains intact and a newly synthesized molecule consists of two daughter strands. Had this been DNA's method of replication, Meselson and Stahl would not have seen the intermediate density band in the first generation of replication. They would have seen a heavy band corresponding to the parental DNA, and a light band corresponding to the daughter DNA. Because they saw an intermediate band, they knew one strand was heavy and one was light; therefore, replication was

semiconservative, with each new molecule consisting of one parent strand and one daughter strand. A third hypothesis was the dispersive model, with each new molecule containing bits and pieces of both old and new strands.

6. See Figure 11.18 in the text.

7. Proofreading occurs during synthesis of the DNA molecule, mismatch repair occurs after synthesis, and excision repair removes any abnormalities in "mature" DNA. DNA damage due to UV exposure is generally repaired via excision repair mechanisms.

8. Refer to Figure 11.20 in your textbook for an explanation of DNA sequencing.

12 From DNA to Protein: Genotype to Phenotype

Important concepts

One gene encodes one enzyme

- Beadle and Tatum studied special mutant (auxotroph) and wild-type (prototroph) strains of *Neurospora*, a bread mold. Auxotrophs are mutant strains that cannot grow on minimal media without the addition of an external mineral, vitamin, or nutrient, because auxotrophs are missing a particular enzyme needed to synthesize these nutrients. Prototrophs can grow on minimal media because in these cells all of the enzymes needed to synthesize these nutrients are functional.
- Auxotrophs for a particular nutrient are missing the enzymes needed to synthesis that nutrient from a biochemical intermediate. For the amino acid arginine, for example, each of these enzymes defines a step in the biochemical pathway of arginine synthesis. These mutant strains can be characterized by supplying them with chemical intermediates in the arginine pathway to determine which enzymes are nonfunctional (see Figure 12.1).
- Beadle and Tatum concluded from their experiments that one gene encoded one enzyme. Because enzymes can consist of several different polypeptides, this conclusion has been modified to the theory that one gene encoded one polypeptide.
- Gene products encode both protein and RNA (such as ribosomal RNA and transfer RNA).

The central dogma of molecular biology

- In a eukaryotic cell, a message (the messenger RNA or mRNA) is needed to carry the genetic information (the DNA) from the nucleus to the cytoplasm. The process of synthesizing mRNA from DNA is called transcription.
- An adapter (the transfer RNA or tRNA) is needed to translate the nucleic acid (in the messenger RNA) into amino acid (protein). Translation is the process of synthesizing protein from mRNA.
- The information flow is unidirectional, from gene (DNA), to message (RNA), to protein.

RNA

- RNA is usually single-stranded (but can base-pair with itself) and contains the bases adenine (A), uracil (U), guanine (G), and cytosine (C) and the sugar ribose.
- Some viruses use RNA as their genome and during infection make more RNA (using viral RNA as a template). These viruses include poliovirus, tobacco mosaic virus, and influenza virus. Other viruses make a DNA copy of their RNA and use the DNA to synthesize more RNA, such as HIV.

Transcription

- RNA polymerase is the enzyme that synthesizes RNA using one strand of the DNA as a template. RNA molecules can be messenger RNA's (which are used to synthesize proteins) or transfer RNA or ribosomal RNA.
- Initiation
  - RNA polymerase initiates transcription at special sequences on the DNA called promoters. The efficiency of transcriptional initiation is related to how tightly RNA polymerase binds those promoter sequences. Some promoters are very effective, and bind RNA polymerase tightly so transcription is initiated very frequently from these promoters. Other promoters bind RNA polymerase less tightly, and transcription is initiated less often from these promoters.
- Elongation
  - RNA polymerase synthesizes a new RNA in a 5′-to-3′ direction, antiparallel to the template DNA strand, using the energy of pyrophosphate hydrolysis. (The two phosphates of the incoming ribonucleotide triphosphate are hydrolyzed as that ribonucleotide is added to the growing RNA chain).
  - RNA polymerase unwinds about 20 base pairs in the DNA as it reads and synthesizes a complementary RNA strand.
  - In prokaryotic cells there is one RNA polymerase. In eukaryotic cells there are three different RNA polymerases, RNA polymerase I, II, and III. Proteins bind at the promoter in eukaryotic cells to help RNA polymerase load onto the promoter sites (see Chapter 14).
- Termination
  - RNA polymerase terminates transcription at special sequences in the DNA. In prokaryotic cells, translation of the mRNA can begin before transcription terminates. In eukaryotic cells, the mRNA is chemically modified and must be transported from the nucleus to the cytoplasm before translation can begin.

Translation

- Genetic code
  - Genetic information in the messenger RNA is read three bases at a time; this three-base sequence is called a codon. Since there are four possible bases that can be used in a codon, there are 43, or 64, different codons. Sixty-one of these codons specify specific amino acids, so the genetic code is redundant, meaning that there is more than one codon for many of the 20 amino acids. Three of these codons, the stop codons UAG, UAA, and UGA, are used to terminate translation since they do not specify any amino acid. The codon used to start translation is AUG, which specifies the amino acid methionine.
  - The genetic code for nuclear genes is almost universally used in all living cells. There are slight differences in the genetic code of one group of protists, and in mitochondria and chloroplast genomes.
  - The genetic code was first determined by experiments in which artificial RNA's (poly U, for example) were added to reaction mixtures containing all of the necessary components needed for protein translation (see Figure 12.6). The resulting polypeptides were analyzed to determine what the protein-synthesizing instructions were in particular RNA sequences. When the synthetic RNA poly U was used, polyphenylalanine was made. Thus, one of the codons for phenylalanine was UUU.
- Transfer RNA
  - Transfer RNA (tRNA) is a small RNA molecule (75–80 nucleotides) that base-pairs with itself to form a characteristic shape, the "cloverleaf," found in all tRNA's (see Figure 12.7).
  - The anticodon of the tRNA (found about midway in the molecule) base-pairs with the codon of the mRNA in an antiparallel fashion. This base pairing occurs on the ribosome during translation.
  - The amino acid is covalently attached to the 3′ end of the tRNA; thus phe tRNA, (with an anticodon of AAA) has phenylalanine covalently attached to its 3′ end. The enzyme that catalyzes the covalent attachment of the amino acid to the correct tRNA is tRNA synthetase (see Figure 12.8).
- Ribosome
  - The ribosome is the site in the cell where protein translation occurs. Ribosomes are composed of a large subunit and a small subunit; both subunits consist of different ribosomal proteins and RNA molecules (ribosomal RNA).
  - The important sites on a ribosome are the T site (where the incoming charged tRNA enters the ribosome), the A site (where the anticodon of the charged tRNA binds the codon of the mRNA), the P site (where the tRNA with the growing polypeptide chain is located), and the E site (where the tRNA exits from the ribosome) (see Figure 12.10).
- Initiation of translation
  - Translational initiation requires the small subunit of the ribosome, the initiator tRNA, the mRNA with the start codon AUG positioned correctly on the ribosome, and the initiation factors.
- Elongation
  - The large subunit of the ribosome joins the small subunit of the ribosome, and the incoming charged tRNA enters the T site and is positioned at the A site. The anticodon of this tRNA base-pairs with the codon on the mRNA that is at the A site of the ribosome. Base pairing between the anticodon and the codon is always complementary (A pairs with U, C pairs with G) in the first two positions. At the third position of the codon (and the first position of the anticodon) base pairing does not have to be exact. This position is called the wobble position and allows the cell to use one tRNA to base-pair with potentially four different codons.
  - Peptidyl transferase, found on the large subunit of the ribosome, forms a peptide bond between the amino acid on the tRNA in the P site and the amino acid on the tRNA in the A site. The catalytic activity of peptidyl transferase can be attributed to the ribosomal RNA in the large subunit (the 23S rRNA) and thus is a ribozyme.
  - Once the peptide bond is formed, the ribosome shifts down the messenger RNA so that the next codon is placed at the A site. The tRNA that was in the P site is moved to the E site and will exit the ribosome.
  - During elongation, the messenger RNA is read 5′ to 3′, with each new codon being moved into the A site during elongation so it can be inspected by the next incoming charged tRNA. Polypeptide synthesis occurs from the N terminus to the C terminus with the polypeptide folding into its tertiary structure as it is synthesized.
  - The elongation factors are proteins that assist in this phase of translation.
- Termination
  - Termination occurs when the stop codon is moved into the A site of the ribosome. Release factor binds the stop codon, and peptidyl transferase hydrolyzes the bond between the amino acid (which will be the C terminus) that is covalently attached to the tRNA in the P site. The polypeptide is released from the ribosome and the small and large subunits of the ribosome dissociate (see Figure 12.12).
  - Antibiotics can block various steps in translation in prokaryotic cells, preventing the bacteria from synthesizing proteins and thus killing the bacteria (see Table 12.2). These antibiotics are effective because they specifically target the bacterial translational machinery without affecting the cells of the individual taking the antibiotic.
  - One messenger RNA is usually bound by many translating ribosomes, which is termed a polysome (see Figure 12.13).
  - Within the amino acid sequence of some proteins in eukaryotic cells are signal sequences, which serve as addresses for the protein's final cellular destination. These signal sequences are recognized by receptor proteins at their destination (in the case of mitochondrial,

plastid, peroxisome, and nuclear proteins), or by signal recognition particles in the cytoplasm that direct the partially translated protein to the endoplasmic reticulum for further addressing (see Figure 12.14).

- Proteins that lack signal sequences remain in the cytoplasm. Some cytoplasmic proteins are phosphorylated.
- Protein modification in the endoplasmic reticulum includes proteolysis and glycosylation. Some proteins are further modified in the Golgi apparatus where sugar groups are added. Proteins that enter the Golgi apparatus are destined for lysosomes, the plasma membrane, or secretion.

Mutations

- Mutations are heritable changes in the genetic information. They can occur in the germ-line cells and thus be passed on to the progeny, or in the somatic cells.
- Some mutations are easy to observe while others are difficult to detect. Conditional mutants are a class of mutations that are only observable under altered conditions, such as high temperature. The inability of certain mutants to grow at high temperature can be related to the unstable tertiary structure of the protein whose gene sequence has been altered.
- Point mutations are changes of one or two base pairs and include silent mutations, missense mutations, nonsense mutations, and frameshift mutations.
- Silent mutations are mutations that do not change the meaning of the codon. For instance, the change of a U to an A in the third position of a codon would change a UUU to a UUA, but both codons specify phenylalanine.
- Missense mutations are single base changes that change one codon to another (UUU, phe to CUU, leu).
- Nonsense mutations are single base changes that change a codon to a stop codon (UGG, trp to UAG, stop).
- Deletions or insertions of one or two bases are frameshift mutations, which alter the reading frame of the ribosome.
- Chromosome mutations are larger deletions, duplications, or inversions that occur in the chromosome.
- Spontaneous mutations can be caused by tautomerization of the bases, errors during DNA replication, and errors during meiosis (leading to unequal crossovers or unequal separation of the chromosomes).
- Induced mutations are caused by chemicals and radiation.
- Mutations for the most part are deleterious, but some mutations may actually benefit the organism by creating an altered gene product, which may improve the survival of the organism.

The Big Picture

- Every protein and RNA in the cell has a DNA blueprint (the gene sequence) that specifies the amino acid or nucleotide sequence of that gene product. Those sequences determine how that gene product will fold up three-dimensionally and ultimately function in the cell. Alterations in those gene products, caused by mutations in the genes, help us to understand how those proteins function in the cell.

Common Problem Areas

- Students often try to understand every detail of gene expression without comprehending the larger picture: accessing genetic information in the DNA and expressing it in the cell so that the cell can function. Familiarize yourself with the details of the central dogma, but then take a step back and understand that all of these details describe the steps required to synthesize gene products.
- Students often confuse transcription and translation. Be able to distinguish between the two processes, and for each one carefully review the template, the product, and the sites of initiation and termination (review Table 12.1).

Study Strategies

- Make diagrams of the process of transcription and translation. Be sure to orient the 5′ and the 3′ ends of the nucleic acid and the N and the C terminus of the protein.
- Pick a hypothetical gene that encodes a peptide of four amino acids. Draw a flow chart describing how that gene in the chromosome is made into protein. Now follow the details of gene expression, the synthesis of the RNA from the DNA, and the synthesis of the protein from the mRNA. Be sure to include the start and stop signals in each of these processes. Alter one of those nucleotides in the DNA sequence. Transcribe and translate the gene. Is the protein sequence changed? Repeat this process using a different mutation in the DNA.
- Review the following tutorials and activity on the website:
  Tutorial 12.1 Protein Synthesis
  Tutorial 12.2 Transcription
  Activity 12.1 Genetic Code

Test Yourself

Knowledge and Synthesis Questions

1. Transcription in prokaryotic cells is
   a. initiated at a promoter using one of three RNA polymerases (RNA polymerase II).
   b. initiated at a start codon with the help of initiation factors and the small subunit of the ribosome.
   c. initiated at a promoter and uses only one strand of DNA, the template strand, to synthesize a complementary RNA strand.
   d. is terminated at stop codons.

2. Which of the following about RNA polymerase is *not* true?
   a. It synthesizes mRNA in a 5′-to-3′ direction reading the DNA strand 3′ to 5′.
   b. It synthesizes mRNA in a 3′-to-5′ direction reading the DNA strand 5′ to 3′.
   c. It binds at the promoter and unwinds the DNA.
   d. It uses the energy of pyrophosphate hydrolysis to synthesize RNA.

3. Translation of messenger RNA into protein occurs
   a. in a 3′ to 5′ direction and from N terminus to C terminus.
   b. in a 5′ to 3′ direction and from N terminus to C terminus.
   c. in a 3′ to 5′ direction and from C terminus to N terminus.
   d. in a 5′ to 3′ direction and from N terminus to C terminus.

4. If a codon were read two bases at a time instead of three bases at a time, how many different possible amino acids could be specified?
   a. 16
   b. 64
   c. 8
   d. 32

5. Translate the following mRNA:
   3′-G A U G G U U U U A A A G U A- 5′
   a. NH2 met—lys—phe—leu—stop COOH
   b. NH2 met—lys—phe—trp—stop COOH
   c. NH2 asp—gly—phe—lys—val COOH
   d. NH2 asp—gly—phe—lys —stop COOH

6. What would happen if a mutation occurred in the DNA such that the second codon of a polypeptide, UGG, was changed to a UAG?
   a. Nothing. The ribosome would skip that codon and translation would continue.
   b. Translation would continue, but the reading frame of the ribosome would be shifted.
   c. Translation would stop at the second codon and no functional protein would be made.
   d. Translation would continue, but the second amino acid in the protein would be different.

7. If the synthetic RNA listed below were added to a test tube containing all of the components necessary for protein translation to occur, what would the amino acid sequence be?
   5′-A U A U A U A U A U A U - 3′
   a. Polyphenylalanine
   b. Isoleucine-tyrosine-isoleucine-tyrosine
   c. Isoleucine-isoleucine-isoleucine-isoleucine
   d. Tyrosine-tyrosine-tyrosine-tyrosine

8. What part of the tRNA base-pairs with the codon in the mRNA?
   a. The 3′ end where the amino acid is covalently attached
   b. The 5′ end
   c. The anticodon
   d. The promoter

9. Peptidyl transferase is an enzyme that is an
   a. enzyme found in the nucleus of the cell which assists in the transfer of the RNA to the cytoplasm.
   b. enzyme found in the large subunit of the ribosome which catalyzes the formation of the peptide bond in the growing polypeptide.
   c. RNA molecule that is catalytic.
   d. b and c

10. Termination of translation requires
    a. release factor, initiator tRNA, and ribosomes.
    b. initiation factors, the small subunit of the ribosome, and mRNA.
    c. elongation factors and charged tRNAs.
    d. a stop codon positioned at the A site of the ribosome, and peptidyl transferase and release factor.

11. Which mutation is the most deleterious?
    a. A missense mutation in the second codon
    b. A frameshift mutation in the second codon
    c. A nonsense mutation in the last codon
    d. A silent mutation in the second codon

12. If the DNA encoding a nuclear signal sequence were placed in the gene for a cytoplasmic protein, what would happen?
    a. The protein would be localized to the lysosomes.
    b. The protein would be localized to the nucleus.
    c. The protein would be directed to the cytoplasm.
    d. The protein would stay in the endoplasmic reticulum.

13. Auxotrophs are mutant strains that
    a. can grow on minimal media.
    b. require the addition of an essential nutrient to grow on minimal plate.
    c. behave like wild-type strains.
    d. can only grow if arginine is added to the growth media.

14. The central dogma of molecular biology states that ________ is transcribed into _______ which is translated into ________.
    a. genes, polypeptides, gene product
    b. protein, DNA, RNA
    c. DNA, mRNA, tRNA
    d. DNA, RNA, protein

15. A gene product is
    a. an enzyme.
    b. a polypeptide.
    c. RNA's.
    d. All of the above.

16. The enzyme that catalyzes the synthesis of RNA is
    a. DNA polymerase.
    b. tRNA synthetase.
    c. ribosomal RNA.
    d. RNA polymerase.

Application Questions

1. What would be the effect of a deletion of the DNA encoding the targeting sequence for that gene product? (Imagine that this protein was targeted to go to the mitochondria and the signal sequence was removed as a result of this deletion.)
2. What would happen if the tRNA synthetase for tryptophan actually added a phenylalanine to the trp tRNA's instead of tryptophan?
3. Mutations can be very harmful to an organism, yet without them life as we know it today would not exist. Explain.

4. Mutations in two different genes cause the same phenotype in a cell. Explain this phenomenon.
5. When fed the same biochemical intermediate, two auxotrophic mutants that had been isolated were able to grow. According to the experiments of Beadle and Tatum, the mutations in each of these auxotrophs should be in the same gene, because they were blocked at the same step in a biochemical pathway. Yet, these two auxotrophs had mutations that mapped in different genes. How do you explain this?
6. Antibiotics block different steps in translation in prokaryotic cells. Why are these chemicals so effective in eliminating bacterial infections?

Answers to Questions

Knowledge and Synthesis Answers

1. **c.** Answer **a** describes transcription in eukaryotic cells; answer **b** describes translation.
2. **b.** RNA polymerase binds at a promoter, unwinds the DNA, and synthesizes mRNA in a 5′ to 3′ (not 3′ to 5′) direction using the energy of pyrophosphate hydrolysis to synthesize the RNA.
3. **b**. Translation of messenger RNA occurs 5′ to 3′ and the polypeptide is synthesized from N terminus to C terminus.
4. **a.** Two bases read three at a time is equal to 23 = 16.
5. **b.** (see the codon table, Figure 12.5).
6. **c.** UAG is a stop codon and translation would terminate at that site.
7. **b.** (see the codon table, Figure 12.5).
8. **c.** Neither the 3′ end nor the 5′ end of the tRNA are part of the anticodon. The promoter is a DNA sequence, which RNA polymerase binds to initiate transcription.
9. **d.** Peptidyl transferase is the enzyme that catalyzes the formation of the peptide bond, and it is located in the large subunit of the ribosome. Its catalytic activity is thought to be due largely to ribosomal RNA found in the large subunit of the ribosome.
10. **d.** Termination of translation requires a stop codon to be positioned in the A site of the ribosome, and peptidyl transferase and release factor. Release factor adds water across the last amino acid attached to the tRNA in the P site, creating the C terminus.
11. **a.** The mutation that is potentially the most dangerous is the frameshift in the second codon. All of the other codons would be out of register and no functional protein could be made. A missense mutation in the second codon could be disastrous or minimal, depending on how crucial that second amino acid was to the folding and functioning of the protein. A nonsense mutation at the end of the coding sequence would have less effect on protein function since almost every amino acid in the protein would have been correctly synthesized. Again, this would depend on how important that last codon was to the proper folding and functioning of the protein. A silent mutation would have no effect on protein function since the codons would be the same.
12. **b.** The nuclear sequence would direct this protein to be delivered to the nucleus.
13. **b.** Answer **d** is true only for arginine auxotrophs.
14. **d.** Genes are not transcribed into polypeptides, protein is not transcribed into DNA, and messenger RNAs are not translated into tRNAs.
15. **d.** Gene products can be RNA's (such as rRNA and tRNA) as well as enzymes and polypeptides. Messenger RNA is translated into a gene product, protein.
16. **d.** DNA polymerase catalyzes the synthesis of DNA, tRNA sythetase covalently attaches amino acids to tRNA's, and ribosomal RNA (in the large subunit) catalyzes the formation of the peptide bond during translation.

Application Answers

1. Since the protein lacked its targeting sequence, it would no longer be able to be localized to the mitochondria and would remain in the cytoplasm after it had been translated.
2. If the tRNA synthetase for tryptophan added phenylalanine to the trp tRNA's, every time a tryptophan codon was read by these trp tRNA's, phenylalanine would be added to the polypeptide. This would create proteins that were nonfunctional, and the cell would die.
3. Mutations very often have damaging effects on a gene, rendering that gene product nonfunctional. Some mutations actually create gene products that are better for the cell or organism and may increase its ability to survive. Mutations over the course of evolutionary time have created new organisms that could survive in different environments or compete more effectively for limited resources. Without mutations evolution does not occur.
4. Mutations that cause the same phenotype in the cell could be auxotrophic mutations in genes involved in the same biochemical pathway. In the biochemical pathway for the synthesis of the amino acid arginine, there are three different enzymes (see Figure 12.1). Mutations in any of these genes would produce the same phenotype, an inability to grow on minimal media lacking arginine.
5. These two genes must encode different polypeptides that are both subunits for the same enzyme in this biochemical pathway.
6. Many of the infections that cause human disease are of bacterial origin, and when antibiotics are applied to these cells, translation stops and the cells die. Eukaryotic organisms such as humans are not as sensitive to these antibiotics, either because their ribosomes do not bind the antibiotics or because the antibiotics cannot get inside their cells. Protein translation machinery in human cells is unaffected by most antibiotics but translation in bacterial cells is, leading to the elimination of the bacterial infection.

13 The Genetics of Viruses and Prokaryotes

Important Concepts

Prokaryotic viruses

- Prokaryotic cells and viruses (also known as bacteriophages) are useful model organisms for genetic studies because they are easy to grow, have a short generation time, and produce large numbers of individual cells or virus-per-milliliter (10^9 *E. coli*/ml, 10^{11} bacteriophage/ml).
- The genome size of bacterial viruses and bacteria are small, making identification and characterization of genes and gene function in these organisms relatively simple.
- Viruses were first identified in the late 1800s using crude filtration methods. In the 1950s direct observation with electron microscopes showed the major components of viruses, which were protein and nucleic acid.
- Viruses are acellular, do not metabolize energy-rich compounds, and thus generate no ATP. They are obligate intracellular parasites, using the host cell protein translation machinery and energy-rich host compounds to express their gene products, to replicate their genome and to multiply. Viruses can infect bacteria, archaea, animal cells, plant cells, fungi, and protists. Many animal viruses cause disease and are not affected by antibiotics since they use the host cell translation machinery to synthesize viral proteins. For a review of how antibiotics inhibit bacterial cells, see Table 12.2.
- Outside of the host cell, viruses exist as individual particles called virions. These virions have a central core of DNA or RNA enclosed in a protein capsid. Some animal viruses have a membrane surrounding the protein capsid, acquired from the host's cell membrane during viral release.
- The nucleic acid in viruses can either be DNA or RNA and can be either double-stranded or single-stranded.
- In prokaryotic cells, viral infection is initiated when proteins on the viral capsid bind receptor proteins on the host bacterial cell. Viral nucleic acid is injected into the cytoplasm through the viral tail assembly.
- Bacterial viruses have two different kinds of life cycles: the lytic cycle, during which the virus multiplies and lyses the cell, and the lysogenic cycle, during which the viral genetic material is harbored in the host chromosome for many generations. When environmental conditions are altered, the viral genome can excise itself from the host chromosome, multiply, and lyse the cell (see Figure 13.2).
- Some bacteriophages (the temperate viruses), are only able to undergo a lytic cycle. Other bacteriophages can alternate between lytic and lysogenic life cycles.
- During the early phase of the lytic cycle, the phage injects its nucleic acid into the host cytoplasm and transcribes and translates early genes whose products shut down host transcription, degrade host DNA, and replicate viral DNA. During the late phase of the lytic cycle, the late viral genes are expressed. These include the proteins that package the viral nucleic acid into mature phage and proteins that lyse the cell. At the end of the lytic cycle, the mature phage lyses the cell (see Figure 13.3). The lytic cycle takes about thirty minutes to produce hundreds of bacteriophage per cell.
- Lysogeny occurs when the viral genome inserts itself into the host genome. The virus is called a prophage at this stage. The prophage prevents the host from being infected by other bacteriophages, providing the host immunity to infections by the same virus. Lysogenesis usually occurs when environmental conditions are limited and the host cell is growing slowly. When environmental resources are abundant, lysogenic viruses undergo a change in gene expression, including excision from the host chromosome and completion of the lytic cycle.

Animal viruses

- Animal viruses infect a wide variety of vertebrates, but among invertebrates only arthropods (crustaceans and insects) are infected. Arboviruses infect insects but do not adversely affect them. Infected insects are vectors for the virus because they transmit the virus to mammals via an insect bite.
- Animal viruses consist of a nucleic acid core (either DNA or RNA) surrounded by protein. Some animal viruses (enveloped viruses) have a membrane surrounding the viral protein core, which is derived from the host plasma membrane. The genome of animal viruses usually encodes only a few proteins.
- Animal viruses, like bacterial viruses, also have early and late stages. Transcription and translation of early genes includes the synthesis of viral proteins that replicate the viral genome and turn off host transcription. Late viral gene products include proteins that package the virus and lyse the cell (see Figure 13.3).

- Animal viruses lacking a membrane are taken into the host cell via the endocytic pathway (refer to the discussion on endocytosis, Chapter 5). Inside the cell, the membrane of the endocytic vesicle is broken down, releasing the virus to the cytoplasm. Enzymes in the cytoplasm degrade the protein capsid of the virus and liberate the viral nucleic acid.
- Enveloped viruses bind host cells using viral glycoproteins embedded in the viral envelope which bind specific protein receptors in the host cell membrane. These viruses can be taken into the host via the endocytic pathway and released from the endocytic vesicle into the cytoplasm. Alternatively, some enveloped viruses release the virion into the cytoplasm of the host cell by fusion of their viral envelope with the cell membrane of the host cell. Enveloped viruses leave the host cell by budding out of the plasma membrane, carrying part of that membrane with them.
- RNA viruses, which include influenza, are endocytosed into a vesicle. Inside the cell, the viral envelope fuses with the endocytic vesicle, releasing the RNA into the host cell cytoplasm. The viral RNA is used to translate viral proteins and is also replicated by viral encoded proteins (see Figure 13.4).
- HIV is an RNA virus which enters the host cell by fusion of its viral membrane with the host plasma membrane. HIV I is a retrovirus, copying its RNA genome into DNA using reverse transcriptase. The DNA copy of the viral genome is then integrated into the host cell's chromosome, and in this form it is termed a provirus (see Figure 13.5).
- Plant viruses can infect horizontally (plant to plant) or vertically (plant to offspring). In a horizontal infection the virus is transmitted from one plant to another by a vector, usually an insect, or by physical contact between a bruised leaf of an infected plant and an uninfected plant. Once inside the cell, the virus reproduces and spreads to other cells in the plant. Viral infection of other plant cells is facilitated by the plasmodesmata, which are special plant structures that form cytoplasmic bridges between cells. Virus travels through these channels to infect adjacent plant cells.
- Viroids are small infectious agents that consist of a circular single-stranded RNA molecule of only a few hundred nucleotides 1/1000th the size of the smallest viruses. Viroids are found only in plant cells, primarily in the nuclei. They are replicated by enzymes in the host cells and may have evolved from introns. They cause a number of plant diseases. It is not well understood how they do so.

Mobile DNA in prokaryotes

- Prokaryotic cells reproduce asexually, generating genetically identical cells (clones). New genetic material can be introduced into bacterial cells in several ways including conjugation, transformation, and transduction.
- Bacteria can pass some of their genetic material to other bacterial cells through the process of conjugation or mating. Donor cells synthesize a thin extension called a pilus, which eventually contacts a recipient cell forming a conjugation tube. DNA is transferred through the conjugation tube to the recipient cell, where it can recombine with the chromosome of the recipient cell.
- During transformation, external DNA can be introduced to cells as seen in experiments in which rough avirulent pneumococci bacteria were transformed to smooth virulent bacteria (see Chapter 11).
- Some viruses can pick up bacterial genes from one cell, pack them in their viral capsid, and transfer them to the next bacterial cell they infect. This process of gene transfer using a viral vector is called transduction.

Plasmids

- Plasmids are small circular DNA molecules (sometimes called episomes) found in the cytoplasm of some bacterial cells. These DNA molecules can move from one bacterial cell to another during conjugation. Plasmids contain a few dozen genes.
  - Genes for metabolic factors encode enzymes with unusual metabolic functions, such as enzymes that digest oil.
  - Genes for mating factors encode proteins that form the pilus and the conjugation tube. These factors are called fertility factors, and the plasmids themselves are called F plasmids (for fertility).
  - Genes for drug resistance encode enzymes that destroy or inactivate antibiotics and heavy metals. These factors are called R factors because they impart drug resistance to the host cells bearing those plasmids.

Transposable elements

- Transposable elements are special DNA sequences that can move to other sites on the chromosome. Transposable elements carry genes whose products can excise these DNA elements and insert them elsewhere in the chromosome. Some transposable elements are replicated during transposition, whereas others are simply excised and placed at a different chromosomal site without DNA replication (see Figure 13.12). Longer transposable elements that have additional genes are called transposons. R plasmids contain parts of transposons, suggesting that transposition might be the mechanism by which plasmids originally gained antibiotic resistance.

Gene regulation

- Gene regulation in prokaryotes is primarily transcriptional. Some regulation occurs at the posttranscriptional level, including hydrolysis of the messenger RNA, blocking of translation, or protein hydrolysis.
- Genes can be expressed constitutively (all the time), or they can be expressed in response to specific environmental conditions.
- Enzyme activity can be repressed by end-product feedback (also known as feedback inhibition) or by regulating gene expression.

Negative regulation of gene expression

- The *lac* operon is an inducible operon.

- The structural genes for the metabolism of lactose are adjacent to one another on the chromosome and are all transcribed from one promoter (review transcription, Chapter 12). The operator overlaps the promoter and is the site where the repressor binds, which prevents transcriptional initiation by RNA polymerase. The regulatory sequences of the *lac* operon are the promoter, and the operator. The structural genes encode enzymes that metabolize lactose or transport lactose into the cell.
- In cells that are not exposed to lactose, the repressor binds the operator and the *lac* genes are not expressed. The *lac* operon is inducible (see Figure 13.13). An inducer of the *lac* operon (lactose and other β-galactosides) binds the repressor, which changes its tertiary configuration so that it can no longer bind the operator site. RNA polymerase initiates transcription and the structural genes are transcribed. The messenger RNA for *lac* structural genes is translated into protein and lactose can be metabolized (see Figure 13.17).
- The repressor is encoded by the *i* gene. The promoter for this gene is constitutive such that there are always about 10 molecules of *lac* repressor per cell.

- The *trp* operon, a repressible operon
  - When RNA polymerase binds the promoter for the *trp* operon, all of the structural genes (*E, D, C, B,* and *A*) are transcribed from the *trp* promoter of the *trp* operon. The *trp* operator site overlaps the *trp* promoter and when the *trp* repressor is bound at the operator, no transcription of the *trp* operon can occur.
  - The *trp* operon is repressible; as the levels of the amino acid tryptophan increase; the *trp* operon is repressed or turned off. Tryptophan is the corepressor. When tryptophan levels are high in the cell, tryptophan binds its repressor, which changes the tertiary configuration of the repressor, allowing it to bind the *trp* operator and preventing transcription (see Figure 13.18).

Positive regulation of gene expression

- Transcription can be positively regulated. The cAMP receptor protein (CRP), when bound to cAMP, will cause the transcription of certain operons to increase (see Table 13.2 for an outline of positive and negative regulation of transcription). The binding site for the CRP protein is just upstream of the promoter. Glucose affects the levels of cellular cAMP by decreasing the activity of the enzyme that synthesizes cAMP. When glucose levels are high in the cell, cAMP levels are low and the CRP protein is not active. Under these conditions, transcriptional activation by the CRP protein does not occur. As glucose levels fall, cAMP levels increase and the CRP protein binds the available cAMP molecules and activates transcription (see Figure 13.19).
- Bacteriophage lambda
  - In bacteriophage lambda, there are two lambda proteins, cI and Cro, which determine whether the phage will complete a lytic or a lysogenic cycle. These two regulatory proteins control a genetic switch. Both cI and Cro compete for two operator promoter sites in the lambda DNA, which control genes involved in lytic and lysogenic functions. If the bacteriophage has infected a host cell that is experiencing limited environmental resources, then lysogeny is favored and the cI-controlled switch is active. If the infected cells are in a rich environment, then the Cro-controlled switch is favored and the bacteriophage undergoes a lytic cycle (see Figure 13.20).

Genomes

- Genomic sequencing and analysis of those sequences is an active area of biological research. Information from these analyses includes open reading frames (which could potentially encode proteins, amino acid sequences of known gene products, and regulatory sequences. Signals for the initiation and termination of translation are used to identify unknown proteins. The field of functional genomics includes assigning roles to genes (annotation) and describing how these gene products work in the organism.
- As more genome sequences have been completed, those from different organisms have been compared to understand the physiology of different organisms.
- Genomic analysis has revealed a new gene for ATP synthesis in bacteria, surface proteins responsible for bacterial infectivity in humans, virulence genes in the bacterium that causes typhus, and bacterial surface proteins that could be used for vaccines. Furthermore, genomic analysis in simple organisms has been used to determine the minimal number of genes required for cellular life.

The Big Picture

- Prokaryotic cells and viruses have small genomes. These have been probed by molecular biologists to understand gene function, regulation, and viral life cycles. During viral infection, sequential gene expression occurs early and late, to facilitate successful viral production.
- Prokaryotic cells use operons to coordinately regulate the expression of several genes. In response to environmental signals (the presence of lactose or the absence of tryptophan) the repressor protein releases the operator site, leaving the promoter accessible to RNA polymerase to initiate transcription. In bacteriophage lambda, other proteins bind promoters to determine whether lysis or lysogeny will occur.
- New genes can be introduced in to prokaryotic cells during transformation, conjugation, and transduction. Transposable elements and transposons are mobile genetic elements which move from one part of the chromosome to another. Plasmids are small circular DNA molecules that contain genes for metabolic factors, drug resistance, heavy metal resistance, and conjugation proteins.
- Genomics is used to study and compare genomes from different organisms and to identify genes, potential genes, and their regulatory sequences. Information from these studies has been used to design new medicines for pathogens and to understand the minimal requirements for a cell to sustain life.

Common Problems

- It is easy to confuse inducible operons (the *lac* operon) with repressible operons (the *trp* operon) since they both use repressors to regulate gene expression. Review the environmental conditions that must exist for each repressor to bind its operator site and what causes each repressor to release the operator site. (In the presence of lactose, the *lac* repressor cannot bind the *lac* operator site; in the presence of tryptophan; the *trp* repressor binds the *trp* operator site.) Review gene expression (Chapter 12) to see how, once transcription is initiated, proteins are produced that will act on those environmental signals.
- Students often think that bacteria reproduce only asexually, but there are several ways that new genetic material is introduced into bacterial cells, including mating, transformation, and transduction. Review transposable genetic elements and plasmids, keeping in mind the advantages that the DNA elements provide their host cells.
- That some viruses insert their chromosome into the host chromosome may seem puzzling at first, since the goal of viral infection seems to be rapid multiplication and infection of adjacent cells. However, lysogeny allows the viral genome to persist until environmental resources are abundant. When resources are abundant, more cells grow and divide, providing the virus with numerous healthy hosts to continue their lytic cycle.

Study Strategies

- Gene expression often occurs in response to environmental signals. Make a list of the environmental signals that a prokaryote receives and then outline the steps the cell initiates to respond to those signals. Include the following signals: lactose in the cell, no glucose or lactose in the cell, high levels of glucose in the cell, low levels of tryptophan in the cell, high levels of tryptophan in the cell, bacteriophage lambda infection under poor environmental conditions, and bacteriophage lambda infection under rich environmental conditions.
- Outline the progress of different animal virus life cycles (Review Figures 13.3, 13.4, and 13.5) from the start of infection through multiplication of virus to viral release. Compare these life cycles to the lytic and lysogenic life cycles of bacteria.
- Review the following tutorials and activity on the website:
  Tutorial 13.1 The *lac* Operon
  Tutorial 13.2 The *trp* Operon
  Activity 13.1 Concept Matching

Test Yourself

Knowledge and Synthesis Questions

1. Viruses consist of
   a. a protein core and a nucleic acid capsid.
   b. a cell wall surrounding nucleic acid.
   c. RNA and DNA enclosed in a membrane.
   d. a nucleic acid core surrounded by a protein capsid and in some cases a membrane.

2. Bacterial cells that are resistant to viruses
   a. lack a cell surface receptor that the virus must bind in order to infect the cell.
   b. harbor a prophage in their chromosome, making the bacterial cell immune to further viral infection.
   c. have enzymes (restriction enzymes) that cut up the viral nucleic acid so that the viral replication cannot occur in the host cell.
   d. All of the above

3. Bacterial viruses
   a. infect the cell, replicate their genomes, and lyse the cell.
   b. infect the cell, replicate their genomes, transcribe and translate their genes, and lyse the cell.
   c. infect the cell, replicate their genomes, transcribe and translate their genes, package those genomes into viral capsids, and lyse the cell.
   d. infect the cell, translate their RNA, and lyse the cells.

4. Animal viruses that integrate their DNA into the host chromosome
   a. are RNA viruses.
   b. are prophages.
   c. copy their RNA genome into DNA using reverse transcriptase.
   d. a and c

5. During conjugation,
   a. DNA from one bacterial cell is transferred to another bacterial cell using a bacteriophage.
   b. mutants that are auxotrophic for one nutrient can be converted to prototrophs when mixed with mutants that are auxotrophic for another nutrient.
   c. a pilus is synthesized and DNA is transferred from one bacterium across the conjugation tube to the recipient bacterium.
   d. a and b

6. Plasmid DNA contains genes that
   a. confer drug resistance to the host cell.
   b. regulate conjugation.
   c. can confer resistance to heavy metals.
   d. All of the above

7. An operon
   a. is regulated by a repressor binding at the promoter.
   b. has structural genes that are all transcribed from same promoter.
   c. has several promoters, but all of the structural genes are related biochemically.
   d. is a set of structural genes all under the same translational regulation.

8. If the gene encoding the *lac* repressor is mutated so that it can no longer bind the operator, will transcription of that operon occur?
   a. Yes, but only when lactose is present.
   b. No, because RNA polymerase is need to transcribe the genes.
   c. Yes, because the operator will not be bound by repressor and RNA polymerase can transcribe the *lac* operon.

d. No, because cAMP levels are low when the repressor is nonfunctional.

9. If the gene encoding the *trp* repressor is mutated such that it can no longer bind tryptophan, will transcription of the *trp* operon occur?
   a. Yes, because the *trp* repressor can only bind the *trp* operon and block transcriptional initiation when it is bound to tryptophan.
   b. No, because this mutation does not affect the part of the repressor that can bind the operator.
   c. No, because the *trp* operon is repressed only when tryptophan levels are high.
   d. Yes, because the *trp* operon can allosterically regulate the enzymes needed to synthesize the amino acid tryptophan.

10. Transcriptional regulation in prokaryotes can occur by
   a. a repressor binding an operator and preventing transcription.
   b. an activator binding upstream from a promoter and positively affecting transcription.
   c. different promoter sequences binding RNA polymerase more tightly, resulting in more effective transcriptional initiation.
   d. All of the above

Application Questions

1. Animal viruses are obligate parasites. What does this mean?
2. Why are antibiotics useless in combating viral infections?
3. Genomics has revealed that the bacterium that causes tuberculosis has over 250 genes that metabolize lipids. What does this finding suggest about the bacterium and about medicinal approaches to this disease?
4. A cell has a mutation that deletes the gene encoding the repressor for a certain operon. A plasmid is introduced into the host cell which carries a wild-type copy of the gene for the repressor. Is normal regulation of this operon restored in the presence of this plasmid?
5. A cell has a mutation that deletes the gene encoding the operator for a certain operon. A plasmid is introduced into the host cell which carries a wild type copy of the operator. Is normal regulation of this operon restored in the presence of this plasmid?
6. Review some of the auxotrophic bacterial strains that Lederberg and Tatum were studying. Describe three different ways that an auxotrophic bacterium can be converted to phototrophic bacterium.

Answers to Questions

Knowledge and Synthesis Answers

1. **d.** Nucleic acids do not form capsids, cell walls are found in bacterial and plant cells, not viruses.Viruses are organized so that the nucleic acid is surrounded by protein (not membranes) and the protein capsid may be surrounded by a membrane.
2. **d.** Bacterial viruses must bind to cell surface receptors to initiate their infective cycle. In bacteriophage lambda, a prophage (a viral chromosome inserted into the host chromosome) prevents other lambda phage from infecting that bacterium. Bacterial cells contain restriction enzymes that cut DNA at specific sites to prevent viral DNA from being replicated in the cell
3. **c.** This sequence includes the most complete details of the viral life cycle.
4. **d.** Answer **a** describes a provirus, which is an animal virus that has inserted its genome into a host chromosome.
5. **d.** Answer **a** describes transduction; answer **b** describes how auxotrophic mutants can be rescued by mating with other bacteria.
6. **d.** All of these genes can be found on plasmid molecules.
7. **b.** An operon is a set of genes that are all transcribed from the same promoter. Answer **a** is not correct. The repressor binds at the *operator* site, which overlaps the promoter. Answer **d** is not correct because the operon is regulated transcriptionally, not translationally.
8. **c.** If the *lac* repressor is nonfunctional, it cannot bind the operator site and transcription of the *lac* operon will occur at all times, whether or not lactose is present.
9. **a.** If the repressor can no longer bind tryptophan, then it cannot bind the operator and transcription of the *trp* operon will always be *on*, whether tryptophan levels in the cell are high or low.
10. **d.** Answer **a** refers to the *lac* and *trp* repressors, answer **b** to the CRP protein, and answer **c** refers to promoters that have different transcriptional efficiencies.

Application Answers

1. Animal viruses cannot express their genes, replicate their genomes, or multiply unless they are in the cytoplasm of the host cell. They use the host cell's components (ribosomes, ATP) to grow and reproduce.
2. Most antibiotics attack bacterial cells by inhibiting prokaryotic translation. Animal viruses use the host translation machinery, which is different enough from the bacterial translation machinery that it is unaffected by antibiotics. Drugs that would inhibit host translation would kill the host cell (by preventing protein synthesis) as well as the virus.
3. The tuberculosis bacterium must use lipids as a source of energy-rich compounds; inhibiting lipid synthesis in this bacterium may inhibit the growth of this bacterium.
4. Yes. The repressor gene can be transcribed and translated from the plasmid DNA and normal regulation will be restored.
5. No. The operator site on the plasmid cannot restore regulation unless it recombines with the host operator site in such a way that it replaces the mutant operator on the host chromosome. The DNA site on the plasmid would bind repressor, but since that site is not adjacent to the promoter or the structural genes on the chromosome, normal regulation of those genes cannot occur.

6. Converting an auxotroph to a prototroph would require the introduction of a wild-type copy of the gene that was nonfunctional in the auxotroph (for example, introducing a gene that encoded an enzyme necessary for leucine biosynthesis to a leucine auxotroph lacking that enzyme). The wild-type copy of this gene could be introduced to the recipient cell by conjugation, transformation, or transduction.

STUDY GUIDE

14 The Eukaryotic Genome and Its Expression

Important Concepts

Genes in eukaryotic cells differ from those of prokaryotic cells in terms of content and organization.

- The eukaryotic genome is larger (more DNA) and more complex (more genes) than the prokaryotic genome. For example, the size of the human genome is 6 billion base pairs (bp) per cell, while bacteria such as *E. coli* have 4.5 million bp per cell, and viral genomes can have as few as 10,000 bp.
- Among eukaryotes, genome size is not related to relative complexity; cells of the lily plant have eighteen times more DNA than human cells.
- Much of the eukaryotic genome is noncoding. Some of this noncoding sequence is important for the structural integrity of the chromosome, specifically centromeres and telomeres.
- The eukaryotic genome (20–100 million bp per chromosome) is located on several linear chromosomes, while the single prokaryotic chromosome (4.5 million bp per chromosome) is circular.
- Each chromosome, consisting of a double helix of DNA, has three essential parts: an origin (the recognition sequence for DNA polymerase to initiate replication), a centromere (which holds the replicated chromosomes together before mitosis and meiosis is completed) and telomeres (the ends of the chromosome).
- Chromosomes are packed tightly into the nucleus (review Figure 9.7) and are separated from the cytoplasm by a nuclear envelope. Replication occurs in the nucleus, as does transcription.

Gene Expression

- During gene expression, transcriptional complexes must be able to access the DNA, including regulatory regions such as regulators, promoters, enhancers, and silencers. Some of these regulatory sequences (enhancers and silencers) can be located quite far from the promoter.
- Pre-mRNA transcripts include noncoding regions that must be removed from the RNA before the mRNA is transported to the cytoplasm for translation.

Model eukaryotes are used to study eukaryotic gene expression and development.

- *Saccharomyces cerevisiae*
  - *Saccharomyces cerevisiae,* brewers or bakers yeast, is a simple eukaryote that has 12 million bp on chromosomes (per haploid genome).
  - *S. cerevisiae* has 6,200 genes, 70 percent of which have been assigned roles (review "Prokaryotic Genomes," Chapter 13).
  - Within the known genes are genes for metabolism (11%), energy production and storage (3%), protein synthesis (12%), DNA replication (3%), and protein targeting and secretion (6%).
  - The genome of *E. coli* has the same number of genes for these basic functions with the exception of genes for protein targeting (see Table 14.2).
- *Caenorhabditis elegans*
  - The nematode *Caenorhabditis elegans* is an extremely useful eukaryotic organism for studying development.
  - In three days it grows from a fertilized egg to a 1000-celled differentiated organism.
  - The genome of *C. elegans* is eight times larger than that of yeast (97 million bp) and four times the number of protein coding genes (19,099 proteins versus 6,000 proteins). Genome comparisons have revealed that 3,000 of these genes in *C. elegans* have direct homologs in yeast.
  - Other gene products in *C. elegans* include proteins used for intracellular communication, cellular differentiation, and cell adhesion.
- *Drosophila melanogaster*
  - The fruit fly, *Drosophila melanogaster*, has ten times more cells than *C. elegans* and 180 million more bp.
  - Using genome annotation, scientists have identified 2,500 *Drosophila* genes and assigned their functions. Protein-coding regions include 13,600 genes (fewer than in *C. elegans*). There are fewer gene families (genes that are related in sequence and function) in *Drosophila*, than there are in *C. elegans.* The genome of *C. elegans* also has a large number of genes devoted to sensing chemicals in the environment.
  - One-third of the genes of *Drosophila* have homologs in the nematode; one-half of its genes have homologs in mammals.

- The fly genome has 177 genes with sequences known to be related to human disease genes, including some genes involved in cancer and neurological disease.

- *Arabidopsis thaliana*
  - *Arabidopsis thaliana* is another eukaryote for which genome analysis and gene regulation are very active areas of research. This plant has 130 billion bp.

There are three types of highly repetitive sequences in the eukaryotic genome: satellites, minisatellites, and microsatellites.

- Satellites are 5–50 bp long, repeated side by side up to a million times in the genome. Satellites are found in the centromeres, and their role is unknown.
- Minisatellites are 12–100 bp long and are repeated several thousand times. The number of repeats of minisatellites varies in individual organisms due to slippage of DNA polymerase during replication.
- Microsatellites are short 1–5 bp sequences that occur in small clusters of 10–50 copies scattered all over the genome.

Moderately repetitive DNA includes telomeres, tRNA and rRNA genes, *Alu* sequences, and transposable elements.

- Telomeres are moderately repetitive sequences located at the ends of the chromosome.
  - Special proteins (telomerases) bind these sequences to help maintain the stability of the ends of the chromosomes during DNA replication. Initially in DNA replication, primers are synthesized and later they are removed.
  - Primer removal from the lagging strands that are synthesized at the ends of the chromosome leave single-stranded parental DNA. DNA repair mechanisms in the cell remove single-stranded DNA (and some double-stranded DNA), causing a shortening of the chromosomes.
  - Telomerase, an RNA-protein complex, binds the ends of the chromosome and facilitates replication of those ends (see Figure 14.3). Telomerase activity is essential to maintain the stability of the chromosomes in cells that continually divide (bone marrow cells and germ-line cells) and also plays a role in cancer cells.
  - In aging cells (i.e., human cells in culture after about 20–30 divisions), telomerase is not active and the chromosomes shorten after each cell division, eventually resulting in cell death.
- The genes for tRNA's and rRNA's also fall within the class of moderately repetitive DNA.
  - Multiple copies of these genes are needed to provide the cell with enough protein translation machinery.
  - In mammals there are four different rRNA molecules: 18S, 5.8S, 28S, and 5S. The 18S, 5.8S, and 28S rRNA genes are all transcribed from one promoter, producing a large precursor RNA which is enzymatically trimmed to yield the mature rRNA molecules (see Figure 14.4).
  - In humans there are 280 copies of these rRNA gene clusters, located on five different chromosomes.
- *Alu* sequences, found in the human genome, are another example of moderately repetitive DNA.
  - These sequences, containing an Alu I restriction site, are 300 bp long, and are found throughout the genome. They are not transcribed.
  - Because there are so many *Alu* copies (300,000 per genome) dispersed throughout the chromosomes, it is thought that these sequences may act as origins for DNA replication.
- Other nongene sequences in the moderately repetitive class of DNA include transposable elements.
  - These DNA sequences are not stably integrated in the chromosome and can relocate to another part of the genome.
  - There are four kinds of transposable elements: SINEs, LINEs, retrotransposons, and DNA transposons.
  - SINEs are short interspersed elements of greater than 500 bp, which are transcribed but not translated.
  - LINEs are long interspersed elements, which are greater than 7000 bp on average, and some are transcribed and translated.
  - SINEs and LINEs (greater than 100,000 copies/genome) make an RNA copy of DNA; this RNA is then used as a template for DNA synthesis which is inserted at a new site in the genome.
  - Retrotransposons are transposons that make an RNA copy when they move to a different site in the genome. These transposable elements are common in animals (except mammals) and yeast.
  - Viral transposons are organized in a manner similar to retroviruses (HIV I), except that they lack genes for the viral protein coat.
  - DNA transposons do not replicate when they move to a new site on the chromosome.
  - Transposons adversely affect the cell by inserting their DNA into different genes and inactivating them, as seen in some forms of muscular dystrophy and hemophilia. Insertions of transposons in genes that control cell division can convert a somatic cell to a cancer cell.
  - If the insertion of transposons occurs in a germ-line cell, it can be passed down to the progeny.
  - Transposons can beneficially affect an organism by picking up adjacent genes during transposition and moving them to a new site on the chromosome, potentially creating a new gene that will be advantageous to the organism's survival.
  - Transposons help explain why some mitochondrial and chloroplast genes are located in the nucleus while other organelle genes remain inside the organelle genome.

Single-copy DNA sequences include structural genes in eukaryotic cells.

- Single-copy genes contain noncoding regions, including the promoter, the transcriptional terminator, and introns.
- Introns are noncoding internal sequences that are transcribed in the pre-mRNA and removed in the nucleus. After intron removal the exons, which contain the coding sequence for the gene, are positioned adjacent to each other.
- Nucleic acid hybridization experiments clearly demonstrated the presence of introns in pre-mRNA molecules (see Figure 14.7).
  - Denatured globin DNA was hybridized to globin messenger RNA and loops of single-stranded DNA representing noncoding intron sequences were observed.
  - Hybridizations with globin DNA and pre-mRNA revealed no single-stranded DNA loops, indicating that in the pre-mRNA the introns were still present.
- In DNA, introns separate gene-coding sequences into distinct parts called exons. All of the exons are in order, and in some cases code for functional domains.
- Pseudogenes are inexact copies of genes and are usually positioned near the real gene on the chromosome.
- Gene families are a set of duplicate or related genes (such as the beta-globin genes).
  - Gene families provide the organism with a functional gene while allowing mutation in the other members of the gene family. Some of these mutations may create new genes that are advantageous to the organism.
  - In humans the globin gene family consists of three alpha-globin genes and three-beta globin genes.
  - Different globins are expressed at different times in development. Gamma-globin, which is expressed in the fetus, binds oxygen more tightly to insure oxygen transfer from the mother across the placenta to the fetus.
  - There are globin pseudogenes which are nonfunctional because they lack promoters or recognition sites for intron removal. However, some pseudogenes are functional.

Modifications of pre-mRNA (RNA processing) occurs in the nucleus of eukaryotic cells.

- A cap of modified GTP is added to the 5′ end of eukaryotic mRNA to facilitate binding of messenger RNA to the ribosome in the cytoplasm, and to protect the messenger RNA from degradation.
- At the 3′ end of the messenger RNA a poly A tail is added.
  - The RNA is first cut by an enzyme that recognizes the sequence A A U A A A at the 3 ′ end of the messenger RNA.
  - The poly A tail (from 100–300 nucleotides) is then added to the 3′ end.
  - The poly A tail assists in the transport of the messenger RNA to the cytoplasm.
- Splicing is the removal of introns from pre-mRNA.
  - Introns are removed from pre-mRNA's by small nuclear ribonucleoprotein particles (snRNP's).
  - SnRNP's bind consensus sequences at the 3′ and the 5′ boundaries of the introns using complementary base pairing.
  - Other proteins bind the snRNP's to form the spliceosome, which uses the energy of ATP hydrolysis to cut the RNA, release the intron, and rejoin the RNA to form the mature messenger RNA.
- Some forms of β-thalassemia (in which there is an inadequate amount of β-globin, resulting in severe anemia) are the result of a mutation in the intron–exon consensus sequence causing inappropriate splicing of β-globin RNA and a nonfunctional gene product.
- Mature RNA is transported from the nucleus to the cytoplasm through the nuclear pores.

The expression of eukaryotic genes must be precisely regulated.

- Multicellular organisms must regulate their genes precisely so that the correct genes are expressed at the appropriate time and development can proceed normally. Differentiated cells have different functions and must express genes specific for those functions.
- Some gene products alter chromosome function and structure, making the DNA more accessible for the transcription complex or making extra copies of certain genes to provide the cell with the necessary gene products.
- All cells must express housekeeping genes, including proteins involved in ATP synthesis and protein translation.
- Transcription of coordinately expressed genes in prokaryotic cells is regulated from one operon, but in eukaryotic cells each gene is usually regulated by its own promoter. Eukaryotic genes that must be transcribed at the same time must have similar control sequences in front of each separate gene.
- Genes are regulated posttranscriptionally at the levels of transport, mRNA stability, translation, protein modification and protein stability.
- There are three RNA polymerases in eukaryotic cells.
  1. RNA polymerase I transcribes rRNA.
  2. RNA polymerase II transcribes mRNA's.
  3. RNA polymerase III transcribes tRNA's and small nuclear RNA's (seen in snRNP's).
- Eukaryotic promoters are diverse and have several kinds of regulatory sequences in front of those genes, including promoters, enhancers, silencers, and regulators.
- Initiation of transcription in eukaryotic cells requires transcription factors, which are proteins that bind the promoter to help RNA polymerase bind the promoter and initiate transcription.
- Prokaryotic promoters have conserved sequences at about 40 bp 5′ from the initaition point of transcription, and 10 bp (the TATA box) 5′ from the initiation point of transcription. RNA polymerase binds these sequences to initiate transcription.

- Eukaryotic promoters have a TATA box positioned about 25 bp 5′ from the start of transcription and one or two recognition sequences about 50–70 bp 5′ from the TATA box.
  - The TATA sequence is common to many eukaryotic promoters.
  - Other promoters have DNA sequences that are recognized by transcription factors, which are only expressed in certain tissues. These transcription factors play an important role in differentiation.
- TFIID, a transcription factor, binds the TATA box, which changes the shape of the DNA and the transcription factor. Other transcription factors then bind the promoter, and finally RNA polymerase binds to initiate transcription.
- There are other important DNA regions that bind proteins and affect transcription, including regulators, enhancers, and silencers.
- Regulator regions, which bind proteins that activate transcription (see Figure 14.15), occur just upstream of the promoter.
  - Enhancers can be located near or far away from the gene they affect, in some cases as far as 20,000 bp away.
  - Activator proteins bind enhancers and have a positive effect on the transcription complex; perhaps by bending the DNA so the activator is in contact with the transcription complex.
- Silencers are sequences that are bound by repressor proteins and negatively regulate transcription.
- Transcription of any eukaryotic gene is the result of the combination of all of these regulatory elements.

Proteins that bind DNA (activators, transcription factors, regulators, and repressors) have characteristic protein motifs.

- Helix-turn-helix motifs are proteins that contain several α-helices with one of those helices making contact with the DNA. Proteins that regulate embryonic development (the homeobox protein, see Chapter 15), and the development of the nervous and immune systems are included in this class.
- Zinc finger proteins fold into fingerlike loops when they bind a zinc ion between a cysteine and a histidine residue. Examples of this kind of DNA binding protein are steroid hormone receptors (see Figure 15.9).
- Proteins that form leucine zippers are polypeptides that have a large number of leucine residues on one side of the protein.
  - These proteins dimerize so that the hydrophobic residues can interact with one another and position positively charged domains of the protein adjacent to the negatively charged DNA.
  - The leucine zipper motif is found in transcription factors, including AP-1, which binds near promoters of genes involved in mammalian growth and division.
  - Overactivity of the AP-1 protein has been linked to cancer.
- Proteins with the helix-loop-helix motif have α-helices separated by a loop. Transcription factors in muscle protein genes and enhancers of immunoglobulin genes are included in this group of DNA binding proteins.

In order for gene expression to occur, the DNA must be made accessible to the transcription complex.

- The DNA in eukaryotic chromosomes is associated with nucleosomes and other chromosomal proteins. (see Figure 14.17), which can inhibit the initiation and elongation of transcription.
  - The first step in gene expression is the recruitment of protein complexes to remodel the chromatin.
  - Upstream of the initiation site, nucleosomes can be disaggregated so that the transcription complex can bind the DNA and transcription can be initiated.
  - Other protein complexes can bind the DNA once transcription has started to help the transcription complex move through the nucleosomes.
- Transcriptionally active chromatin (euchromatin) is less condensed and takes up less nuclear stain than the transcriptionally inactive, highly condensed chromatin (heterochromatin).
- The genotype for the sex chromosomes in mammals is XY in males and XX in females. Yet the expression of X-linked genes is the same in both sexes, due to X-inactivation in females.
  - Early in embryonic development, one of the two X chromosomes is randomly inactivated, producing a highly condensed heterochromatic chromosome, called a Barr body.
  - The number of Barr bodies is equal to the number of X chromosomes minus one (XXY, Klinefelters syndrome, and XXX both have two Barr bodies).
- X inactivation can be related to methylated DNA (methyl groups are added to the 5′ carbon of cytosine), which is more concentrated in inactive chromatin, and to the *XIST* gene.
  - Transcription of the *XIST* gene occurs on the inactive X chromosome.
  - The *XIST* RNA binds the X chromosome from which it was transcribed and inactivates it.

DNA sequences can be regulated by moving them to a new site on the chromosome to activate transcription.

- The mating type locus (MAT) in the budding yeast, *Saccharomyces cerevisiae*, is the site in the yeast genome where the mating type (*a* or α) is determined.
  - Yeast can be either *a* or α mating type, depending on which allele is in the MAT locus.
  - Mating types change with almost every cell division cycle.
  - The *a* and α alleles are located in an inactive region of the chromatin, but during cell division, one of these two alleles is copied and placed into the MAT locus.

- If the yeast is an *a* mating type, than a new DNA copy of α is made. The *a* allele at the MAT locus is removed by enzymes and the α allele is inserted at the MAT and transcribed, resulting in a yeast cell of α mating type.
- DNA rearrangements occur in the genes for human antibodies and in some cancers when inactive genes are moved adjacent to a promoter.

Specific genes can be amplified to meet the needs of the cell, as seen in the amplification of the ribosomal RNA genes.

- Egg cells in frogs and fishes amplify the ribosomal RNA gene clusters so that there are greater than a million copies of ribosomal DNA per egg cell.
- Gene amplification is also seen in cancer cells, specifically in oncogenes (genes that cause cancer) and in genes encoding protein receptors for cancer drugs.

Posttranscriptional control mechanisms include alternate splicing and increasing mRNA stability.

- In mammals, tropomyosin pre-mRNA is spliced five different ways to generate five different forms of this gene product, each for a different tissue: skeletal muscle, smooth muscle, fibroblast cells, liver, and brain.
- When a messenger RNA arrives in the cytoplasm, ribonucleases degrade it. This degradation can be affected by other cytoplasmic factors. When the concentration of free tubulin in the cell's cytoplasm increases; it binds its own mRNA, increasing its susceptibility to ribonuclease degradation.
- Messenger RNA's in the cytoplasm are not always translated, because other factors regulate the accessibility of these mRNA's to the translational machinery. For example, if the G of the 5′ cap is not modified, as seen the embryo of the hornworm moth, the mRNA is not translated. Later during development this G is modified and the mRNA is translated.
- Ferritin is a storage protein for iron. As the concentration of iron ions rises in the cell, the level of ferritin protein increases. Iron ions bind a translational repressor of the ferritin mRNA, preventing that repressor from binding the ferritin mRNA. As a result, ribosomes can translate the ferritin RNA and the cellular levels of ferritin protein increase.
- The translation of the globin messenger RNA increases in the presence of excess heme in the cell. Heme removes a block to translational initiation of globin messenger RNA.

Protein modification includes glycosylation, phosphorylation, and the removal of signal sequences. These modifications can regulate the lifetime of a protein in the cell.

- Protein degradation regulates cellular concentrations of many proteins, including cyclins (which control the cell cycle), transcription factors (which regulate transcription), and abnormal proteins.
  - Proteins that are targeted for degradation are first linked to ubiquitin. This 76-amino acid protein binds a huge protein complex called a proteasome. The proteasome is a hollow cylinder that has ATPase activity.
  - When a protein-ubiquitin complex enters the proteosome, ubiquitin is removed, the protein is unfolded, and three proteases digest the protein (see Figure 14.21).
- Human papilloma virus targets p53 (a cellular protein that inhibits cell division) for proteosome degradation, leading to unregulated cell division and cancer.

The Big Picture

- Eukaryotic genomes are larger and more complex than prokaryotic genomes. The chromosomes in eukaryotic cells are linear and there are more than one.
- Eukaryotic DNA contains many genes and many noncoding regions, including nongene sequences, moderately repetitive sequences, highly repetitive sequences, nontranscribed regions, and introns. Some of the repetitive DNA has a structural function for the chromosomes, including centromeres and telomeres.
- The DNA is highly condensed in the nucleus due to its association with nucleosomes and other chromosomal proteins. Gene expression is a two-step process: The chromatin must be modeled to expose the DNA containing the genes so that the transcriptional complexes (which consist of transcription factors, regulatory proteins, activators, and RNA polymerases) can access those DNA sites to regulate transcription of that gene. Transcription factors and other regulatory proteins have particular motifs that facilitate their binding to DNA.
- In eukaryotic cells, transcription is separated from translation by the nuclear membrane. RNA's are processed in the nucleus before being transported to the cytoplasm. Posttranscriptional regulatory elements affect gene expression, including different mRNA stabilities, translation regulation, protein modification, and protein degradation.

Common Problem Areas

- Sometimes it is difficult to understand why there are so many different ways gene expression is regulated in eukaryotic cells. Examine a picture of a eukaryotic cell to get a feeling for all of the specialized parts of the cell and how gene expression must be tailored for each of these parts. Now think about a multicellular organism and how different genes must be expressed in all of the differentiated tissues.
- Grasping the different of events that are involved in the regulation of gene expression in eukaryotic cells may be overwhelming. Begin by examining how transcription of a eukaryotic gene occurs by reviewing chromatin structure, the promoter, RNA polymerase, and transcription factors. Next examine how regulators, enhancers, and silencers, when bound by proteins, can modulate the levels of transcription of a eukaryotic gene. To complete your analysis, review the details of posttranscriptional regulation.

Study Strategies

- Make a list of the sites on the DNA that affect transcriptional initiation. Next, list the names of the proteins that bind those DNA sites and describe how they affect transcription.
- Draw a simple diagram of a eukaryotic gene.
  - Include the flanking 5′ and 3′ sequences of the gene that are important for regulation and gene expression.
  - Draw a diagram of the pre-mRNA and indicate sites on the RNA where processing occurs.
  - Make another diagram showing the mature mRNA as it appears in the cytoplasm. Indicate sites on the mRNA where posttranscriptional regulation might occur.
  - Finally, describe ways that proteins can be modified to regulate gene expression.
- Give examples of moderately repetitive and highly repetitive sequences in the DNA and describe their function.
- Review model organisms described in this chapter and describe two advantages for studying gene regulation in each of these organisms.
- Review the following tutorials and activity on the website:
  Tutorial 14.1 RNA Splicing
  Tutorial 14.2 Initiation of Transcription
  Activity 14.1 Eukaryotic Gene Expression: Structural Features

Test Yourself

Knowledge and Synthesis Questions

1. Eukaryotic chromosomes
   a. are circular and contain origins and terminator sequences.
   b. are linear and have origins and telomeres.
   c. contain coding and noncoding sequences.
   d. b and c

2. Model eukaryotic organisms have helped biologists understand gene(s)
   a. involved in development.
   b. families.
   c. encoding proteins that are essential for all cells.
   d. All of the above

3. Moderately repetitive DNA includes
   a. only coding sequences.
   b. only noncoding sequences.
   c. coding and noncoding sequences.
   d. satellites, minisatellites, and microsatellites.

4. The ends of eukaryotic chromosomes
   a. are single-stranded and must be bound by transcription complexes.
   b. are highly repetitive sequences that hold replicated chromosomes together.
   c. shorten every time a cell divides.
   d. require telomerase to move to other sites in the chromosome.

5. Transposable genetic elements
   a. always affect the cell adversely, because when they move, they inactivate genes.
   b. are retroviruses.
   c. provide a mechanism for moving genetic material from organelle genomes to the nuclear genome.
   d. always replicate their DNA when they more.

6. Introns are DNA sequences that
   a. code for functional domains in proteins.
   b. are removed from pre-mRNA by spliceosomes.
   c. allow one gene to make different gene products depending on which introns are removed during splicing.
   d. b and c

7. Globin genes are
   a. only transcribed in bone marrow cells.
   b. translationally regulated by excess heme.
   c. part of a family of globin genes, with different genes expressed at different times in development.
   d. All of the above

8. Pre-messenger RNAs must be processed in the nucleus in order to
   a. increase their stability in the cytoplasm.
   b. allow transcription to begin.
   c. permit coding sequences to be joined to adjacent noncoding sequences.
   d. facilitate ribosome recognition in preparation for DNA synthesis

9. Coordinated regulation of genes in eukaryotic cells
   a. is the result of positioning the same regulatory sequence in front of each gene.
   b. results from all of those genes being under the control of one promoter.
   c. occurs because related genes all have the same operons.
   d. occurs because enhancers cause DNA to bend.

10. The transcription complex includes _________ and _________.
    a. transcription factors, promoters
    b. regulator proteins, regulators
    c. repressor proteins, silencers
    d. a and b

11. DNA binding proteins
    a. have distinct three-dimensional structures that allow them to bind to the DNA.
    b. are transcription factors.
    c. inhibit the loss of ends from the chromosome and help DNA condense in the nucleus.
    d. All of the above

12. Chromatin structure must be altered in order for gene expression to occur because
    a. condensed chromatin is replicated but not transcribed.
    b. condensed chromatin makes most DNA sequences inaccessible to the transcription complex.
    c. decondensed chromatin has more nucleosomes per DNA molecule.
    d. heterochromatin is actively transcribed and euchromatin is not transcribed.

13. When DNA sequences are moved to new sites on a chromosome
   a. genes can be transcribed.
   b. genes can be inactivated.
   c. new genes can be created.
   d. All of the above

14. rRNA gene copies are amplified
   a. to ensure that gene expression in somatic cells will continue.
   b. to make more copies of DNA origins for replication.
   c. to ensure that there is enough translational machinery for the rapid development of the embryo.
   d. to help heterochromatin become more accessible to the transcription complex.

15. Posttranscriptional regulation includes
   a. binding of repressor on silencer regions.
   b. transport of messenger RNA into the cytoplasm.
   c. decreasing messenger RNA stability in the cytoplasm.
   d. b and c

16. Genes can be inactivated by
   a. inaccurate removal of introns.
   b. transposable genetic elements.
   c. movement of genes to heterochromatic regions of the chromosome.
   d. All of the above

Application Questions

1. The cells of the lily plant have eighteen times more DNA than human cells, yet human DNA is more complex. Explain this observation.
2. If you were to design a eukaryotic chromosome, what essential elements would have to be included on this chromosome for it to be functional in the cell?
3. When normal cells become transformed into cancer cells, gene regulation and chromosome stability in those cells is altered. Describe five different ways that cancer cells are different from normal cells.
4. A mutation occurs in the promoter of a eukaryotic gene eliminating the TATA box. How will this mutation affect the transcription of this gene?
5. A different mutation eliminates the enhancer sequence on the DNA for this gene. How will this mutation affect the transcription of this gene?

Answers to Questions

Knowledge and Synthesis Answers

1. **d.** Prokaryotic chromosomes are circular and contain termination sequences.
2. **d.** Important developmental genes and gene families have been found in several different eukaryotic organisms. Comparisons of genes in eukaryotes and prokaryotes have revealed many similar genes that provide the same essential functions to those cells.
3. **c.** Moderately repetitive sequences include the ribosomal gene clusters as well as noncoding sequences such as telomeres and Alu I sequences. Satellites, minisatellites, and microsatellites do not contain genes.
4. **c.** The ends of eukaryotic chromosomes are bound by telomerases (not transcription factors). Centromeres (not telomeres) hold the newly replicated chromosomes together. Tranposons do not require telomerase to move to other sites on chromosomes.
5. **c.** Transposable elements do not always have an adverse effect on genes; sometimes their insertion can create new genes. Retrotransposons are like retroviruses, but lack the genes for the viral proteins. DNA transposons do not replicate their DNA before they move to a different site on the chromosome.
6. **d.** Introns are removed by spliceosomes, and differential splicing can produce different gene products as seen in the tropomyosin example (review Figure 14.20). The DNA loops containing introns, seen in hybridization experiments, were done with DNA and mature messenger RNA, not pre-mRNA.
7. **d.** Globin genes cannot be transcribed in the human red blood cells in adults because those cells lack nuclei.
8. **a.** Processing occurs after the initiation of transcription. Intron removal allows the exons (the coding regions) to be joined together. Ribosome recognition does not lead to DNA synthesis.
9. **a.** Each eukaryotic gene is transcribed from a separate promoter and eukaryotic genes do not have operons to regulate them.
10. **d.** When repressor proteins are bound to silencers, transcription is inhibited.
11. **d.** DNA binding proteins must have a three dimensional structure to bind DNA. Some DNA binding proteins are transcription factors, some are telomerases, which bind the ends of chromosomes, and some are nucleosomes, which bind the DNA and help pack it into the nucleus.
12. **b.** Nucleosomes condense the DNA. Heterochromatin is transcriptionally inactive, while euchromatin is transcriptionally active.
13. **d.** Transposable elements can be transcribed if they move adjacent to a promoter. Transposable elements can inactivate a gene if they insert into the middle of a coding sequence. Transposable elements can move parts of genes with them to new chromosomal sites, potentially creating new genes.
14. **c.** Ribosomal RNA amplification occurs in the eggs of fishes and frogs to prepare for the rapid development that immediately follows fertilization. Amplified rRNA is not needed in somatic cells.
15. **d.** Repressors that bind at silencer sequences regulate transcription and thus are not posttranscriptional mechanisms of regulation.

16. **d.** Inaccurate removal of introns can create mRNAs that are missing coding sequence or that have extra non-coding sequences. Transposable genetic elements can move into the coding regions of genes, inactivating those gene products. Moving a gene to a heterochromatic (transcriptionally inactive) region of a chromosome results in that DNA being inaccessible to the transcription complex.

Application Answers

1. The DNA of the lily plant must contain more noncoding and repetitive sequences and less coding sequence than human DNA.
2. A eukaryotic chromosome requires an origin, telomeres, and a centromere.
3. Cancer cells differ from normal cells in the following ways:
   1. Telomerase activity is increased and the stability of chromosomal ends is preserved in cancer cells.
   2. A transposable element can move into the coding sequence of a gene that regulates cell division, inactivating that gene.
   3. Overactivity of the AP-1 transcription factor, which regulates cell division, can cause cells to become cancerous.
   4. Amplification of oncogenes can transform normal cells into cancer cells.
   5. Targeting proteins that control cell division for proteosome degradation (as seen in human papilloma virus and P53) will cause cells to become cancerous.
4. If the TATA box of the promoter for a eukaryotic gene was deleted, transcription factor TFIID could not bind. The transcription complex could not assemble at the promoter and no transcription would be initiated.
5. If the enhancer for a gene was deleted, transcription for that gene would still occur, but at a reduced rate.

STUDY GUIDE

15 Cell Signaling and Communication

Important Concepts

Cells communicate by sending and receiving chemical signals or by detecting signals from their environment. This process of sending and receiving is called signal transduction.

- Signals must bind to a receptor protein on the cell membrane or in the cell's cytoplasm;
- signals cause a conformational change in that protein that relays or amplifies the signal within the cell; and
- signals effect a response or alteration within the cell.

- A signal is not enough to elicit a response. The receptor for a given signal must be present, and the signal must bind or otherwise cause a conformational change in the receptor protein.
- Signals may be environmental (light, chemical) or may be from within an organism (hormones, neurotransmitters, other chemicals).
- Autocrine signals affect the cells that make them.
- Paracrine signals are transmitted and affect other cells.
- Osmotic change in *E. coli* is an example of how a signal (changes in solute concentration within the plasma membrane) causes a conformational change in a receptor protein (EnvZ) that is transduced to other proteins (phosphorylation of OmpR) and ultimately affects the transcription of a gene (*ompF*) that changes the permeability of the cell membrane.

Receptors are specific, and any given cell has different receptors, which allows a given signal to have different effects or no effect on different cell types.

- Binding to the receptor requires a fit of the ligand to the receptor binding site.
- Binding must cause a conformational change on the receptor to have an effect.
- The ligand is not changed in the binding process, and binding of the signal to the receptor is reversible.
- Binding sites may be inhibited by competing chemicals.

There are three main eukaryotic receptor types at the plasma membrane. Often multiple receptor types are activated during a given signal transduction event.

1. Ion channels: Ligand binding causes a conformational change that opens "gates" that ions may pass through.
2. Protein kinases: Ligand binding stimulates the transfer of a phosphate group from ATP to a protein.
3. G protein-linked receptors: Ligand binding to one of the seven regions of this transmembrane protein changes the shape so that a G protein can bind. The G protein transfers GTP to the protein, and it becomes mobile in the plasma membrane until it encounters an effector protein. GTP is hydrolized to GDP, and the signal is transduced. Effector proteins may be inhibited or activated by G proteins.

Cytoplasmic receptors are in the cell cytoplasm.

- Ligand binding causes a conformational change that allows passage into the cell nucleus and activity as a transcription factor.
- Cytoplasm signals are generally autocrine or very small molecules that can enter the cell membrane.

Signal transduction may be direct (occuring at ligand binding) or indirect (requiring second messenger molecules).

- Indirect signal transduction resulting in a cascade of regulated events is common and allows for controlled and varied effects of single signals.
- Secondary messengers are released to the cytoplasm at ligand binding and function to amplify the signal. They also may have multiple effects within a cell. Common secondary messengers include the following:
  - cAMP: This secondary messenger is involved in ion gate opening or in exposing active sites of protein kinases.
  - IP_3 and DAG: These end products of the hydrolysis of PTI interact with G proteins.
  - Free ions: Calcium ions frequently act as secondary messengers either independently or through calmodulin.
  - Nitric oxide: This gas stimulates the formation of cGMP and is the mechanism by which nitroglycerin is effective in treating angina.

Signal transduction is necessarily highly regulated. Protein kinase cascades are common transduction pathways that allow

- amplification of a signal at each step;
- passage of a signal at the membrane to the nucleus; and
- multiple steps that are regulated and provide specificity of response.
- Signal transduction may result in an alteration of membrane channels, changing enzyme activities, and control of gene transcription.

Signals are often transduced between cells.

- Gap junctions in animal cells are "hot spots" of signal relaying. Adjacent cell cytoplasms are connected by protein connexons.
- Plasmodesmata are plasma membrane channels connecting adjacent plant cells.

The Big Picture

- Signal transduction is the means by which cells receive information from the environment or other cells and react to those signals. Transduction is a highly regulated series of events that depend on the binding of a signal ligand to a receptor protein. This binding must cause a change in the shape of the receptor protein which has an effect that is passed on through a cascade to finally effect the cell, or alters it right then. The effects of signals are often mediated by secondary messengers. For signal transduction to be complete, the cell must alter its activity in some manner.

Common Problem Areas

- Avoid getting bogged down in the details of the examples. Each main heading in the text contains a description and an example. Make sure you understand the description before trying to understand the example.
- G protein action frequently gives students trouble. Remember that G protein activity is the interaction of at least three proteins: the receptor, the G protein itself, and the effector protein.
- Gap junctions and plasmodesmata are physical connections between adjacent cells, and although they are similar in function, they are fundamentally different in structure.

Study Strategies

- Anytime you are struggling with understanding how a pathway works, make a flowchart or a diagram.
- The figures in your book are essential for giving you a visual image of what occurs along these pathways.
- Define the components of the process first, then make sure you can link them together.
- Review the following tutorial and activity on the website:
  Tutorial 15.1 Signal Transduction Pathway
  Activity 15.1 Concept Matching

Test Yourself

Knowledge and Synthesis Questions

1. Which of the following is not required for signal transduction to take place?
   a. Ligand binding to receptor
   b. Conformational change to the receptor protein
   c. Conformational change to the signal
   d. Alteration of cellular activity
2. Signals produced in the organism but not in the target cell are referred to as ____________ signals.
   a. paracrine
   b. autocrine
   c. environmental
   d. intrinsic
3. When looking at the osmotic change in *E. coli* as an example of how signals are transduced, the ultimate goal of the pathway is
   a. to change the permeability of the membrane.
   b. to change what DNA is transcribed.
   c. phosphorylation of OmpR.
   d. binding of the ligand to EnvZ.
4. G protein-linked receptors, protein kinases, and ion channels all have which of the following in common?
   a. Ligand binding
   b. Conformational change once the ligand is bound
   c. Amplification of the signal
   d. All of the above
5. Signal binding to a receptor differs from enzyme–substrate binding in what way(s)?
   a. The signal is not altered in any way during the process.
   b. The process is reversible.
   c. The signal is consumed in the binding.
   d. None of the above
6. Secondary messengers function to
   a. amplify the signal.
   b. bind to the active site of the receptor.
   c. result in multiple effects from a single signal.
   d. a and c
7. In the beginning section of this chapter, Carol is relying on caffeine to keep her awake. The caffeine works because it acts as ____________ to the adenosine receptors in Carol's brain, and by stimulating ____________ in her heart and liver that bypass hormonal regulation.
   a. an alloster, a pathway
   b. an inhibitor, cascades
   c. an inhibitor, ligands
   d. a signal, inhibitors
8. Cytoplasmic receptors are available only to
   a. small signals that can permeate the plasma membrane.
   b. secondary messengers like cAMP.
   c. hydrophilic molecules.
   d. All of the above

9. Protein kinase cascades are significant because
   a. amplification can occur at each step in the path.
   b. information at the plasma membrane is communicated to the nucleus.
   c. the multiple steps allow for specificity of the process.
   d. All of the above

10. G proteins provide an example of the many ways in which signal transduction is regulated. Which of the following are mechanisms by which regulation can occur?
    a. The amount of signal present can be regulated.
    b. Enzymes convert active forms of proteins to inactive forms.
    c. Signals are denatured.
    d. None of the above

11. Which of the following statements regarding receptors is *true*?
    a. Only one type of receptor is utilized throughout a cascade.
    b. Cascades may involve many types of receptors including ion channel receptors, protein kinase receptors, and G protein-linked receptors.
    c. Receptors amplify signals in a one-to-one ratio.
    d. All of the above

12. Gap junctions and plasmodesmata differ in that
    a. gap junctions are connected by protein tubules called connexons and plasmodesmata are connected by extensions of the plant plasma membrane.
    b. plasmodesmata allow much larger molecules to pass through them.
    c. gap junctions have no real physical connection but are the space between adjacent cell membranes.
    d. one is of animal origin and the other is of plant origin; otherwise they are physically the same.

Application Questions

1. Based on your knowledge of prokaryotic and eukaryotic cell structure and function, how might signal transduction differ in prokaryotes and eukaryotes?

2. Describe how protein kinases and G protein-linked receptors may interact in a signal transduction cascade.

3. Diagram the Ras signaling pathway, and identify the components by receptor type. Be sure to identify the signal, the receptors, and the resulting effect on the cell.

4. Discuss the role of secondary messengers in a signaling pathway. How are they different from signals and receptors? What roles do they have in common with signals?

5. True or false. A receptor may act as a signal. Explain and justify your answer.

6. Trace how changes in osmotic pressure in the plasma membrane of *E. coli* result in changes in gene transcription. What would be the outcome if OmpR could not cross the nuclear membrane and bind to the OmpF promoter?

Answers to Questions

Knowledge and Synthesis Answers

1. **c.** For signal transduction to take place, the ligand must bind to the receptor, the receptor must undergo a conformational change, and the activity of the cell must be altered.

2. **a.** Autocrine signals function on the cell they are produced in, whereas paracrine signals affect neighboring cells.

3. **a.** The ultimate goal of any signaling pathway is to alter the function of the cell in response to the signal.

4. **d.** All types of receptors bind ligands, undergo conformational change, and amplify the signal.

5. **a.** Substrates are altered in enzyme–substrate complexes. Signals are not altered when they bind to receptors.

6. **d.** Secondary messengers both amplify the signal and have multiple effects within a cell.

7. **b.** Caffeine and adenosine bind to the same receptor protein. The binding of caffeine prevents adenosine from binding. A series of cascades begin in other organs in response to caffeine.

8. **a.** To bind to a cytoplasmic receptor, the ligand must be either produced in the cell or able to pass through the plasma membrane.

9. **d.** Amplification, regulation, and specificity are all roles of protein kinase cascades.

10. **b.** Active forms of proteins may be converted to inactive forms via enzyme action.

11. **b.** A given cascade may utilize ion channel receptors, protein kinase receptors, and G protein-linked receptors.

12. **a.** Both plasmodesmata and gap junctions are physical links between neighboring cells.

Application Answers

1. Signal transduction in prokaryotes is limited to only one cell; there is no nuclear membrane to cross, and cascades tend to be simpler with fewer intermediary steps.

2. G protein-linked receptors frequently expose the protein kinase activities of effector molecules.

3. Refer to Figure 15.11 in your text.

4. Secondary messengers function to amplify and spread signals. They do not bind to receptors and so do not act like signals in the cascade.

5. Receptors may act as signals for the next step in a given cascade. See Figure 15.11 in your text.

6. Refer to Figure 15.3 in your text. If OmpR can not pass through the nuclear membrane and bind to the OmpF promotor, then no change takes place and the cascade is inhibited.

16 Development: Differential Gene Expression

Important Concepts

Once fertilization occurs, the zygote undergoes a series of cell divisions and progressive changes that include growth, differentiation, and morphogenesis. This developmental process produces a mature organism.

- Cell differentiation does not result from the loss of DNA. Somatic cells in the mature organism retain all of the genes present in the fertilized egg.
- Cell differentiation results from the differential expression of genes caused by transcriptional and translational regulation and by signaling mechanisms.

Development includes three stages: growth, differentiation, and morphogenesis.

- The embryo is the earliest stage of development.
- Growth is an increase in size due to cell division and cell expansion. The early growth of plant embryos is characterized by cell expansion, while early growth of animal embryos is characterized by cell division.
- During differentiation, cells become specialized and provide specific structures or functions for the organism.
- The developmental potential for an embryonic cell is totipotent; meaning those embryonic cells can develop into many different kinds of cells in the mature organism. The developmental possibilities narrow later in development as the cell fate becomes determined and the cell becomes fully differentiated.
- Morphogenesis is the creation of form as seen in body shape and organs and is a result of pattern formation during development. Pattern formation helps direct differentiated tissues to form specific structures.
- Plants cells are nonmotile and constrained by cell walls. Formation of differentiated tissue in the mature plant is due to organized cell division and expansion of cells.
- Animal cells are motile, and these movements play a major role in differentiation and morphogenesis.
- Programmed cell death is an important part of development in both plants and animals.

Differentiation and morphogenesis are directed by gene expression responding to extracellular signals.

- Transplantation experiments have shown that early embryonic cells can be redirected by their environment to proceed down a different developmental path.
- Later embryonic cells become committed to a particular developmental fate, even though they are not yet differentiated. When these cells are transplanted to other locations in the embryo, they continue to develop into that differentiated tissue regardless of their environment (see Figure 16.2).
- Cell fate becomes visible as the cells become differentiated.
- Initially, the fertilized egg is totipotent and can give rise to every type of cell in the adult body.
- Cell descendants lose their totipotency as they become committed to a developmental path. A cell's fate becomes determined and then differentiates to form a specialized cell.
- Differentiation is irreversible in cells that have no nuclei, such as mammalian red blood cells in humans and tracheid cells (water conduction cells) in vascular plants.
- In plants, differentiation is reversible, and the nuclei of somatic cells are totipotent.
- Under the appropriate conditions, plant cells in culture can give rise to a new, genetically identical plant, a clone. Thus, differentiated cells contain all of the genetic information needed to express genes in the right sequence and to generate a whole plant.
- Nuclei isolated from frog embryos and fused to enucleated eggs can produce mature frogs. Later experiments demonstrated that somatic nuclei from adult frogs, when fused with enucleated eggs, could develop into tadpoles.
- Whole mammals can be cloned by fusing early embryonic cells (rhesus monkey) or somatic cells (sheep) with enucleated eggs and implanting these eggs in surrogate mothers. The success of cloning mammals from somatic cells has been due to manipulating the donor cells so that they are positioned in G1 of the cell cycle. Upon fusing with the enucleated egg, the donor cells are stimulated to enter S due to cytoplasmic factors in the egg.
- Teratocarcinoma is a type of cancer in which unspecialized cells differentiate to form a tumor containing kidney, hair, and teeth cells.

Normal development is a complex series of signals and transduction of those signals, resulting in the differentiation of cells to form a new organism.

- Stem cells are undifferentiated dividing cells in an adult organism which replace dead cells, typically skin, the lining of the intestines, and the blood system. Stem cells in adults are specific for the kinds of tissue they replace. However, environmental signals can induce stem cells from one tissue to replace cells that have been depleted from another tissue. For example, stem cells isolated from the mouse brain can replace bone marrow cells in mice that have been depleted of bone marrow.
- Embryonic stem cells are totipotent. They can be grown in culture for multiple generations, injected back into blastocysts, and then develop normally. In culture, the developmental fate of stem cells can be directed by the addition of environmental signals such as growth factors or vitamins.
- Future medical applications include the use of embryonic stem cells to replace human cells that have been lost to disease or injury.

Differentiated cells retain all of their original genetic content even though they are expressing only a very small subset of genes.

- Nucleic acid hybridizations have used a DNA sequence containing the gene of interest and nuclear DNA or messenger RNA from the same cells. These experiments have shown that even though a specific gene is not expressed (not present in the mRNA), it is still present in the nuclear DNA.
- Undifferentiated muscle cells (myoblasts) respond to the MyoD1 protein (Myoblast Determining Gene 1) by fusing to form muscle fibers. This protein is a DNA-binding protein with a helix-loop-helix motif. When MyoD1 binds to promoters of muscle determination genes, transcription of those genes increases. It also binds to its own promoter and positively regulates *MyoD1* mRNA production in myoblasts.
- When MyoD1 protein is injected into other precursor cells (for instance, fat cells), those cells are reprogrammed to develop into muscle cells.

Unequally distributed cytoplasmic factors in eggs and zygotes set up the initial pathways for precursor cells in developing embryos.

- Polar distribution of cytoplasmic factors in eggs changes as a result of fertilization. Cell division continues, resulting in an unequal distribution of maternal elements in each of the cells of the embryo.
- Unequal distribution of cytoplasmic factors plays an important role in directing embryonic development. For example, if a sea urchin embryo is cut in half at the eight-cell stage and split from left to right by that cut, the resulting larvae are normal, but dwarfed. If the sea urchin is cut so that its upper half is split from its lower half, the cells from the upper half develop into abnormal ciliated larvae and the cells from the lower half develop into misshapen larvae with an oversized gut (see Figure 16.7).

Fates of particular tissues in developing embryos can induce other tissues to follow a particular developmental path.

- The vertebrate eye lens in surface cells forms due to contact with a double bulging of the forebrain, known as the optic vesicles. If a barrier is placed between these two tissues in the developing embryo, no lens develops. If the forebrain is cut out before it can contact the lens surface cells, no lens develops. The optic vesicles produce a signal, causing the surface cells to develop into a lens. The lens tissue produces embryonic inducers that control the size of the optic cups and the differentiation of the cornea (see Figure 16.8).
- In *C. elegans*, the egg develops into a larva in eight hours at 25°C. The animal has a transparent body, making it is easy to follow the developmental fate of the 959 somatic cells. The adult is hermaphroditic and has both female and male reproductive organs. Eggs are laid through a pore called the vulva. The vulva is induced to form from a single cell called an anchor cell. If this cell is laser ablated, no vulva develops.
- The anchor cell determines the fates of six cells on the animal's ventral surface by producing an inducer that diffuses and interacts with adjacent cells. The closest cell becomes the primary vulval precursor cell. It produces an inducer that causes the next two cells to become secondary vulval precursors and the final two cells (which are the farthest out from the anchor cell) become epidermal cells (see Figure 16.9).
- The primary inducer for the development of the vulva in *C. elegans* is a growth factor, LIN-3, which is homologous to mammalian epidermal growth factor. LIN-3 binds to a receptor on the surface of the vulval precursor cell, which causes a signal transduction cascade in that cell through the Ras protein and MAP kinase proteins (review Figure 15.11). The end result of this cascade is the differentiation of this cell into a vulval cell.

Pattern development is the spatial organization of tissue that results in the appearance of body form (morphogenesis).

- During morphogenesis, some cells are programmed to die. Apoptosis (programmed cell death) is controlled by death genes.
- In *C. elegans*, there are a total of 1,090 cells. During development, 131 of these cells are programmed to die due to the sequential expression of *ced-4* and *ced-3*. *Ced-9* inhibits the actions of *ced-3* and *ced-4* during development.
- In human embryos, apoptosis occurs so that webs of skin that form between the fingers and toes disappear by birth. Capsase, a protein homologous to *ced-3*, stimulates apoptosis. The human homolog to *ced-9* in is *bcl-2*. Over-expression of *bcl-2* leads to the development of a particular kind of cancer known as follicular large cell syndrome.
- The genes for apoptosis are crucial for proper development in many organisms. These genes have been preserved in organisms separated by 600 million years of evolution (nematodes and humans).

- Plants have organs including leaves, roots, and flowers. Flowers have four types of organs: sepals, petals, stamens, and carpels. Flower organs occur in whorls organized around a central axis and derived from meristematic tissue on the plant (see Figure 16.11).
- Plant geneticists have studied the development of flower organs in *Arabidopsis*. There are four whorls of organs in *Arabidopsis*. Three classes of homeotic mutations (class A, B, and C) have been isolated which alter the determination of these whorls. Class A affects whorls 1 and 2 which form the sepals and petals, class B affects whorls 2 and 3 which form petals and stamens, and class C affects whorls 3 and 4 which form stamens and carpels.
- The three genes in class A, B, and C all encode transcription factors that are active as dimers. Gene regulation is combinatorial and the combination of different dimers determines which genes are activated. A dimer of A will activate genes that make sepals. A dimer of A with B will make petals (see Figure 16.11).
- *Leafy* is a transcription factor which regulates the transcription of the genes *A, B* and *C*. A plant bearing a mutation in *leafy* can produce leaves but no flowers.
- These homeotic genes, *A, B*, and *C*, as well as *leafy* may be able to be genetically modified so that the flower organs can produce more fruit and seeds (which come from the carpels).
- In plants, tissue formation is related to the distance that the tissue is from the body surface. For example, tissues that conduct water in the plant are farther away from the plant surface (and thus exposed to lower concentrations of O_2 and CO_2) than tissues that conduct sucrose, the product of photosynthesis. Positional differences may help determine which genes are expressed in the stem and root. Morphogens are signals that establish positional information in the organism. Morphogens diffuse from stem tip to root tip in plants, establishing positional information along the plant's axis.
- In the chick embryo, a round wing bud receives morphogens from three regions that direct its development into the mature wing. The unique concentrations of morphogens at these positions determine proximal–distal, anterior–posterior, and dorsal–ventral axes of the wing.
- Genes that encode developmental morphogens have been highly conserved in diverse organisms such as nematodes and humans.

A cascade of gene expression controls development in *Drosophila*.

- The body pattern of *Drosophila* is segmented, consisting of head (which is formed from three fused segments), three thoracic segments, eight abdominal segments, and one terminal segment.
- The first step of development in *Drosophila* is to establish anterior–posterior and dorsal–ventral polarity. Polarity is based on the distribution of morphogens, which are messenger RNA's and proteins produced in maternal genes. These maternal gene products are distributed to the eggs in a nonuniform manner from transcription in the nurse cells, which nurture the developing egg. Maternal effects on the larvae are not affected by the paternal genotype.
- Mutations in maternal genes (*bicoid and nanos*) create larvae lacking anterior structures (*bicoid*) or lacking abdominal segments (*nanos*).
- Segmental genes are expressed when the embryo is at the 6,000-nuclei stage and determine the number, boundaries, and polarity of the segments.
- Gap genes are segmental genes whose products organize large areas along the anterior–posterior axis of the developing embryo. Gap mutants produce larvae that are missing certain larval segments. Pair rule gene products divide the larva into units of two segments each. Mutations in the pair rule genes produce larvae that are missing every other segment. Segment polarity gene products determine the boundary and the anterior–posterior organization of each segment. Mutant larvae, lacking segment polarity genes, produce larvae in which the posterior structures in the segments have been replaced by anterior structures.
- Differences between segments are encoded by the homeotic genes which give each segment an identity. For instance, thoracic segments produce legs.
- Since early embryos do not transcribe genes, unfertilized eggs have large numbers of messenger RNA's that are made prior to fertilization. These messenger RNA's support the high level of protein synthesis during early stages of development. Some of these messenger RNA's provide positional information; *bicoid* mRNA, for exampale, is localized in the anterior portion of the egg.
- In *Drosophila* there are only nuclear divisions with no division of cytoplasm up until the thirteenth division. This mulinucleate single-celled embryo is called a syncytium. *Bicoid* mRNA's are translated early during development and they diffuse from the anterior (Bicoid protein) and posterior (Nanos protein) end of the egg, creating a gradient. Bicoid encodes a transcription factor that positively affects some genes (*hunchback*) and negatively affects others (*Krüppel*). Nanos protein inhibits the translation of gap genes, including *hunchback*. Gap gene products control the expression of pair rule genes, some of which are transcription factors that control segment polarity.
- Homeotic mutants include *Antennapedia* (producing legs that are positioned in the head, see Figure 16.15) and *bithorax* (producing a fly with an extra set of wings). Homeotic genes are clustered on the chromosome in the order of the segments they determine. The first set of genes in the homeotic cluster are thoracic, the next genes are upper abdomenal, and so on.
- The cluster of homeotic genes that determine the segment identity in the front of the fly *(Antennapedia)* has been predicted to have risen from duplications of an ancestral gene.

Genes that are important for development are conserved in many different organisms.

- Nucleic acid hybridization experiments confirm that genes like *Antennapedia* and clusters next to those genes contain DNA sequences common to all homeotic genes. A DNA sequence (the homeobox), which encodes 60 amino acids (the homeodomain), includes a DNA-binding motif (helix-turn-helix) that has also been found in *bicoid*. *Bicoid* encodes a transcription factor that binds the gap promoters such as *hunchback*. Hunchback is a protein that binds to enhancers to increase transcription of genes involved in head and thorax development.

- In nucleic acid hybridizations using DNA probes for homeoboxes, over 100 proteins in nematodes, frogs, mice, chicks, plants, and humans have been found to contain homeoboxes. Comparative genomics has shown that genes in development are very similar in many organisms, leading to the creation of a new field of evolutionary developmental biology.
- There are 38 homeotic genes in the mouse which control development of specific regions of the mouse embryo. These genes are in the same order on the chromosome as they are expressed in the mouse (from anterior to posterior, see Figure 16.17). There are four clusters of these homeotic genes in all vertebrates and also in the smallest living chordate, the lancelet. Chordates gave rise to vertebrates. These genes must have arisen from repeated duplications of a single ancestral gene followed by small mutations in those genes.
- *Eyeless* is the gene in *Drosophila* that encodes a transcription factor that binds to 1,500 different genes involved in eye formation. The mammalian homolog for *eyeless* is *Pax6*. Mutations in *Pax6* produce a disease known as aniridia, which results in the partial or total absence of the iris. *Eyeless* and *Pax6* can be swapped for one another and still remain functional in the organism's development. *Eyeless* and *Pax6* are examples of the conservation of general instructions for the development of a particular structure that were generated early in evolution and used repeatedly for different eyes in different organisms.
- Pattern information can be flipped in various organisms such as arthropods and vertebrates. The signal for dorsal–ventral determination in both arthropods and vertebrates is the same, but they have opposite meanings. In vertebrates, chordin protein determines dorsal identity while its homolog in arthropods, the Sog protein, determines ventral identity. BMP4 is a protein that directs ventral identity in vertebrates and the arthropod homolog, Dpp protein, determines dorsal identity.

The Big Picture

- The development of a mature differentiated organism from a fertilized egg involves the differential activation of genes in response to environmental signals. Initially in many organisms, positional determinants may be present in different regions of the egg due to maternal factors. The asymmetric distribution of these factors will activate genes differentially, and as cell division proceeds, that asymmetric gene expression continues. As a result, different cells experience different environmental signals based on their position in the developing embryo, and they respond by activating different genes. This differential gene activation determines the fate of the cell.
- Early embryonic cells are totipotent and can develop into a variety of different structures in the adult organism. As development proceeds, the developmental potential of embryonic cells narrows. In the adult organism, differentiated cells express particular genes that give those tissues a particular structure and function, even though all of the genes are still present in the nucleus of those cells.
- Genes that are important for directing developmental pathways are highly conserved in organisms separated by millions of years of evolution.

Common Problem Areas

- Sometimes it is difficult to visualize how a single cell can divide and grow to produce a mature functional organism with highly differentiated tissues. At some very early point in development (either before fertilization or in one of the first set of divisions), different genes begin to be expressed in different cells. This gene expression can be in response to cytoplasmic factors or signals from other cells. Early gene expression sets up positional determinants that activate another wave of genes that further divides regions of the embryo into different developmental regions. As cell division proceeds in the embryo, these developmental genes continue to be differentially expressed, resulting in differentiation and morphogenesis in the organism.

Study Strategies

- Review the genes that are important in the development of *Drosophila*. Make a list of those genes and what their effects are on the phenotype of the *Drosophila* larva.
- Transplantation experiments help explain totipotency. Describe the transplantation experiments in frogs using early embryonic tissues and later embryonic tissues to address the totipotency of those tissues.
- Review the experiments that demonstrated how tissue induction causes the lens in the vertebrate eye to form.
- Apoptosis is an important part of development in all organisms. Give an example of the role apoptosis plays in embryonic development in the nematode *C. elegans* and in humans. Now describe two conditions that result in these organisms when apoptosis fails to occur.
- Review some of the important genes whose products direct development in the organisms discussed in this chapter. Underline homologs from different organisms.
- Review the following tutorial and activity on the website:
  Tutorial 16.1 Pattern Formation in the *Drosophila* Embryo
  Activity 16.1 Stages of Development

Test Yourself

Knowledge and Synthesis Questions

1. Cell differentiation is a result of
   a. the loss of particular genes from the nucleus of the differentiated cell.

b. the differential expression of genes that are responsive to environmental signals.
c. early embryonic cells remaining totipotent in the mature organism.
d. mutations in genes that control the synthesis of DNA.

2. A totipotent cell is a cell
a. whose developmental fate has been decided.
b. that has differentiated into a specialized tissue.
c. that is expressing morphogens to form a particular structure.
d. whose developmental potential is extremely broad.

3. Apoptosis is programmed cell death and
a. only occurs in developing embryos.
b. only occurs in the mature organism.
c. is an important process in both development and in mature organisms.
d. can cause a type of cancer, follicular large-cell lymphoma.

4. Once a cell is committed to a particular developmental fate, it cannot be redirected down a different pathway unless that cell is
a. fused to an enucleated egg cell.
b. exposed to environmental factors that activate genes that control a different developmental pathway.
c. transplanted to a different tissue.
d. All of the above

5. Mammals can be cloned using somatic cells fused to enucleated eggs if the
a. donor cell is in the S phase of the cell cycle.
b. donor cell is in the G1 phase of the cell cycle.
c. mammalian embryo can be propagated in culture.
d. somatic cell is a red blood cell.

6. Identify the function of the protein MyoD1.
a. It is a transcription factor that controls the expression of genes that form muscle.
b. It controls segment identity in mice.
c. It is a transcription factor that regulates its own transcription.
d. a and c

7. Tissue induction occurs because
a. nuclear genes are lost in some tissues.
b. signals are secreted in one part of the developing embryo and are transported via the circulatory system to another site in the embryo to activate tissue differentiation.
c. contact from one tissue allows the environmental signal from those cells to be delivered to the other tissue that must be induced.
d. when two tissues come into contact with one another, transcription factors are released.

8. Apoptosis
a. can cause abnormal development of the nervous system when inhibited in developing *C. elegans*.
b. can cause webbing to be retained between the fingers of developing human embryos.
c. is a natural part of development.
d. a and c

9. It may be possible to genetically engineer plants to produce more seeds and fruits by
a. fertilizing them more in the spring.
b. manipulating their gap genes.
c. altering the homeotic genes that control whorl development.
d. inducing a mutation that eliminates *leafy* gene function.

10. Morphogens
a. include vitamins and growth factors.
b. are expressed as positional signals in developing embryos.
c. direct differentiated cells to form organs.
d. All of the above

11. Maternal genes
a. set up positional axes in the egg prior to fertilization.
b. are mRNA's and proteins that are placed in the egg cell.
c. are masked by the paternal genes.
d. a and b

12. Homeotic genes
a. encode highly conserved protein domains that are important in development.
b. are found in greatly diverse organisms.
c. can produce the wrong structure in the wrong place when mutated, and have been highly conserved over evolutionary time.
d. All of the above

Application Questions

1. Nuclear hybridization experiments were done to show that fully differentiated cells still contained all of their genes even though they were only expressing a small set of those genes. What kinds of hybridizations demonstrated this?
2. Undifferentiated stem cells are totipotent. Describe the experiment that clearly demonstrated this fact.
3. Transcription factors that control the fates of undifferentiated stem cells can be used to redirect other committed stem cells. Describe the experiment that showed this.
4. In cloning whole organisms by fusing nuclei (as seen in frogs) or whole cells (as seen in sheep, cows, and goats) with enucleated eggs, why is the nucleus removed from the egg?

Answers to Questions

Knowledge and Synthesis Answers

1. **b.** Genes are not lost from differentiated cells. There are no embryonic cells in a mature organism. Mutations in cells that affect DNA replication would result in cells that were unable to replicate their DNA.

2. **d.** A totipotent cell is a cell that can develop in any number of ways; its fate has not been determined, it has not differentiated, and it is not expressing morphogens.
3. **c.** Apoptosis is important for embryonic development and in adult tissues such as skin and blood cells.
4. **d.** All of the cells described in this question can be redirected to a different developmental path under the right environmental conditions.
5. **b.** The donor somatic cell must be in G1. Mammalian embryos must be grown in surrogate mothers. Red blood cells in mammals have no nuclei and could not be used for cloning.
6. **d.** MyoDI encodes a transcription factor that controls the expression of genes that form muscle in mice and also regulates its own transcription. MyoDI does not control segment identity in mice. Homeotic genes control segment identity in *Drosophila*.
7. **c.** Induction involves contact between two different tissues, with one tissue providing some kind of environmental signal that is received by the other tissue. Nuclear genes are not lost in developing cells (with the exception of red blood cells and the immune cells). Tissue induction occurs because of tissue contact, not because a signal has been carried from one part of the embryo's body by circulation to another part of its body. Transcription factors are found in the nucleus and are not released to adjacent tissues.
8. **d.** Apoptosis causes the webbing to be removed (not retained) between the fingers and toes during the development of the human embryo
9. **c.** Fertilization is not genetic engineering. Gap genes are found in *Drosophila*, not plants. The leafy protein transcriptionally activates the homeotic genes that control organ development (and subsequent seed and fruit formation). In its absence; no fruit or seeds would develop.
10. **d.** Morphogens (which include vitamins and growth factors) are positional signals in developing embryos that direct differentiated tissues to form organs.
11. **d.** The paternal genotype does not effect the expression of maternal gene products in the egg.
12. **d.** Homeotic genes encode proteins that are highly conserved and important for directing development in many different organisms. Mutation in these genes can produce an abnormal developmental event in the organism.

Application Answers

1. Hybridizations were done using a specific DNA probe to nuclear DNA and messenger RNA. If the DNA probe was complementary to a gene that was not being expressed in the cell, it would not hybridize to the messenger RNA isolated from that cell, but would still hybridize to the nuclear DNA. This indicated that the gene was still present in the nucleus of the cell even though it was not being expressed in the cytoplasm.
2. Undifferentiated stem cells from mouse brain were transplanted to the bone marrow of mice whose bone marrow had been depleted. The brain cells differentiated into bone marrow cells, indicating their totipotency.
3. The MyoD1 transcription factor from muscle cells directs the transcription of several muscle genes, resulting in the fusion of myoblasts to form muscle cells. When MyoD1 was applied to fat cells, it redirected those cells to form muscle cells.
4. The nucleus of the egg contains a full haploid set of chromosomes. Fusing a diploid somatic cell with an egg cell that still contained a nucleus would result in a cell that had three copies of each chromosome and little chance of developing normally.

STUDY GUIDE

17 Recombinant DNA and Biotechnology

Important Concepts

Recombinant DNA technology has allowed scientists to combine DNA molecules from different organisms to create useful gene products and to analyze gene expression and function.

- Gene cloning includes
  - cutting DNA fragments with restriction endonucleases.
  - covalently linking those fragments to vector DNA, creating a new recombinant DNA molecule.
  - introducing the newly recombined DNA into a host cell.
  - growing that cell in conditions that ensure the recombinant molecule will be retained by the cell.
- All of the key techniques in recombinant DNA technology rely on complementary base pairing, including sequencing, rejoining, amplifying, and locating DNA fragments.

Restriction endonucleases are bacterial enzymes that cleave double-stranded DNA's into smaller fragments.

- Bacteria synthesize restriction endonucleases to protect themselves against viral infection. Restriction endonucleases cut viral DNA but leave their own DNA unharmed because that DNA has been modified by bacterial enzymes (usually by the addition of a methyl group).
- Restriction endonucleases cut the DNA at specific sites and leave a 5′ PO_4 and a 3′ OH at the cut ends. For example, *Eco*RI is a restriction endonuclease that cuts the sequence.

  5′ … G A A T T C … 3′

  between the G and the A on both strands of DNA. This type of cutting leaves single-stranded regions on the ends (also called "sticky ends"). These sticky ends can form complementary base pairs with other DNA molecules that have the same sticky ends (see Figure 17.4).
- A different class of restriction endonucleases cuts between the same bases on both DNA strands producing blunt ends.
- *Eco*RI cuts on average about 1 in every 4,000 base pairs in prokaryotic genomes, but sizes of the DNA fragments vary, depending on the actual location of these sites in the chromosome.
- Many restriction endonucleases including *Eco*RI are active as dimers consisting of two identical subunits.

DNA fragments can be separated using gel electrophoresis.

- A gel is a porous molecular sieve, which allows smaller DNA molecules to move faster through it than larger DNA molecules. An electrical field is applied to the gel, causing the negatively charged phosphates of the DNA to be pulled toward the positively charged electrode.
- DNA fragments of known sizes are also separated on the gel as reference.
- DNA fragments containing specific sequences can be identified on a gel using a blot procedure and subsequent hybridization with a radioactively labeled probe. Denatured DNA fragments are transferred to a nylon filter (the blotting part of the procedure) and then exposed to a single-stranded DNA probe that has been labeled with a radioactive isotope, usually P^{32} (radioactive phosphorous). The radioactive DNA probe will hybridize to complementary base sequences on the DNA fragments on the filter (see Figure 17.2).
- The identified DNA fragments can then be removed from gel.

Purified DNA fragments can be mixed with vector (that has been cut with the same restriction endonuclease) to create a new recombinant molecule.

- DNA ligase (the enzyme that covalently joins the lagging strands during DNA replication, see Chapter 11) is used to form a covalent bond on each DNA strand of the recombinant molecule.
- Usually DNA fragments are ligated into vectors (plasmids, viruses, or artificial chromosomes) that contain genetic markers and reporter genes. These reporter genes allow the investigator to follow the recombinant DNA molecule once it is inside the host cell.

Yeast and plants have been used to clone eukaryotic genes.

- Yeast cells have a rapid cell division cycle (2–8 hours), are easy to grow, and have a small genome size (20 million base pairs).
- Plant cells can be grown in culture, coaxed to take up recombinant DNA, and manipulated by alterations in the growth media to form an entire new transgenic plant containing the recombinant DNA molecule.

Once the recombinant DNA has been successfully introduced into the cell, it must be replicated to be maintained.

- Recombinant DNA can be inserted into the host chromosome, where it is replicated when the chromosome is replicated. Alternatively, the vector can provide an origin of replication, so that the recombinant DNA molecule is propagated in the cell.
- Vectors must be small and have restriction sites, origins of replication, and genetic markers.
- Plasmid vectors in *E. coli* are small (from 2,000 to 6,000 base pairs) circular DNA molecules. They have a single set of unique restriction sites, where the cut DNA is inserted, and a drug resistant marker (usually ampicillin resistance) which allows the investigator to confirm the presence of the plasmid in the *E. coli* cell. Plasmids synthesize their DNA independently from their own origins of replication, making many copies of plasmid DNA per cell. DNA fragments of up to 5,000 base pairs can be cloned into plasmid vectors.
- Virus vectors are used to clone larger DNA sequences, especially eukaryotic genes which have introns, exons, and flanking sequences. Genetically engineered bacteriophage lambda can accommodate up to 20,000 base pairs of inserted DNA. The new recombinant virus is still able to infect *E. coli* and produce many viral particles, each containing the cloned DNA.
- Yeast artificial chromosomes can be used to clone genes in a eukaryote, the yeast cell (see Figure 17.5*b*).
- These chromosomes (10,000 base pairs) have a yeast origin of replication, telomeres, a centromere, marker genes, and restriction endonuclease sites where foreign DNA can be inserted. Yeast artificial chromosomes can accommodate 50,000 to 1,500,000 base pairs of DNA.
- *Agrobacterium tumefaciens*, a bacterium that causes crown gall in plants, harbors a Ti plasmid. This plasmid contains TDNA, which is a transposon that makes DNA copies of itself in the chromosomes of infected plant cells.
- There are restriction endonuclease sites within the TDNA which can be used to clone foreign DNA. Once DNA has been inserted into those sites, the tumor causing capabilities of this bacterium are eliminated. However, the recombinant plasmid is still able to act as a transposon. When plant cells are infected by this bacterium, the recombinant plasmid can move into the host DNA via transposition. Individual plant cells bearing this recombinant DNA in their chromosomes can be cultured to produce a new transgenic plant containing the recombinant DNA molecule in every cell.

Recombinant DNA can be introduced into host cells in several ways.

- Bacterial cells can be treated with calcium ions, which coat the negatively charged lipid heads of the cell membrane with positive charges, making those cells more permeable to the electronegatively charged DNA.
- When cloning recombinant DNA into plant or fungal cells, the cell wall must first be removed, leaving a protoplast.
- During electroporation, cells are exposed to rapid pulses of high voltage current, making them temporarily permeable to DNA.
- DNA can be injected into large cells, especially eggs.
- DNA can be coated with lipid to form liposomes. The liposomes are then fused with the plasma membrane of host cell, and the DNA is transferred to the interior of the cell.
- Gene guns can be used to shoot particles of tungsten and gold that have been coated with DNA into host cells.

Reporter genes and genetic markers (such as drug resistance) can be used to determine if a recombinant DNA molecule has been successfully introduced into a host cell.

- If foreign DNA has been inserted into the *Tet* gene (which confers resistance to tetracycline) of a plasmid, that gene will be inactivated and the cell bearing this plasmid will be tetracycline sensitive. The plasmid also has an *Amp* gene, which confers ampicillin resistance to the host cell. (see Figure 17.6). In order to determine if the plasmid is present in the host cell, the investigator will have to screen three kinds of cells: cells that did not receive the plasmid (amp^s, tet^s), cells that received the plasmid but without any inserted DNA (amp^r, tet^r), and cells that received the plasmid with the DNA insert (amp^r, tet^s).
- Other marker genes that are commonly used for detection of recombinant DNA molecules in host cells include the *lac* operon, the genes that code for luciferase and green fluorescent protein (*GFP*).
- In order to favor ligation of the foreign DNA into the plasmid (rather than the plasmid religating with itself), the concentration of foreign DNA must be high relative to the plasmid DNA concentration in the ligation mixture.

DNA molecules can be isolated from genetic libraries, which include random pieces of DNA, from copying messenger RNA (cDNA), or from artificially synthesized DNA.

- Genetic libraries of DNA are usually cloned into bacteriophage lambda since each virus can accommodate 20,000 base pairs of DNA.
- To make a cDNA, the messenger RNA must first be isolated from the cell. Molecules of oligo dT are then hybridized to the poly A tail of the messenger RNA template. These short DNA molecules are then used as a primer on the messenger RNA for reverse transcriptase. A complementary DNA molecule is synthesized by reverse transcriptase, which can then be cloned.
- cDNA clones are used to compare gene expression in different tissues at different stages of development. One-third of all genes in an animal are expressed only during prenatal development.
- To make synthetic DNA, the investigator must know the amino acid sequence of the gene. Sequences for translational initiation and termination as well as transcriptional

initiation can be added to the gene sequence. Regulatory regions and preferred codons can also be used in the synthetic DNA.

- Wild-type and mutant genes can be synthesized to analyze gene function.
- Computational biology is used to study the three-dimensional shape of enzymes and to make predictions about regulatory molecules of those enzymes. Mutant polypeptides can be constructed to examine how enzyme structure is related to enzyme function.
- Gene function can be eliminated by creating knockout genes, which are the result of a homologous recombination event in the cell where a normal gene is replaced by an inactivated form of the gene. The inserted DNA carries a genetic marker so its expression can be followed in the host cell.
- In constructing transgenic mice, a plasmid containing the gene is transfected into a mouse embryonic stem cell (see Figure 17.10). If recombination occurs, the marker gene will be expressed. The stem cell is then inserted into an early mouse embryo, and the resulting phenotype is analyzed.

DNA chip technology allows screening of thousands of DNA sequences at the same time.

- Multiple hybridizations can be done using DNA chips, which are small glass chips containing 10 million copies of different DNA sequences per chip. These sequences (greater than 20 base pairs long) are attached to the chip in a precise order and can contain up to 60,000 different DNA sequences.
- DNA chips have been used to study patterns of gene expression in different tissues at different times in an organism.
- For example, cDNA copies of cellular mRNA can be made and amplified using PCR (see Chapter 11). Those cDNA molecules can be coupled to a fluorescent dye and then allowed to hybridize to the DNA chips. The chip is then scanned for fluorescence to determine which sequences formed a hybrid with the cDNA.
- In a similar manner, DNA chips can be used to detect all of the possible variants of a particular gene.

Antisense messenger RNA is complementary to messenger RNA.

- Antisense RNA will form a double-stranded RNA molecule in the cell's cytoplasm, which cannot be translated and is degraded by the cell.
- A DNA copy of antisense RNA can be cloned into a vector and introduced into the cell's cytoplasm where under the right conditions, it can be transcribed.
- Antisense RNA can be expressed in specific tissues by cloning it in a vector that has a tissue-specific promoter or that has a ribozyme gene, which will cleave the RNA once it is transcribed. Antisense RNA can be used therapeutically to block a gene product that is essential for the growth of cancer cells.

Biotechnology can be used for microbial, plant, and animal cells to produce useful materials for people, including food, medicines, and chemicals.

- In order for cells to produce a cloned gene product, the vector must have DNA sequences that allow the cloned gene to be expressed.
- Prokaryotic expression vectors require a promoter, a termination site for transcription, and a ribosome-binding site.
- Eukaryotic expression vectors require the same elements as well as a poly A addition site, transcription factor binding sites, and enhancers.
- Hormones can induce gene expression from a vector if a hormone response element is positioned in the vector promoter or in the vector enhancer. Tissue-specific promoters placed in front of the cloned gene will yield organisms that express the recombinant gene only in specific tissues. Targeting sequences can be added to expression vectors to direct the gene product to a particular site in the eukaryotic cell.
- Medically useful products that have been cloned in expression vectors include tissue plasminogen activator and human insulin.
- Recombinant DNA offers breeders the opportunity to choose specific genes that will be incorporated into an organism.
- Transgenic plants have been created that express toxins for insect larva, resistance to herbicides, and produce extra nutrients.
- Human genes, such as blood clotting factors, antibodies to colon cancer, and elastase inhibitors have been cloned in sheep and goats next to the lactoglobulin promoter, causing those animals to produce human proteins in their milk.
- Creating transgene plants has raised some concerns about these crops being unsafe for human consumption and about transgenes escaping into other noxious plants. Scientists are examining these issues to promote safeguards.
- DNA fingerprinting is a technique that can be used to identify and distinguish different individuals. Some genes have multiple alleles, and each individual expresses different combinations of these alleles.
- DNA probes have been developed to hybridize to short moderately repetitive sequences that are side by side in the human genome. These sequences, termed VNTRs (variable number tandem repeats), vary from individual to individual.
- At least 1 μg of DNA is needed to complete these DNA fingerprint analyses.
- DNA from single cells can be amplified by PCR and used to analyze the fingerprint.
- DNA fingerprint analyses have been used in forensics, paternity issues, genetic analyses, and breeding selection.
- Blood or tissue can be analyzed for infection by a pathogen using primers whose DNA sequence specifically hybridizes to pathogen DNA.

The Big Picture

- The ability to isolate DNA from any organism, ligate it to vector DNA, introduce that DNA into host cells, and propagate those cells has had an enormous impact on our understanding of genetics, molecular biology, and cell function and development. These techniques have been used to elucidate evolutionary relationships between different organisms and to understand gene regulation and function in greater depth. Applications of these techniques have been used to develop new medicines and diagnostics, agricultural products, and powerful forensic tools.

Common Problems

- There are a variety of ways to clone DNA fragments, and trying to remember all of the different cloning methods can be challenging. Consider the following as you review the different cloning procedures: size of the cloned DNA, host cell in which the DNA should be cloned, and expression of the cloned DNA in the host cell. Different vectors can be used in different host cells to address each of these considerations.

- There are many different experimental questions that can be answered using cloning. Ask yourself what cloning strategies could be used to answer the following questions, and review your text for answers: What is the sequence of a gene? What sequences are important for regulation of that gene? What sequences are important for targeting that gene to a particular site in a eukaryotic cell? What is the difference in function between a mutant gene product and a wild-type gene? What kinds of genes are expressed during development of an organism? How can genes be expressed in plants or in the milk of mammals?

Study Strategies

- Outline the specific steps needed to clone a gene in a bacterial cell. Include how the gene is initially isolated, what kind of vectors can be used, how to introduce that recombinant DNA into the cell, and how to confirm that the recombinant molecule is in the host cell. Now outline the steps needed to clone a gene in a eukaryotic cell.
- Complementary base pairing is important for many aspects of biotechnology. Describe each technique in gene cloning that uses complementary base pairing, detailing specifically how base pairing is involved.
- Review the following tutorial and activity on the website:
  Tutorial 17.1 DNA Chip Technology
  Activity 17.1 Expression Vectors

Test Yourself

Knowledge and Synthesis Questions

1. Cloning a gene involves
   a. restriction endonucleases and ligase.
   b. plasmids and bacteriophage lambda.
   c. yeast artificial chromosomes and complementary base pairing.
   d. All of the above

2. Restriction endonucleases
   a. are enzymes that process pre -RNA's.
   b. are enzymes that degrade DNA.
   c. protect bacterial cells from viral infections.
   d. All of the above

3. DNA fragments are separated using gel electrophoresis
   a. because DNA is pulled through the gel toward the negative end of the field.
   b. because larger DNA fragments move faster through the gel than smaller DNA fragments.
   c. to identify and isolate DNA fragments.
   d. to synthesize DNA for cloning.

4. Complementary base pairing is important for
   a. ligation reactions with blunt-end DNA molecules.
   b. hybridization between DNA and transcription factors.
   c. restriction endonucleases to cut cell walls.
   d. synthesizing cDNA molecules from mRNA templates.

5. In order for a prokaryotic vector to be propagated in a host bacterial cell, the vector needs
   a. telomeres.
   b. centromeres.
   c. drug resistant genes.
   d. an origin of replication.

6. In order for a eukaryotic vector to be propagated in a host eukaryotic cell, the vector needs
   a. telomeres.
   b. centromeres.
   c. an origin of replication.
   d. All of the above

7. Reporter genes
   a. include genes for drug resistance.
   b. include genes for bioluminescence.
   c. include genes for DNA origins.
   d. a and b

8. Vectors include
   a. bacterial and plant plasmids.
   b. viruses.
   c. artificial chromosomes.
   d. All of the above

9. Cloning means that
   a. all of the cells are derived from one cell and are genetically identical.
   b. a gene from one organism has been inserted into a vector and successfully introduced into a host cell.
   c. all of the cells in a particular organism are identical.
   d. All of the above

10. Recombinant DNA can be transferred into host cell by
    a. growing the host cell in growth medium containing ampicillin.
    b. coating the DNA with carbohydrates so that the cells will engulf the DNA.
    c. treating cells with calcium ions or electrical pulses to increase cell permeability.
    d. injecting proteins into host cells to make them more permeable.

11. A cDNA clone contains
    a. mostly cytosine.
    b. a copy of the DNA identical to the nuclear gene.
    c. a copy of noncoding DNA.
    d. a DNA copy of mRNA.

12. Gene expression can be inhibited by
    a. antisense RNA.
    b. knockout genes.
    c. DNA chips.
    d. a and b

13. Expression vectors are different from other vectors because they
    a. contain drug resistance markers.
    b. contain telomeres.
    c. contain regulatory regions that permit the cloned DNA to produce a gene product.
    d. contain DNA origins.

14. DNA fingerprinting works because
    a. genes containing the same alleles make it simple to compare different individuals.
    b. PCR allows amplification of proteins from single cells.
    c. there are multiple alleles for some DNA sequences, making it possible to obtain unique patterns for each individual.
    d. DNA in the skin cells is very diverse.

Application Questions

1. Describe three useful products that have been produced using biotechnology. Outline two specific dangers that could result from producing organisms that contain foreign genes.
2. Your lab assistant has cloned gene *x* into yeast and confirmed that the recombinant DNA molecule is present in the yeast cells. However, the yeast cell is unable to synthe size X protein. Suggest why this part of her experiment is not working and what modifications she needs to make in her cloning procedure.
3. You use a radioactively labeled DNA probe to identify a DNA fragment from an *Eco* RI digestion of *Drosophila* DNA. When you complete your hybridization, you notice that the probe has hybridized to several different fragments in the *Drosophila* DNA sample. How do you explain this result?
4. You are an investigator at a crime scene who is trying to determine if the suspect in custody actually committed the crime. Describe how, using a drop of blood from the crime scene, you would identify the real suspect.

Answers

Knowledge and Synthesis Answers

1. **d.** Cloning a gene requires restriction endonucleases to cut the gene of interest and the vector, ligase to covalently join the DNA to the vector, and vectors including plasmids, bacteriophage lambda or yeast artificial chromosomes. Complementary base pairing is important in forming the recombinant DNA molecule if the DNA fragments have sticky ends.
2. **c.** Enzymes that process pre-mRNA's add 5′ caps and 3′ poly A tails to RNA's, and remove introns. Deoxyribonucleases degrade DNA.
3. **c.** DNA fragments migrate toward the positive end of the electrical field with the smallest fragments migrating the fastest. Synthesizing DNA for cloning is done in a test tube, not in a gel.
4. **d.** No complementary base pairing can occur between blunt end cut DNA molecules. Transcription factors are proteins that bind DNA through interactions with their side chains (which are amino acids) and the nucleotides of the DNA. Restriction endonucleases cut double-stranded DNA, not cell walls.
5. **d.** A prokaryotic vector needs an origin of replication to be propagated in a prokaryotic cell.
6. **d.** A eukaryotic vector requires an origin of replication, telomeres, and a centromere. It does not require a drug resistant marker to be propagated in the cell.
7. **d.** Origins of replication are needed to propagate the vector in the host cell but are not used to detect recombinant molecules in host cells
8. **d.** All of these molecules can serve as vectors for cloned DNA.
9. **d.** If all of the cells are derived from one cell, they are all genetically identical and thus clones of one another. If all of the cells of an organism are identical (derived from a cell isolated from another organism, as seen in plant and mammalian clones), then that organism is called a clone. Cloning also refers to creating a recombinant molecule by inserting a DNA into a plasmid vector and successfully introducing that recombinant molecule into the cell.
10. **c.** Treating cells with ampicillin does not make them more permeable to DNA. Coating DNA with lipids (not carbohydrates) is used to introduce recombinant DNA molecules into host cells. Injecting DNA (not protein) is a method used to place recombinant DNA into host cells.
11. **d.** cDNA clones are not clones that contain mostly cytosine. A cDNA clone is generated by making a DNA copy of a particular messenger RNA using reverse transcriptase. The cDNA clone is not identical to the nuclear gene because in the messenger RNA (which served as a template for the cDNA) the introns have been removed, leaving coding sequence and 5′ and 3′ flanking sequences.
12. **d.** Gene expression can be inhibited by antisense RNA which complementarily base pairs with the target messenger RNA making it inaccessible to the translation machinery in the cell. Knockout genes are genes that have been inactivated by the insertion of DNA into their coding sequences. The Ti plasmid is a vector that is used to clone genes in plants and does not by itself inhibit gene expression. DNA chips are used in hybridization experiments and do not inhibit gene expression.

13. **c.** Expression vectors may contain drug resistance markers, telomeres, and must contain DNA origins but none of these features distinguish them from other vectors. Expression vectors are unique because they contain regulatory sequences that allow the cloned gene to be expressed in the host cell.
14. **c.** Multiple alleles are needed to distinguish differences in individuals. PCR is a technique that amplifies DNA (not protein).

Application Answers

1. Useful products include rice grains that produce beta-carotene, plants that are resistant to herbicides and insect larvae, and sheep that produce human blood clotting factors and antibodies to colon cancer in their milk. Dangers include creating genetically engineered foods that could adversely affect human nutrition and the transfer of herbicide- and insect-resistant genes from crop plants into noxious weeds.
2. The *x* gene can be cloned into a yeast cell, but unless the vector has the appropriate regulatory signals (promoters, poly A addition sites, translational initiation, and termination signals) no expression of gene *x* will occur. Recloning gene *x* in a yeast expression vector will result in the expression of the *x* gene.
3. The DNA sequence of the probe is not unique in *Drosophila* and may be part of a gene family (see Chapter 14) or part of a homeobox sequence (see Chapter 16).
4. DNA from the blood drop would be isolated and amplified for DNA fingerprint analysis using PCR amplification. The DNA would be digested with a restriction endonuclease and electrophoresed through a gel. DNA from a blood sample from the suspect would be similarly digested and run on the same gel. The gel would be blotted to a nylon filter. A radioactively labeled DNA probe containing a VNTR sequence would be hybridized to that filter to determine if the DNA band pattern was the same from the two samples. This process would be repeated with the two DNA samples using several different restriction endonucleases and several different VNTR probes. A match between DNA band patterns from the crime scene and DNA from the suspect would be needed to confirm the suspect's guilt.

18 Molecular Biology and Medicine

Important Concepts

Many diseases in humans are caused by a specific genetic defect.

- Proteins function in cells in many ways, as enzymes, receptors, transporters, structures, and carriers. Mutations in genes encoding these proteins can cause many diseases.
- Patients who inherit an abnormal allele produce an abnormal protein. The environment can have an effect on expression of those genes.
- Both phenylketonuria (PKU) and alkaptonuria are caused by a mutation in genes encoding enzymes in the same biochemical pathway.
- In PKU, the enzyme that converts phenylalanine to tyrosine, phenylhydroxylase, is nonfunctional, leading to build-up of phenylalanine in the blood. As phenylalanine accumulates it is converted to phenylpyruvic acid. Somehow, high concentrations of phenylpyruvic acid prevent normal brain development in infants, leading to mental retardation. In addition, patients with PKU have lighter skin and hair color because melanin, a skin pigment, is made from tyrosine.
- The molecular alteration in patients with PKU is a change in the 408th amino acid (from arginine to tryptophan) of phenylhydroxylase.
- Many proteins are polymorphic. Protein sequencing has revealed a 30 percent variation in amino acid sequences for the same protein from different individuals. A protein is polymorphic if it has a variant one percent of the time. Some mutations in these genes have no effect, producing functional gene products (see Figure 18.2).
- In sickle-cell anemia, which is homozygous in one in 655 African Americans, abnormal alleles produce an abnormal protein (β-globin). The resulting hemoglobin protein forms aggregates in the red blood cells, causing the cells to sickle. Sickle cells block narrow capillaries and damage highly vascularized tissues, especially at low oxygen concentrations.
- The abnormal allele in sickle-cell anemia is caused by a mutation that results in a change at position number 6 (from a glutamic acid to a valine) in the β-globin gene. This altered protein sequence changes the charge of the protein, and the tertiary and quaternary structure of hemoglobin.
- Hemoglobin is easy to isolate and to study. Hundreds of β-globins with alterations in their primary amino acid sequence have been isolated. Some alterations have no effect, while others are quite severe. Hemoglobin C disease results from a change at position 6 of β-globin, altering the glutamic acid to a lysine. The result is anemia that is not as severe as sickle-cell anemia.

Most common genetic diseases result from altered membrane proteins.

- Familial hypercholesterolemia (FH) is a disease in which elevated levels of cholesterol are present in the blood. The excess accumulation of cholesterol on the inner walls of the blood vessels narrows the diameter of the vessel, increasing the probability that clots will lodge in the narrowed vessels, resulting in heart attack and stroke. FH affects about one in 500 individuals, resulting in early heart attacks and strokes before the age of 45.
- Cholesterol is taken in through the diet, and travels in the blood in lipoprotein particles called low-density lipoproteins (LDL). The LDL's carry cholesterol to the liver where they bind LDL receptors and are taken up via endocytosis to be delivered to the interior of the cell (see Figure 18.3). Patients with FH lack normal LDL receptors on the surface of their liver cells, causing cholesterol to remain in the blood.
- Cystic fibrosis occurs in one in 2,500 individuals and results from a defective version of a membrane protein that is a chloride transporter. In normal individuals, chloride is transported to the exterior of the cell by this protein and as a result, water leaves the cell osmotically due to the higher concentration of chloride ions outside the cell. Chloride transport keeps the surface of the lungs (and the mucus secretions) moist. When the chloride transporter is defective, thick dry mucus accumulates in the lungs and the lining of other organs in the body. The dry mucus prevents the lungs from clearing out bacteria and fungal spores and the patient is subject to recurrent respiratory infections and liver, pancreatic, and digestive failures. Patients with cystic fibrosis usually die in their twenties or thirties.

Defects in structural proteins also lead to severe disease.

- Duchenne's muscular dystrophy is caused by a defect in dystrophin, the protein that binds actin to the plasma membrane of muscle cells. This disease occurs in one in 3,000 individuals and results in weaker muscles, respiratory failure, and eventual death.

- Hemophilia is caused by a mutation in a gene encoding a clotting factor in the blood. Lacking this protein, patients can bleed to death from minor cuts.

Abnormal protein conformations can cause disease.

- Transmissible spongiform encephalopathy (TSE) is a degenerative brain disease in which the brain develops holes and the brain tissue appears spongelike. This disease has been identified in sheep (scrapie), cows ("mad cow disease"), and humans (kuru).
- TSE is transmitted from one animal to another via brain extracts from the diseased animal.
- The infectious agent is a nonfunctional conformation of a membrane protein (PrP^{sc}) termed a prion (proteinaceous infective particles). Once these proteins fold inappropriately, they can induce a conformational change in the normal protein (PrP^{c}), so that it too becomes abnormal (see Figure 18.4). Altered proteins pile up as fibers in the brain tissue, causing cell death.
- While some diseases are the result of a mutant allele, most diseases are multifactorial, caused by many genes and proteins interacting with the environment.

Human genetic diseases have several patterns of inheritance.

- Autosomal recessive diseases include PKU, sickle-cell anemia, and cystic fibrosis. Parents who are normal but heterozygous for these alleles have a one in four chance of having a child who expresses the trait (homozygous recessive). Individuals who are homozygous recessive for this trait can synthesize only nonfunctional protein, and express the disease. Heterozygotes can synthesize normal proteins 50 percent of the time, which is usually enough for the cell to function.
- Autosomal dominant diseases are diseases in which the presence of one mutant allele is enough to produce the disease. Familial hypercholesterolemia is an autosomal disease in which heterozygotes only produce one-half of the normal LDL receptors, which is not enough to clear the blood of cholesterol. The pattern of inheritance for autosomal dominant alleles is direct transmission from parent to offspring.
- X-linked diseases include hemophilia and Duchenne's muscular dystrophy. These diseases are more common in males (since males only have one X chromosome) and are passed from mothers (who are heterozygotes and thus carriers) to their sons. The son who inherits the mutant allele from his mother will express that allele; the daughter who inherits that allele from her mother will be a carrier.
- Chromosome abnormalities include an excess or loss of one or two chromosomes (aneuploidy), a loss of a piece of a chromosome (a deletion), or the transfer of a piece of a chromosome to another chromosome (translocation). Chromosomal abnormalities may be inherited or may be the result of a nondisjunctional event in meiosis (see Chapter 9).
- Many spontaneous abortions are due to chromosomal aberrations.
- Fragile-X-syndrome (which occurs in one in 1,500 males and one in 2,000 females) causes mental retardation in some individuals. It is characterized by a constriction in the tip of the X chromosome which sometimes breaks during preparation for microscopy.

Identifying the genes that cause disease has been accomplished in a variety of ways.

- The β-globin gene was cloned using cDNA generated from messenger RNA from red blood cells. β-globin messenger RNA is abundant since it is the major protein in these cells.
- The dystrophin gene was isolated by comparing X chromosomes isolated from two children with muscular dystrophy to X chromosomes from normal individuals. The X chromosomes from these two patients had visible deletions, allowing investigators to position the site of the dystrophin gene on the chromosome.
- Comparing genetic markers in families of individuals who have a particular genetic disease to genetic markers of normal individuals has helped scientists isolate the gene responsible for the disease.
- RFLP's (restriction length fragment polymorphisms, see Figure 18.7) are genetic markers that can be used to distinguish individual variations in chromosomes. If a particular RFLP is positioned adjacent to a mutated gene, the RFLP can be used as landmark to isolate the gene. Comparing RFLP patterns between affected and unaffected individuals can help determine how close these particular markers are to the gene of interest. DNA fragments from those regions can be used in DNA sequencing reactions to look for gene coding sequences. Coding sequences are compared between affected individuals and normal individuals to determine if a particular coding region contains the gene of interest.

Different kinds of mutations can cause human disease.

- Point mutations that change one codon to a different codon (as seen in sickle-cell anemia and PKU) can lead to an alteration in the three-dimensional structure of a protein, eliminating its function.
- Point mutations that change a codon to a stop codon produce truncated proteins, eliminating protein function if the mutation occurs at the beginning of the coding sequence (as seen in some types of cystic fibrosis).
- Hot spots, which are usually sites on the DNA containing methylated cytosines, are more prone to mutation. Loss of the amino group on a cytosine converts that base to uracil, which is removed from the DNA by repair enzymes. Loss of an amino group on a methylated cytosine converts that base to thymidine, which is retained in the DNA, resulting in a base change at that site.
- Deletions can vary in size. Small deletions are generally less severe than large deletions, which can remove more than one gene.
- Expanding triplet repeats are found in fragile-X syndrome (a repeat of CGG in the *FMR 1* gene), in myotonic dystrophy (a repeat of CTG), and in Huntington's chorea (a

repeat of CAG). Triplet repeats are more extensive in affected individuals, and are thought to expand during DNA replication due to slippage of DNA polymerase in those regions.

- Genetic imprinting is observed in genes or groups of genes whose phenotypic effects differ depending on the parent from whom those genes were inherited.
- Studying genetic variation has broadened our understanding of cell physiology, allowing new biochemical approaches to human disease and to medical diagnosis.

Genetic screening is used to identify people who might be carriers of a particular disease.

- In PKU screening blood samples from newborn babies are tested to see if they can support growth of bacteria that are phenylalanine auxotrophs (see Chapter 12). Individuals who test positive for PKU are fed special diets low in phenylalanine to prevent abnormal brain development.
- Tay-Sachs is an autosomal recessive disease that is lethal in childhood. Blood can be screened for the missing enzyme. Parents who are both carriers of this mutant allele have a 25 percent chance of bearing a child with this disease. Screening allows parents to make an informed decision about having a child.
- *In vitro* fertilized human embryos can be screened at the eight-cell stage for mutant alleles. One of these embryonic cells is removed and its DNA is amplified using PCR primers specific for the disease gene. Once a normal genotype has been confirmed, the seven-celled embryo can be implanted into the mother, and will develop normally.
- Newborns can be screened for PKU and other disorders using DNA extracted from their cells and PCR amplification of that DNA which targets the mutant allele. Parents who are potential carriers for a particular mutant allele (such as muscular dystrophy or cystic fibrosis) can be screened for that allele using PCR amplification.
- Once the DNA has been isolated, different kinds of genetic screens can be used. Allele specific cleavage differences can detect changes in restriction endonuclease sites in the DNA containing the gene of interest. Oligonucleotides that are specific for the mutant allele or the normal allele can be used in hybridizations with DNA isolated from different individuals (see Figure 18.12).
- Cancer is a genetic disease in which there is a loss of normal control over cell division. Cells divide continuously, forming a malignant tumor that metastasizes to other tissues in the body.
- Benign tumors are not cancerous, grow slowly, and do not spread to other parts of the body.
- Metastasis is characterized by angiogenesis, secretion of digestive enzymes as cells begin to relocate, and synthesis of cell surface proteins that allow the cancer to invade new host tissue.
- Carcinomas are cancers of surface tissues including skin and epithelial cells that line organs. Carcinomas include lung cancer, colon cancer, breast cancer, and liver cancer. Sarcomas are cancers of tissue such as bone, blood vessels, and muscle. Leukemia and lymphomas affect cells that give rise to blood cells.
- Some cancers (15 percent) are caused by viruses (see Table 18.2), including Hepatitis B and papilloma virus.
- Fifteen percent of all cancers are inherited. Eighty-five per cent of cancer is caused by mutations, either spontaneous or induced by environmental carcinogens, viruses, and various forms of irradiation.
- Two kinds of genes are altered in many cancers, oncogenes (which stimulate cell division when activated) and tumor suppressor genes (which inhibit cell division). Point mutations, translocations, and gene amplifications in these genes can lead to uncontrolled cell division. Proto-oncogenes are genes that can be converted to oncogenes with one mutational hit. Tumor suppressor genes (such as retinoblastoma, Wilm's tumor of the kidney, and inherited forms of breast and prostate cancer) require the inactivation of both alleles (the "two-hit" hypothesis).
- Women who inherit a mutant allele of the *BRCA1* gene are more prone to develop breast cancer than women with the normal alleles for *BRCA1*.
- Tumor suppressors control the progress of the cell through the cell cycle. Retinoblastoma acts during G1 to bind transcription factors that are necessary for progress into DNA synthesis. Inhibition of these transcription factors prevents DNA synthesis and subsequent cell cycle completion.
- P53 is a protein that stops the cells during the G1 phase of the cell cycle. *P53* is a transcription factor for proteins that block the interaction of cyclins and protein kinases (see Chapter 10). *P53* is mutant in many type of cancer cells, including lung and colon cancer cells.
- Multiple mutations transform a normal cell into a cancerous cell (see Figure 18.18). These mutations can be screened using oligonucleotide probes for oncogenes and/or tumor suppressor gene alterations.

Treating genetic diseases includes modifying the phenotype, and gene therapy.

- For individuals with PKU, treatment involves restricting the substrate (phenylalanine) for the missing enzyme (phenylalanine hydroxylase), preventing the buildup of phenylpyruvic acid.
- For individuals with familial hypercholesterolemia, drugs are administered that block the patient's own cholesterol synthesis.
- Drugs that kill rapidly dividing cells are administered to cancer patients (see Figure 18.19). These drugs also affect noncancerous dividing cells in the body, causing digestive upsets, hair loss, and anemia.
- Patients with hemophilia A are given the missing gene product, which is a clotting factor, in purified form.
- In gene therapy, the new gene is inserted into the affected individual. The gene must have all of the important regulatory regions necessary for expression and must be delivered effectively to individual's cells. This requires effective

vectors, efficient uptake, appropriate expression and processing of mRNA and protein, and selection of the target cells for the recombinant DNA.

- Gene therapy has been used to replace a defective gene for adenosine deaminase in the white blood cells of a patient lacking a functional copy of this gene. The disadvantage of this technique is that white blood cells have a set lifetime in the body, eventually dying and eliminating the new gene. Inserting new genes in stem cells will prolong the lifetime of the inserted gene in the patient's body. Other approaches include inserting the gene directly into the cells where the defect exists. For instance, in lung cancer patients, functional alleles of tumor suppressor genes, as well as antisense RNA's against oncogene mRNA's, are provided in aerosols for inhalation.

Sequencing the human genome has been a valuable scientific endeavor.

- Human DNA is first cut into 500 base pair-long fragments and then sequenced. Aligning of the different fragments is done using physical (restriction endonuclease sites) and genetic (such as RFLP's and short simple sequence repeats) map markers.
- Libraries of human DNA have been created using restriction enzymes that recognize 8–12 base pair recognition sequences, resulting in much larger DNA fragments. These large DNA fragments have been cloned into a bacterial artificial chromosome (BAC) which can accommodate 250,000 base pairs of inserted DNA.
- These fragments have been arranged in proper order using sequence-tagged sites (STS) which are unique stretches of DNA (60–100 base pairs long) whose sequences are known or by using genetic map markers (see Figure 18.21).
- Shotgun sequencing involves cutting the DNA into random small fragments which are analyzed for overlapping sequences by computer (see Figure 18.21). Shotgun sequencing was used to sequence the entire 180 million base pairs of the fruit fly genome.
- There are many advantages of sequencing the entire human genome:
  1. Comparison of genes from different organisms that have similar counterparts in humans
  2. Identifying disease related genes using positional cloning
  3. Identifying genes that will react best to particular medications (pharmacogenomics)
  4. Establishing fingerprint patterns of gene expression in cancer cells
  5. Searching for genes that predispose certain groups of people to certain diseases
- Now that complete genetic information for each human being is possible, many individuals including ethicists, psychologists, and geneticists want to establish guidelines regarding how this information will be used.

The Big Picture

- Some diseases in humans can be directly related to a change in the gene, causing the production of an abnormal gene product. Other molecular diseases are affected by the individual's environment. Understanding the molecular alterations in a particular gene product and its effect can lead to therapeutic approaches, new diagnostic tools, and preventative measures in health care.

Common Problems

- Sometimes it is difficult to conceptualize how a mutant allele can be dominant to a wild type allele (as seen in familial hypercholesterolemia). Remind yourself of the function of the wild type gene product and then predict how the individual will be affected with only half of those gene products being expressed.
- RFLP mapping, oligonucleotide hybridization, and allele-specific cleavage analyses are similar techniques used in genetic screening to detect mutant alleles. Individual differences in DNA sequences (including simple base changes, as well as a difference in the number of repeated sequences) can be utilized to distinguish individuals and to map genes. Review Figures 18.7, 18.11, and 18.12 to distinguish these screening techniques.

Study Strategies

- Make a list of the molecular diseases discussed in this chapter. Include in this list the specific proteins that, when altered, cause the disease, and the treatments that are currently available for each of these diseases.
- Review the cell cycle and the signals that are important for causing cells to divide. Predict how mutations in those genes can alter the regulation of the cell cycle and lead to unregulated cell division and cancer. Be sure to include a review of tumor suppressors and oncogenes in this analysis.
- Review the following tutorial and activities on the website:
  Tutorial 18.1 Sequencing the Genome
  Activity 18.1 Allele-Specific Cleavage
  Activity 18.2 Concept Matching

Test Yourself

Knowledge and Synthesis Questions

1. Fragile-X syndrome is
   a. due to a single base change in the DNA.
   b. caused by changing a valine to a glutamic acid.
   c. caused by triplet expansion.
   d. caused by a chromosomal translocation.
2. A nonfunctional membrane protein is responsible for
   a. hemophilia.
   b. sickle-cell anemia.
   c. cystic fibrosis.
   d. PKU.
3. Triplet repeat expansion
   a. occurs during DNA replication due to slippage of DNA polymerase.
   b. is caused by errors in DNA synthesis with reverse transcriptase.

c. differences can be identified using RFLP mapping.
d. a and c

4. In sickle-cell anemia the
   a. sixth amino acid is changed from a valine to a glutamic acid.
   b. sixth amino acid is changed to a stop codon.
   c. hemoglobin concentration builds up in the red blood cells.
   d. structure of β-globin is altered and the hemoglobin protein forms aggregates in the red blood cells.

5. Most cancers are
   a. due to the inheritance of mutant genes.
   b. caused by viruses.
   c. caused by mutations in genes whose products help blood clot.
   d. caused by mutations that are spontaneous or induced.

6. Metabolic disorders
   a. refer to genes that have alterations in their centromeres.
   b. are caused by prions.
   c. are due to abnormal membrane proteins that transport chloride ions.
   d. are the result of mutations in enzymes that are required to synthesize particular compounds in the cell (such as an amino acid).

7. Prions
   a. cause scrapie in sheep.
   b. are caused by abnormally folded proteins interfering with the normal brain cell function.
   c. cause "mad cow disease."
   d. All of the above

8. Human genetic disease can be
   a. due to an autosomal recessive trait.
   b. passed from mothers to their sons (if X-linked) 50 percent of the time and also can be caused by translocations.
   c. due to a mutant allele that is dominant to wild type.
   d. All of the above

9. The mutations that result in human disease include
   a. triplet expansions that occur during DNA synthesis.
   b. point mutations that do not change the amino acid sequence of the gene.
   c. prion-like diseases such as kuru.
   d. All of the above

10. Genetic screening has been used to
    a. identify embryos carrying a mutant allele.
    b. identify parents who are carriers for PKU and Tay-Sachs.
    c. identify fathers who are carriers for X-linked diseases.
    d. a and b

11. Genes that when mutated cause cancer are
    a. only those genes involved in the regulation of the cell cycle.
    b. transcription factors only.
    c. only those genes inactivated by viruses.
    d. None of the above

12. In order for gene therapy to be most effective, genes should be inserted in
    a. white blood cells.
    b. red blood cells.
    c. stem cells.
    d. All of the above

13. The human genome allows scientists to
    a. understand regulatory sequences that are important for gene expression.
    b. locate genes that cause disease.
    c. understand evolutionary relationships by comparing human genes to genes in other organisms.
    d. All of the above

Application Questions

1. Familial hypercholesterolemia (FH) and Duchenne's muscular dystrophy (MD) are both molecular diseases that are caused by mutations in particular genes, yet FH is easier to treat than MD. Why?
2. Two individuals have a mutation in gene *x* but at different sites. The disease adversely affects the first individual, and the second individual experiences no effect. Explain this observation.
3. Why are X-linked diseases more common in males than in females?
4. The β-globin gene (which when mutated can lead to sickle-cell anemia) was much easier to clone than the gene responsible for one of the inherited forms of breast cancer (*BRAC1*). Why?
5. Why do cells with unregulated cell division require a mutation in only one of the two alleles for an oncogene, and in both alleles for a tumor suppressor?

Answers

Knowledge and Synthesis Answers

1. **c.** Fragile-X syndrome is due to a triplet expansion on the X chromosome, not a single base change, a translocation, or an alteration of protein sequence as seen in sickle-cell anemia.
2. **c.** The chloride transporter is a membrane protein that, when nonfunctional, causes cystic fibrosis. Hemophilia is caused by a defect in a clotting factor; sickle-cell anemia is caused by a defect in β-globin. PKU is a defect in an enzyme in a biochemical pathway.
3. **d.** Triplet expansion is thought to be due to slippage of DNA polymerase during DNA replication and will result in an increasing number of triplet repeats in those cells. That increase in the DNA sequence can be distinguished from the DNA sequence of a normal individual using RFLP analysis. Cystic fibrosis is not the result of triplet expansion in that gene. Reverse transcriptase is an enzyme from HIV that copies RNA into DNA.
4. **d.** In sickle-cell anemia, the sixth amino acid is changed from a glutamic acid to a valine (not from a valine to a glutamic acid, and not to a stop codon). The resulting change causes the hemoglobin protein to form aggregates in the red blood cell. Hemoglobin concentration in the red blood cell is not altered in sickle-cell anemia.

5. **d.** Fifteen percent of cancers are due to the inheritance of a mutant allele. The other 85 percent are caused by mutations that are spontaneous or induced by carcinogens and viruses (15 percent). Mutations in genes that encode blood clotting factors result in hemophilia.
6. **d.** Some metabolic disorders are caused by mutations in genes that encode enzymes in a biochemical pathway (such as the pathway for the synthesis of phenylalanine). Mutations in centromeres and genes for chloride transporters do not result in metabolic disorders. Prions are not the result of a mutation; they are the result of abnormal folding of a protein.
7. **d.** Prions result from abnormally folded proteins that interfere with normal brain cell function. They are not caused by a mutant allele for this particular protein.
8. **d.** All of these mutations and chromosome alterations can cause human disease.
9. **a.** Triplet expansions cause a variety of diseases including fragile-X syndrome, Huntington's chorea, and myotonic dystrophy. Point mutations that do not alter the amino acid sequence of a protein are silent mutations and do not cause any disease (in almost all cases). Prions are not the result of a mutation in a gene but are due to the abnormal folding of a particular protein.
10. **d.** Genetic screening has been used to analyze *in vitro*-fertilized embryos (at the 8-cell stage) and DNA from individuals who might be carrying a mutant allele. Fathers are not carriers for X-linked diseases; they express all of their X-linked alleles. Genetic screening is used to identify mutant alleles in a DNA sample, not to transfer a gene into the genome of an individual.
11. **d.** While all of the examples listed are examples of mutations that can result in cancer, none of these mutations is the *only* reason an individual develops cancer.
12. **c.** The most effective approach for genes to reside permanently in the individual's cells is for them to be inserted in stem cells which provide a continual supply of differentiated cells under the correct environmental signals. Red blood cells have no nuclei and thus cannot be used for gene therapy. White blood cells have been used for gene therapy, but because they have a set lifetime in the body, they eventually die, destroying the gene they were carrying.
13. **d.** Human genome sequencing provides all of these advantages.

Application Answers

1. Familial hypercholesterolemia causes an increase of cholesterol in the patient and can be treated with drugs that reduce the patient's own synthesis of cholesterol. Muscular dystrophy is caused by a defect in a structural protein, dystrophin, and results in patients who have dysfunctional muscle tissue. This structural dysfunction cannot be reversed with the application of any drug.
2. The mutation in gene *x* in the first individual must have occurred in an essential region of the gene that is required for its function. The mutation in gene *x* in the second individual must be in a region that is nonessential for the function of that protein.
3. X-linked diseases are more common in males because males have only one X chromosome so that every allele on that chromosome is expressed in the individual. Mutant alleles can be carried on females but masked by a wild type allele on the other X chromosome.
4. β-globin is one of the major proteins found in the red blood cell, and thus isolating messenger RNA from this cell and cloning a cDNA copy of this gene was relatively straightforward. Neither the function nor the location of the *BRAC1* gene was known. Nearby genetic markers had to be identified that were linked to the *BRAC1* gene, and then DNA sequencing of that region had to be done. DNA sequences between affected and unaffected individuals had to be compared to identify the gene.
5. Proto-oncogenes control cell division by responding to an environmental signal and initiating cell division in response to that signal. A mutation in an allele of an oncogene would result in that protein causing the cell to initiate cell division even in the absence of any environmental signal. Half of the wild type proteins would respond appropriately to cell division signals, but the dysfunctional proteins which were now unregulated would still signal the cell to divide. A single wild type copy of a tumor suppressor gene could still provide enough gene product to regulate cell division.

STUDY GUIDE

19 Natural Defenses against Disease

Important Concepts

Animal defense systems are based on the distinction of self from nonself and include innate defenses and specific defenses.

- Innate defenses are general mechanisms that protect the body from pathogens. Innate defenses include physical barriers as well as cellular and chemical defenses (see Table 19.1).
- Specific defenses, which involve two types of lymphocytes, B and T cells, are aimed at a single target.
- Lymphoid tissues, including thymus, bone marrow, spleen, and lymph nodes, are essential to this defense system.
- Phagocytes (neutrophils and macrophages) are white blood cells that engulf and digest nonself materials.
- Lymphocytes are the most abundant class of white blood cells and include B and T cells.
- Immature T cells migrate via the blood to the thymus where they mature. T cells participate in specific defenses against foreign or altered cells, including virus-infected cells and tumor cells.
- B cells circulate through the blood and lymph and make antibodies, which are proteins that bind to nonself or altered self.
- T cell receptors are proteins on the surface of T cells that recognize and bind to nonself substances on the surfaces of other cells.
- MHC proteins protrude from surfaces of most cells in the mammalian body.
- Cytokines are soluble signal proteins released by T cells, macrophages, and other cells. They can activate or inactivate B cells, macrophages, and T cells or limit tumor growth.

Barriers and local agents are nonspecific defenses against invaders.

- Nonspecific defenses include skin, mucus-secreting tissues, lysozyme in tears, digestive enzymes, bile salts, and hydrochloric acid in the stomach.

Innate or nonspecific defenses include chemicals and cellular processes.

- Complement proteins, interferons, phagocytotic cells (neutrophils, monocytes, macrophages, and eosinophils), natural killer cells, and inflammation defend the body against infectious agents in nonspecific ways.

Specific defenses of the immune response occur when immune cells (B and T cells) recognize and destroy nonself substances.

- Immune cells recognize antigenic determinants (i.e., specific amino acids of a protein) on invading pathogens. There can be many antigenic determinants on a pathogen surface.
- Immune cells respond to pathogens by activating lymphocytes of the appropriate specificity. Humans can respond to 10 million different antigenic determinants.
- The immune cells can distinguish self from nonself.
- The immune system "remembers" the pathogen and can respond more rapidly and more effectively the next time it is exposed to it. Immunological memory is the basis of vaccination.
- There are two interactive immune responses, the humoral immune response (which uses B cells) and the cellular immune response (which uses T cells).

Clonal selection accounts for the characteristic features of the immune response.

- Each B cell is able to produce only one kind of antibody. When an antigen that fits the surface antibody binds to the B cell, that cell is activated and divides to produce more B cells which secrete antibody. T cells are clonally selected in a similar manner.

Immunologic memory and immunity result from clonal selection.

- In the primary immune response, activated B and T cells produce effector cells and memory cells.
- Effector cells attack the pathogen by producing specific antibodies (B cells) or cytokines (T cells).
- Memory B and T cells are long-lived and will divide rapidly to produce an effective and more powerful immune response if re-exposed to a pathogen. This rapid response, the secondary immune response, provides a natural immunity to those diseases. Vaccinations are administered to give the recipient artificial immunity, and are effective because of this secondary immune response.

- Animals are able to tolerate their own antigens due to clonal selection which removes any B and T cells (during their differentiation) that recognize self antigens. An immune cell that recognizes self antigen undergoes programmed cell death (apoptosis).
- Clonal anergy is a suppression of the immune response so those T cells that recognize a self antigen on the body do not send out cytokines. This suppression occurs in those body cells which are missing a co-stimulatory signal for the T cell. The co-stimulatory signal for the T cell is only expressed on antigen presenting cells.
- Immunological tolerance occurs when a foreign antigen is introduced into an animal during fetal development, which results in the animal's recognizing that antigen as self.
- Continued exposure to self antigens helps maintain tolerance throughout the lifetime of the organism.

B cells are the basis of the humoral response.

- Activation of B cells results from the binding of antibody to a particular antigenic determinant, and the arrival of a signal from a T cell that has bound the antigen on an antigen-presenting cell.

Antibodies (also called immunoglobulins) are similar in structure, but can be grouped into five different classes.

- Antibody molecules consist of two identical heavy polypeptide chains and two identical light polypeptide chains. Each polypeptide consists of a constant region (which is similar in amino acid sequence from one immunoglobulin to another), and a variable region, which forms the antigen-binding site.
- The constant region determines whether the antibody will be secreted or remain on the cell surface of the B cell. It also determines what type of action will be taken to eliminate the antigen.
- The five immunoglobulin classes (IgG, IgM, IgD, IgA and IgE) are based on differences in the constant region of the heavy chain (see Table 19.2).

Monoclonal antibodies are produced by hybridomas.

- Normal immune responses are polyclonal, generated by the activation of many different B cells and resulting in a complex mixture of antibodies.
- Monoclonal antibodies are produced from a clone of B cells that makes only one antibody that binds a unique determinant.
- Monoclonal antibodies are generated by the fusion of normal B lymphocytes with myeloma cells, which are cancerous tumors of plasma cells.
- Monoclonal antibodies have been used in immunoassays, to detect tiny amounts of molecules in tissues and fluids, to target antigens on the surface of cancer cells (for diagnostic purposes or medicinal purposes), and for passive immunization.

T cells direct the cellular immune response.

- T cells have specific surface receptors that are made up of two different polypeptide chains, both with a variable and a constant region. The variable region provides the specificity of the T cell receptor.
- T cell receptors bind antigen fragments that are displayed on antigen presenting cells.
- Cytotoxic T (T_C) cells recognize virus-infected cells and kill them by causing them to lyse (see Figure 19.15).
- Helper T (T_H) cells assist both the cellular and the humoral immune systems.

The major histocompatibility complex (MHC) encodes proteins that present antigens to the immune system.

- MHC gene products are plasma membrane glycoproteins, which present antigens on an antigen-binding groove of the protein. These antigens are inspected by T cell receptors. MHC proteins can display antigenic peptides of about 10–20 amino acids.
- Class I MHC proteins are present on the surface of every nucleated cell in the animal. Degraded cellular proteins are presented to T_C cells on MHC I complexes. T_C cells have a surface protein called CD8 that can recognize MHC I proteins.
- Class II MHC proteins are found on the surfaces of antigen presenting cells including B cells and macrophages. T helper cells recognize class II MHC proteins using the CD4 cell surface protein.
- Class III MHC proteins include some of the proteins in the complement system that help lyse foreign cells.
- T cell receptors recognize antigenic fragments when they are complexed with the MHC I or MHC II molecules.
- MHC genes, antibody genes, and T cell receptor genes have similar base sequences, suggesting a common ancestral gene.

The humoral response is directed by T helper cells recognizing antigen on MHC II complexes.

- Both macrophages and B cells bind antigen and present antigen fragments on the surface of their cells on MHC II complexes. T_H cells bind both the antigen and the MHC II complex and release cytokines. Those cytokines activate the T_H cells to divide (the activation phase) and activate B cells to divide and produce antibody (the effector phase).
- Cytotoxic T cells destroy cells that are displaying antigens from virus or mutant proteins on their MHC I complexes, eliminating virally infected cells as well as tumor cells. T_C cells must receive a second signal for activation, which involves an interaction between proteins on the Tc cell and proteins on the antigen-presenting cell. The Tc cell then proliferates and divides, releasing cytokines.
- Later, T_C cells are inhibited by the cell surface protein CTLA4.

MHC molecules play a key role in the tolerance of self.

- T cells are tested in the thymus to see if they bind to MHC displaying one of the body's own antigens, or if they are unable to recognize the body's MHC proteins. If T cells have either of these characteristics they are destroyed.
- MHC proteins are specific to each individual, making tissue transplants recognizable as nonself in the donor. Suppression of rejection of organ transplants by the recipient's immune system is accomplished by treating a patient with drugs such as cyclosporin that suppress the immune system.

DNA rearrangements and mutations are used to generate all of the diverse antibodies within an organism.

- DNA fragments are rearranged and joined during B cell development to generate antibody genes. Each B and T cell assembles its own unique set of immunoglobulin and cell receptor genes for that particular cell.
- The variable region of the light chain originates from two families of DNA segments, and the variable region of the heavy chain originates from three families of heavy chains.
- For the constant and variable regions of the heavy chain in mice there are multiple genes coding for each of four kinds of segments: 100 *V*, 30 *D*, 6 *J*, and 8 *C* (see Figure 19.19). Each B cell randomly selects one gene from each of these multiple genes to make the final heavy chain coding sequence *VDJC*.
- Light chains are similarly constructed from DNA segments.
- Light and heavy chains are combined to create billions of possible antibodies.
- Increased diversity is achieved by imprecise joining during DNA rearrangements, addition of extra nucleotides to cut DNA fragments, and an increased mutation rate in immunoglobulin genes.
- In mouse B cells, antibody genes are rearranged by initially joining the *V*, *D*, and *J* segments and then splicing that *VDJ* segment to a constant region after transcription.
- Class switching occurs by rearrangements in the DNA which position the *VDJ* segment (the same variable region on the antibody) adjacent to a constant region farther down the DNA (see Figure 19.21), resulting in the production of an antibody with a different constant region. Class switching in B cells is caused by cytokine release from T_H cells.
- Ancestral antibody genes may have developed from transposons which were able to cut out adjacent DNA sequences and move them elsewhere in the genome.

Other organisms have defense mechanisms.

- Some protists, invertebrates, and vertebrates use phagocytosis as a defense mechanism.
- Invertebrates distinguish between self and nonself but lack immunological memory.
- Invertebrates do not produce antibodies, lymphocytes, or the complement system but do produce proteins similar to vertebrate cytokines.

Disorders of the immune system

- Hypersensitivity occurs when the immune system overreacts to a dose of antigen and may cause inflammation as seen in the allergic response.
- Immediate hyperactivity in allergic reactions is caused by IgE binding the foreign antigen and causing mast cells to release histamines.
- Delayed hypersensitivity is due to T cell responses resulting in cytokine release that can result in macrophages being released and tissue damage (as seen in tuberculosis).
- Autoimmunity occurs when B and T cells direct their response against self antigens. Examples of autoimmune disease include systemic lupus erythematosis, rheumatoid arthritis, multiple sclerosis, and insulin-dependent diabetes mellitus. Some of these diseases may have a genetic component.
- AIDS is caused by HIV, a virus that eventually destroys T_H cells. HIV is transmitted through body fluids including blood and semen.
- HIV initially infects macrophages, T_H cells, and dendrite cells (one kind of antigen-presenting cell). These cells carry the virus to the lymph nodes where they preferentially infect activated T_H cells.
- During the first phase of infection, up to 10 billion viruses are made every day, causing mononucleosis-like symptoms. During the second phase, the virus infects T_H cells and an immune response is mounted, producing HIV antibodies in the blood. Later, the T_H cells are destroyed and the patient becomes susceptible to a wide variety of infections that would normally be eliminated by T_H cells. At this stage Karposi's sarcoma, pneumocystis carinii, and virally caused tumors persist in the system of the HIV infected individual.

HIV infects and replicates in T_H cells.

- HIV is an enveloped virus (see Chapter 13) and has a protein coat that surrounds two identical molecules of RNA as well as the enzymes reverse transcriptase, integrase, and a protease.
- The viral envelope protein gp120 attaches to T_H cells and macrophages by binding their CD4 proteins. The HIV membrane fuses with the host cell plasma membrane by using gp41, a viral protein, and fusin, the host protein.
- A DNA copy of the viral genome is made using three enzymes: reverse transcriptase (an error prone polymerase), RNAse H, and DNA polymerase. Errors made during viral replication create alterations in viral proteins that can now escape the host's immune response.
- Viral cDNA is inserted into the host DNA using the HIV integrase protein.
- When the infected T_H cell is activated, the entire viral genome is transcribed into RNA where it can be used to make new HIV particles or to synthesize new viral structural proteins. These proteins are processed by viral proteases. RNA genomes are packaged into core particles and the mature virus buds from the host cell.

- Therapeutic treatment for HIV includes drug therapies that interfere with the major steps of the life cycle of the virus, such as protease inhibitors and reverse transcriptase inhibitors. Drugs under development include HIV vaccines, integrase inhibitors, inhibitors that block transcription and splicing, and antisense RNA's and ribozymes directed against HIV RNA.

The Big Picture

- Animals have special systems to distinguish self from nonself. Some of these systems are nonspecific, such as innate defenses (complement protein, phagocytotic cells). B and T cells can recognize specific antigens on the surface of pathogens and mount an immune response to destroy those pathogens. When the immune system is compromised, as seen in HIV-infected individuals, the individual becomes susceptible to a variety of pathogens.

Common Problems

- The specificity of the immune response may seem complex initially, but there are really only two major components: B cells (which bind antigen) present antigen and secrete antibody, and T cells, bind antigen on antigen-presenting cells and help direct the immune response.
- MHC proteins may seem complicated at first, but they have only a few essential functions in the immune system. These proteins are specific for each individual and are essential for presenting antigen on antigen-presenting cells (MHC II) to T_H cells and recognizing one's own cells (MHC I).
- The specificity of the immune response may seem complex but it can be understood by reviewing cellular communication. Antibodies of B cells bind with antigen on the surface of pathogens; T cell receptors bind antigen presented on MHC I and MHC II proteins on the surface of cells. Once these membrane proteins interact, cytokines are secreted and cells begin to divide and proliferate. Antibodies are secreted by B cells.
- The rearrangement of DNA in the nucleus of the B cell may seem foreign to students who think the integrity of the genetic information must be preserved. But this change in the genetic makeup of activated B and T cells is essential to provide the organism with a diverse and specific set of antibodies and cell receptors.

Study Strategies

- List the nonspecific (innate) responses that a human organism can make in response to an invading pathogen. Next list cells that are part of the specific responses. Include proteins that are crucial to generating that specificity. Review how genetic rearrangements generate a diverse set of antibodies for B cells.
- Diagram an antibody and label the important components. Highlight those areas where mutations and alterations occur that change the amino acid sequence of the heavy and light chains.
- Review the expression and function of MHC I and MHC II molecules.
- Make a list of the functions of T_H cells and T_C cells.
- Draw a diagram of a chromosome and label it with 4 *V*'s (*V*1–*V*4), 2 *D*'s (*D*1*D*2), 3 *J*'s (*J*1–*J*3) and 8 *C*'s (*C*1*C*8). Imagine that a B cell is maturing to produce antibody on its surface. Select one *V*, one *D*, one *J*, and one *C* segment to make an antibody. Repeat the process using different segments. Predict how many different antibodies can be generated from this set of gene segments.
- List some of the disorders of the immune system and their probable causes.
- Review the following tutorials and activities on the website:
  Tutorial 19.1 Cells of the Immune System
  Tutorial 19.2 Pregnancy Test
  Tutorial 19.3 A B Cell Builds an Antibody
  Tutorial 19.4 Humoral Immune Response
  Tutorial 19.5 Cellular Immune Response
  Activity 19.1 The Human Defense System
  Activity 19.2 Inflammation Response
  Activity 19.3 Immunoglobulin Structure
  Activity 19.4 HIV Reproduction

Test Yourself

Knowledge and Synthesis Questions

1. Phagocytes
   a. are T and B cells.
   b. present antigen on MHC II complexes.
   c. digest nonself materials.
   d. b and c

2. B cells
   a. secrete antibodies.
   b. present antigen on MHC II.
   c. ingest antigens.
   d. All of the above

3. Nonspecific responses in the immune system include
   a. macrophages.
   b. natural killer cells.
   c. complement.
   d. All of the above.

4. When the receptor of a T_H cell binds to a pathogen being presented on a macrophage it
   a. activates itself.
   b. secretes cytokines.
   c. activates B cells.
   d. All of the above

5. Part of the normal immune response includes
   a. the production of B memory cells.
   b. the production of memory macrophages.
   c. antibody secretion by eosinophils.
   d. the production of B cells that attack the individual's own cells.

6. Antibody molecules are
   a. produced by B cells and have a variable and a constant region.
   b. secreted by B cells, once a signal (a cytokine) is received from a T cell.
   c. produced by T cells and have a variable and a constant region.
   d. a and b

7. Cytotoxic T cells
   a. release cytokines that activate B cells.
   b. attack pathogens by binding to cell surface antigens on those pathogens.
   c. destroy pathogens by lysing them.
   d. destroy host cells that are infected with virus.

8. The humoral response
   a. is due to the secretion of antibodies.
   b. occurs when T cells bind antigen-presenting cells.
   c. is due to T cells secreting their receptors.
   d. occurs when natural killer cells engulf cancer cells.

9. Autoimmunity is
   a. active when organ transplantation is successful.
   b. caused by the failure of anergy.
   c. b and d
   d. the response of the immune cells attacking the body's own tissues.

10. Patients with HIV are susceptible to a variety of infections because
    a. the virus produces cell surface receptors that bind to pathogens, making it easier for those pathogens to be infective.
    b. synthesizing a DNA copy of the viral genome makes a person feel sick.
    c. HIV attacks and destroys the T helper cells, which are central to mounting an effective immune response, making those individuals more susceptible to other infections.
    d. HIV destroys B cells so that antibodies cannot be made in response to invading pathogens.

11. DNA rearrangements in the B cell
    a. are responsible for generating single B cells that can express many different antibodies.
    b. lead to mutations in T cells resulting in the elimination of essential T cell genes.
    c. occur only in B memory cells.
    d. are responsible for generating many different antibodies, with each B cell expressing only one set of identical antibodies.

12. Major histocompatibility proteins function in the immune system by
    a. presenting antigen to T cells.
    b. presenting self antigens to T helper cells so those T cells will continue to tolerate self throughout the lifetime of the individual.
    c. generating antibodies to different pathogens.
    d. a and b

Application Questions

1. Imagine that a particular bacterium has surface molecules that resemble membrane proteins found on the heart valves. What will happen to the individual infected with this bacterium if the infection is not treated?
2. Organ transplants are more successful when the donor is a relative of the recipient. Why?
3. Vaccinations are effective against an array of different pathogens. How do they work? Vaccinations against HIV virus have not been effective in preventing the disease. Why not?
4. If the cytotoxic T cells were eliminated from a person's array of immune defenses, what kinds of disease would he or she be susceptible to?

Answers

Knowledge and Synthesis Answers

1. **d.** Phagocytes are nonspecific cells (not B and T cells) that digest nonself materials and present protein fragments of those nonself materials on their surface. They do not have antibodies.
2. **d.** B cells are antigen presenting cells; they bind antigen, digest it, present fragments of that antigen on their MHC II proteins, and secrete antibodies.
3. **d.** Macrophages, natural killer cells, and complement are all part of the nonspecific response of the immune system. B and T cells are part of the specific response of the immune system.
4. **d.** When the T_H cell binds antigen being presented on a macrophage, it secretes cytokines which activate itself and B cells.
5. **a.** In a normal immune response, B memory cells are made so that the organism can mount a faster and more effective response to the pathogen if the organism ever encounters it again. Memory macrophages do not exist. Eosinophils do not secrete antibodies. In an abnormal immune response, B cells that attack the individual's own cells can be activated, as seen in autoimmune diseases.
6. **d.** T cells have cell surface receptors with a variable and a constant region and do not produce antibody molecules. Antibodies are not produced by macrophages. Only B cells produce antibodies in response to cytokines released from the T_H cell. Each antibody molecule has two identical heavy chains and two identical light chains, and those chains each have a variable and a constant region.
7. **d.** Cytotoxic cells bind to virus antigen being presented on MHC I protein and destroy those cells. T_H cells release cytokines to activate B cells. The antibodies of B cells bind cell surface antigens on pathogens. Pathogens are lysed by complement (and not by cytotoxic T cells), which is activated by antibodies bound to those pathogens.

8. **a.** The humoral response refers to that part of the specific response that releases antibodies (B cells) in the lymph and blood (the "humors" of the body). The cellular response is T cell receptors binding antigen on antigen presenting cells. T cells do not secrete their receptors.

9. **c.** Autoimmunity occurs when the immune cells attack the body's own cells. Transplants can be rejected if the transplanted tissue is recognized by the body as nonself, and is not a result of immune cells attacking the body's own tissue. The failure to suppress T cells that recognize self antigens in the individual (anergy) would result in activation of those T cells and autoimmunity. The activation of complement is in response to foreign pathogens and not the cause of autoimmunity.

10. **c.** An HIV-infected individual is more susceptible to a variety of infections because the virus destroys T_H cells, which are essential for mounting an effective immune response. HIV does not bind to pathogens; it binds to T_H cells. HIV does not destroy B cells.

11. **d.** DNA rearrangements occur in B cells and result in each B cell expressing one kind of unique antibody. DNA rearrangements also occur in T cells to generate T cell receptors. This rearrangement does not destroy essential T cell genes. Memory B cells are cells in which the DNA rearrangement has already occurred.

12. **d.** MHC proteins present nonself antigens and self antigens to T cells. T cells that react with self antigens are destroyed, preventing autoimmunity. The cytokines of T cells (and not MHC proteins) activate B cells. MHC proteins do not generate antibodies.

Application Answers

1. If the bacterial infection is not treated, the individual's immune response to this pathogen will include secreting antibodies that can cross-react with the individual's own membrane proteins on the cells of the heart valve. These antibodies will create an autoimmune response and heart disease.

2. Organ transplants are successful if the MHC proteins and other cell surface markers are similar, making the immune system more tolerant of the foreign tissue. Close relatives have more genes in common with the patient and may have more surface antigens on their cells that are similar. Organ transplants from unrelated individuals require the administration of immunosuppressive drugs so that the patient's own immune system does not reject the tissue as foreign.

3. Vaccinations work against foreign pathogens because of B and T memory cells that are made during every immune response. If these memory cells are re-exposed at a later time to the same antigen, they will mount a faster and more efficient immune response to that pathogen. T helper cells, which direct the immune response, including the synthesis of B and T memory cells, are destroyed by HIV.

4. Cytotoxic T cells target virally infected cells and some cancer cells. If these cells were eliminated, the individual would be much more susceptible to viral infections and cancer.

STUDY GUIDE

20 The History of Life on Earth

Important Concepts

Biological evolution is a change over time in the genetic composition of a population.

- Microevolutionary changes occur within the lifetime of a species.
- Macroevolutionary changes involve the occurrence of new species and evolutionary lineages. The fossil record provides evidence of such changes.

Several types of evidence have been used to estimate the age of rocks, of fossils, and of Earth itself. All the evidence indicates that both Earth and life on Earth are very old.

- Fossils, the remains of ancient organisms, are useful for establishing the relative ages of the sedimentary rocks in which they occur.
- The regular pattern of decay of radioactive isotopes provides a means of estimating the absolute age of fossils and rocks. The ratio of radioactive ^{14}C to its stable isotope, ^{12}C, is used to date fossils less than 50,000 years old. The decay of ^{40}Na to ^{40}Ar is widely used to date ancient evolutionary events.
- Sedimentary and igneous rocks can also be aged because they preserve a record of Earth's magnetic field at the time they were formed.
- Earth's geological history is divided into eras, which are subdivided into periods. The boundaries between these divisions are marked by changes in the types of fossils found in successive layers of sedimentary rock.
- The first forms of life evolved early in the Precambrian era, which lasted for more than three billion years. The Cambrian, the earliest period of the Paleozoic era, began about 543 million years ago and was marked by a sudden increase in new life forms.

The history of Earth has been marked by unidirectional and oscillatory changes, as well as by occasional collisions with extraterrestrial objects.

- Two major unidirectional changes have been the cooling of Earth and the increase in concentration of oxygen in its atmosphere. No free oxygen was present in the atmosphere until certain prokaryotes evolved the ability to use water as a source of hydrogen ions for photosynthesis. The gradual increase in oxygen concentration in the atmosphere eventually permitted the evolution of larger eukaryotic cells and of multicellular organisms.
- Earth's crust consists of solid plates floating on a liquid mantle. Convection currents in the mantle result from heat emanating from Earth's core. The currents cause continental drift, which is the gradual shift in the position of the plates and the continents they contain. The drifting of continents has had profound effects on climate, sea level, oceanic circulation, and the distributions of organisms.
- The climate of Earth has alternated between hot/humid and cold/dry conditions. Though most major climatic changes have been gradual, some have occurred within periods of 5,000 to 10,000 years or less.
- The climatic shifts caused by massive volcanic eruptions associated with continental drift are implicated in several mass extinctions, notably the one that occurred in the late Permian period at the end of the Paleozoic era.
- The mass extinction that occurred in the late Cretaceous period at the end of the Mesozoic era was probably caused by the collision of Earth with a large meteorite.

The fossil record reveals patterns in life's history.

- Because an oxygen-rich environment favors rapid decomposition, organisms that become fossils are likely either to have lived in a poorly oxygenated environment or to have been transported to such a site soon after death.
- The 300,000 known fossil species represent only a tiny fraction of the species that have ever lived. Some groups, such as hard-shelled marine animals, are much better represented in the fossil record than others.
- The fossil record shows that an organism of any specific type can predictably be found in rocks of a particular age.
- Living organisms resemble recent fossils more closely than they resemble fossils from more ancient periods.
- According to the fossil record, all species eventually become extinct.
- Contrary to the claim of scientific creationists, many fossil lineages display gradual change over time. Cases in which intermediate forms linking ancestral and descendent organisms are missing from the fossil record are probably the result either of evolution having occurred where fossils have not been discovered, or of very rapid evolution.

The earliest life on Earth appeared about 3.8 billion years ago, but the fossil record of organisms that lived in the Precambrian is fragmentary.

- The first organisms were unicellular prokaryotes. The first eukaryotes evolved about 2.5 billion years ago.
- The best Precambrian fossil deposits, dating from about 600 million years ago, contain diverse soft-bodied invertebrates, some of which may represent lineages with no living descendents.

Beginning with the Cambrian period, the Paleozoic era (543–245 mya) was marked by a dramatic increase in the diversity of life.

- During the Cambrian explosion, the three major lineages of animals separated and began to diversify. Several continental plates began to come together to form the southern landmass known as Gondwana.
- The Ordovician period was marked by a proliferation of marine filter feeders living on the sea floor. Small plants appeared in moist terrestrial habitats. At the end of this period, Gondwana moved over the south pole, massive glaciers formed, the sea level and ocean temperatures dropped, and the majority of animal species became extinct.
- The Silurian period witnessed the aggregation of northern continents, the evolution of swimming marine animals, and the appearance of the first terrestrial arthropods.
- During the Devonian period, all major groups of fishes evolved. On land, the first insects and amphibians evolved, and forests of club mosses, horsetails, and tree ferns appeared. A mass extinction of approximately three-quarters of all marine species occurred at the end of this period, possibly caused by the collision of the northern and southern continents.
- In the Carboniferous period, swamp forests largely consisting of giant tree ferns and horsetails were widespread. The fossilized remains of these plants formed coal. The period also saw the appearance of the first reptiles, which evolved from an amphibian ancestor, and of the first winged insects.
- The Permian period was marked by the formation of a single supercontinent called Pangaea. As the climate cooled, reptiles became the dominant land vertebrates, and the lineage of reptiles from which mammals later evolved became distinct. The occurrence of the most extensive of all mass extinctions brought the Permian to a close.

During the Mesozoic era (245–65 mya), the continents that formed Pangaea slowly separated to form Laurasia and Gondwana, and distinct assemblages of plants and animals evolved on each continent.

- During the Triassic period, seed plants (seed ferns and conifers) were the dominant forms of terrestrial vegetation and reptiles were the dominant vertebrates. A mass extinction occurred at the end of the Triassic.
- The Jurassic period witnessed the radiation of the dinosaurs, the evolution of flying reptiles, and the first appearance of mammals. Bony fishes became dominant in the sea.
- The Cretaceous period was marked by the complete separation of the northern and southern continents. The first flowering plants evolved from gymnosperm ancestors, and the dinosaurs continued to diversify and dominate, despite the presence of many groups of mammals. The collision of a large meteorite with Earth probably caused the extinction of the dinosaurs and of many other animal and plant lineages at the end of the Cretaceous.

Modern groups of plants and animals evolved during the Cenozoic era (65 mya–present).

- During the Tertiary period, the continents drifted toward their present positions. As the climate became cooler and drier, extensive grasslands appeared. Birds and mammals radiated extensively.
- The current geological period, the Quaternary, is divided into the Pleistocene and Holocene (recent) epoches. Modern humans evolved during the Pleistocene, which was a time of severe climatic fluctuations, including four major periods of extensive glaciation. As humans spread geographically, many species of large birds and mammals became extinct.

Rates of evolutionary change have been uneven.

- Harsh, relatively unchanging environments favor evolutionary stability, or stasis, in the species of organisms that are adapted to them. Conversely, changes in the physical or biological environment of a species often trigger rapid evolution of new traits.
- Though species have become extinct throughout the history of life, extinction rates have varied widely, being relatively low during normal times and very high during periods of mass extinction. Some mass extinctions have hit both terrestrial and marine organisms, whereas others have primarily affected either one group or the other.

The history of life on Earth reveals several patterns.

- Almost all evolutionary innovations—new features that adapt an organism to a particular way of life—represent the modification of structures preexisting in ancestral organisms.
- Life on Earth has been dominated by three major animal assemblages resulting from periods of rapid evolutionary diversification. The first was the fauna stemming from the Cambrian explosion; the second fauna dominated the later Paleozoic; and the third originated from the Triassic explosion and resulted in our present-day fauna. Organisms with new body plans appeared in the Cambrian explosion but not in the others.
- A trend favoring greater size and complexity of organisms is apparent in most evolutionary lineages. Insects, which have been small throughout their evolutionary history, are a major exception to this trend.
- Predators have evolved ever more effective means of capturing their prey; in response, prey species have evolved better defenses.

The causes of evolution that have operated in the past continue to operate, but humans are a new and dominant evolutionary agent.

- By changing Earth's vegetation, importing species into regions where they did not previously live, and manipulating the evolution of certain valuable species, humans are having an enormous influence on the future of life on Earth.

The Big Picture

- The evidence provided by nineteenth-century geologists that Earth is very ancient was crucial to Darwin's theory of evolution by natural selection.
- Radiometric dating is an excellent example of how a discovery in one field of science (nuclear physics in this instance) can lead to advances in other fields (geology and paleontology). Prior to the development of this technique, no absolute dates could be accurately assigned to any of the important geological and evolutionary events described in this chapter.
- The concept of continental drift, first widely accepted in the late 1960s, has revolutionized our understanding of geological processes and of the history of life on Earth.
- The accumulating evidence of the sudden and catastrophic effect of the impact of large meteorites and of other rapid changes (e.g., widespread volcanic eruptions) on the environment of Earth has given paleontologists a new perspective on the causes of mass extinctions and on the dynamics of the evolutionary process in general.

Study Strategies

- This chapter contains many names of geological eras and periods and many dates. Instructors vary with respect to the importance they place on memorization of this information. In general, it is best to focus on broad evolutionary patterns and trends in studying this material. Most instructors will also place more importance on your learning phylogenetic sequences (e.g., mammals evolved from reptiles, which evolved from amphibians) than on your memorizing the dates at which these events occurred.
- Making a table or chart to summarize and organize the events that occurred in each geological period is an excellent study strategy.
- Review the following tutorial and activity on the website:
  Tutorial 20.1 Evolution of the Continents
  Activity 20.1 Concept Matching

Test Yourself

Knowledge and Synthesis Questions

1. The half-life of an isotope is best defined as the
   a. time a fixed fraction of isotope material will take to change from one form to another.
   b. age over which the isotope is useful for dating rocks.
   c. ratio of one isotope species to another in a sample of organic matter.
   d. None of the above
2. Mountain ranges are ultimately the result of
   a. plates in Earth's crust that move against one another on top of a liquid mantle.
   b. climate changes and the movement of glacial ice sheets.
   c. leftover debris from ancient collisions with an asteroid or meteor
   d. the breakup of Laurasia and Gondwana.
3. Which of the following would likely be the best setting for fossilization to occur?
   a. The surf zone along a sand beach
   b. A shallow, cool swamp with good deposition rates of mud sediments
   c. The bottom of a hot, dry cave with no running water
   d. A fast-running mountain stream
4. Despite being incomplete as a whole, the fossil record is rather detailed for
   a. soft-bodied insects.
   b. cnidarians and sponges.
   c. most terrestrial animals.
   d. hard-shelled mollusks.
5. When biologists say that Earth's biota became provincialized, they mean that
   a. mass extinctions took place.
   b. continental drift did not occur.
   c. distinctive assemblages of plants and animals arose on different continents.
   d. reductions in species diversity took place.
6. All but one of the following occurred during the Ordovician period. Which statement is *not* correct?
   a. Marine filter feeders flourished.
   b. The number of classes and orders increased.
   c. Modern mammals appeared.
   d. Many groups became extinct at the end of the period.
7. Most of human evolution has occurred during the
   a. Paleozoic era.
   b. Devonian period.
   c. Quaternary period.
   d. Carboniferous period.
8. For terrestrial animals and plants, the most recent mass extinction event that occurred prior to the evolution of humans took place approximately
   a. 10 mya.
   b. 65 mya.
   c. 220 mya.
   d. 400 mya.
9. One of the main factors that distinguishes the Cambrian explosion from all others is that
   a. evolutionarily, it was the most recent explosion.
   b. many novel lineages (particularly phyla) appeared at this time, but not during other explosions.
   c. it was the time when the dinosaurs became extinct.
   d. there was a dramatic drop in species diversity, especially among marine organisms.

10. During which of the following geologic times did the most new kinds of body plans appear?
    a. Paleozoic
    b. Triassic
    c. Jurassic
    d. Cambrian

11. The event that precipitated the sudden disappearance of the dinosaurs some 65 mya may have been the result of
    a. Earth's collision with a large meteorite.
    b. slow climate changes due to planetary cooling.
    c. competition from better adapted organisms.
    d. the rise of birds and mammals.

12. Which of the following statements about evolution is *not* true?
    a. Predators have evolved more efficient methods of capturing prey over time.
    b. Species have become extinct throughout the history of life.
    c. The fossil record of whales provides a good example of gradual evolutionary change within a lineage of organisms.
    d. There is no evidence in the fossil record for periods of rapid evolutionary change.

13. Which of the following groups of organisms has undergone an overall decrease in body size during evolution?
    a. Mammals
    b. Mollusks
    c. Plants
    d. None of the above

14. Which of the following statements about patterns or processes in the evolution of life is *not* true?
    a. ^{14}C can be used to date the age of dinosaur bones.
    b. The supercontinent Pangaea formed during the Permian period.
    c. All of the major groups of animals that have species living today appeared in the Cambrian period.
    d. Rates of evolutionary change can be rapid during times of dramatic change in physical environments.

15. Which of the following pairs of organisms were *not* present on Earth in their living forms at the same time?
    a. Tree ferns and bony fish
    b. Amphibians and birds
    c. Gymnosperms and insects
    d. Humans and dinosaurs

Application Questions

1. ^{14}C decays to ^{12}C with a half-life of about 5,700 years. Suppose you find a fossil in which the amount of ^{14}C is only one-sixteenth of what one would find in a living organism with the same carbon content. Approximately how old is the fossil?

2. For each geological era, make a chart summarizing the major geological and evolutionary events that occurred in each period.

3. Scientific creationists argue that evolution cannot have occurred because transitional forms that would link fossil organisms to living species are missing. How would you refute this argument?

Answers to Questions

Knowledge and Synthesis Answers

1. **a.** Radioactive decay is measured as the time it takes for one-half the amount of a substance to spontaneously convert into another substance.
2. **a.** As the crustal plates move past each other, the friction they cause pushes rocks upward to make mountains.
3. **b.** Fossilization occurs best in areas of low oxygen concentration and rapid sedimentation, where scavengers cannot destroy the body.
4. **d.** Hard-shelled animals such as mollusks are good candidates for fossilization because their shells can withstand decay long enough to become buried. Many also tend to live in quiet, shallow waters.
5. **c.** Coming from the same root word as *province*, which is a specific area within a country or state possessing unique characteristics, provincialized biotas are geographically unique assemblages of organisms produced by a variety of physical and biological isolating factors.
6. **c.** The modern mammals did not appear until the Tertiary, more than 350 million years after the Ordovician.
7. **c.** The Pleistocene epoch, which is part of the Quaternary, was the scene of most hominid evolution.
8. **b.** The mass extinction at the end of the Cretaceous, 65 mya, was the most recent affecting terrestrial life before the evolution of humans.
9. **b.** The Cambrian explosion occurred when many open habitats were available for exploitation. Thus the largest numbers of new phyla were produced then, with later explosions causing increase in diversity only within already existing phyla.
10. **d.** For the reasons outlined in the previous answer, the Cambrian time is the correct answer.
11. **a.** Although there is still some debate, most paleontologists believe that a large meteorite's collision with Earth so altered conditions that the dinosaurs rapidly succumbed during what we know as the great Cretaceous mass extinction.
12. **d.** Evolutionary rates apparently have changed through time. Long periods of stasis may be interrupted by periods of rapid change.
13. **d.** Looking at evolution from an overall viewpoint, paleontologists find that most groups of organisms, with the major exception of insects, have increased in size.
14. **a.** Carbon dating is generally only reliable for fossils less than 50,000 years old.
15. **d.** Dinosaurs disappeared at the end of the Cretaceous, over sixty million years before humans evolved.

Application Answers

1. If only $\frac{1}{16}$ of the ^{14}C remains, then four ^{14}C half-lives have passed since the fossil was formed ($\frac{1}{2} \times \frac{1}{2} \times \frac{1}{2} \times \frac{1}{2} = \frac{1}{16}$). Since the half-life of ^{14}C is roughly 5,700 years, the fossil must be about 22,800 years old ($4 \times 5{,}700 = 22{,}800$).

2. See Table 20.1 in the text.

3. Because fossils are likely to form only under special conditions, they represent a very incomplete record of the history of life. Thus it not surprising that many transitional forms are missing from the fossil record. Furthermore, in cases in which evolution occurred very rapidly, it is even less likely that transitional fossil forms would be found. Nevertheless, contrary to the claim of scientific creationists, there are many cases in which the existence of many intermediate forms in the fossil record enables paleontologists to reconstruct the evolutionary history of a group of organisms in great detail. The fossil record of whales (Figure. 20.7) is a good example.

21 The Mechanisms of Evolution

Important Concepts

An adaptation is a trait that helps its bearer survive and reproduce; the term also refers to the evolutionary process that produces such traits.

- Darwin proposed that adaptations are produced by the process of natural selection.
- The key concepts of natural selection are:
  1. All populations have the capacity to increase in number exponentially.
  2. Since populations rarely increase rapidly, high death rates must be the rule in nature, and each individual organism must face a struggle for existence.
  3. All populations show variations among individuals with regard to numerous inherited characteristics, some of which may affect their chances of surviving or reproducing.
  4. Those individuals whose heritable traits enable them to survive and reproduce more successfully than others in the population will tend to pass those traits on to their offspring.
- In short, natural selection is differential reproductive success among individuals of a population.
- The rediscovery of Mendel's laws of inheritance led to the development of population genetics, which applies these laws to entire populations in order to understand the genetic basis of evolutionary change, including adaptation.

Evolution is any genetic change in a population from generation to generation.

- The genotype is the genetic constitution of a particular trait of the individual. Evolution occurs when individuals with different genotypes survive or reproduce at different rates.
- Agents of evolution actually act on the phenotype, which is the physical expression of the genotype.
- The gene pool of a population is the sum of all the alleles (alternative versions of a gene) found within it. Although a single diploid individual can have at most two different alleles for any gene, a population may contain many alleles. The gene pool contains all of the variation that results in individuals with differing phenotypes on which agents of evolution act.
- Fitness is the contribution of a genotype (or phenotype) to the composition of later generations, relative to the contribution of alternative genotypes (or phenotypes). The fitness of a genotype is determined by the average rates of survival and reproduction of individuals with that genotype, as compared to individuals with other genotypes.
- Differences in fitness among genotypes lead to changes in allele frequencies in the gene pool of later generations. Changes in allele frequencies from one generation to the next constitute evolution.
- Studies of both domestic and wild species show that virtually all populations contain a certain amount of genetic variation.
- Allele frequencies measure the amount of genetic variation in a population, whereas genotype frequencies show how this variation is distributed among its members. Populations that have the same allele frequencies may nevertheless differ in their genotype frequencies.
- The sum of all allele frequencies at a locus is equal to 1, as is the sum of all genotype frequencies. For a locus with two alleles, p and q represent the frequencies of the dominant and recessive alleles, respectively, and thus $p + q = 1$. Biologists estimate allele frequencies by measuring numbers of alleles in a sample of individuals from the population.

A population that is not changing genetically is at Hardy–Weinberg equilibrium.

- To be at Hardy–Weinberg equilibrium, a population must meet five conditions:
  1. Mating is random.
  2. The population is large.
  3. There is no migration between populations.
  4. Mutation can be ignored.
  5. Natural selection is not acting on the population.
- If these conditions are met, allele frequencies at a locus remain the same from generation to generation. Moreover, after one generation of random mating, the genotype frequencies will remain in the proportions $p^2 + 2pq + q^2 = 1$, where p^2, $2pq$, and q^2 represent the frequencies of the homozygous dominant, heterozygous, and homozygous recessive genotypes, respectively.

- Though populations in nature rarely meet the conditions of the Hardy–Weinberg equilibrium, it is important because deviations from it show that evolution is occurring. Moreover, the pattern of deviations is useful in identifying the agents of evolutionary change operating on the population

Microevolution refers to small-scale changes in the genetic makeup of a population produced by evolutionary agents that cause deviations from the Hardy–Weinberg equilibrium.

- The ultimate source of genetic variation in a population is germ-line (heritable) mutation. Because mutations are random changes in genetic material, most are harmful or neutral, but it is the environment that determines whether a particular mutation is disadvantageous or adaptive. Though mutations cannot be totally prevented, their frequency is usually so low as to cause little deviation from Hardy–Weinberg expectations.
- Gene flow occurs when individuals migrate from one population to another and breed in their new location. Gene flow can add new alleles to a population's gene pool, or change the frequencies of alleles already present, or both.
- Random genetic drift is caused by chance events that alter allele frequencies in a population. Though in theory genetic drift acts on all populations, it has its greatest impact on those that are small. In such populations, it may even cause allele frequencies to shift in a direction that is opposed by the action of natural selection and other evolutionary agents.
- During a population bottleneck, when a large population is severely reduced in size, allele frequencies may shift drastically and genetic variation may be reduced as a result of genetic drift.
- The founder effect occurs when a few individuals originate a new population. As in the case of a population bottleneck, some alleles found in the source population will be missing from the founding population, and others will occur with altered frequencies.
- Nonrandom mating occurs when individuals mate assortatively with others of the same genotype, or of a different genotype. The effect of nonrandom mating is to cause either fewer or more heterozygotes to be found in a population than would be expected in a population in Hardy–Weinberg equilibrium. The effect of self-fertilization, another form of nonrandom mating, is to reduce the frequency of heterozygotes. In any type of nonrandom mating, the allele frequencies remain the same despite the changes in genotype frequencies.
- Natural selection, unlike other agents of evolution, adapts organisms to their environment. In cases in which the distribution of a phenotype approximates a bell-shaped curve because it is controlled by many gene loci, selection can produce any one of three results:
  1. Stabilizing selection reduces variation in the population by favoring average individuals.
  2. Directional selection changes the mean value for a trait by favoring individuals that vary in one direction.
  3. Disruptive selection, by favoring both extremes, leads to a population with two peaks in the distribution of the trait.

Biologists use several techniques to study microevolution.

- Natural selection and other evolutionary agents can be studied by field measurements, by experimental alteration of genotypes and phenotypes, and by computer simulations.

Most populations maintain substantial genetic diversity.

- Genetic drift, stabilizing selection, and directional selection tend to lessen the genetic variation of a population.
- By generating new combinations of alleles, sexual reproduction amplifies existing genetic variation.
- Neutral mutations—that is, mutations that do not affect the fitness of an individual—tend to accumulate and thereby increase the genetic variation in a population.
- Geographically distinct subpopulations of a population often vary genetically because they are adapted to different environments.
- Frequency-dependent selection, in which the less common genotype or phenotype is favored by natural selection, preserves variation as a polymorphism.

Genotypes do not uniquely determine phenotypes.

- If one allele is dominant to another, two genotypes can determine the same phenotype. For example, *AA* and *Aa* may be phenotypically identical.
- Phenotypic variability results from the interaction of genetic and environmental factors during development. This phenotypic plasticity is often adaptive and may evolve through natural selection.

There are constraints on the evolutionary process.

- Evolutionary changes, even major ones, generally are based on modifications of previously existing traits.
- Evolutionary theory requires that all modifications be beneficial in every generation. For example, a population cannot temporarily become less adapted in the course of evolving a trait with a new function.

Studies of short-term microevolutionary processes do not allow us to predict the long-term course of evolution.

- Macroevolution is strongly influenced by events that occur so infrequently or so slowly that they are rarely observed during microevolutionary studies. To understand the long-term course of evolution, biologists seek evidence of rare events and search for trends in the fossil record.

The Big Picture

- Most biologists rank Darwin's *The Origin of Species* as the most significant book ever written about their subject. Its importance is twofold. First, it presents a vast amount of evidence for the fact of evolution. Second, it describes a mechanism—natural selection—to explain evolutionary change. Evolution through natural selection remains perhaps the most important unifying concept in biology.

- Population genetics unites the concepts of Darwin with the insights into the mechanisms of heredity provided by Mendel. It enables biologists to study evolution with quantitative rigor.
- The loss of genetic variation that occurs during a population bottleneck is a major concern for biologists working to preserve endangered species. As the discussion of the northern elephant seal makes clear, genetic uniformity in a population renders it more vulnerable to extinction through exposure to a new disease or other sudden environmental change.

Common Problem Areas

- A common mistake made by students in attempting to solve Hardy–Weinberg problems is to use the wrong equation. For example, if you are given phenotypic frequencies for a trait that is at genetic equilibrium and shows dominance, remember that the frequency of the recessive phenotype (and genotype) is equal to q^2, not q. By taking the square root of q^2, you can obtain the frequency of the recessive allele and can then determine the frequency of the dominant allele by subtraction ($p = 1 - q$).
- It is important to bear in mind that for any gene locus with two alleles, the allele frequencies must total unity (i.e., it is always the case that $p + q = 1$). The equation that specifies genotype frequencies ($p^2 + 2pq + q^2 = 1$) is only valid for a population at genetic equilibrium.

Study Strategies

- Population genetics is learned best by doing problems, so be sure you can solve problems similar to those presented in the "Test Yourself" section of this chapter of the *Study Guide*. Your instructor may provide a worksheet with additional examples.
- Review the following tutorials on the website:
  Tutorial 21.1 Natural Selection
  Tutorial 21.2 Hardy-Weinberg Calculator

Test Yourself

Knowledge and Synthesis Questions

1. Evolution occurs at the level of
   a. the individual genotype.
   b. the individual phenotype.
   c. environmentally based phenotypic variation.
   d. the population.

2. What does natural selection act upon?
   a. The gene pool of the species
   b. The genotype
   c. The phenotype
   d. Multiple gene inheritance systems

3. Suppose a particular species of flowering plant that lives only one year can produce red, white, or pink blossoms, depending on its genotype. Biologists studying a population of this species count 300 red-flowering, 500 white-flowering, and 800 pink-flowering plants in a population. When the population is censused the following year, 600 red-flowering, 900 white-flowering, and 1000 pink-flowering plants are observed. Which color has the highest fitness?
   a. Red
   b. White
   c. Pink

4. Which choice would form a statement about Mendelian populations that is *not* true? A Mendelian population must
   a. consist of members of the same species.
   b. have members that are capable of interbreeding.
   c. show genetic variation.
   d. have a gene pool.

5. In comparing several populations of the same species, the population with the greatest genetic variation would have the
   a. greatest number of genes.
   b. greatest number of alleles per gene.
   c. greatest number of population members.
   d. largest gene pool.

6. The ability to taste the chemical PTC (phenylthiocarbamide) is determined in humans by a dominant allele *T*, with tasters having the genotypes *Tt* or *TT* and nontasters having *tt*. If you discover that 36 percent of the members of a population cannot taste PTC, then according to the Hardy–Weinberg rule, the frequency of the *T* allele should be
   a. 0.4.
   b. 0.6.
   c. 0.64.
   d. 0.8.

7. A gene in humans has two alleles, *M* and *N*, that code for different surface proteins on red blood cells. If you know that the frequency of allele *M* is 0.2, according to the Hardy–Weinberg rule, the frequency of the genotype *MN* in the population should be
   a. 0.16.
   b. 0.32.
   c. 0.64.
   d. 0.8.

8. If the frequency of allele *b* in a gene pool is 0.2, according to the Hardy–Weinberg rule, the expected frequency of the genotype *bbb* in a triploid ($3n$) plant species would be
   a. 0.008.
   b. 0.04.
   c. 0.08.
   d. 0.2.

9. A small, isolated population would most likely be subject over time to
   a. assortative mating.
   b. a founder effect.
   c. genetic drift.
   d. gene flow.

10. Allele frequencies for a gene locus are *least* likely to be changed by
    a. assortative mating.
    b. the founder effect.
    c. mutation.
    d. gene flow.

11. Select all of the following evolutionary agents that produce nonrandom changes in the genetic structure of a population.
    a. Self-fertilization
    b. Population bottlenecks
    c. Mutation
    d. Natural selection

12. The following graph shows the range of variation among population members for a trait determined by multiple genes.

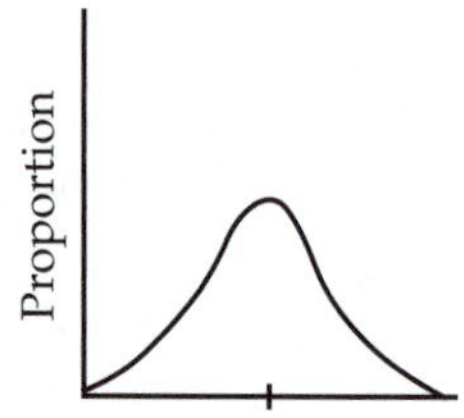

If this population is subject to *stabilizing selection* for several generations, which of the distributions (*a–d*) is most likely to result?

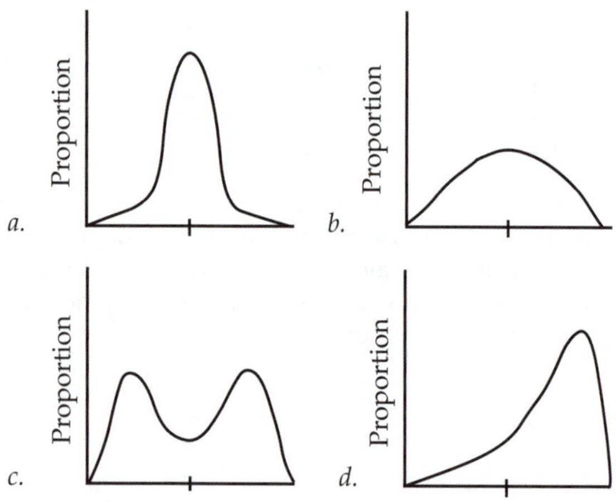

13. In areas of Africa where malaria is prevalent, many human populations exist in which the phenotypes of normal sickle-cell trait (the result of the heterozygous genotype), and severe sickle-cell anemia are constant, despite the fact that sickle-cell anemia frequently causes death at an early age. This is an example of
    a. the founder effect.
    b. a stable polymorphism.
    c. mutation.
    d. assortative mating.

Application Questions

1. Discuss the main use of the Hardy–Weinberg rule in evolutionary biology.

2. The Hardy–Weinberg expression is an expansion of a binomial or $(p + q)^2 = p^2 + 2pq + q^2$. In this equation, the left-hand side represents __________ frequencies, and the right-hand side represents __________ frequencies.

3. In a population with 600 members, the numbers of individuals of three different genotypes are $A1A1 = 350$, $A1A2 = 100$, $A2A2 = 150$.
    a. What are the genotype frequencies in this population?
        1. $A1A1$ =
        2. $A1A2$ =
        3. $A2A2$ =
    b. What are the allele frequencies in this population?
        1. $A1$ =
        2. $A2$ =
    c. What would the expected genotype frequencies be if this population were in genetic equilibrium?
        1. $A1A1$ =
        2. $A1A2$ =
        3. $A2A2$ =
    d. Is this population in genetic equilibrium? Explain.

4. The following data were collected in a study in which dark and light moths of the peppered moth (*Biston betularia*) were released and later recaptured in several different areas (*a,b,c,* and *d*). Based on these data, in which area is the light phenotype most advantageous?

5. Peppered moths are subject to predation by birds that eat them as the moths rest on bark surfaces during the day. Assuming that moths that are more camouflaged will be preyed on less, what can you say about the color of bark surfaces in area *d*?

Answers

Knowledge and Synthesis Answers

1. **d.** Evolution is defined as changes in the genetic structure of a population over time, so evolution occurs at the level of the population.

2. **c.** Natural selection acts on phenotypes, not genotypes. For example, a harmful recessive allele is "invisible" to natural selection when it occurs in a heterozygote, where its harmful effect is masked by the dominant allele.

3. **a.** Fitness measures the relative contribution of a genotype or phenotype to subsequent generations. The red-flowering plants, which doubled in number, had the greatest percentage increase of any of the plants, and thus had the highest fitness.

4. **c.** Although most populations show genetic variation, this condition is not part of the definition of a Mendelian population.

5. **b.** Genetic variation is related to the number of different alleles per gene. Regarding choice a, recall that all species members have the same number of genes. Population size per se has little to do with genetic variation, so choices c and d are also incorrect.

6. **a.** Recall that q^2 = the frequency of the *tt* genotype, so = the frequency of the *t* allele or = 0.6, and $1 - 0.6 = 0.4$ is the frequency of the *T* allele.

7. **b**. Since $p = 0.2$ and therefore, $q = 1 - 0.2 = 0.8$, the frequency of the *MN* genotype is $2pq$ or $2(0.2)(0.8) = 0.32$.
8. **a**. The probability of one allele *b* in a genotype is equal to its frequency, or 0.2, so the probability of 3 *b* alleles in a genotype (*bbb*) would be $(0.2)^3$ or 0.008.
9. **c**. Genetic drift is always most prevalent in small populations. The founder effect may be important in the act of creating a small population, but after it is established, random variation due to small size is called genetic drift.
10. **a**. Unlike the founder effect, mutation and gene flow, all of which change allele frequencies in a population, assortative mating only causes a deviation from the frequency of heterozygotes predicted by Hardy–Weinberg equilibrium.
11. **a**. and **d**. Self-fertilization leads to increased numbers of homozygous individuals and natural selection is also nonrandom.
12. **a**. Stabilizing selection results when individuals that are intermediate in phenotype make a larger contribution to future generations than individuals of more extreme phenotype. This leads to reduced variation for the trait and causes the curve to be higher and narrower. Curve *b* shows greater variation, curve c would result from disruptive selection, and curve *d* from directional selection.
13. **b**. Polymorphism in a population is the existence of two or more alleles at a particular gene locus. If phenotypes are stable through time, then the underlying alleles will also be constant. In this instance, the polymorphism is stable because malaria is a significant cause of mortality in some parts of Africa, and heterozygotes have greater resistance to this disease than individuals with a "normal" phenotype. Thus the superior fitness of the heterozygotes maintains both alleles in the population.

Application Answers

1. Most populations do not meet the Hardy–Weinberg conditions, and the main use of the rule is to determine if evolutionary change is occurring within a population.
2. allele, genotype
3. a. $A1A1 = 350/600 = 0.58$
   $A1A2 = 100/600 = 0.17$
   $A2A2 = 150/600 = 0.25$

   b. $A1 = (350 + 50)/600 = 0.67$
   $A2 = (150 + 50)/600 = 0.33$

   c. $A1A1 = p^2 = (0.67)^2 = 0.45$
   $A1A2 = 2pq = 2(0.67)(0.33) = 0.44$
   $A2A2 = q^2 = (0.33)^2 = 0.11$

   d. No. The observed genotypic frequencies differ from those predicted by the Hardy–Weinberg rule using the observed allele frequencies for *p* and *q*. The population is *not* in genetic equilibrium.
4. Area *c*. Relatively more light moths are recaptured in this area than dark moths, indicating that this phenotype has greater fitness.
5. Dark moths seem to survive best in area *d*. If the increased survival of the moths is due to cryptic coloration, then the bark surfaces in that area should also be dark.

STUDY GUIDE

22 Species and Their Formation

Important Concepts

Species are the fundamental units of evolutionary diversification and of the biological classification system.

- The morphological concept of species used by early biologists grouped organisms into species on the basis of their appearance. For some groups of organisms, morphological similarity is still the major criterion.
- The biological species concept widely used by modern biologists regards a species as a genetic unit, the members of which can interbreed among themselves but are unable to interbreed successfully with members of other such units because of their genetic differences.
- Determining whether two populations constitute different species may be difficult because speciation is frequently a gradual process.

Speciation is the process by which one species splits into two species.

- Evolutionary change can occur without speciation. A single lineage may change through time without diverging into two species.
- Speciation requires that the gene pool of the original species divide into two isolated gene pools. After separation, the agents of evolution cause sufficient genetic differences to accumulate in the populations for them to be unable to interbreed successfully if they come together again.
- In allopatric speciation, the population is initially divided by a geographic barrier. If the barrier results from a geological or climatic change, the two isolated populations are often large and genetically similar but diverge because the places in which they live are, or become, different. Alternatively, separation may result from a founder event, in which a relatively small group of individuals crosses a barrier and founds a new population. Because it is small, the founding population will usually lack many of the alleles found in the parent population. Natural selection and other agents of evolution will then cause additional genetic divergence of the populations.
- The effectiveness of a barrier in preventing gene flow depends on the size and mobility of the species.
- Evidence suggests that allopatric speciation is the most common mechanism of speciation among animals.
- Sympatric speciation occurs without geographic subdivision of the gene pool of the original species. Polyploidy, an increase in the number of sets of chromosomes, produces new species because the polyploid organisms cannot interbreed with members of the parent species. Polyploid species that have a single ancestor are called autopolyploids, whereas those that have resulted from the hybridization of two species are referred to as allopolyploids. Speciation by polyploidy is very common in plants.
- Among animals, sympatric speciation is rare, but it may result from individual choices of mating sites and habitats.
- In parapatric speciation, strong environmental differences prevent gene flow among individuals living in adjacent habitats even though a geographic barrier is absent. The frequency of parapatric speciation is unclear.

Speciation requires that reproductive isolating mechanisms evolve to prevent previously allopatric populations from exchanging genes if and when they become sympatric.

- Prezygotic barriers prevent members of different species from mating. The four prezygotic isolating mechanisms are:
  1. spatial (the two species live and mate in different locations);
  2. temporal (they mate at different times);
  3. mechanical (their reproductive organs are structurally incompatible); and
  4. gametic (the sperm and egg are chemically incompatible).
- Postzygotic barriers prevent effective gene flow between species even if mating occurs. The four postzygotic isolating mechanisms are:
  1. hybrid zygote abnormality (the hybrid embryo develops abnormally);
  2. hybrid infertility (the hybrids are viable but sterile);
  3. low hybrid viability (hybrid offspring survive less well than nonhybrid offspring);
  4. absence or sterility of one sex.
- A hybrid zone may form where two populations come into contact if complete reproductive isolation failed to develop during the period of allopatry.
- If two species can hybridize, they are very similar genetically, but even species that do not hybridize may display few genetic differences.

Several factors affect the rate of speciation of different evolutionary lineages.

- Characteristics of a species that make it prone to undergo speciation include membership in a large evolutionary lineage, an extensive range, relative immobilility, complex behavior, and a short generation time.
- Periods of environmental change favor speciation.

In an evolutionary radiation, the speciation rate greatly exceeds the extinction rate within a lineage, and it gives rise to many daughter species.

- The native biota of the Hawaiian islands illustrates that populations colonizing environments that have relatively few species are particularly likely to produce evolutionary radiations.
- A lineage may radiate if it acquires a new adaptation that enables it to exploit the environment in new and varied ways.

The Big Picture

- The significance of speciation is that it is the process that has produced the millions of life forms—each adapted to a particular environment and way of life—that constitute life on Earth.
- Archipelagos such as the Galápagos and Hawaiian Islands have been called natural laboratories of evolution. Studies of the evolutionary radiations of the Galápagos finches and of several Hawaiian groups, such as *Drosophila* and the silverswords, have provided crucial insights to evolutionary biologists since the time of Darwin.
- Chapter 58, "Conservation Biology," discusses the reasons for the rapid pace of human-caused species extinctions, including the disappearance of many members of groups that have undergone evolutionary radiations, such as Australian marsupials and Hawaiian honeycreepers. It also describes some strategies for the preservation of Earth's precious biological diversity.

Study Strategies

- Since most instructors consider the heart of this chapter to be the discussions of allopatric (geographic) speciation and reproductive isolating mechanisms, you should pay particular attention to these topics.
- Review the following tutorial and activity on the website:
  Tutorial 22.1 Speciation Mechanisms
  Activity 22.1 Concept Matching

Test Yourself

Knowledge and Synthesis Questions

1. Select all the correct answers that complete the following sentence: It is difficult to apply the biological species concept to groups of organisms that
   a. are asexual.
   b. produce hybrids only in captivity.
   c. show little morphological diversity.
   d. exist only in the fossil record.
2. Which of the following statements about geographic speciation is *not* true?
   a. Geographic speciation can sometimes involve small populations.
   b. Geographic speciation only occurs in species that are widely distributed.
   c. Geographic speciation always involves a physical barrier that interrupts gene flow.
   d. Geographic speciation can sometimes involve chance events.
3. The activities of a mining company result in deposition of a new soil type within the range of a widespread plant species. Which of the following phenomena is likely to occur as a result?
   a. Geographic speciation
   b. Sympatric speciation
   c. Parapatric speciation
   d. Allopatric speciation
4. Plant species *A* ($2n$=20) and *B* ($2n$=14) hybridize to produce species C, an allopolyploid. How many chromosomes would be present in the cells of species *C*?
   a. 17
   b. 28
   c. 34
   d. 40
5. Which type of speciation is most common among flowering plants?
   a. Geographic
   b. Sympatric
   c. Parapatric
   d. Allopatric
6. Which of the following would *not* be considered an example of a prezygotic reproductive isolating mechanism?
   a. One bird species forages in the tops of trees for flying insects while another forages on the ground for worms and grubs.
   b. The males of one species of moth cannot detect and respond to the sex attractant chemicals produced by the females of another species.
   c. Sperm of one species of sea urchin are unable to penetrate the egg plasma membrane of another species.
   d. Mosquitos of one species are active in foraging and searching for mates at dusk, whereas those of another species are active at dawn.
7. Which of the following factors would *not* be expected to increase the rate of speciation in a group of organisms?
   a. A species range consisting of fragmented populations
   b. A diet consisting of food items whose abundance varies widely
   c. High birth rates
   d. Increased behavioral complexity
8. Which of the following is *not* a suggested reason for the evolutionary radiation of silverswords on the Hawaiian archipelago?
   a. Water is an effective barrier for many organisms.

b. Because islands are small compared with mainland areas, you would expect more species to develop there.
c. Frequently, competition is reduced on islands.
d. More ecological opportunities exist on islands.

9. Studies of species of Hawaiian *Drosophila* show that
   a. sympatric speciation via polyploidy is common in insects.
   b. few genes need be involved in establishing reproductive isolation.
   c. species with nonoverlapping ranges can be very similar.
   d. a geographic barrier is not always necessary for the establishment of reproductive isolation.

10. Which of the following statements about speciation is *not* true?
    a. A small founding population can be involved in speciation.
    b. Speciation always involves interruption of gene flow between different groups of organisms.
    c. The rate of speciation can vary for different groups of organisms.
    d. Speciation cannot occur in a single generation.

11. Which of the following observations constitutes conclusive evidence that two overlapping populations that had been geographically separated have *not* diverged into distinct species?
    a. Matings between members of the two populations produce viable hybrids.
    b. A stable hybrid zone exists where their ranges overlap.
    c. Interbreeding is common between members of the two populations.
    d. None of the above.

Application Questions

1. Discuss the conditions on the Galápagos Islands that led to the evolution of the birds known as Darwin's finches.
2. You look at several polymorphic loci in land snails removed from the areas A through E of four adjacent city blocks, as shown in the following map.

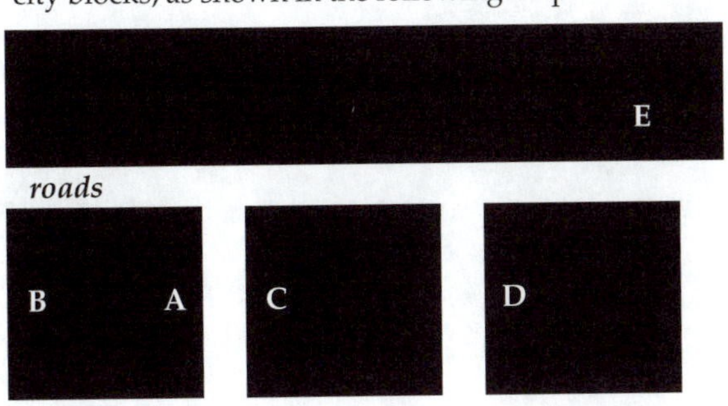

Snails from area A should be genetically most like individuals from area __________ and least like individuals from area __________.

3. Discuss what is known about the evolutionary radiations on islands, based on studies of Hawaiian silverswords and tarweeds.

Answers

Knowledge and Synthesis Answers

1. **a.** and **d.** The key criterion of a biological species is that its members be reproductively isolated from other such groups. This criterion is impossible to evaluate in asexual and fossil species.
2. **b.** A wide distribution is not a requisite for geographic speciation.
3. **c.** Differences in soil type and other environmental discontinuities frequently lead to parapatric speciation.
4. **c.** The hybrid offspring of a cross between between species *A* (n=10) and *B* (n=7) would be expected to have 17 chromosomes, but would be sterile because of the inability of the chromosomes to synapse properly during meiosis. If the chromosome number were later to double (perhaps owing to nondisjunction during meiosis), Species *C* would be an allopolyploid with 34 chromosomes. It would be able to carry out meiosis normally because homologous pairs of chromosomes would be present.
5. **b.** Sympatric speciation is most common among flowering plants. It has been estimated that about 70 percent of all flowering plant species are polyploid.
6. **a.** Provided that the two species are active in the same locality at the same time, a difference in the habitat in which they forage would not in itself be a barrier to interbreeding. All of the other choices describe reproductive barriers that would act prior to fertilization.
7. **c.** Birth rates per se do not seem to affect the rate of speciation in organisms. All other factors have been shown to increase speciation rates in the lineages of some organisms.
8. **b.** Actually, biogeographers have found that larger islands tend to have more species than smaller islands, so you might expect the reverse effect.
9. **b.** Studies of the Hawaiian *Drosophila* show that these species have only slight genetic differences, despite being morphologically diverse.
10. **d.** New species formed by polyploidy can arise in a single generation.
11. **d.** Interbreeding, production of viable hybrids, and establishment of a hybrid zone do not necessarily mean that speciation is not complete. If, on the other hand, the hybrids were successful, fertile, and bred freely with members of both original populations, their gene pools would merge and you would conclude that speciation had not taken place.

Application Answers

1. The relatively great distance between the Galápagos Islands and the South American mainland and also between each of the islands in the archipelago insured that once immigrants had arrived on an island, they would be genetically isolated for a substantial period of time. Also, because the islands differ greatly in climate

and vegetation, the resident birds were subject to different selection pressures. This, in combination with reduced gene flow between the islands, led to a rapid evolutionary radiation of finches.

2. Snails from area A would be genetically most like individuals from area B and least like individuals from area D. Snails from area A are in the same block of habitat as area B snails, but must cross two barriers (roads) to interbreed with area D snails. They need only cross one road to interbreed with area C or E snails.

3. Studies of the silverswords of the Hawaiian archipelago show that taxa that evolve on islands frequently show great morphological diversity because of the reduced competition that immigrants encounter on islands. Thus Hawaiian silverswords have evolved tree- and shrub-like species because there were few resident tree and shrub species with which they had to compete, unlike their mainland tarweed relatives.

STUDY GUIDE

23 Reconstructing and Using Phylogenies

Important Concepts

Phylogenetic trees show the evolutionary patterns of life on earth.

- Phylogenetic trees display the order in which lineages are split. They are constructed by analyzing evolutionary changes in the traits of organisms.
- Homologous traits are keys to reconstructing phylogenetic trees because they are features that are shared by members of a lineage owing to their descent from a common ancestral trait. A trait that differs from its ancestral form is called a derived trait.
- Homoplastic traits create confusion when reconstructing the evolutionary history of a lineage because they are features that are similar for some reason other than descent from a common ancestral trait.
- Three processes generate homoplastic traits: convergent evolution, parallel evolution, and evolutionary reversals.
- In convergent evolution, features that were independently evolved become superficially similar.
- In parallel evolution, developmental processes cause similar traits to occur in distantly related organisms.
- In an evolutionary reversal, a character reverts from a derived state to an ancestral one.
- The focal group is the assemblage of organisms whose phylogeny is to be determined. Any feature of the organisms in the focal group that will be used to analyze its evolutionary history is called a character (e.g., flower color); a trait is a particular form of the character (e.g., white flowers). To determine the group's evolutionary history, systematists must distinguish between ancestral and derived traits, and between homologous and homoplastic traits.
- One method of distinguishing ancestral and derived traits is to compare the focal group to an outgroup—a lineage that is closely related to the focal group but which split off from the lineage of the focal group below its base on the evolutionary tree. Ancestral traits are expected to be found in both the focal group and the outgroup, whereas derived traits should occur only in the focal group.
- Inclusion of as many traits as possible in the phylogenetic analysis helps to detect homoplasies because they may be inconsistent with the single phylogenetic pattern revealed by the analysis of the other traits.
- In reconstructing a phylogeny, systematists assume that organisms that share unique derived traits have a common ancestor not shared with organisms lacking those traits.
- A phylogenetic tree does not describe the ancestors or date the splits between lineages; it simply shows the sequential order of the splits.
- In reconstructing phylogenies, systematists use the parsimony principle, which states that one should prefer the simplest hypothesis capable of explaining the known facts. In practice, this means that the phylogenetic reconstruction that assumes the smallest number of evolutionary changes is the best hypothesis. Complex computer programs are now employed by systematists to determine the shortest possible phylogenetic tree—that is, the one with the fewest homoplasies. In some cases, several phylogenies may be equally likely on the basis of the parsimony principle. A consensus tree based on all the most parsimonious trees retains the branch points (nodes) common to all the trees and leaves the others "unresolved."
- The maximum likelihood method uses computer analysis for the reconstruction of phylogenies based on molecular data. It takes into account differences in the likelihood of different kinds of nucleotide-substitution mutations.

To reconstruct phylogenies accurately, systematists consider the widest possible array of characteristics, including molecular, structural, developmental, and behavioral traits

- Morphology—the sizes and shapes of body parts—is an important source of traits for phylogenetic analysis. The morphology of fossils is particularly useful in helping to distinguish ancestral and derived traits.
- Structures in early developmental stages sometimes reveal evolutionary relationships that are not evident in adults.
- Like morphological characters, molecules are heritable characteristics that may diverge among lineages. Thus, comparisons of the amino acid sequences of homologous proteins, or of the base sequences of comparable regions of DNA provide excellent evidence of evolutionary relationships.
- Phylogenetic trees provide important information to biologists trying to answer many kinds of questions. One important use of phylogenetic trees is to determine how

many times a particular trait has evolved within a lineage. In combination with fossil evidence, phylogenetic trees can also help to determine when lineages diverged.

Classification systems are important for several reasons.

- They improve our ability to explain relationships among organisms.
- By grouping organisms into categories based on shared features, they are an aid to memory, and provide a basis for predictions about characteristics to be found among members of a group.
- They provide unique, universally used names for organisms.

Taxonomy is the theory and practice of classifying organisms.

- A taxon is a group of organisms that is treated as a unit in a biological classification system.
- The system of binomial nomenclature developed by Linnaeus in 1758 assigns each species two names, one identifying the species itself, and the other the genus to which it belongs. The generic name (always capitalized) is followed by the species name (lower case) and both are italicized.
- In the Linnaean system of classification, species are grouped into higher order taxons. The hierarchy of taxons, ranked from most to least inclusive, is: kingdom, phylum, class, order, family, genus, species. Thus, a genus includes one or more species, a family comprises one or more genera, and so forth.
- All taxonomic categories above the species level are mental constructs. There are no rigorous criteria by which to decide, for example, whether a particular lineage should be ranked as an order or a family.

Biological classification systems are designed to express evolutionary relationships among organisms.

- Although early classification systems were nonevolutionary, most present-day taxonomists believe that evolutionary relationships should be the basis for classifying organisms.
- A monophyletic taxonomic group (also called a clade) contains all the members of a lineage descended from a common ancestor and no other organisms. Ideally, all taxa should be monophyletic.
- A polyphyletic taxon consists of members that do not all share a common ancestor. Because they do not reflect phylogenetic relationships, such taxonomic units are eliminated by systematists as soon as they are discovered.
- A paraphyletic taxon includes some, but not all, descendents of a particular ancestor. Many systematists believe that some paraphyletic taxa should be retained to emphasize that members of some lineages have undergone rapid evolutionary change and diversification.
- In the future, the reconstruction of phylogenies will depend increasingly on computer analysis and on molecular methods for the determination of biochemical and cellular traits.

The Big Picture

- The parsimony principle, which holds that one should prefer the simplest hypothesis capable of explaining the known facts, is fundamental not only to the reconstruction of phylogenies but also to every other field of scientific research.
- The concept of evolution revolutionized taxonomy. Prior to Darwin's time, Linnaeus and others had developed "natural" systems of classification based primarily on similarity of morphology. Modern biologists realize that similarities of organisms (other than homoplasies) are the result of descent from a common ancestor and believe that a truly natural system of classification should reflect evolutionary relationships.
- Knowing that organisms are evolutionarily related enables biologists to make predictions about their characteristics. As this chapter points out, this knowledge can provide important hints in the search for organisms with valuable properties, such as the ability to produce medically useful drugs.

Study Strategies

- Here is a mnemonic for the hierarchy of taxa (kingdom, phylum, class, order, family, genus, species) in the Linnaean system of classification: "**K**indly **P**rofessors **C**annot **O**ften **F**ail **G**ood **S**tudents."
- Review the following activities on the website:
  Activity 23.1 Constructing a Cladogram
  Activity 23.2 Types of Taxa

Test Yourself

Knowledge and Synthesis Questions

1. Which of the following would not be expected to result in homoplasy?
   a. Convergent evolution
   b. The independent evolution of similar structures in different lineages
   c. Selection for traits that perform similar functions
   d. The inheritance of ancestral traits

2. A derived trait is one that
   a. differs from its ancestral form.
   b. is homologous with another trait found in a related species.
   c. is the product of an evolutionary reversal.
   d. has the same function, but not the same evolutionary origin, as a trait found in another species.

3. Which of the following statements about reconstructing phylogenies is *false*?
   a. Traits found in the outgroup as well as in the focal group are likely to be ancestral traits.
   b. Shared traits are generally assumed to be homoplastic until they can be proven to be homologous.
   c. Phylogenies do not include ancestors of modern groups, or date the splits between lineages.
   d. Nodes (branching points) in phylogenetic trees have only two branches because during speciation a lineage normally splits into only two daughter species.

4. The *most* important attribute of a biological classification scheme is that it
   a. avoids the ambiguity created by using common names.
   b. reflects the evolutionary relationships among organisms.
   c. helps us remember organisms and their traits.
   d. improves our ability to make predictions about the morphology and behavior of organisms.

5. Suppose you are writing a scientific paper about a unicellular green alga called *Chlamydomonas reinhardi.* What would be the proper way to refer to this species after you had used the full binomial earlier in the same paragraph?
   a. *Chlamydomonas reinhardi*
   b. *Chlamydomonas* spp.
   c. *Chlamydomonas* sp.
   d. *C. reinhardi*

6. Organisms in a higher taxon are _________ similar, usually have diverged from a common ancestor _________ recently, and include _________ species than organisms in a lower, included taxon.
   a. less, less, fewer
   b. less, more, fewer
   c. less, less, more
   d. more, more, fewer

7. Which of the following incomplete lists of taxonomic categories ranks them properly from most inclusive to least inclusive?
   a. Phylum, order, family, genus
   b. Class, phylum, order, species
   c. Order, class, family, genus
   d. Family, order, class, kingdom

Application Questions

1. Based on the following phylogenetic tree showing the evolutionary relationships of five species (A – E) relative to five traits (1 – 5), fill in the table using 1 to indicate the presence of a derived trait, and 0 to indicate the presence of an ancestral trait.

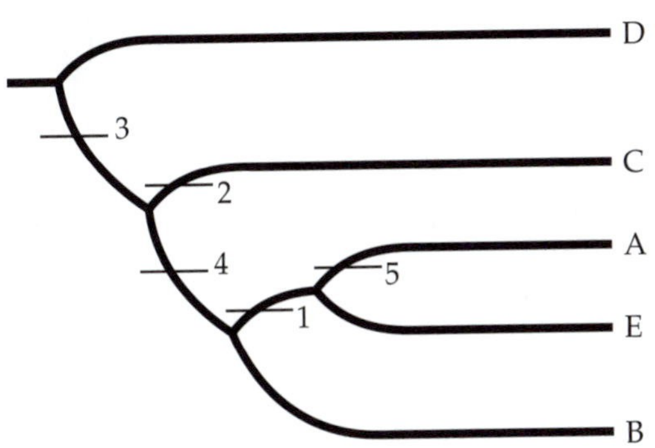

	TRAIT				
SPECIES	1	2	3	4	5
A					
B					
C					
D					
E					

2. In the phylogenetic tree shown in question 1, which species is considered the outgroup? Explain your answer.

3. Based on the following table showing the ancestral and derived traits of five species (A – E), construct a phylogenetic tree that represents the evolutionary relationships of this group using conventions presented in the textbook.

	TRAIT				
SPECIES	1	2	3	4	5
A	1	1	1	0	0
B	0	0	0	0	0
C	1	1	0	1	0
D	0	1	0	0	0
E	1	1	0	1	1

The ancestral state of each trait is indicated by 0 and the derived state is indicated by 1.

4. Discuss the application of the parsimony principle in the construction of phylogenetic trees.

5. Discuss the implications the following statement has for systematics: "DNA is the genetic material for all eukaryotes."

6. Why are paraphyletic groups (taxa) acceptable to many systematists whereas polyphyletic taxa are not?

Answers to Questions

Knowledge and Synthesis Answers

1. **d.** Homoplasy is the appearance of similar structures in different lineages that were not present in the common ancestor.

2. **a.** Derived traits are those that have undergone a change during evolution from the ancestral (original) character state. Whether a trait is ancestral or derived depends on the size of the lineage being examined.

3. **b.** Most shared traits, especially in species with a recent common ancestor, are likely to be homologous, not homoplastic. Therefore it is more consistent with the parsimony principle to assume that traits are homologous until proven homoplastic than to assume the reverse.

4. **b.** Although all of the statements listed are important attributes of biological classification schemes, the most important attribute is that biological classification reflects evolutionary relationships.

5. **d.** After a scientific name is referenced once in a paragraph, typically the genus is abbreviated, but the species name is given in full.

6. **c.** Organisms in a higher taxon are *less* similar, have diverged from a common ancestor *less* recently, and include *more* species than organisms in a lower, included taxon.

7. **a.** The complete hierarchy of taxonomic categories, from most to least inclusive, is: kingdom, phylum, class, order, family, genus, species.

Application Answers

1.

SPECIES	TRAIT 1	2	3	4	5
A	1	0	1	1	1
B	0	0	1	1	0
C	0	1	1	0	0
D	0	0	0	0	0
E	1	0	1	1	0

2. Species D. The outgroup has the ancestral form for all traits.

3.

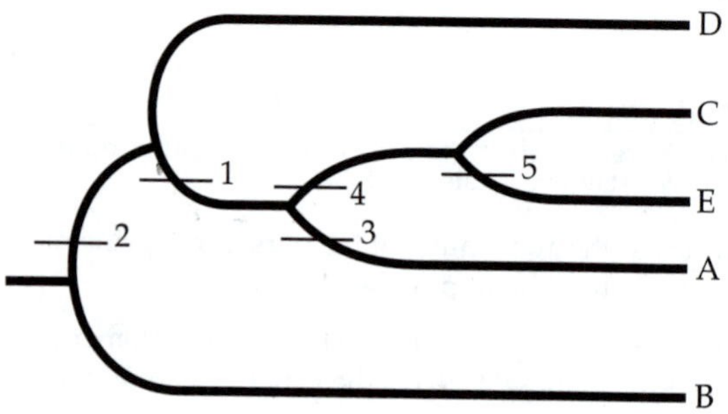

4. In the construction of a phylogenetic tree, you initially assume that derived traits appear only once and never disappear. Given a set of traits for a group of species, sometimes these restrictions must be relaxed to produce a phylogenetic tree for the group. Parsimony involves arranging the species so that you minimize the number of required reversals and multiple origins. Generally, the simplest explanation is most likely to be correct.

5. Some of the implications of this statement are that DNA evolved as the genetic material before eukaryotes had diverged from prokaryotes, that DNA is an ancestral and general homologous trait, and that all surviving eukaryotes use DNA as the genetic material.

6. Paraphyletic taxa are tolerated as a way to emphasize groups that have undergone rapid evolution. For example, birds are put in their own class, Aves, even though most biologists agree that the birds share a common ancestor with the reptiles. This makes the class Reptilia paraphyletic. Polyphyletic taxa are not acceptable, because they include members with more than one ancestor. This is antithetical to the purpose of constructing phylogenies.

STUDY GUIDE

24 Molecular and Genomic Evolution

Important Concepts

The field of molecular evolution is the study of the evolution of nucleic acids and proteins and molecular phylogenetics.

- Nucleic acids and proteins are investigated to determine how rapidly and why they have changed. Molecular phylogenetics is the reconstruction of the evolutionary history of genes and organisms.
- Molecular evolution differs from phenotypic evolution in that its rate and direction are strongly influenced by random genetic drift and mutation.
- A synonymous (silent) substitution is a mutation that replaces a nucleotide base in a codon but does not change the amino acid specified by the codon. Because synonymous substitutions are unlikely to be affected by natural selection, they accumulate in populations at rates determined by rates of mutation and genetic drift.
- A nonsynonymous substitution is a mutation that changes the amino acid specified by the codon. Though such mutations are often harmful, they are sometimes selectively neutral (or nearly so) and thus can accumulate in populations in the manner of synonymous substitutions.
- The neutral theory of molecular evolution postulates that most evolutionary change in macromolecules, as well as much of the genetic variation within species, is the result of random genetic drift, rather than natural selection.
- Most polymorphisms at specific gene loci are transitory rather than stable because the frequencies of neutral alleles slowly change over time. On the other hand, advantageous mutations are rapidly fixed in a population and harmful mutations are quickly lost.

The structures of macromolecules can be determined and compared.

- Biologists use X-ray crystallography and computer graphics to determine the precise three-dimensional structure of macromolecules.
- The polymerase chain reaction (PCR) technique makes it possible for biologists to determine the base sequence of regions of DNA from both living tissues and preserved remains.
- By first aligning amino acid sequences of homologous proteins or nucleotide sequences of homologous regions of DNA and then constructing a similarity matrix, biologists can determine the similarity of these macromolecules in different species. The assumption is that the longer the molecules have been evolving separately, the more differences they will have.
- The rate of evolution of macromolecules is extremely variable. Regions of a molecule where any change is likely to affect function tend to evolve much more slowly than areas where changes are functionally insignificant.
- Because the rate of nucleotide or amino acid substitutions is reasonably constant in many macromolecules, differences in these molecules can be used as a molecular clock to estimate the approximate time of divergence of lineages of organisms.

The many thousands of genes found in modern organisms have evolved by gene duplication from a relatively small number of ancestral genes.

- Polyploidy, in which the entire genome is duplicated, is one important source of duplicate genes.
- After a gene has been duplicated, one copy may retain its original function while the other accumulates mutations until it can perform a different task. Successive rounds of gene duplication and mutation can result in a gene family —a group of homologous genes with related functions.
- Groups of genes (such as the homeotic gene complex that is important in the control of development) arranged in the same order on chromosomes of distantly related species are probably homologs of one another.

Duplicated proteins can acquire new functions during evolution via mutation and natural selection.

- The globin gene family illustrates that gene diversification can produce molecules with different functions (hemoglobin and myoglobin) as well as functionless pseudogenes.
- The evolution of lysozyme shows how a protein can become adapted to a new function through mutation and natural selection. While serving in almost all animals as an important first line of defense against invading bacteria, lysozyme also has taken on an essential role in the digestive process of several groups of foregut-fermenters. The independent evolution in these groups of a type of lysozyme adapted to its new environment and function shows also that convergent evolution can occur at the molecular level.

There is tremendous variation in the size and composition of the genomes of various species.

- The size of the coding portion of the genome is larger in more complex organisms. Thus, eukaryotes have many times more genes than prokaryotes; vascular plants and invertebrates have more genes than single-celled organisms, and vertebrates have more genes than invertebrates.
- Most of the variation in genome size of various organisms is not due to differences in the number of functional genes, but in the amount of noncoding DNA.
- Most of the noncoding DNA appears to be nonfunctional, but some of it may have undetected functions.

Biological molecules can be used to reconstruct phylogenetic trees.

- Molecules that have evolved slowly, such as ribosomal RNA, are useful for determining ancient lineage splits. Molecules that have evolved rapidly can be used to estimate relationships among organisms with more recent common ancestors.
- For ancient lineage divisions, molecular data are often the only evidence available for reconstructing phylogenies. In other cases, molecular data are used in combination with morphological and fossil evidence.
- The division of living organisms into three major domains—the Bacteria, the Archaea, and the Eukarya—is strongly supported by comparisons of rRNA.
- DNA extracted from fossils is being used to determine the evolutionary relationships between those organisms and their living relatives.
- Mitochondrial DNA is useful for studying the recent evolution of closely related species because it accumulates mutations rapidly and is maternally inherited. The mitochondrial DNA sequences of modern human populations indicate that all modern humans shared a common African ancestor about 200,000 years ago.

The Big Picture

- The growing importance of the study of macromolecules in the reconstruction of phylogenies is an example of the far-reaching impact that advances in molecular biology have had in the decades since the discovery of the structure of DNA by Watson and Crick in 1953.

Study Strategies

- Study Figure 24.3 carefully to understand how a similarity matrix is constructed. Similarity matrices are essential to reconstructing phylogenies based on molecular data.
- Review the following activities on the website:
  Activity 24.1 Amino Acid Sequence Alignment
  Activity 24.2 Similarity Matrix Construction
  Activity 24.3 Gene Tree Construction

Test Yourself

Knowledge and Synthesis Questions

1. Which of the following statements about mutations is *false*?
   a. A silent substitution results in no change in the amino acid sequence of a protein.
   b. Since most substitution mutations are selectively neutral, most evolutionary changes in macromolecules are the result of random genetic drift.
   c. A substitution mutation in the third codon position is more likely to be neutral than a substitution at the first or second codon position.
   d. Nonsynonymous substitutions are rarely, if ever, selectively neutral.
2. Which of the following statements about gene families is *false*?
   a. Gene families evolve via gene duplication.
   b. Pseudogenes are quickly removed from gene families by deletion.
   c. Members of a gene family can include several functional genes.
   d. Examples of gene families include the homeotic gene complex in *Drosophila* and the alpha hemoglobin gene family in humans.
3. Which of the following is *not* a valid conclusion based on studies of the structure of proteins, such as cytochrome *c*, in different species?
   a. The rate of accumulation of neutral substitutions is based on the mutation rate.
   b. Many nucleotide substitutions result in neutral changes in the amino acid sequence of the protein.
   c. Fewer differences were observed in the amino acid sequences of a protein if the organismal sources of the proteins were closely related.
   d. Functionally important regions of a protein can be discovered by identifying the regions with the most amino acid substitutions.
4. Which of the following statements about molecular clocks is *false*?
   a. The molecular clock of cytochrome *c* ticks more slowly than that of rRNA.
   b. The tick rate of a clock for a molecule with great functional constraints would be very slow.
   c. The tick rate of a molecular clock that mostly reflects neutral changes is determined by the mutation rate.
   d. The tick rate of a molecular clock that mostly reflects neutral changes is directly related to the slope of the line showing the relationship between the time of divergence of different taxa and amino acid sequence differences for the protein.
5. Select the correct statement about the enzyme lysozyme.
   a. A small group of closely related mammals have evolved a special form of lysozyme that functions in digestion.
   b. The lysozymes found in the foregut fermenters resulted from convergent evolution.
   c. Lysozyme could not have evolved a secondary function if it had been an enzyme with a vital primary function.
   d. A higher mutation rate in the foregut fermenters allows their lysozymes to evolve rapidly.
6. Which of the following sequences of organisms ranks them from least to most, in terms of the expected total

amount of coding DNA in their genomes?

a. Bacterium, single-celled eukaryote, roundworm, insect, bird
b. Bacterium, roundworm, bird, insect, single-celled eukaryote
c. Single-celled eukaryote, bacterium, roundworm, insect, bird
d. Roundworm, single-celled eukaryote, insect, bird, bacterium

7. Which of the following molecules would be best to use in a study of the domains of the phylogenetic tree of life?
   a. Lysozyme
   b. Mitochondrial DNA
   c. rRNA
   d. Myoglobin

Applications Questions

1. Discuss the importance of the polymerase chain reaction (PCR) to studies of molecular evolution.
2. What is the most important conclusion about our origins that has been obtained by a study of the mitochondrial DNA (mtDNA) from over 100 ethnically distinct modern human groups?
3. Determine the amino acid sequence for an eight-residue section of a small protein from five different species (1–5).

Position	1	2	3	4	5	6	7	8
Species 1:	Arg	Cys	Leu	Leu	Ser	Thr	Asn	Met
Species 2:	Arg	Cys	Phe	Leu	Leu	Ser	Thr	Asn
Species 3:	Arg	His	Leu	Leu	Ser	Thr	Asn	Met
Species 4:	Arg	Cys	Leu	Ser	Ser	Thr	Asn	Met
Species 5:	Arg	His	Leu	Leu	Ser	Gln	Asn	Met

Complete the following similarity matrix from these sequences:

SPECIES (columns; *Differences* above the diagonal, *Similarities* below)

SPECIES	1	2	3	4	5
1	■				
2		■			
3			■		
4				■	
5					■

Which species differ because of an amino acid insertion?

Which species differ because of amino acid substitutions?

Answers to Questions

Knowledge and Synthesis Answers

1. **d.** A nonsynonymous substitution mutation can be selectively neutral if, as sometimes happens, it results in an amino acid change that has no significant effect on the shape (and hence the functional properties) of the protein.
2. **b.** Pseudogenes (nonfunctional DNA) may be removed via deletion, but numerous pseudogenes persist in many gene families.
3. **d.** Functionally important regions of a protein can be discovered by identifying the regions with the fewest amino acid substitutions. Since the sequences in these regions have been optimized by natural selection to accomplish the function of the molecule, random (neutral) changes in these areas are not tolerated.
4. **a.** The molecular clock of cytochrome *c* would tick more rapidly (i.e., the rate of neutral changes-per-unit time would be greater) than the clock of rRNA. rRNA has changed very little because it is a molecule with great functional restraints.
5. **b.** Since the animals that evolved a similar lysozyme do not share a recent common ancestor, the mechanism involved is convergent evolution. All other statements are false.
6. **a.** Recall that there is a rough relationship between the amount of coding DNA and organismal complexity.
7. **c.** Unlike the other molecules, rRNA is found in all living organisms and also has evolved very slowly.

Application Answers

1. Since the study of molecular evolution depends on obtaining fairly large amounts of pure DNA for sequencing purposes, the PCR is commonly used to amplify and purify DNA. This is especially important in studying DNA from fossil organisms where the amount of DNA available may be very small and degraded.
2. The most important conclusion is that all humans share a recent common mitochondrial ancestor from about 200,000 years ago. This supports the "out of Africa" hypothesis that our species evolved in Africa and dispersed to the rest of the world.
3.

SPECIES (columns; *Differences* above the diagonal, *Similarities* below)

SPECIES	1	2	3	4	5
1	■	*1*	*1*	*1*	*2*
2	*7*	■	*2*	*2*	*3*
3	*7*	*6*	■	*2*	*1*
4	*7*	*6*	*6*	■	*3*
5	*6*	*5*	*7*	*5*	■

Species 2 shows an amino acid insertion (Phe) at position 3. There are no deletions.

Species 3, 4, and 5 all show amino acid substitutions. Species 1 shows no changes.

STUDY GUIDE

25 The Origin of Life of Earth

Important Concepts

Life first appeared on Earth about 4.5 billion years ago.

- As our solar system came into being, Earth formed about 4.5 billion years ago by gravitational attraction of rocks of various sizes.
- Earth consists of a fluid iron-nickel core, a mantle of dense silicates, and a crust of lighter materials.
- The early atmosphere of Earth consisted mainly of methane, carbon dioxide, ammonia, hydrogen, nitrogen, and water vapor. As Earth cooled, most of the water vapor condensed to liquid water and formed the oceans.
- The existence of fossils of complex, unicellular life dating from 3.5 billion years ago (bya) suggests that life first appeared on Earth about 4 bya.

Although the origin of life was probably a unique event, it can be studied scientifically.

- Three principles guide the study of the origin of life.
  1. The principle of continuity—the origin of life was probably a continuous, gradual process.
  2. The signature principle—traces of prebiotic processes should still exist in contemporary biochemistry.
  3. The no-free-lunch principle—all organisms require energy obtained by the oxidation of some material.
- Any plausible and comprehensive hypothesis for the origin of life must explain all of life's essential properties, including a number of universal biochemical characteristics.

Conditions on early Earth differed from those of today and favored the origin of life.

- Radiation from the sun and the chemical potential of reduced compounds in magma are two sources of energy that could have provided energy for the origin of life.
- At the time of life's origin, Earth had a reducing atmosphere that lacked the large quantity of free oxygen found in the present-day atmosphere. Experiments have demonstrated that the gases in such an atmosphere, given an energy source, will react to form many of the chemical building blocks of life.
- As small molecules (monomers) accumulated, energy-requiring condensation reactions used monomers to build polymers such as polysaccharides, proteins, and nucleic acids.

The evolution of life required self-duplicating molecules subject to error, a constant supply of free energy, and partial isolation of the replicating system from its surrounding environment.

- The property of partial isolation is exhibited by artificially produced aggregations of organic molecules called protobionts. Both major categories of protobionts—coacervates and microspheres—can maintain internal chemical environments that differ from their surroundings. The protobiont precursors of cells probably had a lipid-based outer membrane.
- The lipid membranes of some protobionts probably contained light-absorbing molecules called chromophores. Because such membranes are able to transduce light energy to drive chemical reactions, it is possible that solar energy was the main source of free energy for the earliest cells.
- The first genetic code was probably based on RNA that catalyzed its own replication as well as catalyzing other chemical reactions. This view is supported by the discovery in present-day cells of ribozymes, which are RNAs that catalyze chemical reactions. As proteins evolved, they replaced RNA as the principal catalysts in cells.
- Because DNA is more stable than RNA in the intracellular environment, it replaced RNA as the hereditary molecule, and RNA took on its present roles in the translation of the genetic code into protein.

Photosynthesis is the source of atmospheric O_2.

- Slightly more than 2 billion years ago, noncylic photophosphorylation evolved in sulfur bacteria, which later evolved into cyanobacteria. Because this process uses water as a source of electrons for the reduction of CO_2, it liberates O_2 as a waste product that is released into the atmosphere.
- As O_2 accumulated in the atmosphere, it poisoned most anaerobic prokaryotes. Some cells evolved to tolerate O_2, and the presence of O_2 in the atmosphere made possible the evolution of aerobic metabolism.

Life is not being formed from nonliving matter on today's Earth.

- Because of the conditions present on the early Earth, the origin of life was almost inevitable.

- Louis Pasteur performed definitive experiments in the nineteenth century that demonstrated that spontaneous generation—the origin of life from nonliving substances—does not occur today.
- Spontaneous generation is impossible today because simple biological molecules that form in today's environment are quickly consumed by existing life or oxidized.

Whether life exists elsewhere in the universe is unknown.

- For a planet to support simple life, it would require liquid water on its surface and a relatively constant output of energy from an associated star. Many such planets may exist.
- The more stringent requirements for multicellular life include a planet with a nearly circular orbit, a rapid rate of spin, nearby planets to minimize comet and meteor impacts, and a large moon that stabilizes the planet's orbit. Such conditions may occur very rarely.
- Notwithstanding periods of mass extinction, multicellular life has existed on Earth for almost a billion years.

The Big Picture

- Pasteur's disproof of the theory of spontaneous generation was a major landmark in our fundamental understanding of life. It established firmly the principle that on today's Earth, all life comes from previously existing life.
- The work of Oparin and Miller was instrumental in reawakening the interest of biologists in how life could have originated in the conditions found on early Earth.
- The progress made in recent decades in the field of molecular biology has provided knowledge crucial to our advancing understanding of how life began on Earth. For example, the discovery of ribozymes (catalytic RNAs) lends strong support to the hypothesis that RNA was the original genetic material.

Study Strategies

- The origin of life on Earth is essentially an historical narrative. As you study this material, focus on understanding how each event in the story set the stage for the next.
- Review the following tutorial on the website:
  Tutorial 25.1 Pasteur's Experiment

Test Yourself

Knowledge and Synthesis Questions

1. Which of the following is *not* evidence of the conservatism of biochemical evolution?
   a. The storage and release of energy in the living world is associated with the formation and hydrolysis of phosphate bonds.
   b. All proteins are made from the same group of amino acids.
   c. The first genetic code was based on DNA.
   d. Ribosomes are the sites of protein synthesis in all cells.
2. Which of the following gases was probably *least* abundant in Earth's early atmosphere?
   a. CO_2
   b. NH_3
   c. H_2
   d. O_2
3. Experiments to synthesize organic compounds under simulated early Earth conditions have demonstrated that
   a. organic compounds can be formed in atmospheres containing a variety of gases, provided oxygen is absent.
   b. a variety of organic compounds can be formed in these experiments, including amino acids found in most organisms.
   c. organic compounds cannot be formed in experiments that simulate an aquatic environment.
   d. a and b
4. Concerning the earliest catalysts, scientists studying the origin of life believe that
   a. the first catalysts were nonproteins and lacked the specificity shown by enzymes.
   b. DNA had no catalytic function, either before or after the evolution of enzymes.
   c. RNA had important catalytic functions before the evolution of enzymes.
   d. All of the above
5. Which of the following statements about coacervates is *false*?
   a. Coacervates are clumps of protein and polysaccharide in an aqueous environment.
   b. If artificial coacervates contain nucleic acids, they can accurately reproduce themselves.
   c. Oparin was able to create coacervates with the ability to absorb substrates and catalyze reactions.
   d. Because Oparin's artificially produced coacervates lacked lipid outer membranes, they differed from the probable precursors of the first cells.
6. Which of the following was *not* presented in the text as evidence that catalysis by RNA preceded catalysis by enzymes?
   a. RNA is only stable within membrane-enclosed cells, where the water concentration is lower than the surrounding environment.
   b. RNA can excise its own introns and splice together its exons.
   c. Ribosomes contain a catalytic RNA that functions in protein synthesis.
   d. RNA can catalyze the polymerization of RNA.
7. The earliest organisms were probably
   a. able to use the small quantity of oxygen available in the atmosphere for their metabolism but also able to survive without oxygen.
   b. unable to live in the absence of oxygen.
   c. unable to live in the presence of oxygen.
   d. like present-day cyanobacteria.

8. As a result of the evolution of photosynthesis using water as an electron donor,
   a. Earth developed an oxidizing atmosphere.
   b. organisms using aerobic metabolism became predominant.
   c. the atmospheric concentration of CO_2 increased.
   d. a and b

9. Which of the following represents a correct (though not necessarily complete) sequence of events in the origin of life, from earliest to latest?
   a. Organic monomers accumulate; protobionts evolve into cells; DNA becomes the genetic material; oxygen accumulates in the atmosphere.
   b. Monomers are polymerized; ribozymes appear; oxygen accumulates in the atmosphere; protobionts evolve into cells.
   c. Organic monomers accumulate; cyanobacteria appear; oxygen accumulates in the atmosphere; DNA becomes the genetic material.
   d. Monomers are polymerized; DNA becomes the genetic material; ribozymes appear; protobionts evolve into cells.

10. Any planet that supports life will almost certainly have
   a. liquid water on its surface.
   b. a constant supply of radiant energy from a star.
   c. an atmosphere containing oxygen.
   d. a and b

Application Questions

1. How did Stanley Miller demonstrate that the organic building blocks of life could have been produced in the environment that probably existed on early Earth?
2. Describe the guiding principles used by biologists in the scientific investigation of the origin of life on Earth.
3. Discuss why life does not evolve from nonliving matter on Earth today.
4. How did Louis Pasteur disprove the theory of spontaneous generation?
5. Why is the evolution of complex, multicellular organisms in the universe likely to be far less common than the evolution of single-celled life forms?

Answers to Questions

Knowledge and Synthesis Answers

1. **c.** The first genetic code is believed to have been based on RNA. Once cells evolved, DNA replaced RNA as the genetic material.

2. **d.** Being highly reactive, oxygen would have been bound into compounds such as water, carbon dioxide, and iron oxides. Thus, free oxygen (O_2) probably did not exist in the early atmosphere.

3. **d.** For organic molecules to be produced in these experiments, oxygen must not be present. Amino acids are among the molecules produced. But organic molecules can also be produced in simulated aquatic environments.

4. **d.** The first biological catalysts were probably RNA molecules, not proteins. Protein enzymes are superior catalysts to RNA because they can catalyze a wider variety of specific reactions. There is no evidence that DNA has ever had a catalytic function.

5. **b.** Protobionts in general, including coacervates, lack the ability to reproduce.

6. **a.** Catalysis by RNA probably originated before the evolution of the first cells. In an aqueous solution, RNA is more stable than DNA, but the reverse is true in the intracellular environment.

7. **c.** Little or no oxygen was present in the atmosphere of the early Earth, and until organisms evolved adaptations to use it, oxygen was poisonous.

8. **d.** With the evolution of noncyclic photophosphorylation, which releases O_2 as a waste product, free oxygen began to accumulate in the atmosphere. This event paved the way for the evolution of aerobic metabolism. But the atmospheric concentration of CO_2 was *reduced* as more became fixed in organic form or deposited in ocean bottoms.

9. **a.** The most probable complete sequence (i.e., including all the events mentioned in the answer choices) would be the following: organic monomers accumulate; monomers are polymerized; ribozymes appear; protobionts evolve into cells; DNA becomes the genetic material; cyanobacteria appear; oxygen accumulates in the atmosphere.

10. **d.** Liquid water and a constant supply of free energy are essential for life. Oxygen, however, was not necessary for the origin of life on Earth, and even today there are many anaerobic prokaryotes.

Application Answers

1. See Figure 25.2 in the text.

2. According to the principle of continuity, any stage in life's origin should be derivable from preexisting stages; sudden major changes are unlikely to have occurred. According to the signature principle, the biochemistry of contemporary organisms should give us some clues about prebiotic processes. Finally, according to the no-free-lunch principle, all living organisms must oxidize some material to obtain energy for their maintenance and growth.

3. The present-day oxidative atmosphere prevents the accumulation of abiotically formed simple organic compounds. Even if these materials could accumulate, they would be consumed by living organisms.

4. See Figure 25.6 in the text.

5. The evolution of complex, multicellular organisms would probably require relatively stable conditions for an extremely long period of time. Here on Earth, with conditions that favored the eventual evolution of multicellular organisms, unicellular prokaryotes were the only life forms for almost 3 billion years. The specific conditions that would be necessary for stability include a planet with a nearly circular orbit, a rapid rate of spin, nearby planets to minimize comet and meteor impacts, and a large moon that stabilizes the planet's orbit. Such conditions are probably very rare.

STUDY GUIDE

26 Bacteria and Archaea: The Prokaryotic Domains

Important Concepts

Domains attempt to reflect evolutionary relationships among organisms.

- All living things can be placed in the following three domains: Archaea, Bacteria, and Eukarya. This system reflects evolutionary relationships among the organisms.
- All organisms have similarities, but differences allow classification. Eukarya have nuclei, membrane-bound organelles, and cytoskeletons. Neither Archaea nor Bacteria have these characteristics and thus are both prokaryotic. Differences between Archaea and Bacteria have been resolved through biochemistry and genetic studies.
- All three groups arose from a common ancestor possessing a circular chromosome and machinery for transcription and translation of the DNA code to protein. Divergence from this ancestor occurred billions of years ago. The kingdoms of Archaea are more closely related to Eukarya than to Bacteria.

Prokaryotes have many features in common.

- Though members of two different domains, Archaea and Bacteria have several structural similarities that are indicative of evolutionary adaptation to similar environments.
- Prokaryotes are prolific and on a per-number basis are the most common of all organisms. Because they are single-celled (there are a few exceptions) and very small, millions exist without our being aware of them. They play extremely important roles in the biosphere, for instance in nutrient cycling. They can also metabolize many things that eukaryotes cannot.
- Prokaryotes generally exist in one of three shapes; cocci are spherical, bacilli are rod-shaped, and spirilli are spiral. All of the different shapes may aggregate to form chains or plates of cells. Advanced associations may result in filaments that are encased in a protective sheath.
- Prokaryotes do not have membrane-bound organelles; therefore, the organization and replication of DNA proceeds differently in prokaryotes than it does in eukaryotes. Also, the processes of cellular respiration and photosynthesis are carried out differently and rely on infoldings of the cell membrane. Finally, prokaryotes lack a cytoskeleton; therefore, true mitosis does not occur. Prokaryotes go through a process called fission after replicating their DNA.
- Many prokaryotes are mobile and move by rolling, regulating gas vesicles, or by means of flagella. Bacterial flagella are whiplike and are composed of a single protein called flagellin. In contrast, the flagellum of eukaryotes is complex.
- The cell wall of prokaryotes is thick and rigid and composed of peptidoglycans (bacteria) or protein (archaea). This cell wall is very different from that of plant and fungal cell walls. Cell wall composition is one of many ways of differentiating bacteria and archaea.
- Cell wall structure within the bacteria also serves as a diagnostic characteristic among different bacteria. This difference is used in the Gram staining technique. Whether a cell is Gram-negative or Gram-positive relates to the amount of peptidoglycan in the cell wall.
  - Gram-positive cells stain purple and have a thick layer of peptidoglycan.
  - Gram-negative cells do not retain stain and appear red due to a thin layer or peptidoglycan.
- Antibiotics work by attacking the cell wall structure of prokaryotes. Eukaryotic cells are not harmed because they lack this structure.
- Prokaryotes reproduce asexually but may exchange genetic information through transformation, conjugation, or transduction.

Metabolic processes in prokaryotes are varied and diverse.

- Obligate anaerobes are oxygen sensitive and can only survive in oxygen-poor conditions. Facultative anaerobes can shift between aerobic and anaerobic modes of respiration. Some facultative anaerobes do not respire aerobically but are not harmed by an oxygen-rich atmosphere as obligate anaerobes are. Obligate aerobes require oxygen and cannot survive in the absence of it.
- Photoautotrophs are photosynthetic and convert light energy to chemical energy. Some photoautotrophs use chlorophyll *a* and give off oxygen as a by-product. Others use bacteriochlorophyll and produce sulfur as a by-product.
- Photoheterotrophs use light for energy but also require outside carbon sources.

- Chemoautotrophs oxidize inorganic substances for energy. Some chemoautotrophs fix carbon dioxide into carbohydrate. Others fix nitrite or ammonia, hydrogen gas, hydrogen sulfide, and other materials.
- Chemoheterotrophs obtain energy and carbon from consuming organic molecules.
- Some bacteria carry out respiratory electron transport using nitrogen or sulfur compounds as an electron acceptor. These organisms are extremely important in nutrient cycling.

Prokaryotes make diverse contributions to their environments.

- The nitrogen cycle depends on nitrification by bacteria. Nitrifying bacteria use nitrogen to make ATP that is used to fix carbon dioxide into glucose and other food molecules. Through this process, unusable nitrogen sources are converted into biologically available nitrogen sources. See Figure 26.10.
- Cyanobacteria, using photosynthesis, enriched the atmosphere with oxygen. Oxygen levels became too high for many existing organisms; many others evolved to fill those niches and new organisms arose.
- Prokaryotes live in many mutualistic associations. Many of your digestive processes rely on the bacteria that inhabit your intestine.
- A small minority of bacteria are pathogenic. Bacteria cause disease because they produce toxins that are harmful to host tissues. Endotoxins are released when bacterial cells burst, but are rarely fatal. Exotoxins are secreted by bacteria and may be fatal.
- The German physician Robert Koch came up with criteria for determining if a bacterium was a disease causing agent.
  - The organism must always be found in the individual with the disease.
  - The organism must be taken from the host and grown in pure culture.
  - A sample of the culture must produce disease if injected into a healthy individual.
  - The newly infected individual must yield a new pure culture of the same organism.
- In order to be pathogenic an organism must be able to reach a new host, invade the host, evade the host's defenses, multiply, and infect a new host. Very few organisms can do this.

Evolutionary relationships among prokaryotes are determined at the genetic level.

- Physical characteristics are used for classification for identification purposes, but to understand the true evolutionary relationships among organisms it is necessary to look at genetic sequences and biochemistry.
- Ribosomal RNA gives us information about relationships among organisms. It is useful for this purpose because it is evolutionarily ancient, found in all organisms, has the same function in all organisms, and is highly conserved (it evolved slowly). Scientists compare signature sequences to make evolutionary inferences.
- rRNA sequences have not provided clear data. Lateral gene transfer—genetic transfer between species—makes these data hard to understand.
- Mutations are important in prokaryotic variation. Because prokaryotes have single circular DNA pieces, mutations cannot be masked and are passed on through fission. This is in contrast to diploid organisms in which mutation plays a small role in genetic variation.
- Prokaryotes evolve comparatively rapidly because of rapid reproduction, persistent mutation, and selection. Mutations leading to antibiotic resistance are of extreme concern and can occur in a matter of just a few years.

The majority of known prokaryotes are bacteria.

- Bacteria can be subdivided into a series of monophyletic groups. These groups are not classified the same way by all scientists, and are an active area of research.
- Proteobacteria contain the largest number of known species. This group is believed to be the origin of the eukaryotic mitochondria. The early ancestor of this group was probably a photoautotroph, but today, members of the group have multiple metabolic schemes. This group contains important nitrification bacteria and several pathogens. Another member, *Agrobacterium tumefaciens*, has been important in understanding genetic transformation in plants.
- The Cyanobacteria undergo photosynthesis similar to eukaryotic photosynthesizers. They contribute to the current oxygen-rich environment. These organisms frequently exist in colonies with distinct divisions of labor. Some cells are vegetative and photosynthesize. Others form resting stages called spores that can withstand harsh environmental conditions. Still others, called heterocysts, fix nitrogen.
- Spirochetes are corkscrew-shaped prokaryotes characterized by the presence of axial filaments that run through their periplasmic space. Many spirochetes are parasites of humans.
- Chlamydias are the smallest bacteria and exist only as parasites. They have a two-stage life cycle and are responsible for several human diseases.
- Most, but not all, Firmicutes are Gram-positive. They produce endospores that are capable of surviving in extremely harsh conditions. Many medically important bacteria fall within this group including *Clostriduim botulinum* (botulism), *Staphylococcus* (many skin infections), and *Mycobacterium tuberculosis* (tuberculosis). However, one of our most potent antibiotics comes from this group. Streptomycin is produced by the actinomycete *Streptomyces*. Most of currently used antibiotics come from the actinomycetes which are firmicutes. Another subgroup, the mycoplasmas, may represent the smallest life form.

The Archaea live in some of the harshest known environments.

- Very little is known about the phylogeny of the Archaea and much research still needs to be done, but there are shared common characteristics. Archaeans lack peptidoglycan cell walls and have very distinctive lipids in their cell membranes. These are linked by ether linkages. No other organisms possess this type of lipid.
- Crenarchaeota live in hot, acidic environments, and are thermophilic and acidophilic. They require temperatures above 55° C and can survive in an environment with a pH as low as 0.9 while maintaining an internal pH of 7.
- Some Euryarchaeota produce methane gas and are obligate anaerobes. These organisms are prevalent in the guts of herbivores. Others live near volcanic vents. Another group of euryarchaeota live in extremely salty conditions and are referred to as halophiles.

The Big Picture

- All organisms fall within one of three domains. This chapter focuses on the Archaea and Bacteria. Subsequent chapters focus on Eukarya. Archaea and Bacteria are prokaryotic organisms. They have no membrane-bound organelles, single circular chromosomes, and cell walls of peptidogylcans or proteins. Their structures are simple and they exist as single cells. Some prokaryotes are stationary and others move by way of flagella. Their metabolic processes are highly varied, as are their modes for procuring energy. Many prokaryotic organisms make valuable contributions to their environment through nitrogen fixation, photosynthesis, and other metabolic processes. The majority of prokaryotes are free living, but some live in mutualistic relationships and yet others are pathogenic to other organisms.
- The evolutionary relationships between prokaryotes are still highly disputed and best understood at the gene level. The most ubiquitous extant prokaryotes are the bacteria. They are grouped according to physical and biochemical characteristics and these groupings do not necessarily reflect their evolutionary relationships. Archaeans are thought to be an older lineage and exist in some of the harshest known environments.

Study Strategies

- This chapter is an overview of two domains of organisms. A single chapter can only hit the high points of prokaryotic diversity. Focus your study on what makes prokaryotes different from eukaryotes, and then move on to learning about the example organisms.
- The temptation is to focus on the pathogenic organisms, but remember that only a very small percentage of eukaryotes are pathogenic.
- Review the following tutorial and activities on the website:
  Tutorial 26.1 Lateral Gene Transfer
  Activity 26.1 Gram Stain and Bacteria
  Activity 26.2 Nitrogen Cycle

Test Yourself

Knowledge and Synthesis Questions

1. Which of the following characteristics are unique to prokaryotes?
   a. Lack of membrane bound organelles
   b Presence of cell walls
   c. Plasma membranes
   d. All of the above
2. A tooth scraping yields large numbers of corkscrew shaped bacteria. These bacteria are referred to as
   a. bacilli.
   b. cocci.
   c. spirilli.
   d. helici.
3. Which of the following statements concerning prokaryotes are true?
   a. Because prokaryotes do not contain organelles they cannot photosynthesize or carry out cellular respiration.
   b. Prokaryotes have no chromosomes and therefore lack DNA.
   c. Prokaryote flagella are similar in structure to eukaryote flagella.
   d. None of the above
4. Gram-negative cells stain pink because
   a. they have specialized lipids in their cell walls.
   b. their peptidoglycan layer is thin.
   c. their peptidoglycan layer is thick.
   d. they are receptive to antibiotics.
5. Which of the following statements concerning bacteria are *not* true?
   a. Most bacteria are pathogenic.
   b. Bacteria live in diverse environments.
   c. Bacteria are derived from Archaea.
   d. a and b
6. Determining evolutionary lineages of bacteria is difficult because
   a. bacteria evolve rapidly due to short life spans.
   b. the evolution of bacteria is greatly influenced by mutation due to their single chromosome.
   c. Much of the data appears conflicting.
   d. All of the above
7. You would find several pathogenic bacteria in which of the following groups of bacteria?
   a. Proteobacteria
   b. Cyanobacteria
   c. Chlamydias
   d. None of the above
8. Parasitic spirilli are found in which of the following bacteria groups?
   a. Proteobacteria
   b. Cyanobacteria
   c. Spirochetes
   d. None of the above

9. Autoclaves must pass a "spore test" in many states to be considered sterile. The spore producing bacteria they use for this test is most likely a member of which of the following groups of bacteria?
   a. Firmicutes
   b. Proteobacteria
   c. Cyanobacteria
   d. Chlamydia

10. Archaeans often live in harsh environments. An archaean that lives in extremely salty conditions is referred to as a
    a. thermophile.
    b. halophile.
    c. salinophile.
    d. None of the above

11. Methane gas is thought to be a threat to ozone. The primary source of methane emission is from grazing cattle. This is because cows harbor a methane producing archaean from which of the following groups?
    a. Crenarcheaota
    b. Euryarcheaota
    c. Anarchaeota
    d. Proteobacteria

12. One of the major diagnostic characteristics that distinguishes archaea from bacteria is lack of
    a. peptidoglycan cell walls in archaea.
    b. peptidoglycan cell walls in bacteria.
    c. ribosomes in archaea.
    d. chemoautotrophic bacteria.

13. A bacterium that requires a carbon source other than carbon dioxide, yet can convert light energy to chemical energy is called a
    a. photoautotroph.
    b. photoheterotroph.
    c. chemoautotroph.
    d. chemoheterotroph.

14. A bacterium that cannot live in the presence of oxygen is called a(n)
    a. obligate aerobe.
    b. facultative aerobe.
    c. obligate anaerobe.
    d. facultative anaerobe.

Application Questions

1. You work for a waste water treatment plant. An organism is getting past your treatment process and fish are dying in a stream that the runoff enters. You are able to isolate the organism. How do you know if you have isolated a bacterium or protist?

2. You work with the Centers for Disease Control in Atlanta. You have been called to a remote area of Uganda to study a mysterious disease that is causing respiratory ailments in a small village. You do not have your bag of high-tech equipment. You suspect the disease is bacterial in nature. How do you determine if a bacterium is causing the illness?

3. As a U.S. Department of Agriculture field representative, you counsel a young farmer to plant alfalfa in his fields because there are high nitrite levels in his soil. You know alfalfa roots are hosts for nitrogen fixing bacteria. Explain, in terms the young farmer can understand, how the alfalfa and its associated bacteria will help "fertilize" his soil.

4. A patient comes to your medical practice with a bacterial infection. You are terribly concerned because the bacteria he is harboring produce an exotoxin. Why are exotoxins often more dangerous than endotoxins?

5. Assume you are teaching a high school class. How would you explain the difference between archaea and bacteria to your students?

6. The book talks about the different groups of bacteria. Discuss two ways in which the different groups are distinguished from one another.

7. Describe how Gram staining depends on bacteria cell wall structure. Would Gram staining technique work for archaea?

8. Differentiate between the following terms:
   a. obligate anaerobe/facultative anaerobe
   b. photoautotroph/photoheterotroph
   c. chemoautotroph/chemoheterotroph

Answers

Knowledge and Synthesis Answers

1. **a.** Prokaryotes are differentiated from eukaryotes by lacking membrane-bound organelles. They also have a single circular piece of genomic DNA and lack a cytoskeleton.

2. **c.** Spirilli are corkscrew shaped.

3. **d.** Bacteria do carry out both photosynthesis and multiple forms of cellular respiration. All life forms must have DNA. The flagella of bacteria consist of a single protein called flagellin rather than multiple proteins as found in eukaryotes.

4. **b.** Their peptidoglycan layer is thin and does not retain crystal violet stain like Gram-positive bacteria do.

5. **d.** Most bacteria are *non*pathogenic, and, though archaea and bacteria have a common ancestor, one was not derived from the other.

6. **d.** Rapid evolution and the strong influence of mutation make data misleading. Also, transformation leads to data that appear conflicting.

7. **c.** Chlamydias are all parasitic and cause several human diseases.

8. **c.** The spirochetes are all spirilli bacteria. Some of them are disease-causing.

9. **a.** Only the Firmicutes produce spores. These spores can withstand extreme environmental conditions, but by law, autoclaves must be able to destroy spores.

10. **b.** The term "halophile" means "salt loving."

11. **b.** Methanogenic bacteria that reside in the guts of cows belong to the Euryarcheota bacterial group.

12. **a.** Archaeans lack peptidoglycans and have a unique lipid in their cell walls.

13. **b.** Photoheterotrophs require an outside carbon source other than carbon dioxide, yet are able to harvest light energy.

14. **c.** Oxygen gas is toxic to obligate anaerobes.

Application Answers

1. If the organism is a bacterium you will note the absence of membrane organelles, a circular chromosome, and absence of a cytoskeleton. All of these would be present in a eukaryote.

2. You would use Koch's postulates. These postulates stipulate that the organism must always be found in the individual with the disease, the organism must be taken from the host and grown in pure culture, a sample of the culture must produce disease if injected into a healthy individual, and the newly infected individual must yield a new pure culture of the same organism.

3. Nitrogen fixing bacteria can convert nitrites that are unusable by plants to nitrates that are usable by them. Refer to Figure 26.10.

4. Exotoxins are continuously produced by the infecting bacteria. Endotoxins are produced only at cell death. There are a limited number of endotoxin molecules released. Exotoxins can be released until the host dies.

5. There are important differences in the cell walls of archaea and bacteria. Archaea occur in a range of extreme environments that bacteria cannot.

6. Different types of bacteria can be distinguished through Gram staining, an important technique for observing morphology and for chemical analysis.

7. Gram staining results depend on the arrangement of peptidoglycans in the bacterial cell wall. This technique does not work with archaea because they do not have similarly structured cell walls.

8. Obligate anaerobes die in the presence of oxygen gas. Obligate aerobes die without oxygen gas for cellular respiration. Photoautotrophs convert light energy to chemical energy using carbon dioxide as their carbon source. Photoheterotrophs rely on an outside carbon source other then carbon dioxide but still can convert light energy to chemical energy. Chemoautotrophs can produce what they need from inorganic molecules. Chemoheterotrophs depend on nutrients from external sources.

27 Protists and the Dawn of the Eukarya

Important Concepts

Protists are a nonmonophyletic group of very diverse organisms.

- All protists are eukaryotic. This is the single characteristic that they all have in common.
- Protists are organisms that do not fit into the kingdoms of Animalia, Planta, or Fungi. Many are unicellular but some are multicellular.

The origin of the eukaryotic cell was pivotal in evolution.

- The origin of the eukaryotic cell is still largely debated by biologists. The changes leading to the eukaryotic cell were vast and occurred at a time of great environmental change.
- The eukaryotic cell evolved a flexible cell surface, a cytoskeleton, a nuclear envelope, digestive vesicles, and organelles (possibly through symbiosis).
- The first step in the path toward a eukaryotic cell may have involved loss of the rigid cell wall. Loss of this rigidity allows infoldings of the cell surfaces to increase the surface area-to-volume ratio. This allows cells to become larger. It also allows vesicles to form from bits of infolded membrane that pinch off. This may have begun the nuclear envelope. The prokaryotic chromosome is attached to the cell membrane. Infolding may have begun to encase the nuclear material.
- New structures had to form during the evolution of the eukaryotic cell. Ribosomes became associated with internal membranes to form the endoplasmic reticulum, a cytoskeleton of actin and microtubules formed, and digestive vesicles developed. The cytoskeleton opens up modes of locomotion with actin/myosin-based flagella. How these structures formed is currently conjecture.
- It is thought that these early motile cells may have been phagocytes that engulfed prokaryotic cells. Since the atmosphere was shifting from anaerobic to aerobic, it is possible that these phagocytes engulfed prokaryotes that could metabolize the toxic by-products of the oxygen-rich atmosphere. These prokaryotes were retained as symbionts and may have led to modern peroxisomes.
- Subsequent endosymbiotic events may have led to the evolution of mitochondria and plastids.
- See Figure 27.3 for a diagram of events leading to the eukaryotic cell.
- Organisms such as *Giardia* lack some common structures of eukaryotic cells and once were thought to support the sequence of events listed above; however, current research indicates that they may be organisms that have lost components through later evolution.
- An understanding of the evolution of eukaryotes is complicated by lateral gene transfer. There are so many bacterial genes in many prokaryotes that lateral transfer cannot explain them all, nor can endosymbiosis. Additional hypotheses are being tested.

Protists have diverse ways to accomplish the same functions.

- Most protists are aquatic. Even those that seem terrestrial live in soil water or other extremely moist environments.
- Metabolism mechanisms are relatively uniform among protists but nutritional acquisition varies. Some protists are autotrophic, some are heterotrophic, and some have the ability to switch between the two modes. Many protists are involved in symbioses from mutualism to parasitism.
- Most protists are motile and move via pseudopodia, flagella, or cilia.
- Many protists are microscopic though some algae literally form aquatic forests of very large organisms. Those single-celled protists are able to increase their cell size with the presence of vesicles, which effectively increase surface area. Vesicles assist with water regulation (contractile vacuoles) and nutrient procurement (food vacuoles).
- Many protists have cell walls or "shells" to protect their cell membranes.
- Many protists contain endosymbionts. Some harbor other protists that photosynthesize or carry out other metabolic processes.
- Most protists undergo both sexual and asexual reproduction but some lack the ability to reproduce sexually. Sexual and asexual reproductive practices are as diverse as the protists themselves.

Accepted groups of protists may be monophyletic groups or phenetic groups (groups of classification for convenience).

- The Euglenozoa is a monophyletic group.
  - Euglenozoa are single-celled flagellates that reproduce asexually by binary fission.
  - This group consists of two subgroups, the euglenoids and the kinetoplastids.
  - Euglenoids have anterior flagella. Many are photosynthetic and are characterized by a complex cell structure (Figure 27.10). They are flexible in nutritional requirements and may switch between autotrophism and heterotrophism. Euglenoid chloroplasts have triple membranes.
  - Kinetoplastids are parasitic and are characterized by having a single large mitochondrion. The mitochondrion is unique in having a kinetoplast containing DNA in minicircles and maxicircles. The minicircles encode "guides" for RNA editing in the mitochondrion. Many tropical human diseases are caused by kinetoplastids including African sleeping sickness.
- The Alveolata are a monophyletic group.
  - All members of Alveolata are characterized by cavities called alveoli just below their cell surfaces.
  - This group consists of dinoflagellates, apicomplexans, and ciliates.
  - Dinoflagellates are mostly marine organisms with two flagella. They are yellowish in color due to photosynthetic and accessory pigments and are primary producers in marine environments. Many live in symbiosis with other organisms. Dinoflagellates are responsible for "red tides" in warm marine waters and can damage fish. Toxins produced by these organisms can accumulate in shellfish and be fatal to people eating the shellfish. Dinoflagellates are also bioluminescent and produce light when disturbed.
  - Apicomplexans are parasitic protists. Their name comes from the cluster of organelles at the apex of their spores which assist with invasion of their host organism. The life cycles of these parasites are complex. The most well noted of these organisms, *Plasmodium*, causes malaria. See Figure 27.13 for the *Plasmodium* life cycle.
  - Ciliates move via cilia and are characterized by multiple macronuclei and micronuclei. Most ciliates are heterotrophic. *Paramecia* are common and well-studied ciliates.
    - Paramecia move by cilia, their membranes are protected by a pellicle, and they protect themselves with trichocysts.
    - Paramecia reproduce asexually by binary fission. Genetic recombination is accomplished through conjugation whereby equal amounts of genetic material are exchanged (Figure 27.17). The resulting recombined paramecia then go through binary fission.
- Stramenopils, most likely a nonmonophyletic group, are referred to as the brown algae.
  - All members are either flagellated or have lost their flagella over the course of evolution.
  - This group consists of diatoms, brown algae, oomycetes, and a few smaller groups.
  - Diatoms (Bacillariophyta) are yellowish brown and store chyrsolaminarin and oils as photosynthetic products. They are most noted for their silica (glass) containing cell walls. The cell walls occur in two "valves" and fit together like a petri plate. All diatoms are symmetrical. Diatoms reproduce asexually and sexually (See Figure 27.19). Diatom cell walls have significant human uses from filters to reflective paint.
  - Brown algae (Phaeophyta) are multicellular and are the largest of the protists. They are almost exclusively marine and make up large oceanic kelp forests. They produce leaflike growths called thalli. They are brown because of chlorphylls *a* and *c* and fucoxanthin. They have specialized regions called holdfasts which anchor them to a substrate and allow them to resist water movement. The brown algae are noted for specialization of structures. Brown algae are commercially important for the presence of alginic acid, used as a binder in many food and cosmetic products. Brown algae, like many other protists and all plants, go through alternation of generations in which a diploid sporophyte gives rise to haploid spores through meiosis to form the haploid gametophyte which, in turn, produces haploid gametes. See Figure 27.22. Heteromorphic alternation of generations occurs when the two generations differ morphologically and isomorphic alternation of generations occurs when the two generations do not differ morphologically.
  - Oomycetes (oomycota) make up the nonphotosynthetic group of stramenopiles. These are commonly known as water molds and downy mildews but they are not fungi. Oomycetes are coenocytic and have multiple nuclei per cell. Many are saprobic and feed on dead material. Others are infectious to plants. The Irish potato famine was caused by an oomycete.
- Rhodophyta, the red algae, are mostly multicellular and are part of a monophyletic group.
  - Rhodophyta are characterized by the pigment phycoerythrin, which gives them their characteristic red color. Depending on light intensity, red algae alter the ratio of pigments in the chloroplast. In high light conditions, red algae may appear very green.
  - Most red algae are marine, large and have specialized structures.
  - Red algae store the products of photosynthesis as floridean starch and have no motile cells at any point in their life cycles.
  - Red algae are the source of agar, which is important in food products as well as the study of biology.

- The green algae are not a monophyletic group. Chlorophyta, the largest lineage of green algae, is a monophyletic group and will be discussed here.
  - Chlorophytes contain chlorophylls *a* and *c*, store photosynthetic products as starch in plastids and are mainly aquatic.
  - Morphology of the chlorophytes is diverse and ranges from simple single-celled organisms to very complex multicellular organisms. Colonial and aggregate forms exist. Colonial forms show the first indications of cell specialization without multicellularity. Individual cells range from single-nucleated to coenocytic.
  - Life cycles within the chlorophytes are diverse. Figures 27.26 and 27.27 illustrate the life cycle of *Ulva* and *Ulothrix* respectively. Some chlorophytes are isogamous, indicating that their gametes are identical. Others are anisogamous and have female gametes much larger than the male gametes. Some chlorophytes have isomorphic life cycles and others are heteromorphic. In one variation of the heteromorphic life cycle—the haplontic life cycle—a multicellular haploid individual (gametophyte) produces gametes that fuse to form a zygote. Some chlorophytes exhibit diplontic life cycles much like animals, in which meiosis of sporophytes produces gametes directly.
- Choanoflagellida are thought to be most closely related to animals.
  - These organisms are flagellated and the majority are colonial.
  - Choanoflagellates are structurally similar to sponges, the simplest of all animals.

Endosymbiosis is the origin of chloroplasts

- Some chloroplasts are double membrane-bound and others are triple membrane-bound. This phenomenon can be explained by endosymbiosis.
- All chloroplasts can be traced to the engulfment of an ancestral cyanobacterium with a single membrane. Its engulfment in a vesicle of the phagocyte's membrane resulted in a double membrane. This initial event is called primary endosymbiosis. Primary endosymbiosis gave rise to the green and red algae.
- Photosynthetic euglenoids, with their triple membranes, arose from secondary endosymbiosis. In this case, a chlorophyte was ingested with its housed chloroplast, resulting in three membranes. Ultimately all of the constituents of the chlorophyte except the chloroplast were lost. Another line is thought to have arisen by means of a red algal endosymbiont.

Some body plans recur in several protist lineages.

- The amoeboid body plan is seen in several groups of protists. This body plan is characterized by pseudopods, cytoplasmic extensions that are used in locomotion. This body plan is highly effective in nutrient-rich environments. Some organisms have the amoeba body plan at different stages of their life cycle.
- Actinopods have very thin psuedopods reinforced by microtubules. The pseudopods increase surface area, help the organisms float, provide locomotion, and are used in feeding. Radiolarians, a specific type of actinopod, secrete glassy endoskeletons. Other types, called, heliozoans, lack an endoskeleton.
- Foraminiferans secrete shells of calcium carbonate and live either as plankton or on the sea bottom. Their psuedopods are used to trap food. Limestone deposits are often the result of discarded foraminiferan skeletons.
- Slime molds occur in three distinct groups. All groups are motile, feed by endocytosis, form spores on erect fruiting bodies, and undergo drastic changes during their life cycle.
  - Acellular slime molds form multinucleate masses. During the vegetative stage an acellular slime mold is a wall-less mass of cytoplasm with multiple diploid nuclei. It oozes over a network of strands called plasmodium. This phenomenon is called cytoplasmic streaming and is used to move and engulf food. If exposed to adverse environmental conditions the slime mold will form a dormant sclerotium from which it can turn back into a plasmodium when conditions are right. It may also transform into a fruiting structure called sporangiophores.
    - Sporangiophores form rigid walls between the nuclei. The walled nuclei go through meiosis to become spores. Spores germinate into swarm cells which can divide to produce more swarm cells. They can also release flagellated swarm cells, which fuse to form a zygote which can then form a new plasmodium.
  - Cellular slime molds are made of large numbers of cells called myxamoebas with single haploid nuclei. This life cycle can continue as long as conditions are favorable. Once conditions become unfavorable, myxamoebas secrete cAMP, aggregate, and form a motile slug that eventually produces fruiting bodies. Spores from the fruiting bodies germinate to form more myxamoebas. Sexual reproduction occurs with the fusion of two myxamoebas which then undergoes meiosis to release haploid myxamoebas.

The Big Picture

- Protists as well as all fungi, animals, and plants are eukaryotic. Their cells are characterized by membrane bound organelles, the presence of a cytoskeleton, and a nuclear envelope. The evolution of the eukaryotic cell is a major evolutionary milestone. The eukaryotic cell allows for compartmentalization of processes and increased adaptability. Though the evolutionary lineage is still debated, it is thought that eukaryotes evolved via multiple symbiotic events.
- Protists as a group are not monophyletic. Organisms in this kingdom are placed in it because they do not fit into other organism groupings. The unifying characteristic is the eukaryotic cell. Protists exhibit vast diversity and range from microscopic single-celled organisms to complex multicellular organisms. Several subgroups of protists are overviewed in this chapter. Some groups are monophyletic and others are phenetic.

- Certain body plans and modes of nutritional acquisition are repeated throughout the protist groups and are good examples of form following function. Protists as a group play a vast role in the environment and medicine, and are economically valuable.

Common Problem Areas

- There is a tendency for students to become overwhelmed with organism names and characteristics. Focus on the trends outlined in the chapter and use lab opportunities to understand the organism groupings.
- Endosymbiotic theory is often confusing upon first approach. Think of the great numbers of endosymbiotic organisms around you, from the flora of your own digestive tract to lichens on trees. Realizing that endosymbiosis is a common occurrence makes it easier to understand how it could give rise to eukaryotic cells.

Study Strategies

- This chapter has lots of information in it. It is best to break up your time and study small sections at a time.
- In order for you to learn the organism groups, it is best to create a chart with key characteristics. This will help you to compare and contrast the groups.
- Review the following tutorial and activities on the website
  Tutorial 27.1 Family Tree of Chloroplasts
  Activity 27.1 Ulva Life Cycle
  Activity 27.2 Ulothrix Life Cycle
  Activity 27.3 Anatomy of a Paramecium

Test Yourself

Knowledge and Synthesis Questions

1. Which of the following modes of reproduction can be found in at least some protists?
   a. Binary fission
   b. Sexual reproduction
   c. Spore formation
   d. All of the above

2. During the evolution of eukaryotes from prokaryotes, which of the following did *not* occur?
   a. Infolding of the flexible cell membrane
   b. Loss of the cell wall
   c. A switch from aerobic to anaerobic metabolism
   d. Endosymbiosis of once free-living prokaryotes

3. Which of the following statements about protists is *false*?
   a. Apicomplexans are the only protist group without parasitic representatives.
   b. Foraminiferans and radiolarians are shelled protists.
   c. Ciliates have great control over the direction their cilia beat.
   d. Although they appear structurally simple, amoebas are not primitive organisms.

4. Ciliates are characterized by protective structures called
   a. trichonympha.
   b. trichocysts.
   c. trichomes.
   d. throclea.

5. A major difference between the vegetative states of cellular and acellular slime molds is that
   a. acellular slime molds have haploid nuclei while cellular slime molds have diploid nuclei.
   b. acellular slime molds produce fruiting bodies and cellular slime molds do not.
   c. acellular slime molds use cAMP as a chemical cue for aggregation and cellular slime molds do not.
   d. acellular slime molds exist as a coenocytic mass and cellular slime molds exist as individual myxamoebas.

6. Isogamy is best characterized as
   a. indistinguishable male and female gametes.
   b. female gametes being significantly larger than male gametes.
   c. male gametes being significantly larger than female gametes.
   d. a diplontic mode of reproduction.

7. Red tides often cause massive fish kills and human illness in those eating shellfish. Which group of protists is responsible for red tides?
   a. Chlorophyta
   b. Rhodophyta
   c. Phaeophyta
   d. Chrysophyta

8. Holdfasts and alginic acid are characteristic of which group of protists?
   a. Chlorophyta
   b. Rhodophyta
   c. Phaeophyta
   d. Chrysophyta

9. A water mold is discovered growing on the surface of a freshwater green algae. Which two groups of protists are involved?
   a. Myxomycota and Rhodophyta
   b. Oomycota and Chlorophyta.
   c. Foraminifera and Chrysophyta.
   d. Apicomplexa and Ciliophora.

10. Which of the following statements concerning colonial organisms is *false*?
    a. Colonial organisms have division of labor among cells.
    b. Colonial organisms are multicellular.
    c. Colonial organisms are aggregates of single cells.
    d. None of the above

11. Which of the following statements regarding conjugation in *Paramecium* is *false*?
    a. Conjugation results in genetic recombination.
    b. Conjugation results in clones.
    c. Conjugation results in offspring.
    d. b and c

12. Which of the following regarding protists in general is *false*?
    a. Protists are always parasitic.
    b. Protists are single-celled.
    c. Protists are all heterotrophic.
    d. All of the above

Application Questions

1. What role did environmental conditions play in the evolution of the eukaryotic cell?
2. Explain alternation of generations. Why are the protists the first group in which this process could be seen?
3. Differentiate between asexual and sexual reproduction. Select one mode of protist sexual reproduction and explain it fully.
4. You have identified a new organism. What would the criterion be for placing it with the protists ?
5. Explain one line of thought regarding the origin of the eukaryotic cell. Why are scientists unsure of the evolutionary relationships among the protists?
6. Most protists are motile. Describe some of the means of motility.
7. Describe the origin of double membrane-bound and triple membrane-bound chloroplasts.
8. Many protists are significant human pathogens. Select one pathogenic protist, describe its life cycle, identify the protist group to which it belongs, and discuss the disease it causes.

Answers

Knowledge and Synthesis Answers

1. **d.** Sexual reproduction modes are quite varied among the protists.
2. **c.** The environment was becoming oxygen-rich. There was not a switch from aerobic to anaerobic metabolism but vise versa.
3. **a.** Malaria is caused by *Plasmodium*, which is an apicomplexan. All apicomplexans are parasitic.
4. **b.** Trichocysts are defensive barbs ejected from ciliates when they are disturbed.
5. **d.** The organization of the vegetative state differentiates cellular and acellular slime molds.
6. **a.** "Isogamy" means "same gametes." The gametes appear identical.
7. **d.** Dinoflagellates (Chrysophyta) are the cause of red tides.
8. **c.** Holdfasts and alginic acid are characteristics of the brown algae (Phaeophyta).
9. **b.** Water molds are Oomycota and green algae are Chlorophyta.
10. **b.** Colonial organisms are not truly multicellular though they do exhibit division of labor among cells.
11. **d.** Conjugation does result in genetic recombination but does not result in the production of clones or offspring.
12. **d.** Protists really do not have a unifying characteristic other than being eukaryotic; they do not fit into the kingdoms of Planta, Animalia or Fungi.

Application Answers

1. The switch from an oxygen-poor to an oxygen-rich environment was a selective pressure for organisms that could maintain a symbiosis with a tolerant organism. It also selected for organisms that could convert the harmful products of an oxygen-rich atmosphere to nontoxic products.
2. In alternation of generations an organism exists in a haploid gamete-producing form and a diploid spore-producing form. Bacteria have a single chromosome. Only in eukaryotes, with multiple copies of chromosomes, is a diploid portion of a life cycle possible.
3. Asexual reproduction results in clones of the original organism and there is no genetic recombination or variation possible. Sexual reproduction allows for new genetic combinations. Protists undergo varied types of sexual reproduction, from conjugation, to haplontic alternation of generations, to a diplontic type of reproduction.
4. The organism must be a eukaryote and not fit the criterion for being plant, animal, or fungus.
5. See Figure 27.3. The evolutionary relationships are difficult to understand due to limited fossilization, lateral gene transfers, and the contributions of symbiotic prokaryotes.
6. Protists move by flagella, cilia, and pseudopodia. Since many are aquatic, additional modifications allow for floating and movement along currents.
7. Double membrane-bound chloroplasts most likely arose from the endosymbiosis of a cyanobacteria in a process called primary endosymbiosis; whereas triple membrane-bound chloroplasts most likely arose from the endosymbiosis of a chlorophyte or a rhodophyte.
8. You may select from a variety of pathogenic protists: malaria caused by one of the apicomplexans, African sleeping sickness caused by a kinetoplastid, or others.

28 Plants without Seeds: From Sea to Land

Important Concepts

The plant kingdom has specific unique characteristics.

- Plants are photosynthetic eukaryotes that utilize chlorophylls *a* and *b* to produce and store carbohydrates as starches.
- Plants develop from an embryo that is protected by tissues of the parent plant.
- Plants undergo alternation of generations, with haploid gametophytes producing haploid gametes through mitosis and diploid sporophytes producing haploid spores through meiosis. The diploid sporophyte arises from the fusion of gametes to produce a diploid zygote that develops into the mature sporophyte. The extent of the development of the two life stages is characteristic of the different phyla. Nontracheophytes typically have a dominant gametophyte stage and highly reduced sporophyte stage. Progression through the tracheophyte phyla yields to groups in which the sporophyte stage is dominant and the gametophyte stage is highly reduced.
- Plants have either motile sperm or sperm that have lost flagella through evolution.
- There are twelve extant (surviving) groups of plants. These groups either lack vascular tissue and are called nontracheophytes (Hepatophyta, Anthocerophyta, and Byrophyta), or have vascular tissues and are called tracheophytes. Tracheophytes can further be subdivided into nonseed tracheophytes (Lycophyta, Sphenophyta, Psilotophyta, and Pterophyta) and seed tracheophytes. The seed tracheophytes are classified as gymnosperms (Cycadophyta, Ginkgophyta, Gnetophyta, and Coniferophyta) or angiosperms (Angiospermae). See Table 28.1 for characteristics of the phyla. Plants as a whole are monophyletic, but the classification into tracheophytes and nontracheophytes is phenetic, not monophyletic.
- Molecular and fossil evidence indicate that plants arose from the green algal lineage. We know that green plants most likely arose from a group of green algae called Charophytes which lived at edges of marshes and ponds. The cycles of wet and dry allowed the genetic modifications that would eventually lead to life on land.

Survival on land poses a series of challenges.

- Survival on land meant that organisms had to reduce their dependence on water. This posed a series of difficulties, for which adaptations were necessary.
- Terrestrial plants needed to support themselves against gravity if they were to grow upward.
- Terrestrial plants needed a mechanism to move water to all of the cells in the plant.
- Terrestrial plants needed a mechanism to prevent losing the water in their tissues to evaporation.
- Fertilization was dependent on motile sperm. Some mechanism of moving sperm from one plant to another in a dry environment was necessary.
- Terrestrial plants had to develop a mechanism to prevent dessication of the developing embryo.

Specific modifications were necessary for plants to make the initial transition to land.

- The earliest land plants developed characteristics that allowed life on land and led to the plants we now see. The earliest plants developed the following modifications for terrestrial life: waxy cuticle, gametangia, embryos, protective pigments, and thick spore walls.
- Waxy cuticles, though minimal in early plants, did prevent water loss from tissues as they were exposed to dry air. This waxy covering functions to prevent water loss from surface evaporation. The extent of the cuticle varies among plant groups and is most well developed in the more recently evolved gymnosperms and angiosperms.
- Multicellular gametangia mark the transition from algae to plant. Gametangia enclose the gamete-producing tissue and protect the developing gametes.
- Embryos are significant in that they contain all of the developing sporophyte and are housed within protective tissues. Recently evolved angiosperms have a complex protective mechanism.
- Terrestrial plants are without a UV screen of water; therefore, to protect against UV damage, pigments are necessary to screen out UV radiation.
- Spore walls allow for dispersion of spores without dessication. This is particularly important in the nonseed plants.

The development of vascular tissue was key to terrestrial existence.

- Not all land plants have vascular tissue; therefore, it is not an absolute requirement for terrestrial life. However, if vascular tissue were not present, plants could never be very large or very far away from a water source. The development of vascular tissue has allowed plants to occupy many more niches in the biosphere.
- Vascular tissue provides for the transport of water and food through a plant. The fundamental unit of a vascular system is a tracheid, a strawlike cell that conducts water in the plant. Though some mosses have cells that are precursors to tracheid development, nontracheophytes are said to lack developed vascular tissue.
- Plants without vascular tissue depend on abundant water sources, thin tissues that can pass water, mechanisms for capturing water vapor in the air, and capillary action to move water through the plant. They lack the support of lignin and closely hug the ground.

The nontracheophytes were the first plant colonizers of land.

- Most nontracheophytes live in dense mats that hug the ground and are in areas of high moisture content.
- Though small, nontracheophytes occupy more diverse habitats and, number-for-number, are far more abundant than any other plant type on earth.
- The life cycle of the nontracheophyte is dominated by the gametophyte generation. When you look at a nontracheophyte, it is the haploid gametophyte that you are observing. The sporophyte is very tiny and is completely dependent on the gametophyte. Figure 28.3 illustrates the life cycle of a moss as an example of the nontracheophyte life cycle. You should be able to follow this life cycle and understand which stages are haploid, which are diploid, and where meiosis takes place.
- Liverworts (Hepatophyta)
  - Liverworts are the oldest of the plant lineages. They are characterized by low, thin thalli that may be lobe-shaped or dissected into small leaflike structures. The thallus of a liverwort is only a few cells thick and has one-cell thick rhizoids on the underside for anchorage and minimal water absorption.
  - The life cycle of the liverwort is dominated by the gametophyte. The gametophyte raises up the gametangia on special reproductive structures called antheridiophores that bear antheridia (sperm-producing structures) and archegoniophores that bear archegonia (egg-producing structures). See Figure 28.5. Water droplets literally splash swimming sperm from the antheridia to the archegonia. After fertilization, the developing sporophyte remains attached to the archegoniophore. Specialized elater cells in the sporophyte help disperse the spores away from the mother plant.
  - Liverworts also reproduce asexually through small clumps of shed cells called gemmae. See Figure 28.5.
- Hornworts (Anthocerophyta)
  - Hornworts appear very similar to liverworts but differ in that they have actual functioning stomata that regulate gas exchange and prevent water loss during gas exchange.
  - Hornworts also have continually growing sporophytes. Though the sporophyte is dependent on the gametophyte, it can continue to grow until its water transport ability becomes limiting. See Figure. 28.6.
  - The structure of the chloroplasts of hornworts is also significantly different from the other nontracheophytes and the tracheophytes.
- Mosses (Bryophyta)
  - Mosses have specialized cells called hydroids. These hydroids are thought to be precursors to the tracheid seen in tracheophytes, but they lack lignin for structural support and waterproofing. Like tracheids, the hydroid cell dies and leaves a tiny channel that can transport water through the plant.
  - Mosses also differ because they exhibit apical cell division. This characteristic is seen in "higher" plants and allows for regular, upright growth.
  - Review the moss life cycle in Figures 28.3 and 28.4.

The tracheophytes have specialized vascular tissues.

- All tracheophytes are characterized by the presence of water-conducting tracheids. This is significant, because water can be transported throughout the plant in this tubular network and the lignin lining the tissue provides enhanced support for taller plants. This support network also allowed tracheophytes to fossilize much better then nontracheophytes. Thus our morphological knowledge of early tracheophytes is much more extensive than our knowledge of nontracheophytes.
- Tracheophytes are also characterized by branching, independent sporophytes.
- Figure 28.8 illustrates the relationships between the different groups of tracheophytes.
- The first tracheophytes were members of a now extinct group of plants called the rhinophyta. *Rhynia* have characteristics of extant tracheophytes and show the earliest features of this group, including independence of a branched sporophyte and tracheids, but they lacked more advanced characteristics such as true leaves and roots. See Figure 28.10.

Significant new features arose in early tracheophytes.

- Roots are believed to have evolved over time from branches that grew underground. Because branches aboveground and below-ground would have been exposed to different selective pressures, underground branches might have given rise to what we know today as roots.
- True leaves possessing vascular tissue arise late in the evolution of the tracheophytes. Leaves may be simple and consist of a single protrusion of vascular tissue. This type of leaf, thought to have arisen from sterile sporangia,

serves a protective function for the fertile sporangia. Leaves may be complex and arise from webbing between branched vascular tissue. This is thought to be a modification of a flattened stem system. See Figure 28.12.

- Most early tracheophytes exhibit homospory. This means that the spores developed by the sporophyte are all of the same type and may develop into "male" or "female" gametophytes. Later evolving sporophytes exhibit heterospory, whereby one spore type develops into a "megaspore" that gives rise to the female, egg-producing gametophyte and another spore type called a "microspore" develops into a male, sperm-producing gametophyte . Typically megaspore production is limited and microspores are abundant. Figure 28.13 differentiates between homospory and heterospory.

Extant nonseed tracheophytes are varied and abundant.

- Club mosses (Lycophyta) diverged early and are the most "primitive" of the extant nonseed tracheophytes. They have simple leaves arranged spirally on the stem. Their sporangia are held in apical strobili which are simple branching structures of fertile sporangia and sterile leaves. See Figure 28.14.
- Horsetails (Sphenophyta) are represented by few current species. They have true roots. Their sporophytes are large and independent and their gametophytes are highly reduced but also independent. The leaves are simple and form whorls around the stem. This group is one of the few plant groups that exhibits basal growth. See Figure 28.15.
- Whisk ferns (Psilotophyta) once were thought to be the "missing link" to *Rhynia*, but molecular evidence indicates that they evolved much later from more highly complex plants. This simple plan works well and indicates that evolution does not necessarily move toward greater complexity.
- Ferns (Pteriophyta) are the largest group of nonseed tracheophytes currently in existence. Ferns are characterized by large complex leaves with branching vascular strands. Like all other nonseed tracheophytes and nontracheophytes, ferns continue to be dependent on water to carry motile sperm. Ferns have advanced vascular structures and may reach great heights but are not considered "woody." Their root systems are also limited.

The fern life cycle has independent sporophytes and gametophytes.

- The fern life cyle is dominated by the sporophyte, but the gametophyte is an independent photosynthetic structure.
- Ferns have maximized spore dispersal by developing a complex spore dispersal system consisting of elaters and highly evolved sporangia that cast spores into the wind where they may be carried for long distances.
- Trace the fern life cycle in Figure 28.19. Keep in mind that some fern groups are heterosporous, even though the illustration shows a homosporous life cycle.

The Big Picture

- Plants are distinguished from other organisms by being photosynthetic eukaryotes that utilize chlorophylls *a* and *b*, undergoing alternations of generations, and develop from multicellular embryos that are protected by the parent plant. The twelve extant groups can be classified as nontracheophytes (those without highly developed vascular tissue) and tracheophytes (those with highly developed vascular tissue). This chapter focuses on the nontracheophytes and the nonseed tracheophytes.
- In order to colonize land, plants had to evolve strategies for dealing with desiccation and gravity. This involved mechanisms for extracting water from soil, means of transporting water throughout the plant, methods of insuring fertilization, and modes of protecting developing embryos. The plants discussed in this chapter have developed simple but effective means for dealing with desiccation and gravity.
- This chapter introduces and differentiates the liverworts, the hornworts, and the mosses, which are all considered nontracheophytes, and the club mosses, horsetails, whisk ferns, and ferns of the nonseed tracheophytes. Emphasis is on the adaptations each have for survival of conditions and life cycles.

Common Problem Areas

- It is very tempting to memorize plant types and characteristics, but it is more important to focus on the trends and relationships among the plants. Your goal is to understand the evolutionary relationships among the plants.

Study Strategies

- Focus on relationships among the plants and how to differentiate one group from another. You are interested in how they have adapted to survival on land.
- Try to get a visual picture of what each plant type looks like and what characteristics they have that contributed to their success on land, using the figures in your book.
- When studying life cycles, be sure to note whether the sporophyte or the gametophyte is dominant and make note of where mitosis and meiosis take place.
- Review the following tutorial and activities on the website:
  Tutorial 28.1 Life Cycle of a Moss
  Activity 28.1 Fern Life Cycle
  Activity 28.2 Homospory
  Activity 28.3 Heterospory

Test Yourself

Knowledge and Synthesis Questions

1. Plants may be differentiated from green algae by which of the following characteristics?
   a. Chlorophyll type
   b. Presence of multicellular gametangia
   c. Presence of roots
   d. Swimming sperm

2. In order to colonize land, plants needed to acquire which of the following characteristics?
   a. A mechanism for moving water throughout the plant
   b. A mechanism to prevent desiccation of tissues
   c. An ability to screen ultraviolet radiation
   d. b and c

3. The main differentiating factor between nontracheophytes and tracheophytes is
   a. lack of gametophytes.
   b. spore production.
   c. the presence of tracheids.
   d. All of the above

4. In alternation of generations, the sporophyte generation is _________ and the gametophyte generation is _________.
   a. haploid, diploid
   b. diploid, haploid
   c. haploid, haploid
   d. diploid, diploid

5. Which of the following characteristics can help differentiate between Hepatophytes and Byrophytes?
   a. Presence of hydroids
   b. Gametophyte dominance
   c. Sporophyte dominance
   d. Swimming sperm

6. During alternation of generations, meiosis takes place in the _________ to produce haploid _________.
   a. gametophyte, gametes
   b. sporophyte, gametes
   c. sporophyte, spores
   d. gametophyte, spores

7. Assexual reproduction in liverworts is accomplished by
   a. gametophytes.
   b. spores.
   c. gemmae.
   d. tracheids.

8. Hornworts have continually growing sporophytes that rely on the gametophyte for sustenance. Which of the following limit the size of the sporophyte?
   a. Water transport ability
   b. The gametophyte cannot produce enough nutrients
   c. Cells divide a finite number of times
   d. All of the above

9. You are walking along a roadside and find a plant with the following characteristics: very thin waxy cuticle, stomata, simple leaves in whorls around a central stem, independent sporophyte and gametophyte, sporangia in strobili. This plant is most likely a member of which of the following phyla?
   a. Bryophyta
   b. Sphenophyta
   c. Pterophyta
   d. Lycophyta

10. Match the phyla with their diagnostic characteristics.

____ Bryophyta	a. no true stomata
____ Hepatophyta	b. terminal sporangia
____ Anthocerophyta	c. complex branching leaves
____ Lycophyta	d. unique chloroplasts
____ Sphenophyta	e. first strobili of sterile leaves and sporangia
____ Psilotophyta	f. presence of hydroids
____ Pterophyta	g. basal growth patterns

Application Questions

1. Explain why the largest of mosses are less than a meter tall.
2. Diagram the evolutionary relationships among the nontracheophytes and nonseed tracheophytes. Label the major differentiating characteristics at each branch of your tree.
3. Discuss how the relationship between the sporophyte and gametophyte changes as you move from the first nontracheophytes to later nonseed tracheophytes.
4. Explain how leaves and roots are thought to have arisen. Are their origins similar or different?
5. Compare and contrast the Bryophyte life cycle and the Pterophyte life cycle.
6. Motile sperm are retained in all the plants discussed in this chapter. Describe how land plants cope with the issue of motile sperm.
7. Discuss the challenges of land dwelling that plants needed to address for survival.
8. Compare and contrast homospory and heterospory. Why is microspore production abundant but megaspore production limited?

Answers

Knowledge and Synthesis Answers

1. **b.** Multicellular gametangia differentiate plants from green algae. Green algae and plants have the same type of chlorophyll. Not all plants have roots or swimming sperm.
2. **d.** This is correct because there are several successful groups of plants that lack mechanisms for moving water throughout the plant.
3. **c.** Tracheids are found only in the tracheophytes.

4. **b.** Meiosis occurs in the diploid sporophyte to produce haploid spores that develop into the gametophyte.
5. **a.** Bryophytes are the only group with hydroids, the precursors to true vascular tissue.
6. **b.** See answer to number 4.
7. **c.** Liverworts have specialized asexual reproductive structures called gemmae.
8. **a.** The sporophyte eventually reaches a size at which water cannot be transported to its tissues.
9. **b.** You have come across a horsetail. These plants are very common along roadsides in damp ditches, particularly in the Midwest.
10. **f.** Bryophyta
    a. Hepatophyta
    d. Anthocerophyta
    e. Lycophyta
    g. Sphenophyta
    b. Psilotophyta
    c. Pterophyta

Application Answers

1. Mosses have very rudimentary water transport cells called hydroids. Hydroids lack the waterproofing and support molecule lignin. Because of this, they can carry water only short distances and cannot support tall plant parts.
2. See Figure 28.8 and Table 28.1.
3. Early nontracheophytes have reduced sporophytes that are highly dependent on the gametophyte for nutrition. Advanced nonseed tracheophytes have independent sporophytes and gametophytes and the gametophyte is highly reduced.
4. Complex leaves and roots are thought to be modifications of stems. Simple leaves are thought to be derived from sterile sporangia. See Figure 28.12.
5. Pay particular note to the relative dominance of the sporophyte versus the gametophyte. Use Figures 28.3 and 28.19.
6. Motile sperm mean that there is dependence on water for fertilization. These plants have derived a variety of mechanisms for taking advantage of what little water is available. Some rely on splashing of sperm from plant to plant (Bryophytes, Hepatophytes), others have elaborate spore dispersal mechanisms and hormonal attractants that insure gametophytes are close and are of the correct sex (Pterophytes).
7. Land dwelling poses problems of support, water conduction, UV radiation, water loss, and embryo protection.
8. See Figure 28.13.

STUDY GUIDE

29 The Evolution of Seed Plants

Important Concepts

Seed plants have characteristics that set them apart from nonseed plants.

- Seed plants consist of two groups of tracheophytes, the gymnosperms and the angiosperms. Gymnosperms consist of the following phyla: Cycadophyta, Ginkgophyta, Coniferophyta, and Gnetophyta. Angiosperms currently consist of only one phylum, angiospermae.
- Seed plants have highly reduced gametophyte generations that are dependent on the sporophyte for survival. When you look at a seed plant, what you see is the sporophyte.
- Very few seed plants have retained swimming sperm. In most, the male gametophyte has evolved for dispersal.
- All seed plants are heterosporous. Microspores and megaspores develop in specialized cones or in flowers. Megaspores are not shed but retained within the megasporangia. The megasporangia, surrounded by sterile sporophyte tissues (integument), forms the ovule and will be the basis of the seed. The presence of this ovule separates seed plants from nonseed plants. Microspores develop into pollen grains that are shed and used for dispersal.
- Only one of the products of meiosis in the megasporangium is retained. This resulting cell divides by mitosis to form the multicellular gametophyte. Pollen grains consist of sperm nuclei and supporting cells again formed from mitosis and subsequent mitosis. Pollen produces pollen tubes that deliver the sperm nuclei to the egg nuclei of the female gametophyte for fertilization. The resulting embryo grows to a defined state and then goes dormant within the surrounding tissues. This dormant, protected embryo and its surrounding tissues form the seed.
- Seeds protect the embryo until conditions are right for germination. Many seeds have adaptations that allow dispersal by wind or by vector.

Gymnosperms are naked seed plants.

- Gymnosperms are seed plants but they do not form flowers or true fruits. They are often called "naked seed plants."
- Gymnosperms are varied and live in a variety of habitats, from sparse deserts to vast forests. The Coniferophyta are the most abundant of the gymnosperms.
- Gymnosperms have significant vascular tissue and are capable of woody secondary growth.
- The gymnosperm life cycle can be illustrated by the pine, a member of the Coniferophyta. Conifers differ from the other phyla in that they produce cones, which are specialized structures for reproduction. There are separate microspore-producing cones and megaspore-producing cones.
- Pines do not have swimming sperm and the pollen is modified for wind dispersal. Pollen lands on the female cone and lodges in the pollen chamber. The pollen tube grows down through the micropyle to the female gametophyte where it fertilizes the egg nucleus within the archegonium. The resulting zygote develops into an embryo that remains encased in the tissues of the megasporangium and gametophyte. The seed is additionally protected by the sterile scale of the cone until it is mature and ready for dispersal. Dispersal is aided by modifications of the seed coat. See Figure 29.5.

Angiosperms are characterized by flower formation.

- Angiosperms are a diverse group of plants that all produce flowers and are thus called flowering plants. They are currently the dominant plant form on Earth.
- Angiosperms have the most reduced gametophyte generation and the most highly developed sporophyte generation of all plant types.
- Angiosperms differ from all other plants in that they produce triploid endosperm, have double fertilization, their ovules are enclosed in an ovary, they may have multiple carpels, they produce flowers and fruits, and their xylem and phloem have multiple modified cell types including vessel elements, fibers, and companion phloem cells.

Flower structure is varied among plants but consists of similar structural elements.

- Flowers are the result of modification of leaves. All flower parts can be traced to leaf structures. Though all flowers are structurally different, they have common parts. See Figure 29.6 for a diagram of floral structures.
- The reproductive structures of the flower consist of the stamens, which produce microspores, and the pistil, which houses the ovule and receives the pollen. The additional floral structures are supporting structures and help to insure pollination. The stamen is composed of the filament, which supports the anther. The anther is the site of microspore production. The pistil can be subdivided into the stigma, which forms a sticky surface and exudes materials

necessary for pollen adhesion and pollen tube germination; the style, which provides nutrition for the developing pollen tube; and the ovary, which houses the ovule. A flower may have multiple ovaries, each housed in its own modified leaf called a carpel.

- Nonreproductive floral structures include petals, sepals, and the receptacle. The receptacle is important in some fruit formation.
- Flowers that have both microsporangia-producing structures and megasporangia-producing structures are called perfect. If either structure is missing it is called imperfect. Plants that have both megasporangiate flowers and microsporangiate flowers on the same plant are called monoecious. If they are on separate plants they are called dioecious.
- Natural selection has led to the evolution of many varied flower forms. Each form serves to best insure pollination of a particular plant.
- Many plants and animals have coevolved such that the nutrition of the animal and pollination of the plant are interdependent. The products of this coevolution can be highly specific in terms of species-to-species interaction or may be more generalized.

The angiosperm life cycle is dominated by the sporophyte.

- The angiosperm life cycle is different from that of all other plants in that double fertilization occurs. Double fertilization results in a diploid embryo and triploid endosperm. The triploid endosperm provides nutrition for the developing embryo. Review Figure 29.10 for the angiosperm life cycle. Be sure you understand where the endosperm comes from.
- Angiosperm embryos have one or two seed leaves called cotyledons. The fate of the cotyledons serves to differentiate between two main lines of flowering plants.
- Fruits develop from the ovary and its supporting tissues. Fruit types depend on the number of carpels associated with the fruit and the extent of support tissue incorporated into the fruit structure.

Evolutionary relationships among angiosperms are not completely understood.

- The evolutionary relationships and the origin of angiosperms are still a significant area of study for many botanists. Originally thought to trace to the magnolia line, the basis of angiosperm evolution is now traced to a line consisting of only one extant species called Amborella.
- Angiosperms can be divided into two monophyletic groups, the eudicots with two embryonic seed leaves and the monocots with only one seed leaf. Differentiation among groups beyond this division is still an active area of study without clear results.

The Big Picture

- Seed plants can be characterized into two main groups: the gymnosperms and the angiosperms. Seed plants have developed complex modes of protecting the embryo, dispersing gametes, and moving water, nutrients, and food throughout the plant. Both groups are characterized by the presence of seeds for protecting the embryo until conditions are correct for germination. The main difference between gymnosperms and angiosperms is the presence of the flower and the ovary in angiosperms. The flower insures fertilization and allows for the development of protected seeds. Both groups have highly developed vascular tissue, may exhibit secondary woody growth, and show significant diversity.
- There is significant variation among flower types, but the basic flower consists of the structures diagrammed in Figure 29.6. Many plants and animals have coevolved such that the flower structure and the animal are integrally linked for nutrition of the animal and pollination of the plant. The angiosperm life cycle is different from that of all other plants in that double fertilization occurs. This allows for nutrient formation that is used during germination.
- Both the angiosperm and gymnosperms life cycles are dominated by the sporophyte. Both groups are heterosporous. Megasporogenesis and microsporogenesis are similar in angiosperms and gymnosperms. Review the similarities and note the differences in the two life cycles.

Common Problem Areas

- Not all gymnosperms are pines. The examples used in this chapter focus on the pine, but gymnosperms are a varied group and cones are not a diagnostic characteristic of this group. Cones define the conifers only.
- Angiosperms are extremely varied as well. Not all flower parts are found on all plants, and some are highly modified. Also note that flowers coming from florists are often sterile, have aberrant parts, and may be altered through selective breeding and treatment.

Study Strategies

- Recall that you are interested in evolutionary trends among the plant groups. Focus on how gymnosperms and angiosperms have met the challenges of desiccation and gravity. You should connect the knowledge gleaned from this chapter to that of Chapter 28.
- Be sure that you truly understand what sets angiosperms and gymnosperms apart. Focus on their life cycles for definitive characteristics.
- Review the following activities on the website:
  Activity 29.1 Flower Morphology
  Activity 29.2 Life Cycle of a Pine

Test Yourself

Knowledge and Synthesis Questions

1. You are enjoying a stroll in the botanical gardens. You notice a plant with a beautiful flower. Upon closer inspection you find that the flower has a pistol but no anthers. You look at several more flowers on the same

plant and they are also missing anthers. This plant is ____________ and its flowers are ____________.
a. monoecious, perfect
b. monoecious, imperfect
c. dioecious, perfect
d. dioecious, imperfect

2. Your sweetheart sent you an incredible rose with many petals on it. When you look closely at the rose, it appears that the petals have sterile anthers attached to them. Which of the following best explains this phenomenon?
   a. Both anthers and petals are derived from leaf structures; therefore, the genes of the anther could be altered to form a petal.
   b. Anthers are derived from petals.
   c. The anthers and the petals probably fused.
   d. None of the above

3. Why are gymnosperms referred to as "naked seed plants?"
   a. They lack ovules.
   b. They lack ovaries.
   c. They don't protect their embryos.
   d. They have no seed coats.

4. Which of the following functions make seeds useful to plants?
   a. Seeds provide a mechanism for dispersal.
   b. Seeds protect the embryo.
   c. An embryo may remain dormant until optimum growth conditions are available.
   d. All of the above

5. Many angiosperms and animals have coevolved. What roles do animals play in the life cycle of plants?
   a. They act as pollinators.
   b. They assist in dispersal of seeds.
   c. They insure fertilization.
   d. All of the above

6. The product(s) of fertilization in gymnosperms is (are) ____________ and in angiosperms is (are) ____________.
   a. endosperm and embryo, embryo
   b. embryo, endosperm and embryo
   c. embryo, embryo
   d. embryo, endosperm

7. In gymnosperms there is(are) ____________ functional sperm nuclei and in angiosperms there is(are) ____________ functional sperm nuclei.
   a. one, two
   b. two, one
   c. four, one
   d. one, one

8. The diagnostic characteristic of an angiosperm is the presence of
   a. multiple carpels.
   b. woody growth.
   c. the flower.
   d. All of the above

9. Which of the following statements regarding gymnosperms is true?
   a. All gymnosperms produce cones.
   b. Gymnosperms are heterosporous.
   c. Gymnosperm seeds have no protection.
   d. Gymnosperms are all woody.

10. Match the following flower structures with their function.

____ Ovule	a. Leaf modification that protects the ovary
____ Anther	b. Colorful display that assists in attracting pollinators
____ Stigma	c. Secretes sticky material to help pollen adhere
____ Style	d. Site of ovule development
____ Carpel	e. Holds the pistol, forms some fruits
____ Ovary	f. Surrounds petals
____ Petal	g. Houses the egg
____ Sepal	h. Nourishes the elongating fertilization tube
____ Receptacle	i. Found in plants in which the petals and sepals cannot be differentiated
____ Tepal	j. Site of microsporogenesis

11. What is the significance of the fruit?
    a. It aids in dispersion of seeds.
    b. It protects seeds until they are mature.
    c. It attracts pollinators.
    d. a and b

12. Vascular tissue in angiosperms is highly developed. The purpose of this vascular tissue is to
    a. move water.
    b. move food.
    c. move nutrients.
    d. All of the above

Application Questions

1. Discuss the significance of double fertilization in angiosperms.
2. Discuss the differences between the following terms: ovary, ovule, carpel. Which of the structures allows you to differentiate between gymnosperms and angiosperms?
3. Refer back to the obstacles plants had to survive on land. How have gymnosperms surmounted those obstacles?
4. Pollen is found only in seed tracheophytes. What is the significance of pollen, and from what structures is it derived? Is it sporophyte or gametophyte?
5. Where would you find a female gametophyte in an angiosperm? Explain your answer.
6. Using a conifer as an example, describe heterospory in gymnosperms.
7. What is a cotyledon? How is it used to classify angiosperms into two monophyletic groups?
8. What is the significance of the angiosperm fruit? What types of tissues does the fruit arise from?

Answers

Knowledge and Synthesis Answers

1. **d.** If upon inspection you find only female flowers then you are most likely looking at a dioecious plant in which male and female flowers appear on separate plants. The fact that the flower is missing anthers makes it an imperfect flower.
2. **a.** All floral structures are derived from leaves. The genes necessary for any other part are there, but are "turned off."
3. **b.** Gymnosperms lack ovaries and thus lack the ability to produce fruit. Their embryos are protected and they have ovules and seed coats.
4. **d.** Seeds protect the embryo, provide a mechanism for dispersal, and allow an embryo to remain dormant.
5. **d.** By acting as pollination vectors, animals insure fertilization of the plant. Animals also assist with dispersal of seeds.
6. **b.** Angiosperms differ from gymnosperms by having double fertilization and producing endosperm.
7. **a.** Double fertilization occurs only in angiosperms.
8. **c.** Angiosperms are different from all other plants because they have flowers. It is true that only angiosperms can have multiple carpels, but multiple carpels are not a necessary characteristic of an angiosperm.
9. **b.** All gymnosperms are heterosporous. Only conifers are cone producers.
10. **g.** Ovule
    j. Anther
    c. Stigma
    h. Style
    a. Carpel
    d. Ovary
    b. Petal
    f. Sepal
    e. Receptacle
    i. Tepal
11. **d.** Fruit does not attract pollinators.
12. **d.** Vascular tissues in angiosperms move food, water, and nutrients throughout the plant.

Application Answers

1. Double fertilization allows for a diploid embryo and a triploid endosperm. The endosperm provides nutrition to the developing embryo at the time of seed germination.
2. An ovule is a composite of sporophyte and gametophyte tissue and is the basis of seed development. The ovary is the site of megasporogenesis. The carpel may house multiple ovaries and ovules.
3. Gymnosperms have a sophisticated vascular system that allows movement of nutrients and also support for large plants competing for sunlight. Wind-borne pollen removes the dependence on water for fertilization and increases the range of gymnosperms. Seeds protect embryos from dessication and provide a mechanism for dispersal.
4. Pollen is the male gametophyte of the gymnosperm. It is significant in that it provides a means of dispersal and eliminates the need for water for fertilization. The sporophyte undergoes meiosis to produce the haploid microspore. Each microspore goes through mitosis to produce the resulting gametophyte that is then shed.
5. The female gametophyte is highly reduced and is part of the ovule of the angiosperm.
6. Microspores go on to develop into pollen. Microspores along with sporophyte tissue develop into ovules. See Figure 29.5.
7. A cotyledon is the first "seed leaf" produced in an angiosperm embryo. Monocotyledons have only one seed leaf whereas eudicotyledons have two. The monocotyledons are believed to have arisen later during evolution through reduction.
8. Angiosperm fruit arises from tissues of the sporophyte and the gametophyte. How that fruit comes to develop depends on fruit type. Different fruits encompass more or less of the parent sporophyte tissues.

30 Fungi: Recyclers, Killers, and Plant Partners

Important Concepts

General features of fungi

- Fungi absorb the nutrition needed for their survival from dead matter (saprobes), from living hosts as parasites, and from mutually beneficial symbiotic relationships with other organisms.
- Fungi produce spores and reproduce sexually.
- The cell walls of fungi are made up of the polysaccharide chitin.
- The phyla Chytridiomycota, Zygomycota, Ascomycota, and Basidiomycota make up the kingdom Fungi (see Table 30.1). About 25,000 species of fungi lack sexual stages and have yet to be placed in one of the four phyla; these are known as the imperfect fungi or deuteromycetes.

The unicellular fungi

- All of the fungal phyla contain unicellular fungi commonly called yeasts.
- Yeasts reproduce by budding, by fission, or sexually.

The body of a fungus is composed mainly of hyphae.

- Most fungi are multinucleated, rather than multicellular.
- The mycelium is the vegetative body of the fungus composed of tubular filaments. The tubular filaments, or hyphae, have no separate cells, which allows organelles to move about freely (see Figure 30.3).
- Partial cross-walls known as septa subdivide some hyphae, while those with no septa are known as coenocytic.
- Rhizoids are hyphae that anchor fungi from the phylum Chytridiomycota to the substrate.
- In fungi that are plant parasites, hyphae enter a leaf from a spore on the surface and form mycelium within the leaf (see Figure 30.4).
- The hyphae provide a large surface area-to-volume ratio providing a large surface area for absorption to occur. The hyphae are all close to the food source.
- Fungi are tolerant to hypertonic environments and temperature extremes.

Fungi acquire nutrients in a variety of ways.

- Saprobes secrete enzymes into the environment that help in absorption of dead matter and the decomposition and recycling of elements used by living organisms.
- Many fungi can use both nitrate and ammonium as a source of nitrogen.
- Parasitic fungi are either facultative, possessing the ability to grow by themselves, or obligate, needing their living host for growth.
- Predatory fungi may trap microscopic animals in a sticky substance, while other species form a constricting ring around their prey.
- Fungi form associations with plants known as mycorrhizae and with bacteria known as lichens.

Fungi reproduce sexually and asexually.

- Fungi reproduce asexually by the production of haploid spores and conidia spores, cell division, and breakage of mycelium.
- Sexual reproduction involves two mating types rather than a morphological male and female.
- Self-fertilization is prevented because individuals of the same mating type cannot mate with one another.
- The zygote nuclei are the only diploid nuclei in the life cycle of a fungus.
- A dikaryon stage with two nuclei occurs in some fungi as the first stage of sexual reproduction. The cytoplasm of two individuals fuses, but the nuclei do not fuse at this time. A fruiting structure is eventually formed and the two nuclei fuse to form a zygote. This form of reproduction has no gamete cells and no true diploid tissues ($2n$), but are instead $n + n$.

The phylum Chytridiomycota

- The chytrids are the earliest diverging fungi.
- They are aquatic with chitinous cell walls and are the only fungi with flagella.
- The genus *Allomyces* employs alternation of generations. A haploid zoospore is produced and becomes a small organism with both female and male gametangia that produce gametes. When the gametes fuse they form a diploid zygote and a diploid organism. The diploid organism produces diploid zoospores. Finally, the diploid organism produces sporangia that give rise to haploid zoospores.

The phylum Zygomycota

- Over 900 species of zygomycetes have been described. The hyphae spread and reproduce asexually by producing stalked sporangiophores that contain sporangium and spores.
- Sexual reproduction of Zygomycota occurs between adjoining individuals with different mating types (see Figure 30.9).
- Branches from each individual grow toward each other until they fuse to produce gametangia. A thick-walled zygosporangia with a zygospore forms at the fusion site. The sporangium sprouts from the zygospore.

The phylum Ascomycota

- The ascomycetes are distinguished by their ascus, a sexual reproductive structure. There are approximately 30,000 species divided into two groups. The euascomycetes have an ascocarp, while the hemiascomycetes lack an ascocarp.
- Most hemiascomycetes are unicellular and reproduce asexually by fusion or budding.
- Sexual reproduction occurs between two mating types and involves the forming of a diploid cell population in some species, or direct meiosis of the zygote. The occurrence of meiosis results in the whole cell becoming an ascus with either eight or four ascospores.

The euascomycetes produce a dikaryon during sexual reproduction.

- Euascomycetes include molds and the cup fungi and have a number of uses for humans. Mold from the genus *Penicillium* produces the antibiotic penicillin and others provide flavor for cheese. Brown mold is used in brewing and soy sauce.
- Asexual reproduction involves the production of conidia at the tip of hyphae.
- A dikaryon is produced during sexual reproduction by the fusing of two mating structures (see Figure 30.13). From the dikaryon hyphae ($n + n$), dikaryotic ascus form and the nuclei fuse. The zygote ($2n$) undergoes meiosis and then produces ascus that are shed to produce a new individual (n).

The phylum Basidiomycota

- Puffballs and mushrooms make up some of the 25,000 species of fungi in the phylum Basidiomycota.
- In basidiomycetes meiosis and nuclear fusion occurs in the basidium, found at the tips of hypha (see Figure 30.15). After meiosis, four basidiospores are formed as tiny stalks. The basidiospores are released into the environment and germinate to form haploid hyphae. The hyphae grow and fuse with different mating types to form the dikaryotic mycelium. The fruiting structure, or basidiocarp, grows from the dikaryotic mycelium. The cap of a mushroom is the fruiting structure. Basidia form on the underside of the cap along gills and discharge their spores which will form the new developing basidium.

Fungus forms mutually beneficial symbiosis with plants.

- Tracheophytes form associations with fungi. The root hairs of these plants become infected with fungi forming an association called mycorrhiza. The fungus helps to increase the water and nutrient uptake of the plants, while obtaining nutrients and amino acids from the plants.
- Plants that live in nitrogen-poor areas form mycorrhizae.
- Lichens are associations between algae and fungi. Ascomycetes and basidiomycetes form the fungi component of lichens. The algal cells leak photosynthetic products to the fungi. Lichen reproduces by the formation of a vegetative body called a thalus or by producing algae surrounded by hyphae in soredia.
- On new rock, lichens are the first organisms to invade, and they are very sensitive to the environment.

The Big Picture

- The kingdom Fungi are comprised of heterotrophic organisms that absorb nutrients from the environment. Fungi can be unicellular or multicellular. The mycelium is the body of a multicellular fungus and is composed of many tubular filaments known as hyphae. The hyphae are not totally divided; they are actually multinucleated cells. Some fungi form symbiotic relationships with both plants and algae.
- Reproduction in the fungi occurs both sexually and asexually. Asexual reproduction involves the production of spores, breakage, or budding. Sexual reproduction requires the union of hyphae with two different mating types.
- The four phyla in the kingdom Fungi are Chytridiomycota, Zygomycota, Ascomycota, and Basidiomycota. The Chytridiomycota are aquatic fungi that may display alternating generations and motile gametes.
- The Zygomycota have hyphae that spread throughout the environment. Sexual reproduction occurs when hyphae of two mating types join to form a zygosporangia from which a sporangium will eventually grow.
- The Ascomycota produce an ascus as their reproductive structure. The species of Ascomycota are separated into the euascomycetes if the ascus is contained in a fruiting structure (an ascocarp), or the hemiascomycetes if the ascocarp is absent. The hemiascomycetes reproduce by budding, fission, or the production of an ascus. The euascomycetes have a dikaryotic ($n + n$) stage during reproduction.
- In the Basidiomycota the basidium is the reproductive structure. The fusion of haploid hypae form a dikaryotic mycelium that develops the basidiocarp.

Common Problem Areas

- Students are often confused by the numerous ways that fungi reproduce. Create flow charts to help you understand the ploidy level at each stage of the life cycle in the different phylum of fungi. Make sure to include both the sexual and asexual stages in your charts. Comparisons of your flow charts for the different phyla will help you learn the differences between the groups.

Study Strategies

- Fungi are placed in particular systematic groupings because they share unique characteristics. Try to learn these features, along with the names of the groups, and look for patterns that distinguish major groups from each other. Many of the fungi are grouped in the same phylum based on the reproductive structures that they produce.
- Review the following tutorial and activities on the website:
  Tutorial 30.1 Life Cycle of a Zygomycete
  Activity 30.1 Fungal Phylogeny
  Activity 30.2 Euascomycete Life Cycle

Test Yourself

Knowledge and Synthesis Questions

1. The phyla of fungi are distinguished based on
   a. methods of sexual reproduction.
   b. methods of asexual reproduction.
   c. presence or absence of cross-walls separating the cell-like compartments.
   d. All of the above

2. Which of the following would you *not* expect to find in the life cycle of a typical fungus?
   a. Haploid cells
   b. Diploid cells
   c. Spores
   d. Chloroplasts

3. All fungi are absorptive heterotrophs. Which of the following is an adaptation shown by many fungi that greatly aids this mode of nutrient procurement?
   a. Dikaryosis
   b. Large surface-to-volume ratio
   c. Conjugation
   d. Complex life cycle

4. Assume that two normal hyphal cells of different fungal mating types unite. After a period of time the cell walls between these cells will dissolve, producing a
   a. mycelium.
   b. fruiting body.
   c. zygote.
   d. dikaryotic cell, which is also heterokaryotic.

5. The dikaryotic hyphae are
   a. $2n$
   b. n
   c. $n + n$
   d. n/n

6. Which of the following statements about sexual reproduction in fungi is *false*?
   a. Motile gametes are present in all fungal species.
   b. Water is not required for fertilization to occur in fungi.
   c. There is no true diploid tissue in the life cycle of most sexually reproducing fungi.
   d. Sexual reproduction often begins with contact between opposite mating types.

7. A lichen is
   a. an imperfect fungi.
   b. the fruiting structure of the Basidiomycete
   c. a symbiotic association of fungus and algae.
   d. a symbiotic association of fungus and plants.

8. Which of the following is not a life form of fungi?
   a. Saprobe
   b. Parasite
   c. Mutualist
   d. Herbivore

Application Questions

1. Complete the following diagram, which shows the generalized life cycle of a fungus, by adding the following terms in their appropriate locations: spores, dikaryotic hyphae, meiosis, mitosis, fertilization. Also, circle and label the portions of the life cycle corresponding to the haploid (n), diploid ($2n$), and the dikaryotic ($n + n$) stages.

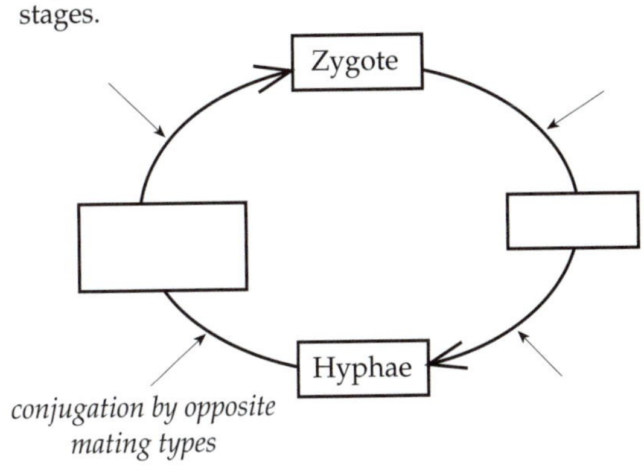

Answers

Knowledge and Synthesis Answers

1. **d.** All of the traits can be used to help classify the fungi.
2. **d.** Fungi have haploid and diploid cells at different stages of their life cycle. They also produce spores. The chloroplasts are not found in the fungi, but in plants.
3. **b.** The large surface area of the hyphae increases their ability to uptake nutrients.
4. **d.** When two hyphal cells of different mating types fuse, they form a dikaryotic cell that is heterokaryotic.
5. **c.** The dikaryotic hyphae are $n + n$.
6. **a.** Not all of the phyla in the kingdom Fungi have mobile gametes. The gametes of the Chytridiomycota are motile.
7. **c.** Lichens are symbiotic relationships between a fungus and an algae. The algae provides nutrients for the fungus.
8. **d.** Fungi are not herbivores.

Application Answers

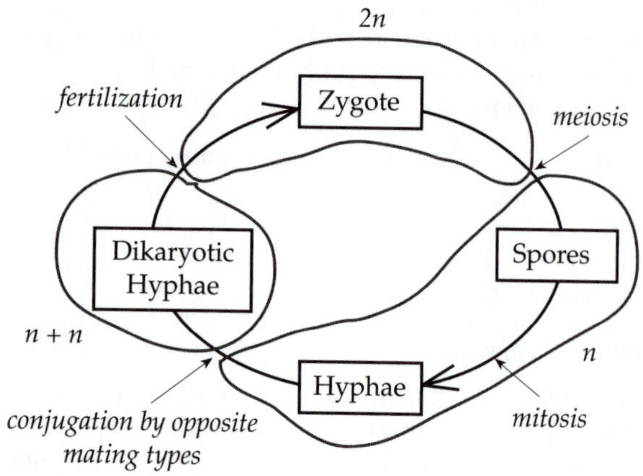

31 Animal Origins and Lophotrochozoans

Important Concepts

Animals are descendants of a common ancestor.

- The animal kingdom is a monophyletic group with all animals descended from a single lineage.
- All animals have similarities in their 5S and 18S ribosomal RNA. They all have cell-to-cell junctions that are tight junctions, desmosomes, and gap junctions. Animals share a set of extracellular matrix molecules.
- Colonial flagellated protists appear to be the common ancestor of modern animals.
- Animals are characterized as multicellular organisms that must take organic molecules in with their food. They also have differentiated cells that coordinate their activities.

Animals are classified based on morphology and genetics.

- The classification of animals has typically been based on an animal's developmental pattern and morphology. With the advent of modern molecular techniques, small subunit rRNA and mitochondrial genes have become common tools used for classification.
- The sponges, cnidarians, and ctenophores diverged from the rest of the animal lineage early in evolutionary history (see Figure 31.2). The remaining animals are categorized as either protostomes or deuterostomes.
- Most of the protostomes undergo complete radial cleavage, but one lineage has spiral cleavage. The blastopore of the protostome embryo becomes the mouth.
- Deuterostomes have indeterminate, radial cleavage. The anus of the deuterostome will be formed by the blastopore.
- Diploblastic animals have two cell layers during development (ectoderm and endoderm), and triploblastic animals have three layers during development (ectoderm, mesoderm, and endoderm).
- Animals have three different types of body cavities (see Figure 31.1). Acoelomates have no enclosed body cavity, pseudocoelomates have a liquid-filled space known as the pseudocoel where many of the internal organs are located, and the coelomates have a true body cavity, a coelom, developing within the mesoderm. In coelomates, the internal organs are in pouches of the peritoneum.

All animals have a body plan.

- A body plan describes how the body of an animal is laid out. The symmetry of a body is a key descriptor of an animal body plan.
- An animal that can be divided along at least one plane into similar halves is said to be symmetrical. Asymmetrical animals, such as the sponges, have no plane of symmetry.
- Spherical symmetry is defined as a body plan in which all body parts radiate from a central point.
- The sea anemone has radial symmetry, meaning that the body can be divided by any plane running along a main axis of the body.
- The sea star has biradial symmetry; the body can be divided into two planes that are at right angles to each other (see Figure 31.3).
- Animals exhibiting bilateral symmetry can be divided into two mirror images by a single plane (see Figure 31.3). A second plane at a right angle divides the animal into two dissimilar halves. A common feature of bilateral symmetry is cephalization, or the presence of a head.
- Animals have evolved a skeleton for support that can be internal or external and rigid or flexible.

Sponges lack a symmetrical body plan.

- About 10,000 marine species and 50 freshwater species of sponges are found in the phylum Porifera. They are groups of cells loosely organized around a water canal system. The Porifera are the only animals that do not have distinct cell layers or symmetrical bodies.
- Sponges filter small organisms and nutrient particles out of the water as it flows thorough pores in the walls of the inner cavity. Choanocyte cells use flagella to divert water through the pores and filter out the food particles. The water exits the sponge through large openings called oscula (see Figure 31.4).
- Sponges have a supporting skeleton composed of branching spines (spicules) or an elastic network of fibers.
- A single sponge reproduces sexually by producing both egg and sperm and asexually by budding and fragmentation.

Cnidarians have two stages in the life cycle, the polyp and the medusa.

- The 10,000 or so species in the phylum Cnidaria are composed of three classes: Hydrozoa, Scyphozoa, and Anthozoa (see Figure 31.6). All of the Cnidariians are diploblastic and have biradial symmetry.
- Cnidarians' low metabolic rates, coupled with their ability to catch prey, allow them to live in environments with little food. The tentacles of the cnidarians have specialized cells, cnidocytes, which inject toxins into their prey with the help of stingers, nematocysts (see Figure 31.7).
- A gastrovascular cavity has a single opening used for digestion, respiratory gas exchange, and circulation.
- The life cycle of cnidarians has a sessile polyp stage and a motile medusa stage (see Figure 31.8). The polyp stage usually reproduces asexually and the medusa stage reproduces sexually, with the fertilized egg becoming a planula larva and eventually a polyp. A layer of jellylike material, the mesoglea, is found between the two cell layers of all cnidarians.
- Species from the class Hydrozoa are found in fresh water and are dominated by the polyp stage of the life cycle. The polyps tend to be colonial, with many polyps sharing a common gastrovascular cavity (see Figure 31.9).
- The Scyphozoa include the marine jellyfish which have medusae with thick mesoglea. These species spend most of their life in the medusa stage. Polyps are produced sexually by adult medusae. The polyps reproduce by budding, finally budding off small medusae that grow into adult medusae.
- The class Anthozoa (about 6,000 species of sea anemone and coral) lack a medusa stage altogether, with the polyps reproducing sexually and asexually. Sexual reproduction produces a planula that will develop into a polyp. Sea anemones are solitary and motile, and corals are sessile and colonial. The structures of the coral skeleton are composed of calcium carbonate, are species-specific, and diverse. Corals live in symbiosis with dinoflagellates which provide nutrients for the coral colony through photosynthesis, allowing corals to do well in clear tropical waters.

Ctenophores

- The marine comb jellies (phylum Ctenophora) have bodies similar to cnidarians. The main difference between the two phyla is that ctenophores have a complete gut with a mouth and two anal pores.
- Ctenopores get their name from the eight rows of comblike plates of cilia known as ctenes. The cilia are used to propel the animal in the aquatic environment. Prey is caught on sticky filaments on the tentacles or body (see Figure 31.12).
- Ctenopores have a simple life cycle and reproduce sexually, with the externally fertilized egg becoming a miniature ctenophore.

Protostomes and deuterostomes evolved bilateral symmetry.

- All animals with bilateral symmetry had a common ancestor, the urbilateria, and share intercellular signaling systems and development regulating homeobox genes.
- The urbilateria appear to have been complex animals with circulatory systems, antagonistic muscles, and a fluid-filled body cavity.
- The deuterostomes have a dorsal nervous system and larvae with a food-collecting system consisting of cells with a single cilium.
- The protostomes have a central nervous system with an anterior brain and a ventral nervous system consisting of paired or fused longitudinal nerve cords and free-floating larva.
- Most of the living animals are protostomes. The lophotrochozoans are protostomes that use cilia for locomotion, may have free-living larva (trochophore), and grow by adding skeletal elements (see Figure 31.14).

Flatworms exhibit bilateral symmetry.

- The simplest lophotrochozoans are the bilaterally symmetrical flatworms (phylum Platyhelminthes).
- Flatworms lack respiratory organs, have only a simple organ for waste removal, and have a closed body cavity. The flat shape of the animal helps in oxygen transport and waste removal (see Figure 31.15).
- The digestive tract has only one opening and has many branches that aid in nutrient absorption.
- Many of the 25,000 species of flatworms are parasites of other animals (see Figure 31.16). The parasitic flatworms have nutrients brought to them and disperse their eggs in the feces of the host. Many parasitic flatworms have more than one host species in their life cycle. Two common flatworms are the tapeworms and the flukes.

Rotifers have a pseudocoel.

- The phylum Rotifera contains 1,800 species that are bilaterally symmetrical, unsegmented animals (see Figure 31.17). They are pseudocoelomates and use the pseudocoel as a hydrostatic skeleton.
- They have a complete gut with a mouth and anus. The corona on the head of many rotifers has cilia that direct water and organic matter into the mouth and the mastax. Food is ground in the mastax for digestion.

Lophophorates are named after the lophophore, a food collecting and gas exchange structure.

- The next group of animals to diverge were the lophophorates from the phyla Phoronida, Brachiopoda, Ectoprocta, and Pterobranchia.
- Animals from these phyla have three body parts, each with a separate coelomic compartment: the anterior prosome and protocoel, the middle mesosome and mesocoel, and the posterior metasome and metacoel (See Figure 31.18).
- The lophophorates have a lophophore surrounding the

mouth with two rows of ciliated tentacles used for food collection and gas exchange.

- The 20 species of the phylum Phoronida are sedentary worms that live in chitinous tubes.
- The Pterobranchia are solitary or live in colonies and live in a tube secreted by a proboscis (see Figure 31.19).
- The Ectoprocta are colonial with many individuals living together connected by strands of tissue that reproduce sexually and asexually. The colonies of Ectoprocta are produced asexually by the founding member. They are the only phylum of lophophorates that can completely retract the lophophore.
- The phylum Brachiopoda are marine and are surrounded by a divided shell with two halves (see Figure 31.21). The lophophore is located in the shell and the cilia help draw water and food into the shell. Gas exchange occurs across the body.

The spiralian lineage contains wormlike phyla.

- The spiralians are wormlike organisms with bilateral symmetry and soft bodies.
- The Nemertea, or ribbon worms, have a complete digestive tract with two openings. Small ribbon worms use their cilia for movement and large ribbon worms use muscle contractions to move.
- The feeding organ or proboscis of the ribbon worms is kept in a rhynchocoel, or fluid-filled cavity (see Figure 31.22). The proboscis, with a sharp stylet can be forcefully ejected out of the body to catch prey.

The spiralian lineage also contains the segmented worms.

- Segmentation has evolved a number of times in the spiralians. Segmentation allows for greater control of body movements.
- The phylum Annelida is composed of approximately 15,000 species of segmented worms living in marine, freshwater, and terrestrial environments. Each segment of the worm has an isolated coelom and a ganglion nerve center connected to a nerve cord. Gas exchange occurs across the thin moist surface of the animal.
- The class Polychaeta contains more than half of the Annelida species. The Polychaeta tend to be found in marine environments. Outgrowths used in gas exchange, the parapodia, extend from segments over much of the body. Setae extending from the parapodium help to attach the animal to the substrate and aid in movement. The joining of male and female gametes produces trochophore larvae.
- Unlike the Polychaeta, the Oligochaeta live mainly in freshwater and terrestrial environments. The Oligochaeta are hermaphroditic, containing both male and female reproductive organs on the same individual. The earthworm is a typical oligochaete (see Figure 31.23).
- The class Hirudinea, or leeches, are all hermaphroditic species that live either in freshwater or on land. Parasitic leeches attach externally and feed on a host's blood.
- The Vestimentiferans lack a digestive tract. The vestimentiferans are able to take up nutrients quickly from the environment. Species from this class are found in the deep ocean near hydrothermal vents and use hydrogen sulfide to obtain energy.

All mollusks share three main morphological features.

- Mollusks have evolved into a morphologically diverse group, with all of the classes including the same three basic parts in their body plan. All species have a large muscular foot, a tissue fold covering the visceral mass known as the mantle, and a feeding structure, the radula. The mantle is extended in some species to create a mantle cavity holding the gills used in respiration. Mollusks have a reduced coelom. Some groups, such as the clams, have lost the radula.
- Only a limited number of Monoplacophora survive today. They are unlike other mollusks in that they have repeated body segments containing gills, muscles, and excretory structures.
- Polyplacophora are marine mollusks that feed on algae using their radulae. They have simple internal organs, multiple gills, and bilateral symmetry.
- The Bivalvia are composed of the clams, oysters, scallops, and mussels. Bivalves create a current of water flowing into their two-part shell through a siphon. Large gills inside the shell are used to extract the food and also function as respiratory organs.
- The Gastropoda use their foot either to crawl or swim. The land-dwelling slugs and snails are the only terrestrial mollusks.
- Cephalopods are composed of the octopuses and squids which first appeared around 600 mya. They have evolved the ability to control the water movement into the mantle, allowing them to use ejected water as a jet for propulsion. They are also able to control the gas movement in the mantle, which helps in buoyancy control. The gills are found within the mantle cavity.

The Big Picture

- All animals share a common ancestor. There are a number of morphological and genetic traits that are shared by all animals. All animals are multicellular organisms that cannot synthesize organic molecules. All animals, except the sponges, cnidarians, and ctenophores, are either protostomes or deuterostomes. Animals are also distinguished by their number of cell layers during development, their type of body cavity, and their body symmetry.
- The phylum Porifera contains the sponges, which are loosely organized groups of cells. The phylum Cnidaria are diploblastic animals that have evolved a gut. The cnidarian life cycle has an asexually reproducing polyp stage and a sexually reproducing medusa stage. The three classes of Cnidarians are the Hydrozoa, the Scyphozoa, and the Anthozoa. Animals in the phylum Ctenophora are similar to the cnidarians, except they have a complete gut.

- The lophotrochozoans are all protostomates with a anterior brain, ventral nervous system, and free-floating larva. The simplest lophotrochozoans are the Platyhelminthes and Rotifera which both have bilateral symmetry.
- The lophophorates, composed of the Phoronida, Brachiopoda, Ectoprocta, and Pterobranchia phyla, all have a lophophore. A lophophore is an anatomical feature used in food collection and gas exchange.
- The spiralian lineage contains the wormlike animals, including the unsegmented phylum Nemertea and the segmented phylum Annelida. The Annelida phylum is further divided into the Polychaeta, Oligochaeta, Hirudinea, and Vestimentiferan classes. The evolution of a segmented body allowed for better control of movement.
- The five classes in the phylum Mollusca share three morphological characters: a muscular foot structure, a mantle covering the internal organs, and a radula used to rasp for food.

Common Problem Areas

- The difference between deuterostomes and protostomes is often difficult to remember. This chapter, along with the next, is concerned with the protostomes, while Chapter 33 discusses the deuterostomes. Create a table to be used with the upcoming chapters containing all of the characters of a protostome and a deuterostome. To this table you can add the different phyla that are included in each group. Use this table as a tool to help you organize and remember the differences between the groups.

Study Strategies

- The phylogenies that are presented in your book are good skeletons on which you can add more information about the different phyla. This will provide you with a framework and a point of reference when you are studying the differences between phyla. Remember that groups of animals have a set of traits they share with earlier lineages and a new set of traits not shared with the earlier lineages.
- Review the following activities and tutorial on the website:
  Tutorial 31.1 Life Cycle of a Cnidarian
  Activity 31.1 Animal Classification
  Activity 31.2 Animal Features

Test Yourself

Knowledge and Synthesis Questions

1. Which of the following is not a derived trait that is shared by all animals?
   a. Similarities in the 5S and 18S ribosomal RNA of animals
   b. The extracellular matrix molecule collagen
   c. Tight junctions, desmosomes, and gap junctions
   d. Bilateral symmetry

2. Which of the following traits do the Ctenophora share with the Lophotrochozoans?
   a. Complete gut
   b. Three body layers
   c. Indeterminate cleavage
   d. Internal skeleton

3. An animal is known to be a deuterostome. Which of the following does not apply to this animal?
   a. Three distinct layers of tissue were present during development.
   b. If a coelom is present, it formed from the embryonic gut.
   c. Its early embryonic cleavage pattern was radial.
   d. It is diploblatic.

4. The primary reason animals are more diverse than plants is the considerable variation in animal
   a. multicellularity.
   b. embryology.
   c. methods of food acquisition.
   d. symmetry.

5. Which of the following statements about the body cavity of animals is true?
   a. The body cavity of coelomates develops from the embryonic ectoderm.
   b. The acoelomates' body cavity is filled with liquid.
   c. The pseudocoel of the pseudocoelomates have a peritoneum.
   d. The acoelomates do not have an enclosed body cavity.

6. Sponges have a very simple body plan. Which of the following statements about sponge structure or function is false?
   a. Choanocytes are flagellated cells that play a role in feeding.
   b. Large species are found in areas of heavy wave action, where food is most abundant.
   c. Individual sponges are both male and female.
   d. Water enters a sponge through pores and exits via one or more oscula.

7. Which of the following traits is not shared by both sea anemones and jellyfish?
   a. A medusa as the dominant stage in the life cycle
   b. Possession of a gastrovascular cavity
   c. Sexual reproduction
   d. Nematocysts on the tentacles

8. Which of the following traits is not shared by the Ctenophora and the Cnidaria?
   a. Both are diploblastic.
   b. Both have radial symmetry.
   c. Both have complete guts.
   d. Both have feeding tentacles.

9. The rhyncocoel is a body plan feature found in which of the following groups?
   a. Trematoda only
   b. Rotifera and Nematoda
   c. Rotifera only
   d. Nemertea only

10. Which of the following phylum is not a lophophore?
    a. Bryozoa
    b. Phoronida
    c. Annelida
    d. Brachiopoda

11. Which of the following statements is not true of the Rotifera?
    a. They have a complete gut with an anterior mouth and posterior anus.
    b. They are coelomates.
    c. The corona is a ciliated organ used in acquiring food.
    d. They use a hydrostatic skeleton.

12. A meal containing snails, clams, and octopus represents which of the following classes of mollusk?
    a. Bivalvia only
    b. Gastropoda and Cephalopoda
    c. Bivalvia and Gastropoda
    d. Bivalvia, Gastropoda, and Cephalopoda

13. The combination of a true coelom and repeating body segmentation allows the annelids (unlike the anatomically "simpler" worms) to do which of the following?
    a. Attain complex body shapes and thus locomote more precisely
    b. Move through loose marine sediments
    c. Be hermaphroditic
    d. Inject paralytic poisons into their prey

14. Which class of Cnidarian contains species that live symbiotically with dinoflagellates?
    a. Hydrozoa
    b. Ctenophora
    c. Scyphozoa
    d. Anthozoa

Application Question

1. The mollusks are a very diverse group morphologically. However, they all share three common morphological traits. Describe the traits and their functions.

Answers

Knowledge and Synthesis Answers

1. **d.** Not all animals have bilateral symmetry. The sponges, cnidaria, and ctenophora do not exhibit bilateral symmetry in their body plan.
2. **a.** They share a complete gut.
3. **d.** The deuterostomes have thee layers, the ectoderm, mesoderm, and the endoderm, making them triploblastic.
4. **c.** Animals eat foods from almost all of the kingdoms. Thus, they have evolved a wide variety of methods for acquiring diverse types of food.
5. **d.** The body cavity of coelomates develops from the mesoderm and contains a peritoneum. The acoelomates lack a body cavity.
6. **b.** Because they are not structurally robust, heavy wave action would destroy large, upright sponges.
7. **a.** While the jellyfish have both a medusa and polyp stage in their life cycle, the sea anemones have lost the medusa stage and spend the life cycle as polyps.
8. **c.** The Ctenophora have evolved a complete gut with a mouth and two anal pores.
9. **d.** The rhynchocoel is a fluid-filled cavity containing a proboscis feeding organ found in the phylum Nemertea.
10. **c.** The Annelida phylum belongs to the spiralian lineage, not the lophophorates.
11. **b.** The phylum Rotifera are pseudocoelomates and have a pseudocoel.
12. **d.** Snails are found in the phylum Gastropoda, clams are in the phylum Bivalvia, and octopus are in the phylum Cephalopoda.
13. **a.** The segmentation of the annelids allows for more complex coordinated movement.
14. **d.** The Anthozoa include the corals which live symbiotically with the photosynthetic dinoflagellates.

Application Answer

1. The mollusk body plan has three basic morphological components shared by all of the classes. The first trait is a muscular foot. This foot is used for locomotion in all of the classes. In the cephalopods the foot has been modified into arms and tentacles. The second feature is a mantle which covers the internal organs. Inside the mantle covering are the gills used for gas exchange. The third character is a radula. The radula in the gastropods is used for scraping algae. In some species the radula has been modified into a drill and in others it has been lost.

STUDY GUIDE

32 Ecdysozoans: The Molting Animals

Important Concepts

Marine ecdysozoans have a thin exoskeleton.

- Ecdysozoans have developed an exoskeleton that ranges from thin and flexible to hard and rough.
- A large number of marine ecdysozoans have a cuticle, a thin exoskeleton designed for protection. The cuticle allows for exchanges with the environment, but offers little protection.
- The priapulids are 16 species of unsegmented, tiny wormlike animals that make up the phylum Priapulida.
- The Kinorhyncha phylum is composed of 150 species of 13 segmented marine worms, all less than 1 millimeter in length, which live in mud and feed on decaying organic matter (see Figure 32.2).
- The arrow worms of the phylum Chaetognatha have no circulatory system, no larval stage, and are major predators of small prey in the open ocean (see Figure 32.3). All respiratory gas exchange occurs across the body surface.

A tough exoskeleton has evolved in some worms.

- Species of both the Nematomorpha and Nematoda phyla have tough cuticles.
- The horsehair worms (phylum Nematomorpha) are composed of 230 species and are long and very thin, as the name suggests. They live near the edge of ponds and streams and feed only in the larval stage as parasites of terrestrial and aquatic insects and crabs (see Figure 32.4).
- The nematodes, or roundworms (phylum Nematoda), can reach up to 9 meters in length. They comprise from 20,000 to an estimated 1,000,000 species. Nematodes use their gut for both gas exchange and nutrient uptake (see Figure 32.5). The nematodes live in the soil as scavengers on the bottom of lakes and streams, and as parasites of plants and animals.
- Parasitic roundworms have a life cycle that helps to transfer them from host to host. The larvae of *Trichinella spiralis* infect mammals when they eat flesh containing the roundworm. The gut of the host activates the larvae to attach to the intestinal wall and feed. They bore through the intestinal wall and make cysts in the muscles.

The arthropods have segmented external skeletons.

- The Onychophorans and Tardigrada are close relatives of the arthropods that have unjointed legs and thin cuticles with chitin which acts as a hydrostatic skeleton (see Figure 32.7).
- The rest of the species that are considered arthropods have a hard exoskeleton of chitin that provides protection, waterproofing, and muscle attachment sites. They also have appendages that allow for movement.
- The extinct phylum Trilobites, which lived in the Cambrian and Ordovician time period, had heavy armor and jointed appendages (see Figure 32.8).
- The chelicerates (phylum Chelicerata) are broken into three classes, all with two body regions, each with four walking legs. The pycnogonids are a rarely seen marine species (see Figure 32.9). The Merostomata class contains only the horseshoe crab. The class Arachnida is made up of the predatory spiders, scorpions, mites, and ticks. They have a simple life cycle, hatching miniature adults from eggs.
- Spiders are web-spinning predators. Protein threads are used to make a web that is used to catch prey items.
- The dominant marine arthropod, composed of 40,000 species, is the phylum Crustacea, containing the decapods, isopods, amphipods, and copepods. The body is divided into the head, thorax, and abdomen, with the thorax and abdomen having one pair of appendages. A carapace extends dorsally from the head to protect and cover the body. The larva of a crustacean, a nauplius, has three pairs of appendages and one eye (see Figure 32.12).

The phylum Uniramia is composed of the myriapods and insects.

- Animals in the phylum Uniramia have a body with either two or three regions.
- The subphylum Myriapoda are terrestrial and have many segments in their posterior region with many pairs of legs and are composed of the centipedes and millipedes. They are scavengers.
- There have been 1.5 million species described in the subphylum Insecta. There are three basic body parts: the head with a pair of antennae, the thorax with three pairs of legs and the wings, and the abdomen. Gas exchange occurs in the tracheae system composed of a series of air sacs and tubular channels. The two classes of insects are the wingless Apterygota and the winged Pterygota.

- The Apterygota include firebrats and collembolans.
- The Pterygota typically have two pairs of wings that are attached to the thorax.
- The Pterygota are further divided into 28 orders with three major lineages: those that cannot fold back their wings (dragonflies and damselflies), those that can fold their wings and undergo incomplete metamorphosis, and those that can fold their wings and undergo complete metamorphosis.
- Incomplete metamorphosis occurs in Orthoptera, Isoptera, Plecoptera, Dermaptera, Thysanoptera, Hemiptera, and Homoptera. The hatchlings are like small adults and gradually molt from one stage to the next. Complete metamorphosis occurs in Neuroptera, Coleoptera, Lepidoptera, Diptera, and Hymenoptera. In these, the larvae and adult forms are different, with the developing animal becoming a pupa before it is an adult.

Protostome evolution occurred heavily in the oceans.

- Dissolved organic matter and very small organisms were the only food during much of animal evolution.
- The sessile lifestyle has evolved repeatedly with the advantages of access to local resources and sharing of resources. The major disadvantages are the necessity to rely on external fertilization and the competition for space on which to grow.
- Specialized body coverings have evolved to help provide protection and increase the ability to move as a predator or to escape predators in a terrestrial environment.

The Big Picture

- The Ecdyozoans have more species than all other lineages combined. The earliest groups had a flexible, unsegmented exoskeleton. Wormlike animals with a thin cuticle are the Priapulida, Kinorhyncha, and Chaetognatha phyla. The unsegmented worms with a tough cuticle belong to the Nematomorpha and Nematoda phyla.
- A segmented body with a hard exoskeleton and jointed appendages for movement characterize the Arthropods. A related phylum, the Onychophora, have soft bodies and unjointed appendages. The phylum Trilobita is extinct. The Chelicerata are composed of species that have invaded the land, while the Crustacea tend to be marine.
- The phylum Uniramia is composed of the millipedes and centipedes. The subphylum Insecta comprises well over 1.5 million species. They are broken down into the Apterygota without wings and the Pterygota with wings. The Pterygota are further broken down into three groups, according to their development and how their wings fold.

Common Problem Areas

- Learning systematics of any large animal group can be challenging, particularly when they are organisms with which you are unfamiliar. This is particularly true for the ecdysozoans, represented by a large number of animal groups.

Study Strategies

- In this chapter there are many facts about a large number of phyla. Try to use the phylogeny presented in Figure 32.1 as a skeleton for learning about the different groups. Understanding that certain characteristics are shared between related groups can take you a long way toward remembering what characteristics define each group.
- Sometimes learning the Latin root of the group you are studying can help organize your thinking. Although the names are unfamiliar at first, they all "make sense" if you understand what they mean.
- Review the following activities on the website:
  Activity 32.1 Animal Classification
  Activity 32.2 Animal Features

Test Yourself

Knowledge and Synthesis Questions

1. Lobsters, millipeds, and butterflies all share which of the following traits?
   a. Parapodia
   b. Setae
   c. Gas exchange across the skin
   d. Jointed appendages
2. Which of the following phyla or subphyla have a specialized internal gas exchange system?
   a. Priapulida
   b. Chaetognatha
   c. Insecta
   d. Nematomorpha
3. Which of the following insect order undergoes complete metamorphosis and can fold their wings back against the body?
   a. Coleoptera (beetles)
   b. Ephemeroptera (mayflies)
   c. Isoptera (termites)
   d. Odonata (dragonflies)
4. The bulk of arthropods belong to the phylum Uniramia, and of these, the majority of species belong to the
   a. Myriapoda.
   b. Insecta.
   c. Cirripedia.
   d. Tardigrada.
5. The typical crustacean larva is called a
   a. nauplius.
   b. pupa.
   c. instar.
   d. cyst.
6. Which class of the phylum Chelicerata is considered a living fossil?
   a. Pycnogonida (sea spiders)
   b. Arachnida (mites, ticks, and spiders)
   c. Merostomata (horseshoe crabs)
   d. None of the above

7. Which phylum or subphylum of the arthropods is extinct?
   a. Chelicerata
   b. Crustacea
   c. Onychophora
   d. Trilobita

8. Which characteristic is not part of the arthropod body plan?
   a. Segmentation
   b. Jointed appendages
   c. Soft exoskeleton
   d. Hard exoskeleton

Application Question

1. A thick cuticle provides an animal with protection, but it also can seal it off from the world, obstructing the exchange of gases. How would you design a simple, hypothetical animal with a covering that provided for both protection and exchange?

Answers

Knowledge and Synthesis Answers

1. **d.** All of these species have jointed appendages.
2. **c.** The Insecta have an internal trachea system that is for respiratory gas exchange.
3. **a.** The Coleoptera, or beetles, have complete metamorphosis and are able to fold their wings against the body.
4. **b.** The Insecta contain about 1.5 million species.
5. **a.** The larva of crustaceans are known as nauplius.
6. **c.** The horseshoe crabs have changed little over their evolutionary life span, making them living fossils.
7. **d.** The Trilobita are extinct.
8. **c.** Arthropods have a hard exoskeleton, a segmented body, and jointed appendages.

Application Answer

1. An animal with a thick covering would benefit from the protection, but there are problems that must be overcome. The animal would need special mechanisms for gas exchange with the environment. The insects have overcome this problem with their tracheae system. This network of tubes is found throughout the body and opens to the environment. The network is in close contact will all of the cells of the animal, allowing for gas exchange to take place and providing enough oxygen. In an aquatic environment, the animal could have an internal chamber with gills used for respiratory gas.

   The one drawback to a thick outer covering is that it limits the size of an animal. As the size of the animal increases, the thickness of the covering would increase to the point that it would be too heavy for the animal to carry around.

STUDY GUIDE

33 Deuterostomate Animals

Important Concepts

Deuterostomes and protostomes evolved in the same environments.

- The deuterostomes and protostomes evolved similar solutions to similar environmental conditions. While the protostomes invaded land from the oceans, the deuterostomes invaded from fresh water.
- The bodies of both groups are divided into compartments.
- The deuterostome ancestral traits are indeterminate cleavage, the anus formed from the blastopore, the gut formed from the mesoderm, and a well-developed coelom.

The Echinodermata evolved an internal skeleton.

- Of the 20,000 species of Echinodermata known to have lived, about 7,000 are living today. They are found in two lineages, the Pelmatozoa and the Eleutherozoa.
- Two features allowed the Echinodermata to evolve into active species from sluggish ancestors. They have evolved a calcified internal skeleton, and they have a water vascular system used for gas exchange, locomotion, and feeding (see Figure 33.2).
- The larvae and adults show bilateral symmetry.
- The only class of Pelmatozoa surviving today is the Crinoidea, with 80 species including the sea lilies and feather stars. The body is formed of a cup-shaped structure and has anywhere from five to several hundred arms. The arms catch food particles with the tube feet on the arms. Cilia on the tube feet move the food to a groove that leads to the mouth. The tube feet also provide a surface for respiratory gas exchange and nitrogenous waste excretion.
- The eleutherozoans are comprised of the classes Echinoidea, Holothuroidea, Concentricycloidea, Asteroidea, and Ophiuroidea (see Figure 33.3).
- The Echinoidea are the sand dollars and sea urchins, which both feed on algae and organic debris. Neither has arms, but the sea urchin has spines that are attached to the skeleton.
- The class Holothuroidea comprises the sea cucumbers. The tube feet of the sea cucumber are used for anchoring the animal to the substrate rather than for moving. The anterior tube feet have evolved into tentacles that catch food particles.
- Little is known about the sea daisies of the class Concentricycloidea other than that they feed on decaying oceanic driftwood.
- The sea stars in the class Asteroidea are predators of many animals. Some species use their tube feet to pry open the shells of mollusks. Once the mollusks are open, the sea star inverts its stomach through the shell of the clam and digests it. The tube feet are also sites of gas exchange.
- The 2,000 species of the class Ophiuroidea have five flexible arms and feed on the surface sediments and the accompanying organic material through their single opening to the digestive tract.

Hemicordata have pharyngeal slits.

- The animals of the phylum Hemichordata are composed of a trunk, a collar, and a proboscis (see Figure 33.4). The sticky proboscis is used for digging and catching prey. Cilia move the food from the proboscis to the mouth.
- A pharynx that opens to the outside through pharyngeal slits is found behind the mouth. The pharyngeal slits are used as respiratory surfaces; water is moved through the slits and oxygen is extracted.

Chordates have a notochord.

- The chordates are evolved from a sister lineage of the acorn worm that had enlarged pharyngeal slits used to catch prey.
- All chordates share five anatomical features. They all have pharyngeal slits at some point in their life, a hollow dorsal nerve cord, a ventral heart, a tail beyond the anus, and a notochord. In the tunicates the notochord is lost, while in vertebrates it is replaced by skeletal structures.
- The subphylum Urochordata contains the 2,500 species of marine tunicates. Tunicate larvae have pharyngeal slits, a nerve cord, and a notochord. The adults lose the notochord and nerve cord but have an enlarged pharynx, the pharyngeal basket, for catching food (see Figure 33.5).
- The lancelets in the subphylum Cephalochordata retain the notochord for their entire life and use their pharyngeal baskets to catch prey items (see Figure 33.6).

Vertebrates evolved an internal skeleton.

- The lineage from which the vertebrates evolved had enlarged pharyngeal baskets used to collect prey from mud.

- Vertebrates can be categorized by the following traits: vertebral column replacing the notochord, an internal skeleton with two pairs of appendages, an anterior head with a large brain and protective skull, a closed circulatory system, and a large coelom containing the internal organs.
- The animals that gave rise to the fishes lacked jaws. The Ostracoderms were a class of early jawless fish that were able to swim above the substrate and exploit the environment. The hagfish and lamprey are the only jawless fishes surviving today.
- The jawed fishes evolved in the Devonian period. The evolution of jaws provided access to new food sources. The early jawed fishes such as the placoderms were the major predators of the Devonian time.
- The Chondrichthyes fishes have a skeleton composed of cartilage and pairs of unjointed appendages known as fins. The fins allow for better control of movement in the water.
- Osteichthyes are comprised of the bony fishes with a skeleton of bone and a protective covering of scales. A gas-filled sac to aid in respiration evolved in early bony fishes. In the ray-finned fishes, the sac has become a swim bladder and aids in the control of buoyancy. With the aid of the swim bladder and fins, many fish can control their position in the water column.
- The lung sacs may have been used to supplement oxygen uptake in low oxygen water and may have allowed some fishes to survive out of water. The evolution of jointed fins may have helped the evolution onto land. The subclasses Crossopterygii have jointed fins and were thought to be extinct. One species, *Latimeria chalumnae*, composed mainly of cartilage, has been found in the waters of the Indian Ocean (see Figure 33.13).

Amphibians ventured onto land.

- The class Amphibia evolved from a common ancestor shared with the lobe-finned fishes. Amphibians have jointed walking legs, like reptiles, birds, and mammals.
- Amphibians live in moist environments and spend at least part of their life cycle in the water (see Figure 33.16). Many species come back to the water to lay their eggs, producing aquatic larvae.
- The skin surface is a major site of respiratory gas exchange. They also have lungs used in gas exchange.
- Extant amphibians are found in the orders Gymnophiona (caecilians), Anura (frogs and toads), and Urodela (salamanders).

Reptiles invaded the dry environments.

- Amphibians are confined to moist environments. Two major morphological adaptations occurred to allow reptiles and birds to colonize dry land: a shelled egg and waterproof skin.
- The animals that evolved a protective calcium-based shell are known as amniotes. The shell inhibits dehydration of the developing egg and allows eggs to be laid in dry environments.
- The class Reptilia branched off during the Carboniferous period about 250 mya. The Reptilia is not a monophyletic group because it does not contain the birds. This classification has been used in the classical system and is the one used in your book.
- The skin of reptiles is impermeable to water, decreasing water loss and at the same time decreasing the ability to exchange gases. Gas exchange occurs in ventilated lungs.
- The subclass Squamata contains the lizards, snakes, and amphisbaenians. Most lizards and snakes are carnivores, but some lizards are herbivores.
- Tuataras are a subclass that contains only two extant species.
- The Chelonia (turtles and tortoises) and the Crocodylia (crocodiles and alligators) are closely related and their ancestors shared the trait of armored plates.
- Dinosaurs evolved from the reptiles around 215 mya and survived for about 150 million years.

Birds evolved from a dinosaur ancestor.

- *Archaeopteryx* lived 150 mya and is the oldest known fossil from the subclass Aves. It was covered in feathers and had well-developed wings and a wishbone like present-day birds (see Figure 33.20).
- One theory about the evolution of birds is that they evolved from bipedal dinosaurs with the feathers evolving to aid in insulation or courtship display.
- Both the lightweight bones and feathers of birds have evolved to aid in flight. Along with the ability to fly came a high metabolic rate needed to fuel flight.
- Birds typically lay eggs in nests and incubate the eggs until they hatch. Development of the embryos occurs rapidly, producing either chicks that are well developed with feathers or featherless chicks that require parental care.
- The 9,000 species of birds make them the largest subclass of terrestrial vertebrates (see Figure 33.21)

Mammalia evolved from mammal-like reptiles.

- The mammals evolved from a group of mammal-like reptiles around the Mesozoic era. During the dinosaur extinction, mammals flourished in size and number.
- A number of unique traits have evolved in the mammal line. The jawbone is comprised of a single bone, they have mammary glands that provide nutrients for their young, and they have insulating external body hair.
- The subclass Prototheria is composed of the Monotremata, an egg-laying order of mammals (see Figure 33.22).
- The subclass Theria contains the Marsupialia and eutherian mammals. The Marsupialia development occurs mainly in a pouch on the front of the mother (see Figure 33.23). The eutherian mammals develop in the mother's uterus.

Humans belong to the primates.

- Primates share the following traits: opposable thumbs, nails on the fingers and toes, eyes on the front of the face, and production of only a few offspring which are provided with parental care.

- The earliest split in the primates occurred in the late Cretaceous, splitting into the prosimians (comprised of lemurs, bush babies, and lorises) and the anthropoids (comprised of the monkeys, apes, and humans).
- The anthropoids further split into the arboreal New World monkeys and the arboreal or terrestrial Old World primates. The apes split off about 20 mya.

Early human ancestors evolved bipedalism.

- The ardipithecines and their descendants, the australopithecines, were adapted for bipedalism, freeing up their hands for other tasks and elevating the eyes.
- The *Australopithecus afarensis* appears to be the ancestor of the humans. From this species a number of *Australopithecus* species evolved, including the tool using *A. garhi* which gave rise to the genus *Homo*.
- The genus *Homo* evolved larger body and brain sizes compared with the *Australopithecus*. The oldest *Homo*, *Homo habilis*, lived 2 mya. Tools have been found with the bones of *H. habilis*. *Homo erectus* lived from 1.6 mya to 200,000 years ago and used fire and stone tools. *Homo erectus* may have played a role in the extinction of *H. habilis*.
- A number of types of *Homo sapiens* lived from 1.5 mya to about 30,000 years ago. Neanderthals, a type of *H. sapiens*, lived between 75,000 and 30,000 years ago. The Cro-Magnons were most likely responsible for the demise of the Neanderthals. The Cro-Magnons used sophisticated tools, spread across Asia, and were in North American by about 20,000 years ago.

Humans have developed culture.

- With the evolution of larger brain sizes in our *Homo sapiens* ancestors, the capacity for language increased.
- Humans have developed culture and pass knowledge and traditions from generation to generation.
- Human societies were transformed from communities of hunters and gatherers to those of farmers.
- About 11,000 years ago agriculture developed in the Middle East and reached the British Isles around 4,000 years ago.
- There have been three major times of human population growth. The first occurred during the period of first tool use. The second was brought about by the domestication of plants and animals. We are currently living in the third growth which started with the industrial revolution.

The Big Picture

- The deuterostomes share a number of traits including being triploblasts, indeterminate cleavage, a well-developed coelom, and the anus formed from the blastopore. They also evolved an internal skeleton that allowed for a large increase in size.
- The phylum Echinodermata are the first major lineage of deuterostomes. They have an internal skeleton and a special water vascular system used in feeding. The Echinodermata include the sea stars, sea lilies, sea cucumbers, and sea daisies, which all exhibit biradial symmetry.
- The next major phylum, the Hemichordates, have pharyngeal slits, a trait they share with the chordates. The Hemichordates are thought to be similar to the ancestors of the chordate lineage.
- The phylum Chordata have several shared traits: pharyngeal slits, a nerve cord, a ventral heart, a tail beyond the anus, and a notochord. The Chordata are composed of the tunicates, lancelets, and vertebrates. The vertebrates have evolved a rigid endoskeleton for support and a dorsal vertebral column that replaces the notochord during development.
- The Vertebrates include the subclasses cartilaginous fishes (Chondrichthyes), bony fishes (Osteichthyes), Amphibians (Amphibia), reptiles (Reptila), birds (Aves), and mammals (Mammalia).
- Humans evolved from bipedal ancestors in Africa. The *Australopithecus afarensis* from 3.5 mya appears to be an ancestor of the modern human. The most recent ancestor of humans was Cro-Magnon. During the evolution of the humans, brain size has increased, allowing humans to develop culture and learn languages.

Common Problems

- Learning systematics is a difficult task. Luckily, in this chapter you should run across a number of phyla that you are familiar with. As you work your way through the phylogenies presented in the book, try to recall the representative organisms that you know from each group and what their special traits are. When learning about the evolution of amphibians, recall the frogs you used to catch in the summer. If you base your learning on information you already know, you will have an easier time learning the facts.

Study Strategies

- Many students become confused with the evolution of the chordates and vertebrates and the traits that separate them. Create a time line with all of the major chordate classes. Insert the specific traits that distinguish one class from another. This will help you organize the evolution of the chordates so that memorizing it becomes a manageable task.
- Review the following tutorial and activities on the website:
  Tutorial 33.1 Life Cycle of a Frog
  Activity 33.1 Deuterostome Phylogeny
  Activity 33.2 Amniote Egg
  Activity 33.3 Deuterostome Classification

Test Yourself

Knowledge and Synthesis Questions

1. Which of the following is not a characteristic of species in the phylum Echinodermata?
   a. They have a water vascular system.
   b. They have an internal skeleton.
   c. They are protostomes.
   d. They have bilateral symmetry.

2. All but one of the following classes of echinoderm belongs to the subphylum Eleutherozoa. Select the exception.
   a. Asteroidea
   b. Crinoidea
   c. Asteroidea
   d. Holothuroidea

3. Which of the following traits is not shared by both the hemichordates and the chordates?
   a. Pharyngeal slits
   b. Deuterostomes
   c. Notochord
   d. Triploblasts

4. The primary anatomical characteristic that preadapted pharyngeal slits for use as feeding structures was their
   a. large surface area.
   b. location anterior to the mouth.
   c. bilateral symmetry.
   d. ability to provide support for the internal skeleton.

5. Which of the following groups of chordates evolved prior to the appearance of cartilaginous fishes?
   a. Bony fishes and sea squirts
   b. Tunicates, lancelets, and agnathans
   c. Tunicates, lancelets, and bony fishes
   d. Agnathans and bony fishes

6. The vertebrate body plan includes all of the following traits except
   a. a ventral spinal cord.
   b. an internal skeleton.
   c. a well-developed circulatory system.
   d. organs suspended in the coelom.

7. The swim bladder of many fishes evolved from a lung-like sac. What important function does the swim bladder provide?
   a. Aids in prey capture
   b. Controls swimming speed
   c. Controls buoyancy
   d. Aids in reproduction

8. Which one of the following traits do the Condrichthyes and Osteichthyes share?
   a. The gills are the major site of gas exchange.
   b. The skeleton is composed of cartilage.
   c. The outer surface is covered with scales.
   d. They have a swim bladder.

9. The transition from aquatic to terrestrial lifestyles required many adaptations in the vertebrate lineage. Which of the following is *not* one of those adaptations?
   a. Switch from gill respiration to air-breathing lungs
   b. Improvements in water resistance of skin
   c. Alteration in mode of locomotion
   d. Development of feathers for insulation

10. The amniotes evolved the ability to reproduce by laying eggs. What is the major advantage to laying eggs?
   a. The embryo needs only a small amount of yolk for development.
   b. The shelled egg does not have to be laid in a moist environment.
   c. Evaporation from the egg is increased.
   d. The nitrogenous wastes can be excreted across the shell.

11. How do birds differ from reptiles?
   a. Birds have a lower metabolic rate.
   b. Reptiles lay eggs and birds give birth to live young.
   c. Birds can breathe and run at the same time, while reptiles cannot.
   d. They do not share a common ancestor.

12. Which of the following is *not* a trait that you could use to identify an animal as a mammal?
   a. Mammary glands
   b. Hair
   c. Sweat glands
   d. Kidneys

13. Dinosaurs dominated the Earth for nearly 150 million years before becoming extinct. Which of the following vertebrate groups is arguably a living representative of the dinosaur lineage?
   a. Amphibians
   b. Birds
   c. Lobe-finned fishes
   d. Snakes

14. The now extinct hominid most like ourselves is
   a. *H. habilis*.
   b. *H. erectus*.
   c. Cro-Magnon.
   d. Neanderthal.

15. Which of the following statements about human evolution is *false*?
   a. Bipedalism was a hominid adaptation for life on the ground.
   b. Increases in the size of hominid brains preceded the appearance of language and culture.
   c. The extinction of the Neanderthal people was caused by the emergence of *H. habilis*.
   d. Humans are not the direct descendants of modern-day chimpanzees.

Application Questions

1. You have been working in the field as an anthropologist for the past year. You have come upon the bones of a hominid that was buried along with some tools and clothing. Using these clues, when do you think this hominid could have been alive, and to what species might it belong?

2. The evolution of a hinged jaw resulted in an extensive radiation of the agnathan fishes into the many modern jawed forms. What was the main advantage and significance of the hinged jaws in this radiation?

Answers

Knowledge and Synthesis Answers

1. **c.** Species from the phylum Echinodermata are deuterostomes.
2. **b.** The class Crinoidea belongs to the Pelmatozoans and not the Eleutherozoans.
3. **c.** The Chordates have evolved a notochord. The Hemichordates do not have a notochord.
4. **a.** Morphological structures that are used for filter feeding need a large surface area in order to catch sufficient food for a large animal.
5. **b.** The tunicates, lancelets, and agnathans all evolved before the cartilaginous fishes.
6. **a.** Along with an internal skeleton, well-developed circulatory system, and organs suspended in a coelom, the vertebrates have a dorsal spinal cord.
7. **c.** The swim bladder of modern-day fish are involved in controlling buoyancy.
8. **a.** In both the bony fishes and cartilaginous fishes the major site of gas exchange is the gills. The bony fishes have a skeleton of bone, are covered with scales, and have a swim bladder. These traits are not shared with the cartilaginous fishes.
9. **d.** In the move onto land, the development of feathers for insulation was not required. The amphibians and reptiles do not have an insulation layer and the mammals have hair for insulation.
10. **b.** The shelled egg of the birds and reptiles allowed them to occupy dry terrestrial habitats because the shell decreases the water loss from the egg.
11. **c.** Birds and reptiles differ in the morphology of their muscles that control movement and breathing. Birds can run and breathe at the same time, while reptiles must stop running to take a breath.
12. **d.** The presence of kidneys is shared with the reptiles, birds, and mammals. Therefore, it is not a trait that could be used to identify a mammal.
13. **b.** The birds are thought to be direct descendants of the dinosaurs.
14. **c.** Both Cro-Magnon and Neanderthal were *Homo sapiens*, but the Cro-Magnon was anatomically more like us and survived into the more recent past.
15. **c.** The extinction of the Neanderthal is thought to have been due in some part to the presence of Cro-Magnon. *Homo habilis* was extinct for more than a million years by the time the Neanderthals were on Earth.

Application Answers

1. The bones that you have found could be as old as 2.5 million years. The *Australopithecus garhi* were among the first humanlike species to use tools. However, it was the early species of *Homo sapiens* that first began to bury their dead with their possessions. These early *H. sapiens* lived between 1.5 million to 300,000 years ago.
2. The hinged jaw of the fishes evolved during the Devonian period. The evolution of the jaw opened up a new untapped food source for these animals. With a jaw, fish could now grasp and kill larger living prey. It allowed fish to chew and tear the body parts of prey.

STUDY GUIDE

34 The Plant Body

Important Concepts

Vegetative structures serve a variety of roles including nutrient procurement, support, water uptake, and transport.

- Vegetative organs are involved in all plant functions except reproduction.
- Angiosperms consist of two lines of plants with distinct characteristics. See Figure 34.1 for a review of the differences between monocots and eudicots.
- Plants are composed of root systems and shoot systems (Figure 34.2). Roots are responsible for mineral and water uptake and support. Shoot systems consist of leaves (and leaf derivatives) and stems. The shoot system is responsible for photosynthesis and support.
  - Root systems are generally one of three types:
  - Taproots are large single roots that grow deep within the ground and serve as storage tissue. Smaller lateral roots absorb water as the taproot enlarges.
  - Fibrous roots are numerous thin roots that spread far out from the plant. They hold soil in place and rarely grow very deep.
  - Adventitious roots occur along the stem and assist with re-rooting, or support.
- Stem systems produce buds, leaves, and flowers. Buds are embryonic shoots. At each node where leaves meet stem, a lateral bud is produced. This bud may produce a new branch. At each stem tip is an apical bud. The apical bud is responsible for elongation of that stem.
- Stems may be modified into tubers and runners. Stems may remain photosynthetic or become nonphotosynthetic woody material.
- Leaves are the site of photosynthesis but may have modifications to allow for specific jobs. The blade of the leaf is attached to the stem by a petiole. The leaf blade, with help from the petiole, is able to maintain a constant orientation toward sunlight for maximum photosynthesis. Leaves may be simple and consist of a single blade, or may be compound with leaflets arranged along a central axis (see Figure 34.5). Vein patterns may be netted as in eudicots or parallel as in monocots.
- Some leaves are modified for storage of nutrients or water. Many have color modifications to attract pollinators. Some are modified into "climbing" structures like tendrils and suckers. Humans have taken advantage of these characteristics by breeding for these modifications.

Plant cells are specifically suited for their functions.

- Plant cells have the characteristics of all eukaryotic cells but differ in that some have plastids, many have vacuoles, and every plant cell is bounded by cellulose containing cell wall. Some plant cells are most functional to the plant after death.
- Cell walls play important roles in plants. They provide turgor pressure, are chemically active, allow for transport, and become modified for specific functions. Cell walls are formed after cytokinesis. The daughter cells secrete the middle lamella and then begin laying down the primary cell wall. Once the cell is full size, it may begin secreting other substances to form the secondary wall.
- Cell walls do not block communication between cells. Small cytoplasmic strands called plasmodesmata extend through the walls allowing movement of substances from cell to cell. Pits in cell walls (tiny holes) also allow materials to pass from cell to cell.
- Plant cells are often specialized to carry out specific functions
  - Parenchyma cells are frequently photosynthetic or used for storage. They are thin-walled and have large central vacuoles. They may also continue dividing and serve as "parent cells" in a wounded area.
  - Collenchyma cells are support cells with special thickenings at the cell wall corners. They are elongated and allow for support without rigidity which is an important structure for plants in windy areas.
  - Sclerenchyma cells provide rigid support after they die. They have highly thickened cell walls. Sclerenchyma cells may be elongated fibers or variously shaped sclereids. Fibers strengthen bark and woody stems. Sclereids produce the gritty texture of pears.
  - Xylem conducts water. Tracheary elements are functional after death. Vessel elements are found in angiosperms. They secrete a specialized waterproofing substance that keeps them from "leaking" as water is moved.
  - Phloem moves carbohydrates and nutrients. Phloem cells are living. Individual phloem cells are called sieve tube members. Plasmodesmata enlarge where sieve tube members join, making sieve plates. Sieve tube members

may lose nuclei and other organelles to prevent clogging of the sieve plates. These cells are filled with sieve tube sap which is high in carbohydrates. Adjacent companion cells regulate the function of the sieve tube members.

Plant cells are organized into tissues.

- Simple tissues are composed of only one cell type; complex tissues are composed of multiple cell types. Xylem and phloem are examples of complex tissue types that perform multiple functions by means of multiple cell types.
- Tissue systems extend through the plant and are either vascular, dermal, or ground.
- The vascular tissue system is made of xylem and phloem and is the conductive tissue of the plant.
- The dermal tissue system provides the outer covering of the plant. This includes the epidermis and the layer of cuticle it secretes.
- The ground tissue system composes the rest of the plant. This tissue functions in storage, support, and photosynthesis.

Plants develop in specific patterns from embryonic tissue.

- From the beginning, the plant is laid out in an apical–basal pattern which establishes the cells along the main axis, and in a radial pattern that establishes the layout of the tissue systems.
- Plants frequently lose parts during growth. To counter this loss, plants are laid out in modules or units. A unit is a node and its attached leaf, the internode below that node, and the lateral buds at the base of the internode. New units are formed as the plant grows. Leaves in some respects can be thought of as units themselves.
- Plants differ from animals in having indeterminate growth. They continue to grow throughout their life span. This is possible because plants have regions of continual cell division called meristems.
- Apical meristems give rise to the primary plant body, and are located at the tips of roots and stems and in buds. They are responsible for elongating the plant body. Apical meristems give rise to primary meristems that produce the primary plant body. These are called the protoderm, the ground meristem, and the procambium. All plant parts arise from division of the apical meristems.
- Lateral meristems lay down wood and bark. This does not occur in all plants. The vascular cambium arises from the lateral meristem and allows the production of new xylem and phloem. The cork cambium produces new dermal tissues to accommodate increasing diameter. The action of the vascular and cork cambiums is called secondary growth. Wood results from secondary xylem, and bark is everything outside of the cork cambium.
- Root primary meristems (protoderm, ground meristem, and procambium) produce root tissues. The protoderm gives rise to the epidermis and root hairs. Ground meristem gives rise to the cortex and endodermis. The endodermis is specialized with a waxy coating of suberin to assist with water movement. The procambium gives rise to the stele which houses three tissues, the pericycle, the xylem, and the phloem. In eudicots, the very center of the root is xylem but in monocots the center is pith tissue. (see Figures 34.16 and 34.17).
- The shoot primary meristem also gives rise to three primary meristems that produce the three tissue systems. Leaves arise from leaf primordia. Vascular tissues are arranged in bundles. In eudicots, the vascular bundles are arranged in a cylinder. In monocots they are scattered. Therefore, only eudicots can have woody growth.
- Stems and roots may undergo secondary growth from the lateral meristems. The orientation of the vascular cambium gives rise to cylinders of secondary xylem (wood) and secondary phloem. Connections of living cells to secondary phloem occurs via vascular rays. Vascular rays are perpendicular to the vessel elements for conduction. Only eudicots have secondary growth.
- Annual rings seen in wood are a result of climate shifts in temperate zones, particularly water availability. You do not see rings in tropical trees. Heartwood in trees appears darker because the xylem has been clogged with minerals. Sapwood is still conducting at the time of cutting and appears lighter and less dense. Knots are the result of branches becoming buried in the expanding trunk.
- Bark results from the stretching, breaking, and flaking off of epidermis and cortex. This leaves the secondary phloem at risk. Cells at the surface of the phloem produce cork that is thickened and reinforced with suberin. New cork is produced as secondary growth proceeds.

Leaves are designed to maximize photosynthesis

- Review leaf structure in Figure 34.23.
- Leaves have two zones of photosynthetic cells called mesophyll. The upper level of cylindrical mesophyll is called palisade mesophyll. The lower level is called spongy mesophyll. Air space around mesophyll cells is necessary for carbon dioxide to reach the photosynthesizing cells.
- Vascular tissue extends throughout leaves in veins. Veins carry water to cells and transport carbohydrates to sink tissues.
- The entire leaf is covered by a protective epidermis and is waterproofed by a waxy cuticle. Gas exchange occurs through guarded stomata. Guard cells open and close stomata to limit water loss.

The Big Picture

- A plant can be thought of as having vegetative structures that carry out the major functions of day-to-day life, and reproductive structures that are responsible for reproducing the plant. You will see that there are cases in which the vegetative portions of the plant are quite adept at reproducing asexually. The vegetative plant body consists of the root system, which anchors the plant and absorbs water and nutrients, and the shoot system, which photosynthesizes and supports the plant against gravity. Modifications of the root or shoot system lead to specialization of the plant.

- Plant cells are uniquely suited for support, transport, and for carrying out the cellular functions of the plant. Groups of cells form tissues that have specific roles within the plant. Most plants cells are indeterminate, and under appropriate conditions can become different cell types. Patterns of development are under the control of hormones and regulatory genes. Indeterminate growth and the modular organization of plants allows for regeneration of parts lost to damage and disease.
- Plant growth occurs from meristems. Apical meristems allow for elongation of the plant, whereas lateral meristems allow for secondary or woody growth. Not all plants exhibit secondary growth. All tissue types arise from the meristems and go through a process of elongation, then differentiation. Review stem, root, and leaf anatomy using the figures in your book.

Common Problem Areas

- This chapter covers plant anatomy. Basic plant anatomy is not difficult, but it is new to most students. The more time you spend looking at diagrams and live specimens, the easier it will be to understand the anatomy. As you study the anatomy, think about the function of each of the parts. You will find that "form follows function."
- This chapter exposes you to new vocabulary words. Resist the temptation to merely memorize. By looking at the words and understanding their roots, you will be better able to understand the terms.

Study Strategies

- The best study strategy for this material is to use the figures, pictures, and live material. Plant anatomy is structure. Structures are three-dimensional and best understood visually.
- Make a vocabulary list. Though you don't want to memorize the terms, a list will help you organize your study.
- If your school has three-dimensional models make use of them. There are 3-D models available now on the Internet. Using these will make it easier to understand the orientation of the structures.
- Review the following tutorial and activities on the website:
  Tutorial 34.1 Secondary Growth: The Vascular Cambium
  Activity 34.1 Monocot Root
  Activity 34.2 Monocot Stem
  Activity 34.3 Monocot Leaf
  Activity 34.4 Dicot Root
  Activity 34.5 Dicot Stem
  Activity 34.6 Dicot Leaf

Test Yourself

Knowledge and Synthesis Questions

1. You are studying tropical plants in a Costa Rican cloud forest. You have identified a plant that has not been studied before. It has compound leaves with netted veins, vascular bundles arranged in a cylinder, and five petals on its very large and showy flower. This plant is most likely a(n)
   a. monocot.
   b. eudicot.
   c. gymnosperm.
   d. None of the above

2. The plant described above grows to approximately 40 m at maturity and is in a forest ecosystem. This plant most likely has which of the following root types?
   a. Fibrous
   b. Taproot
   c. Adventitious
   d. Rhizoids

3. You planted sweet peas along your back garden fence. You note that they are very effective at climbing the fence. Upon closer inspection you notice that the plants are attached to the fence by tendrils. These tendrils are modifications of
   a. stems.
   b. roots.
   c. branches.
   d. leaves.

4. Plant cells are easily distinguished from animal cells because they have
   a. rigid cell walls.
   b. plastids.
   c. large vacuoles.
   d. All of the above

5. Plant cells that are photosynthetically active are found in the ___________ layer of the leaf and are ___________ cells.
   a. mesophyll, parenchyma
   b epidermis, parenchyma
   c. mesophyll, sclerenchyma
   d. epidermis, sclerenchyma

6. Water is conducted in ___________ tissue, and carbohydrates and nutrients are transported in ___________ tissue.
   a. xylem, phloem
   b. phloem, xylem
   c. parenchyma, phloem
   d. parenchyma, xylem

7. Plants are capable of indeterminate growth because of
   a. meristems.
   b. regions of continually dividing cells.
   c. their modular nature.
   d. All of the above

8. Which of the following best describes the origin of wood?
   a. Xylem cells enlarge and deposit large amounts of lignin.
   b. Primary meristems increase the amount of xylem deposited.
   c. Secondary meristems allow for continuous increases in vascular tissue.
   d. None of the above

9. Which of the following statements concerning wood is true?
   a. All woody plants show annual growth rings.
   b. Patterns of both cylindrical secondary growth and lateral vascular rays are visible in wood.
   c. Sapwood and hardwood result from patterns in primary xylem.
   d. All of the above.

10. Which of the following best describes the function of the cork cambium?
   a. It lays down a protective cork covering over exposed phloem tissue.
   b. It allows for the sloughing of epidermal tissue.
   c. It allows for diameter expansion in stems and roots.
   d. All of the above

11. Sieve tube members have sieve plates where they join other sieve tube members. Which of the following best describes the sieve plates?
   a. Sieve plates are enlargements of the plasmodesmata.
   b. Sieve plates are necessary to allow conduction between sieve tube cells.
   c. Sieve plates allow joining of cytoplasm between adjacent sieve tube cells.
   d. All of the above

12. Plants regulate gas exchange and water loss via
   a. the cuticle.
   b. guarded stomata.
   c. coated pits.
   d. sieve plates.

Application Questions

1. Draw a typical eudicot plant. Label the root system and shoot systems. Indicate on your drawing where you would find lateral buds and where you would find apical buds. Label the following structures: internode, petiole, leaf blade, taproot, lateral roots.
2. Differentiate between apical and lateral meristems. Do all plants have apical meristems? Do all plants have lateral meristems?
3. Trace the origin of a xylem cell in a monocot back to the apical meristem.
4. Diagram a growing root. Label the primary meristems, the root cap, cortex, stele, and epidermis. Discuss the function of each of these structures.
5. Compare and contrast monocots and eudicots. Identify the following plants as monocot or eudicot: corn, oak tree, day lily, rose, dandelion.
6. Look at the leaf in Figure 34.23 in your book. Which surface of that leaf faces the sun? How do you know? Why are the stomata opposite the palisade layer?

Answers

Knowledge and Synthesis Answers

1. **b.** See Figure 34.1 for a comparison of eudicots and monocots.
2. **b.** A taproot would be necessary to anchor a plant of that size. Most forest trees have taproots.
3. **d.** Tendrils are modified leaves. Some climbing plants produce tendrils and others produce suckers.
4. **d.** Plant cell shape is maintained by their cell walls. No animal cells have cell walls. Plants have plastids and vacuoles as well. Though these characteristics are seen in some protists, no animal cells have them.
5. **a.** The mesophyll, both palisade and spongy, is photosynthetically active and derived from parenchyma cells.
6. **a.** Xylem tissue transports water from the roots throughout the plant. Phloem tissue transports carbohydrates and nutrients from source tissue to sink tissue.
7. **d.** The presence of continually dividing regions called meristems allows plants to have modules. Because plants are rooted in place, there must be mechanisms for surviving the loss of parts and for continued growth. Meristems and modules allow for this.
8. **c.** Secondary meristems, also known as lateral meristems, allow for production of new xylem and phloem. The secondary xylem gives rise to wood.
9. **b.** Wood grain is a result of patterns in the vascular cylinders. Wood porosity results from the lateral phloem rays.
10. **d.** The cork cambium is a meristematic region outside the secondary phloem. As girth increases and splits the epidermis causing loss of those protective layers, the cork cambium produces new cells to cover the expanding vascular tissue.
11. **d.** The sieve plates allow for the conduction of the sieve tube sap from sieve tube cell to sieve tube cell. The pores are due to enlargements of the plasmodesmata. These cells may also lose nuclei and organelles so that carbohydrates can more easily pass through the sieve tubes.
12. **b.** Guard cells at the edges of stomata respond to changes in osmotic pressure. These changes lead to the opening and closing of the stomata to regulate gas exchange and water loss.

Application Answers

1. See Figure 34.2 for a labeled diagram.
2. Apical meristems are responsible for elongation of the plant body. Lateral meristems are responsible for an increase in girth. Lateral meristems are found only in woody eudicots and are responsible for creating "wood." All plants have apical meristems.
3. The apical meristem gives rise to the protoderm, the ground meristem, and the procambium. The procambium gives rise to the vascular tissue system of which xylem cells are part. Because the plant is a monocot, you know that only primary xylem is possible because monocots do not have secondary xylem.

4. See Figure 34.15 for a diagram of root growth. The protoderm gives rise to the epidermis for protection. The ground meristem gives rise to the cortex for storage. The procambium gives rise to the stele for transport. The root cap protects the meristem as it pushes through the soil.
5. See Figure 34.1 for a comparison of monocots and eudicots. Corn is a monocot. The easiest way to be sure of this is to look at the parallel veins in its leaves. An oak tree is a eudicot because only eudicots have woody secondary growth. Day lilies are monocots. They have the characteristic parallel veins and classically three petaled flowers. Roses are eudicots. Many roses have woody secondary growth, petals in fives, and netted veins. Dandelions have netted veins but the clearest indication that they are eudicots is their well developed taproot (see Figure 34.15).
6. The "top" of the diagram in your book would be the surface that faces the sun. This is the surface where photosynthetic cells are aggregated. They need maximum sun exposure. The stomata are on the opposite side to reduce water loss due to evaporation during photosynthesis.

35 Transport in Plants

Important Concepts

Water and nutrients are taken up in the roots of plants.

- Water moves via osmosis. For osmosis to occur across a semipermeable membrane there must be a solute potential great enough to initiate a movement, and a pressure potential small enough to allow for movement. The overall tendency of a solution to take up water across a membrane is called water potential and is the sum of the negative solute potential and the positive pressure potential. All three parameters can be measured in megapascals (MPa). Water always moves to a region of more negative water potential.
- The structure of plants is maintained by osmotic phenomena. If a plant loses turgor by a decrease in pressure potential it wilts. Movement of water from cell to cell depends on water potential. Long-distance movement depends on pressure potential and is referred to as bulk flow.
- Mineral uptake requires active transport via proteins. When mineral concentrations are greater in the soil solution than in the plant, they are taken up by facilitated diffusion. If minerals are in smaller concentrations outside the plant than inside the plant, or if they must be moved against an electrical gradient, then the plant must rely on active transport.
- Plants rely on a proton pump for active transport of minerals into cells. Plants actively pump protons out of cells, which causes the area outside the cell to be more positive. This assists facilitated diffusion of positive ions through protein channels. It also produces energy to drive the movement of negatively charged ions into the cell by active transport (see Figure 35.2). The results of this pumping action make the internal environment of the plant cell highly negative compared to its environment. This difference in charge is called membrane potential.
- Water moves into a root because the root has a more negative water potential than the soil solution it is bathed in. Movement inside the root takes place because the stele (vascular tissue) has a more negative water potential than the cortex. Minerals dissolved in water are moved via bulk flow of water once it is in the vascular system.
- Minerals follow two paths for reaching the vascular tissue: the apoplast or the symplast.
  - The apoplast is formed by the cell walls and intercellular spaces. Water and minerals may move through this space without ever having to cross a membrane.
  - The symplast is the living portion of the plant and is enclosed in membranes. Water and mineral movement in the symplast is highly regulated.
  - Water and minerals can travel through the apoplast as far as the endodermis. At the endodermis, the water and minerals are stopped by the Casparian strips which are waxy structures surrounding the endodermal cells.
  - Because of this, water can reach the stele only via the symplast. The transport proteins in the endodermal cells determine which minerals enter the stele.
  - Once past the endodermal barrier, water and minerals can again leave the symplast and move back to the apoplast. Some cells (transfer cells) are specialized for doing this.
- Transfer cells are parenchyma cells specialized with knobby projection to become transport cells. This increased surface area allows for increased transport of proteins. Transfer cells also have large numbers of mitochondria to manufacture ATP needed to drive active transport.
- Ultimately, water and minerals from the soil solution end up in xylem cells and are referred to as xylem sap.
- Specialized membrane channel proteins in plant cells called aquaporins can increase the rate of water movement by preventing interaction with the plasma membrane that slows water flow down. Though they can increase the rate, they cannot influence the direction of flow.

Water and nutrients are moved through the plant in the xylem.

- It was originally thought that a pumping mechanism might be active in plants. This was shown to be false in experiments. Plant trunks immersed in poison showed progressive death of all living cells as the poison progressed through the plant. This experiment indicated that movement continued even as cells were killed, so no "pumping" cells were active; that the leaves were critical to transport because transport continued until the leaves died; and that transport did not depend on the roots because it occurred in the absence of roots.
- Root pressure based on guttation (the forcing of water out of openings in leaves) was another theory for water movement. It was thought that the pressure exerted by root tissue might be sufficient to force water up the xylem. In actuality, xylem sap is under negative pressure as it is ascending.

- Pulling forces from transpiration at the leaf surface is what actually moves water through the xylem. Transpiration generates tension on the water molecules which pulls them through the apoplast of the leaves. Water molecules are cohesive enough to resist the tension and pull other water molecules along. This results in bulk flow.
- The transpiration–cohesion–tension mechanism pulls water from the roots up through the plant. Water evaporates from mesophyll cells during transpiration. This puts tension on the film of water associated with the mesophyll cell wall. The tension at the mesophyll cell draws water from the xylem of the nearest vein. This creates tension in the entire xylem column and the column is drawn upward from the roots. None of this would be possible without the cohesive nature of the water molecule. All of this movement requires no energy on the part of the plant.
- Minerals are drawn passively along with the water column. They may be redistributed via the phloem, but initial movement is through the xylem.
- Transpiration also assists with temperature regulation through evaporative cooling.
- Tension in the xylem sap can be measured with a pressure bomb. By measuring tension in stems, the mechanism of water transport was tested and understood. If xylem sap is under tension and a stem is removed from a plant, the xylem sap should be drawn up in the xylem away from the cut. If the cut portion is placed under pressure, the sap should rise back to the surface of the cut end. The pressure that it takes to raise this is equal to the tension of the xylem sap. Using this mechanism, Per Scholander showed that all xylem is under tension.

Water loss in a plant must be controlled.

- Leaf surfaces are modified to prevent excessive water loss, but gas exchange must occur. Any time a plant surface is open enough to allow gas exchange, water is also lost.
- Stomata with guard cells regulate gas exchange and water loss. Guard cells, in response to osmotic differences, shrink and swell to open and close the stomata. Guard cells open when light is sufficient to maintain photosynthesis and carbon dioxide levels are low. Guard cells are also regulated by water potential. If the water potential in mesophyll cells is low, mesophyll cells release abscisic acid, which causes the guard cells to close.
- Guard cells open when potassium ions are actively transported into them as a result of the action of the proton pump. High potassium levels cause water to move in by osmosis. Turgor builds in the cells and they are pulled apart to reveal the stoma. For them to shut, the proton pump stops and potassium ions move back across the membrane. Water follows and the cells go limp and seal off the stoma. See Figure 35.9.
- Certain wavelengths of light stimulate the proton pump and thus regulate guard cell activity.
- Agriculturalists are looking for ways to reduce water loss to transpiration without reducing carbon dioxide availability to plants. This has been attempted through foliar sprays of antitranspirants like abscisic acid and currently through genetic modification of plants to make them more sensitive to abscisic acid.
- Plants that use crassulacean acid metabolism open and close their stomata on a cycle opposite that of other plants.

Phloem moves materials from sources to sinks.

- Sources are organs that produce more sugars than are required to carry out metabolism. Sinks are organs that do not make enough sugar for their own growth or storage needs. Sugars, amino acids, minerals, and other substances are translocated between sources and sinks.
- Translocation stops if phloem tissue is killed. Translocation proceeds in both directions along the stem. Translocation is inhibited whenever respiration and the availability of ATP are limited.
- Scientists can sample the sieve tube sap of a single sieve tube member by using aphids. An aphid drills into a single cell and sap is forced out. Once the aphid begins eating it is frozen and its feeding organ is used as a tap to collect the sap from a single sieve tube. From this, scientists can analyze the contents of a single sieve tube over time.
- The pressure flow model explains how materials move through the phloem. Once in the sieve tubes, sieve tube sap moves via bulk flow and requires no energy. Energy is required for the loading of the sieve tubes at the sources and the unloading of solutes when the sink is reached. Sucrose is actively transported into the sieve tube members at the sources. This causes water to move into surrounding sieve tube members by osmosis, thereby increasing the pressure potential at the source end and pushing the contents toward the sink end of the tube, resulting in bulk flow.
- For the pressure flow model to work, the sieve plates must be open to allow uninterrupted flow of sieve tube sap from one sieve tube member to another. Microscopic analysis indicates that the sieve tube plates are open in undamaged cells.
- There must also be a mechanism for loading and unloading sucrose and amino acids. Neighboring cells assist with the loading and unloading of sucrose at sources and sinks.
- Plasmodesmata assist in moving substances from cell to cell and help with the loading and unloading of sieve tube members. Plasmodesmata in source areas differ from those in sink areas. They are larger and more abundant in sink tissues for unloading. Scientists are exploring the regulation of plasmodesmata in crops to enhance production.

The Big Picture

- In terrestrial plants water must be acquired from the soil, transported through the plant, and used in the leaves for photosynthesis. At the same time, the nutritional products of photosynthesis must be transported throughout the plant to nonphotosynthetic organs. This two-way transport is achieved through specialized cells that make up the vascular tissue of the plant. Water and mineral nutrients are absorbed through the roots and transported to cells in the xylem of the plant, where it is pulled up the stem to the leaves via the transpiration–cohesion–tension mechanism. Sugars and solutes are moved out of the leaves and to the rest of the plant through cells in the phloem.

- Water uptake is regulated by osmotic and water potentials in the root cells, and the rate of transport is controlled by the rate of evaporation at the leaf surface. Guard cell activity in the leaves regulates the opening and closing of stomata to match water availability, light, and drying conditions. Sucrose movement is regulated by active transport and facilitated diffusion in the phloem tissue. The rate of sucrose transport depends on the rate of loading and unloading at source and sink tissues.

Common Problem Areas

- Students often fail to look at the anatomy of the structures that are used in transport. Each structure is uniquely suited to its function. If you couple form and function, the process is much easier to understand.

Study Strategies

- Water and sucrose transport are pathways that can be understood by visually tracing a molecule of water or sucrose through the plant. Use the figures in your book to follow these pathways.
- In order to understand water transport, you need to understand osmosis and the properties of water molecules. If you are struggling with this chapter, review these concepts from earlier chapters.
- The basis of nutritional transport in plants is cell-to-cell transport. Review the sections in your book on diffusion, osmosis, and active transport.
- Review the following tutorial and activity on the website:
  Tutorial 35.1 The Pressure Flow Model
  Activity 35.1 Apoplast and Symplast of the Root

Test Yourself

Knowledge and Synthesis Questions

1. The primary function of the Casparian strip is to
   a. force water and minerals through the membranes of endodermal cells.
   b. prevent entry into the stele solely through the apoplast.
   c. provide regulation for water and mineral movement in the plant.
   d. All of the above

2. The primary difference between the apoplast and the symplast is that the
   a. apoplast is nonliving spaces and cell walls.
   b. apoplast relies on active transport.
   c. symplast is nonliving spaces and cell walls.
   d. apoplast prevents passive diffusion.

3. Which of the following regarding water transport is true?
   a. Root pressure is sufficient to drive xylem sap movement.
   b. Bulk flow is not a mechanism by which water and minerals are transported.
   c. The cohesive nature of water is central to water movement in a plant.
   d. None of the above

4. Tension is a result of which of the following?
   a. Transpiration at the leaf surface
   b. The cohesive nature of water
   c. The narrowness of the xylem tube
   d. All of the above

5. The fact that water transport continues as long as leaves are alive and active indicates that
   a. leaves pump water.
   b. leaves are necessary for transport of water.
   c. roots are active.
   d. water is not needed for leaves to remain alive.

6. Which of the following is true regarding transport in phloem?
   a. Transport in phloem is always in the direction of leaves to roots.
   b. Transport in phloem is from source tissue to sink tissue.
   c. Transport in phloem cells requires energy.
   d. None of the above

7. If the pressure potential is + 0.16 megapascals (MPa) and the osmotic potential is – 0.24 megapascals, then the water potential would be
   a. + 0.04 MPa
   b. + 0.08 MPa
   c. – 0.08 MPa
   d. – 0.24 MPa

8. If you were to order the water potential of the following root cells/regions from least to most negative, which cell/region would be third?
   a. Xylem
   b. Soil next to root
   c. Cortex apoplast
   d. Stele apoplast

9. The movement of water up the stems of tall plants is least dependent on which of the following factors?
   a. Guttation
   b. Transpiration
   c. Cohesion of water molecules
   d. Tension within water columns

10. Which of the following characteristics applies to both xylem and phloem transport?
    a. It is a passive, nonenergy requiring process.
    b. It involves only living cells.
    c. It follows a water potential gradient.
    d. It can occur in both directions.

11. Which of the following regulates stomatal opening and closing?
    a. Abscisic acid levels
    b. Light levels
    c. Carbon dioxide concentrations
    d. All of the above

12. Regulators of stomatal opening and closing work by activating the
   a. proton pump in guard cells.
   b. proton pump in stomata.
   c. sodium–potassium pump.
   d. All of the above

Application Questions

1. Differentiate between source and sink tissues. What happens relative to phloem in each?
2. Under what conditions does transpiration occur most rapidly? What effect will increased transpiration have on water flow in a plant? What happens if adequate water for the plant is not available?
3. Trace the a water molecule as it moves from the soil solution to the stele of a plant. Identify where the molecule is traveling through the apoplast and where the molecule is traveling through the symplast.
4. Describe the role of the proton pump in moving minerals into the root.
5. Explain how transpiration, cohesion, and tension work together to move water in a large plant.
6. Describe the pressure flow model of phloem transport.

Answers

Knowledge and Synthesis Answers

1. **d**. The endodermal cells of the root are the primary regulatory cells for water entry into the stele. They prevent water and minerals from reaching the stele solely through the apoplast.
2. **a**. The intercellular spaces and cell walls of the plant constitute the apoplast.
3. **c**. Water movement depends on the cohesive nature of water to withstand the tension placed on the water column by transpiration.
4. **a**. Transpiration causes tension.
5. **b**. Leaves are necessary for transpiration to take place.
6. **b**. Transport in phloem can move in both directions, but it is always in the direction of source tissue to sink tissue.
7. **c**. Water potential is equal to pressure potential plus osmotic potential.
8. **c**. The xylem would be the most negative, followed by the stele, then cortex, then the area outside the root.
9. **a**. Guttation occurs under extremely humid conditions when water is plentiful.
10. **c**. Both xylem transport and phloem transport depend on water potential.
11. **d**. Abscisic acid, light, and carbon dioxide levels all regulate stomatal opening and closing.
12. **a**. Stomatal regulators work by activating and deactivating the proton pump in guard cells.

Application Answers

1. Source tissues produce sugars in excess of what can be used and stored. Phloem loading occurs in source tissues and creates a pressure potential that results in bulk flow of sieve tube sap toward sink tissues. Sink tissues produce fewer sugars than can be stored or used, and unload phloem through active transport.
2. Transpiration will occur most rapidly in high light conditions when stomata are open, along with high wind conditions and low humidity when evaporation is greatest. This will result in faster bulk flow through the xylem and increased water demands by the plant. If water is not available, plant cells will lose turgor and the plant will wilt.
3. The water molecule enters the epidermis due to water potential. It travels through the apoplast until it reaches the endodermis and the Casparian strip. There it enters the symplast and crosses the membrane of the endodermis. Once across both sides of the endodermis it may again enter the apoplast in its journey to the stele (see Figure 35.3).
4. Plants rely on a proton pump for active transport of minerals into cells. Plants actively pump protons out of cells, which causes the area outside the cell to be more positive. This assists facilitated diffusion of positive ions through protein channels. It also produces energy to drive the movement of negatively charged ions into the cell by active transport (see Figure 35.2).
5. The transpiration–cohesion–tension mechanism pulls water from the roots up through the plant. Water evaporates from mesophyll cells during transpiration. This puts tension on the film of water associated with the mesophyll cell wall. The tension at the mesophyll cell draws water from the xylem of the nearest vein. This creates tension in the entire xylem column and the column is drawn upward from the roots.
6. The difference in solute concentration between sources and sinks allows a pressure potential along sieve tubes, resulting in bulk flow. For this to occur, sugars must be loaded at the source tissue and unloaded at the sink tissue through active transport, and the sieve plates must remain open and unclogged along the phloem column.

STUDY GUIDE

36 Plant Nutrition

Important Concepts

All organisms must make nutrients through the cycling of other compounds or acquire nutrients from their environment.

- The basic nutrient requirements for all living things are carbon, hydrogen, oxygen, and nitrogen. These elements are the fundamental building blocks of all macromolecules.
- Carbon, oxygen, and hydrogen enter biological organisms through photosynthesis.
- Nitrogen's entry into biological organisms is dependent on nitrogen-fixing bacteria in the soil.
- Mineral nutrients are essential to life. Sulfur, phosphorous, magnesium, and iron are all essential components of many macromolecules.
- Mineral nutrients enter biological organisms through soil solutions which plants take up through their roots.
- Plants form the basis of the terrestrial food chain because they make organic compounds that are available to other organisms. The ability to make organic compounds from inorganic compounds is called autotrophism. Heterotrophs require preformed organic compounds.
- Plants cannot move around and therefore cannot "search" for nutrients. They overcome this problem by growing to new resources. A plant grows taller to procure more sunlight for itself and to shade other plants, and its roots spread to acquire new mineral nutrients in the soil.

Mineral nutrients are vital to proper plant growth.

- Every plant requires specific essential nutrients. Essential nutrients are necessary for growth and development, cannot be replaced by another element, and must be directly required for the day-to-day functioning of the plant. A deficiency in any essential nutrient leads to unhealthy plants.
- Macronutrients are required at a rate of 1 g per 1 kg of dry plant matter, and micronutrients are required at a rate of 100 mg per 1 kg of dry plant matter.
- See Table 36.1 for a list of macro- and micronutrients, their sources, and their functions.
- Though many deficiencies ultimately lead to plant death, specific symptoms are displayed in the plant before it dies. Nitrogen deficiency is the most common plant deficiency. Many farmers use chemical fertilizers to supplement nitrogen in soils so that crop plants have adequate nitrogen sources.
- Plant growth will slow to match the availability of nutrients until the nutrient loss becomes severe.
- Chlorosis, or the yellowing of leaves, is the most common symptom of nutrient deficiency. This occurs when chlorophyll cannot be synthesized because of inadequate nitrogen. Chlorosis is also evident in iron deficiencies for the same reason. Chlorosis of young leaves is characteristic of iron deficiency because iron cannot be moved through the plant easily.
- Many essential elements have multiple roles in plants. Essential nutrients were identified by depriving plants of specific nutrients and observing what the result was in the plant. Missing nutrients that prevented the plant from completing its life cycle were deemed essential.

Soils are nature's nutrient sink.

- Plants and soils interact. Plants change soils, and soils influence the growth of plants.
- Soils are composed of living and nonliving matter. The living portion of soil contains roots, protists, bacteria, fungi, and many small animals. The nonliving portion consists of rock fragments, water, air spaces, and dead organic matter.
- All soils have a soil profile consisting of two or more horizons (layers). Water-soluble nutrients are leached to deeper horizons through rainfall. Three major horizons (also called zones) can be identified in soils. Zone A is topsoil. Topsoil is organically rich and very biologically active. This is the most agriculturally important layer. Loam soil with sand, silt, and clay, high nutrient content, plentiful water and air spaces is ideal topsoil for agriculture. Sands typically do not hold nutrients or water well, and clays are too dense for root growth. Zone B is the subsoil and holds many leached nutrients. Zone C is the parent rock which roots cannot penetrate.
- Soils form from weathering rocks. Rocks are broken down by mechanical weathering and chemical weathering. Mechanical weathering occurs through freeze/thaw cycles, rain, and drying. Chemical weathering can actually change the composition of the rock. Chemical weathering is very important to clay formation.

- Mineral nutrients are dependent on clay particles in the soil. Because many nutrients are cations, clays with a negative charge can hold these nutrients and make them available to the plants. Roots release protons into the soil which bind to the clay particles and the cation minerals are released. This process is called ion exchange. Nutrients that are negatively charged and therefore do not participate in ion exchange are rapidly leached from the soil. Thus nitrate and sulfate are often not available to plants.
- Agricultural fertilizers replace nitrogen, phosphorous, and potassium in soils and are rated by their "N-P-K" percentages. A 10-10-10 fertilizer contains 10 percent nitrogen, 10 percent phosphate, and 10 percent potash (potassium source). These nutrients need to be replenished in soils because they are negatively charged and leached from soils rapidly.
- Fertilizers may be organic in nature (manures, compost) or inorganic (chemical fertilizer). Organics release nutrients much more slowly and do not leach as quickly as inorganics. Inorganics provide a much more rapid release of nutrients.
- The pH of soils affects nutrient availability. Specific plants have specific pH needs. Soils tend to get slightly acidic from leaching and rain. The practice of liming raises soil pH. This has the secondary effect of making calcium available to plants. Adding sulfur reduces soil pH.
- Foliar spraying of nutrients works to add copper, iron, and manganese.
- Plants alter the pH of soil, adding decaying matter in the form of humus, and altering soil temperature.

Nitrogen is vital to plant growth.

- Nitrogen gas is readily available in the atmosphere but plants cannot use it directly. Only a few bacteria species can convert nitrogen gas into biologically usable ammonia through nitrogen fixation.
- The most biologically important nitrogen fixers are associated with plant roots and release up to 90 percent of the nitrogen they fix to the soil. *Rhizobium* bacteria are closely associated with the roots of plants in the legume family. These bacteria and the associated legumes are important agriculturally and are the basis of crop rotation.
- Nitrogenase catalyzes the fixation of nitrogen. It is inhibited by oxygen and therefore is active only under anaerobic conditions. The process of reducing nitrogen gas to ammonia is energy intensive. See Figure 36.7.
- The anaerobic *Rhizobium* bacterium resides in a root nodule of the legume where it is protected from oxygen. This relationship is symbiotic. The bacteria are provided with a growing environment and the plant benefits from the released ammonia. This dual benefit is called mutualism. To establish the symbiosis the plant releases flavonoids to attract the bacteria. In response to the flavonoids the bacterial *nod* genes are turned on. The resulting gene products cause formation of the nodule by the plant. Once housed in the nodule, the bacteria form swollen bacteriods capable of nitrogen fixation. The plant surrounds the bacteriods with leghemoglobin to support respiration. See Figure 36.8.
- Bacterial nitrogen fixation is not adequate to supply all the needed nitrogen for agriculture. Currently, nitrogen fertilizers are produced via industrial fixation through the Haber process. This is an energy-expensive process. Much research is going into the possibility of engineering plants that have their own nitrogenase.
- Ammonia itself is toxic to plants. Therefore, soil bacteria that convert ammonia to nitrate are necessary. These bacteria are called nitrifiers.
- Plants, through their metabolism, reduce nitrate to ammonia. This is accomplished by the plant's own enzymes.
- The entire nitrogen cycle is completed by denitrifiers which convert the nitrogen from waste and dead organic matter back to nitrogen gas. See Figure 36.9 to review the nitrogen cycle.

Sulfate reduction in plants is important for amino acid production.

- Cysteine and methionine are sulfur-containing amino acids. All animals must acquire these two amino acids by consuming them. Plants make them after reducing sulfate and incorporating the sulfur into cysteine. All other sulfur-containing compounds are made from cysteine.

Some plants are heterotrophic.

- Some plants have lost the ability to photosynthesize. These plants must acquire their nutrients from other sources. Some of these plants are parasitic and acquire some or all of their nutrients from a host plant at the host plant's expense.
- Some plants that are deficient in particular nutrients are carnivorous. These plants generally grow in very nitrogen- and phosphorous-poor areas. Carnivorous plants acquire nitrogen from trapped decaying animals.

The Big Picture

- Plants require specific macro- and micronutrients in order to sustain life. Deficiencies in any of these produce consequences for the health of the plant. Essential nutrients must be present, and no substitutions will sustain the plant. Nutrients are procured from the soil solution which bathes the roots of the plant. The availability of nutrients depends on quantity, solubility, and structure of soil. Many agricultural practices deplete soils of nutrients, which must be added back through fertilizing.
- Nitrogen availability is essential to plant growth. Bacteria and plants are intrinsically linked in the nitrogen cycle (see Figure. 36.9). Biological nitrogen fixation is utilized in agriculture by rotating crops with those that harbor nitrogen-fixing bacteria in their root nodules. Commercial nitrogen fixation for chemical fertilizers is extremely energy demanding, and alternatives through genetic engineering are being sought.
- Animals are incapable of synthesizing sulfur containing amino acids. Plants are able to reduce sulfur and produce cysteine and methionine, thus adding to their importance in the food web.

- A small number of plants do not photosynthesize. These heterotrophic plants are often parasites of other plants and acquire their nutrients solely through that host plant.

Study Strategies

- The most difficult component of this chapter is the nitrogen cycle. Use the figures in your book to visualize this process, and understand how the organisms involved interact with the environment.
- Be sure you understand the consequences of lack of nutrient availability in plants.
- Review the following tutorial and activity on the website:
  Tutorial 36.1 Nitrogen and Iron Deficiencies
  Activity 36.1 Nitrogen Cycle

Test Yourself

Knowledge and Synthesis Questions

1. Which of the following nutrients are not considered essential for plant growth?
   a. Cadmium
   b. Nitrogen
   c. Manganese
   d. Potassium

2. Macronutrients are __________ than micronutrients.
   a. larger molecules
   b. needed in greater quantities
   c. more essential
   d. more important for growth

3. Carbon and oxygen are acquired through __________ and nitrogen and potassium are acquired from __________.
   a. consumption, soil solution
   b. photosynthesis, nitrification
   c. photosynthesis, soil solution
   d. photosynthesis, fertilizer

4. You notice that the young leaves of your tomato plants are very yellow. What type of deficiency does this suggest?
   a. Nitrogen
   b. Carbon
   c. Water
   d. Iron

5. Years of cotton farming in the South stripped away much of the A horizon of the soils. Subsequent agriculture has been difficult for farmers because the
   a. A horizon contains most of the available nutrients.
   b. B horizon contains significantly less available nutrients.
   c. A horizon is most conducive to root growth.
   d. All of the above

6. Clay particles in soils are important for
   a. holding soil together.
   b. ion exchange.
   c. holding water.
   d. All of the above

7. Most clays form from the __________ of rock.
   a. mechanical weathering
   b. chemical weathering
   c. heaving
   d. All of the above

8. You purchase a commercial fertilizer at your local garden center. The label says that it is 10-20-10. This label refers to the
   a. percentages of nitrogen, phosphate, and potassium.
   b. percentages of nitrogen, carbon, and oxygen.
   c. rate at which nitrogen is released from the fertilizer.
   d. ratio of organic to inorganic matter in the fertilizer.

9. The relationship between *Rhizobium* and the roots of legumes can best be described by which of the following terms?
   a. Parasitic
   b. Symbiotic
   c. Mutualistic
   d. Carnivorous

10. Nitrogen gas is reduced to ammonia by which of the following enzymes?
    a. Rhizobium
    b. Nitrogenase
    c. Nitrification
    d. Denitrification

11. Plants are unique in that they can synthesize the amino acids __________ and __________, while all animals must consume these amino acids.
    a. cysteine, methionine
    b. threonine, methionine
    c. uracil, guanine
    d. cytosine, methionine

12. Carnivorous plants are often found in acidic and nutrient-poor environments. The main selective pressure for carnivory is
    a. lack of nitrogen and phosphorous sources.
    b. lack of iron and calcium sources.
    c. incomplete ion exchange.
    d. All of the above

Application Questions

1. Explain how it was determined whether a nutrient was essential or not. What were the drawbacks of the early experimentation?
2. Sketch the nitrogen cycle. What components of the cycle are impacted by plants and how?
3. Describe how plants and bacteria interact to form nitrogen-fixing root nodules. Why do many farmers plant crops such as alfalfa and soybeans without harvesting them?
4. Most carnivorous plants are found in boggy, wet, acidic environments. What effect would an environment such as this have on nutrient availability?
5. Explain how plants "grow into their nutrients."

6. Differentiate between micro- and macronutrients. Where are most of these nutrients acquired?
7. Discuss the impact agriculture has on plant nutrition. Which procedures benefit nutrition and which procedures rob nutrients from the soil?

Answers

Knowledge and Synthesis Answers

1. **a**. Cadmium is not one of the 14 essential micro- and macronutrients.
2. **b**. The main difference between micronutrients and macronutrients is in how much of them a plant needs to survive.
3. **c**. Carbon, oxygen, and hydrogen are all acquired through photosynthesis. Nitrogen and mineral nutrients are acquired in soil solution (dissolved in water).
4. **d**. An iron deficiency is suspected in a plant. If the older leaves are yellow, a nitrogen deficiency would be suspected because nitrogen is easily relocated in the plant while iron is not.
5. **d**. Topsoil, or the A horizon, is most conducive to root growth. Ample nutrients are available, as are air spaces and water for ease of root growth.
6. **d**. Clay particles are critical for ion exchange. They are also important for retaining water and for the integrity of the soil.
7. **b**. Chemical weathering leads to clay formation.
8. **a** Your fertilizer is 10 percent nitrogen, 20 percent phosphorous, and 10 percent potash.
9. **c**. The relationship is mutualistic in that both the plant and the bacteria benefit from the association.
10. **b**. Nitrogenase catalyzes the reduction of nitrogen gas to ammonia. This is an energy-expensive process.
11. **a**. Sulfur is reduced and incorporated into cysteine. From this all other sulfur containing compounds are made.
12. **b**. Carnivory supplements insufficient nitrogen and phosphorous availability.

Application Answers

1. Plants were grown in cultures that lacked specific nutrients. If the plant could not complete its life cycle, the missing nutrient was said to be essential. Early experiments were limited due to impurities in the chemicals used to grow the plants. These experiments were ultimately verified through well-controlled hydroponic studies.
2. See Figure 36.9. Plants are actively involved in nitrogen fixation through the symbiosis with *Rhizobium*. *Rhizobium* fix nitrogen only in close association with the roots of plants in the legume family.
3. See Figure 36.8. By rotating crops (and even more by rotating and plowing legume crops under) farmers can organically add nitrogen to depleted soils. This results in significantly larger yields to offset the years when the land is not utilized for harvested crops.
4. Acidic environments limit decomposition and thus nitrogen sources. Acidic environments also limit ion exchange. Both conditions result in reduced nutrient availability which can be offset by carnivory.
5. Plants cannot physically move to areas with greater nutrients. They are limited to extending their roots into soils that contain a larger nutrient reserve. When nutrients are limited, plant growth matches the availability of resources.
6. Macronutrients are needed at a rate of one g/one kg of dry plant tissue. Micronutrients are needed at a rate of 100 mg/one kg of dry plant tissue. The majority of these nutrients are in soil solution and are taken up as water is drawn into the roots.
7. Agriculture has increased plant yields-per-unit land through the use of fertilizers and advanced farming techniques, but at a cost. Much of the nation's topsoil has been depleted through poor farming practices, thus limiting the nutrient resources available and forcing dependence on fertilizer. The industry is attempting to find new methods of farming that are more sustainable and less harsh to the environment.

STUDY GUIDE

37 Plant Growth Regulation

Plant development is influenced by multiple regulatory factors.

- Plant development depends on the interplay of environmental cues, receptors that sense those cues, and hormones that mediate the effects of the environment. Enzymes influence development at all stages.
- Hormones are regulatory compounds that are produced in one area of the plant and translocated throughout the plant. The effects of the hormones result from their relative concentrations, and they play multiple regulatory roles in plants.
- Photoreceptors are proteins that sense light and are altered by light quality to induce changes within a plant.
- A plant's genome ultimately regulates its growth and development, under the influence of hormones and photoreceptors.
- Signal transduction mediates hormone and photoreceptor action. A receptor receives an environmental cue, a signal transduction pathway is initiated, and ultimately a cellular response is generated.

Plant development depends on cell division, cell expansion, and cell differentiation.

- Plant seeds are dormant and remain so until seed germination. Germination is triggered by one or more mechanical or environmental cues. To germinate, a plant must imbibe water and draw nutrients from the endosperm or cotyledons.
- Since most seeds germinate underground, there must be rapid elongation of the stem to reach light. This rapid growth is halted by signals from a photoreceptor. Subsequent plant growth is under hormonal control.
- Flowering is initiated by plant age, size, or season. Seasonal flowering is controlled by light duration and is sensed by photoreceptors. Flower formation is most likely induced by hormones, as is subsequent flower development.
- Perennials, which live year after year, have buds that enter dormancy during winter. This dormancy is maintained by hormones. In plants that lose their leaves, senescence is controlled by hormone interactions. Plant death may follow senescent activity.

Defined events regulate the breaking of seed dormancy and germination.

- Dormancy, lasting for different amounts of time depending on the plant, involves exclusion of water or oxygen from the embryo by the seed coat, mechanical restraint of the embryo by the seed coat, and chemical inhibition of the embryo.
- Dormancy can be broken by mechanical abrasion or fire. Prolonged exposure to water may leach chemical inhibitors away from the seed and induce germination.
- Dormancy assures survival through adverse conditions. Some seeds rely on other environmental cues before germination, such as exposure to cold or the passage of a specific amount of time.
- The break in dormancy begins with a plant's imbibing water. Water results in metabolic changes. DNA synthesis is halted until the radical emerges from the seed coat. Breakdown of starch and protein reserves in the cotyledons and endosperm begins to provide nutrients for the developing embryo. In cereal seeds, the embryo secretes gibberellins which promote enzymes that break down starch and proteins.

Gibberellins regulate growth from germination to fruiting.

- Gibberellins are produced in both plants and fungi. The first gibberellin was isolated from a fungus that infects rice plants and causes them to grow tall and spindly. Initial experiments indicated that the medium that the fungus was grown in was sufficient to cause rice plants to be spindly; therefore, the agent that caused the phenomenon was a chemical produced by the fungus.
- Experiments using mutant and normal corn plants showed that plants have innate gibberellins. Normal corn plants exposed to gibberellins showed no alteration in appearance, but dwarf plants exposed to gibberellins showed shoot elongation to near normal lengths. These experiments proved that the chemical was present in plants and that its addition did nothing to alter the plant when it existed in sufficient quantities. In plants lacking the chemical, however, application of gibberellin induced stem elongation.
- Gibberellins also have roles in fruit development. Developing seeds produce gibberellins which enhance development of fruit tissue. It is common agricultural

practice to spray seedless fruits (especially grapes) with gibberellins to enhance fruit growth. Gibberellins also induce bolting and seed set in biennials under appropriate environmental cues.

- Additional roles of gibberellins include fruit development from unfertilized flowers, germination of some seeds, and breaking of winter dormancy.

Auxin affects plant growth and form.

- The discovery of auxin (indoleacetic acid) was the result of work of Charles and Francis Darwin in the 1880s. They were interested in phototropism and which part of an emerging plant coleoptile was responsible for sensing light. They found that the tip was the light-receptive portion, but that the growing region (responsible for the bending to or away from light) was some distance below the receptor region. From this they reasoned that some chemical must be transmitted from the tip to the growing region. Additional experiments with removed tips and tips on gelatin blocks indicated that a chemical was indeed moving from the tip to the growing region. Subsequent experiments by other researchers showed that gelatin exposed to the tips was sufficient to cause altered growth. The chemical was later isolated and determined to be auxin.
- Auxin movement in plant tissues is unidirectional and polar from apex to shoot. Auxin carrier proteins are responsible for the import and export of auxin from cells, and contribute to the unidirectional movement of auxin. Because a gradient is established via transport, auxin acts as a morphogen and directs how the cells differentiate along the gradient.
- Redistribution of auxin laterally is responsible for phototropism and gravitropism. Figure 37.12 illustrates how the lateral distribution of auxin affects the curvature of coleoptiles in response to light and gravity. Higher auxin concentrations correspond to increased rates of growth.
- Auxin affects vegetative growth by initiating root growth in cuttings, causing abscission of leaves, maintaining apical dominance, and promoting stem elongation and inhibiting root elongation. Auxin also can stimulate parthenocarpy in fruits.
- The effects of auxin are on cell walls. The cell wall determines the rate and direction of cell growth. Cells grow by taking up more water. The amount of water that can be taken up is restricted by a rigid cell wall. Cell walls must loosen, stretch, and add polysaccharides and cellulose to maintain structure. Cell walls exhibit elasticity when their shape is altered but return to normal. Plasticity means the shape is altered and does not return to normal. Plasticity is necessary for cell growth. It is thought that auxin affects plasticity and "loosens" the cell wall so that the plant can grow. Auxin stimulates a cell wall loosening factor that results in a pH change. The pH change allows proteins called expansins to alter polysaccharide bonding so that they slide past one another during expansion.
- Because auxin has multiple functions, it stands to reason that there would be multiple auxin receptors and transduction pathways. Research supports the notion of multiple receptors and pathways.
- Auxins and cytokinins regulate organ development based on the relative concentrations of the two. High auxin levels favor root formation and high cytokinin levels favor bud formation.

Cytokinins have multiple roles within plants.

- Cytokinins initiate bud formation, aid in seed germination, inhibit stem elongation, and delay leaf senescence. They are powerful cell division stimulators. They are synthesized primarily in the roots and are translocated throughout the plant.
- Each of these activities is regulated by the relative ratio of cytokinen to auxin.

Ethylene promotes senescence.

- Ethylene is a gaseous hormone that promotes senescence of leaves and ripening of fruit. Commercial use of ethylene spray to promote ripening is widely used for a variety of fruits. Ethylene scrubbers are used in fruit storage to prevent ethylene from accumulating and causing fruit to spoil. Other chemicals are used in the flower industry to inhibit ethylene's affects on flower senescence.
- Ethylene also has a role in maintaining the apical hook on emerging eudicots by inhibiting cells on the inner portion of the hook. Ethylene also inhibits stem elongation and promotes lateral swelling of stems.
- Figure 37.20 illustrates the signal transduction pathway initiated by ethylene. The process begins when ethylene binds to its receptor. This binding initiates a cascade that leads to the expression of genes that result in the physiological changes stimulated by ethylene.

Abscisic acid is responsible for maintaining dormancy.

- Abscisic acid inhibits seed germination and promotes storage during embryo formation. It accumulates during times of stress and is responsible for maintaining winter dormancy.
- Plants deficient in abscisic acid often show vivipary.
- Abscisic acid regulates guard cell activity by opening calcium channels. Increased calcium levels eventually lead to the opening of potassium channels and the closing of the stoma as guard cells sag together.

Hormones regulate reactions to pathogens.

- Upon invasion by pathogens, plants release oligosaccharins from damaged cell walls which trigger plant defenses.
- Auxin also may cause the release of oligosaccharins through its effect on cell walls.
- Other hormones including jasmonates, salicylic acid, and systemin serve as signals during plant defense.

Brassinosteroids are not completely understood, but have many effects.

- Brassinosteriods were isolated from a member of the Brassicaceae. They have been shown to stimulate cell elongation, pollen tube elongation, vascular tissue differentiation, and to inhibit root elongation, all of which are similar to the effects of auxin. It took the isolation of a mutant for scientists to determine definitively that brassinosteriods were themselves a group of hormones.
- Research is being carried out on the effectiveness of brassinosteriods as agricultural growth regulators.

Light and photoreceptors interact to stimulate a variety of plant events.

- Photoreceptors interpret light, its duration, and its wavelength. Light regulates a wide variety of plant processes including germination and flower production.
- Phytochromes mediate the effects of red and dim blue light. Blue and red light promote germination. Far-red light reverses the effect of prior exposure to red light. Far-red light is 730 nm, whereas red light is 660 nm. Exposure to red and far-red light have different effects on plants. The "switching" effects occur because phytochrome can be shifted from one form (red P_r) to the other (far-red P_{fr}) upon absorption of light. When exposed to red light, P_r is converted to P_{fr}. When exposed to far-red light, P_{fr} is converted to P_r.
- Phytochromes regulate etiolation, which functions to conserve resources until plants can begin photosynthesizing. In seedlings that have not been exposed to light, all the phytochrome is the P_r type. Once exposed to light, it shifts to P_{fr} and begins a series of events that reverses etiolation.
- There are multiple phytochromes with multiple regulatory functions. They may be responsible for changes in the plant throughout the day in response to shading. It is thought that they interact with G proteins in a variety of cascades.

Cryptochromes and phototropin are blue-light receptors.

- Cryptochromes absorb blue and ultraviolet light. These receptors are found both in plants and animals. In plants, they carry out many of the same regulatory functions as phytochrome, but they are active in the nucleus rather than in the cytoplasm.
- Phototropin, under blue light, initiates the signal responsible for phototropism.

The Big Picture

- Plant growth is controlled by interactions between the plant's environment, hormones, and its genetic makeup. Changes in plant growth are dependent on receptors to receive information, cascades of events involving enzymes and mediated by hormones, and alteration of gene expression. Receptors are often highly specific and respond to select signals. Hormones are produced in a specific region of the plant body and translocated throughout the plant; therefore, their effects are often concentration-dependent. All growth in a plant is a result of changes in cell division, cell expansion, and cell differentiation.
- Seed germination, growth of the vegetative structures, reproduction, and senescence all proceed in defined patterns. Seed dormancy is broken and development begins when the seed coat is abraded, inhibitory chemicals are diluted, and the seed imbibes water. This begins a series of events that mobilize nutrients and induce growth. Once the seedling emerges from the soil, light begins to influence subsequent development under the control of hormones.
- Hormones exist in several classes, each with their own effects on growth and development. Some of those effects are antagonistic in nature and therefore control is maintained via relative concentration of several hormones. Hormones influence every step of development, from breaking of seed dormancy to senescence.
- Light regulates plant processes through photoreceptors. Photoreceptors respond to very specific wavelengths of light and induce cascades that lead to changes in the plant.

Common Problem Areas

- Understanding how phytochrome shifts from the red to far-red forms frequently gives students difficulty. This is understandable, since phytochromes have given researchers a difficult time for many years.
- The tendency when studying hormones is simply to memorize functions. You need to understand what relative ratios of hormones will do in a plant, not what one individual hormone does.
- It is easy to get overwhelmed with information in this chapter. Chances are the material is completely new to you. Take your time in learning how growth is regulated and the effects of the various hormones.

Study Strategies

- Avoid focusing too much on details. Try to assimilate the big picture of how the environment, receptors, hormones, and the genome interact. From there, begin to work toward the detail. The biggest mistake is to attempt to jump in and memorize functions of hormones or sequences of development without understanding the broad perspective.
- Reviw the following tutorial and activities on the website:
  Tutorial 37.1 Tropisms
  Activity 37.1 Monocot Shoot Development
  Activity 37.2 Eudicot Shoot Development
  Activity 37.3 Events of Seed Germination

Test Yourself

Knowledge and Synthesis Questions

1. Plant growth is ultimately regulated by which of the following?
   a. Environmental cues
   b. Hormones
   c. Signal transduction pathways
   d. The plant's genome

2. Which of the following schemes best represents how environmental cues are transduced to changes in a plant?
   a. Receptors receive environmental cues, a signal transduction pathway is initiated, there is an alteration in the particular genes that are transcribed and translated, and a cellular response is generated.
   b. Receptors are triggered, hormones are released, signal transduction pathways are initiated, there is an alteration in expression of genes, and a cellular response is generated.
   c. Neither a nor b
   d. a and b

3. Which of the following triggers seed germination?
   a. Seeds imbibe water.
   b. Seeds are released from fruit.
   c. Seeds undergo chemical changes.
   d. All of the above

4. Which of the following may function to break dormancy in seeds?
   a. Penetration of the seed coat
   b. Leaching of inhibitory compounds by water
   c. Exposure to fire
   d. All of the above

5. Which of the following hormones are responsible for bud break in the spring in deciduous trees?
   a. Auxins
   b. Cytokinens
   c. Gibberellins
   d. Ethylene.

6. Auxin regulates cell growth by which of the following mechanisms?
   a. Altering the elasticity of cell walls
   b. Altering the plasticity of cell walls
   c. Synthesizing new cell walls
   d. Breaking down cell walls in growing cells

7. Which of the following hormones are responsible for maintaining bud dormancy in deciduous trees?
   a. Auxins
   b. Cytokinens
   c. Gibberellins
   d. Abscisic acid

8. You have installed an outdoor gas burning grill on your back patio next to your favorite camellia bush. After the first few chilly nights of using your grill you notice that your camellia, which does not normally lose its leaves, is beginning to do so. Which of the following is the best explanation for what is happening?
   a. The bush is getting too warm next to your grill.
   b. Ethylene is a by-product of the gas you are burning and is causing senescence in your plant.
   c. Abscisic acid is a by-product of the gas you are burning and is causing senescence in your plant.
   d. The plant is a biennial and is bolting.

9. You are slicing a green pepper for the pizza you are making at home. As you slice into it you notice lots of tiny pepper plants emerging from the seeds of the pepper. This pepper is exhibiting __________ and may be lacking in __________.
   a. parthenocarpy, gibberellins
   b. parthenocarpy, abscisic acid
   c. vivapary, gibberellins
   d. vivapary, abscisic acid

10. Which of the following light receptors is responsible for absorbing blue and ultraviolet light?
    a. Phytochrome P_r
    b. Photochrome P_{fr}
    c. Cryptochrome
    d. Photropin

11. Etiolated seedlings are produced by germinating seeds and keeping them in total darkness. Under which of the following conditions will plants kept in the dark begin to synthesize chlorophyll?
    a. After being given a pulse of blue light
    b. After being given a pulse of red light
    c. After being given a pulse of red light followed by a pulse of far-red light
    d. After being given a pulse of far-red light followed by a pulse of red light

12. Match the major site of production on the left with the plant growth substance produced at that site on the right. A substance may be produced at multiple sites.

____ Abscisic Acid	a. Embryo
____ Auxin	b. Buds
____ Cytokinin	c. Leaves
____ Ethylene	d. Fruit
____ Gibberellin	e. Root tips
	f. Shoot tips

Application Questions

1. You have planted pea seeds in your garden. Trace the steps that occur between your planting of the seed, and the emergence of the pea plant from your garden.
2. In the pea plant in question one, what hormonal influences are affecting it from the moment you put it in the ground until its emergence?
3. Discuss how gibberellins were discovered. How did researchers determine they were chemical in nature?
4. Explain how auxin distribution regulates phototropism and gravitropism.
5. Auxins and cytokinens appear to cancel out the effects of each other. Why would a hormone that affects bud growth be produced in the roots and one that affects root growth be produced in the shoots? How does this relate to polar distribution of hormones?
6. Genetic engineering has produced fruits that are deficient in the ability to produce ethylene. How is this deficiency useful in the storage and marketing of fruits?

7. You are experimenting in the laboratory with tissue culture methods. You have isolated pith tissue and grown an undifferentiated mass of cells (called a callus). You divide up the tissue and place it in the following culture media:
   a. necessary nutrients plus indoleacetic acid.
   b. necessary nutrients plus zeatin (a cytokinin).
   c. necessary nutrients plus equivalent concentration of indoleacetic acid and zeatin.
   d. necessary nutrients plus an excess of zeatin and minimal indoleacetic acid.

   Explain what happens to the mass of tissue under each condition.
8. Commercial bananas are triploid; therefore, their seeds abort without development. Under normal conditions the fruits would not develop. Explain how commercial banana growers use hormones to produce sweet fruit with no seeds. Which hormones could they be using?

Answers

Knowledge and Synthesis Answers

1. **d.** Though the environment influences development, final regulation is at the genomic level.
2. **d.** Receptors receive environmental cues and trigger hormone release or a signal transduction pathway that involves hormones. The signal alters gene expression by altering which genes are transcribed and translated. Cellular response depends on expression of those genes.
3. **a.** The uptake of water by the seed begins the processes that lead to seed germination.
4. **d.** Mechanical abrasion, leaching of inhibitors by water, and exposure to fire may all trigger germination, but actual germination cannot begin until the seed imbibes water.
5. **c.** Gibberellins are responsible for bud break in deciduous trees.
6. **b.** Altering the plasticity allows for permanent changes in cell wall shape. The cell wall must increase in size in order for cell growth to occur.
7. **d.** Abscisic acid is responsible for maintaining winter dormancy in deciduous plants.
8. **b.** Ethylene gas promotes senescence and is one of the by products of burning your gas grill. You should move your grill or your camellia bush.
9. **d.** Vivapary is the germination of seeds before they leave the parent plant and is caused by a deficit of abscisic acid.
10. **c.** Cryptochromes respond to blue and ultraviolet light wavelengths.
11. **c.** The pulse of red light converts P_r to P_{fr}. Subsequent pulses of far-red light stimulate changes that lead to chlorophyll synthesis.
12. **b.** Abscisic acid
    c.,f. Auxin
    e. Cytokinin
    d. Ethylene
    a. Gibberellin

Application Answers

1. Upon planting a pea seed you watered it, promoting the leaching of germination inhibitors. At this point the seed also begins to imbibe water, resulting in metabolic changes. DNA synthesis is halted until the radical emerges from the seed coat. Breakdown of starch and protein reserves in the cotyledons and endosperm begin to provide nutrients for the developing embryo. The apical hook begins to push through the soil. Upon its emergence and exposure to light, chlorophyll synthesis begins.
2. Dormancy of the seed is maintained by abscisic acid until it is leached from the seed with water. Once water is imbibed cytokinins begin to influence germination. Gibberellins assist with the mobilization of storage products to the growing embryo. Ratios of auxins and cytokinins balance root and bud formation. The apical hook is maintained by ethylene.
3. The first gibberellin was isolated from fungus that infects rice plants and causes them to grow tall and spindly. Initial experiments indicated that the medium that the fungus was grown in was sufficient to cause rice plants to be spindly; therefore, the agent that caused the phenomenon was a chemical produced by the fungus.
4. Lateral distribution of auxin controls phototropism and gravitropism. Auxin accumulates in the shaded portions of a stem and stimulates cell growth. This uneven cell growth results in a bending of the stem toward the light. The same mechanism works in response to gravity. An accumulation of auxin occurs where the gravitational pull is the strongest. Cells grow in response to auxin, and stems bend upward, away from the gravitational force.
5. Roots need shoots and vice versa. Increases in roots require increases in shoots for photosynthesis. Regulation of the development of one by the other keeps growth in tandem. Because the distribution of the molecules is polar, a concentration gradient can be established. It is this relative gradient that controls development.
6. Fruits can be kept from ripening until they reach their destination. Once in markets they can be sprayed with ethylene to stimulate ripening. The result is fruit that can be shipped more easily yet can be ripe at any time in the market.
7. a. Roots will develop from the callus.
   b. Buds will develop from the callus.
   c. Both roots and buds will develop from the callus.
   d. Buds will develop from the callus.
8. Commercial banana and grape growers frequently use gibberellins to assist with fruit formation. Gibberellins, and to some extent auxins, can be used to stimulate parthenocarpy or fruit development without seeds

38 Reproduction in Flowering Plants

Important Concepts

Plants may reproduce asexually or sexually.

- Asexual reproduction results in plants with the same genetic makeup as the parent plant. In agriculture, this may be desirable.
- Sexual reproduction is necessary for genetic recombination. This promotes genetic variability and is used in agriculture to reap the benefits of "hybrid vigor."

Sexual reproduction provides for genetic variability.

- Sexual reproduction is advantageous because it provides for greater adaptability within a population; however, it is energy-consuming and risky.
- The flower is the basis of sexual reproduction in angiosperms. Review Figure 29.6 for basic flower structures, including: carpels, pistil, ovule, ovary, stigma, style, stamen, anther, filament, petals, sepals, tepals, calyx, corolla, and receptacle.
- The flower is produced from the sporophyte generation. The flower produces spores which develop into highly reduced gametophytes and produce gametes. The female gametophytes, called embryo sacs, develop in the megasporangia. Male gametophytes, called pollen grains, develop in microsporangia.
- Within the ovule, a megasporocyte produces four haploid megaspores through meiosis. Only one of these megaspores survives. It divides mitotically to produce eight nuclei within a single large cell. The nuclei migrate to either end but two remain in the middle of the cell. Cell walls form, isolating the three nuclei at either end into individual cells while the two nuclei in the middle remain together in one cell. At one end, the three cells become two synergid cells and one egg cell; at the other end-cells are the antipodal cells, and in the large central cell are the two polar nuclei. This seven-celled embryo sac is the gametophyte. See Figure 38.1.
- Pollen grains are developed when a microsporocyte undergoes meiosis. All products of meiosis are retained and undergo mitosis to form a two-celled pollen grain. Further development is halted until after pollination.
- Pollination allows for fertilization without water. Pollen grains are carried to female flowers via wind or another vector. Pollen grains must attach to the stigma for fertilization to occur.
- The stigma allows plants to practice "mate selection." Pollen of the right species sticks readily to the stigma via cell-to-cell signaling; however, pollen of other species does not adhere and does not germinate. Specific self-incompatibility genes prevent self-fertilization in many plants. Other plants self-fertilize without problem.
- Sperm nuclei are delivered to the embryo sac via a growing pollen tube. The pollen tube grows down through the tissue of the style.
- The pollen grain consists of two cells at the time of pollination, the tube cell and the generative cell. The tube cell controls the growth of the pollen tube. As the tube is growing, the generative cell undergoes meiosis and produces two sperm cells. Once the pollen tube enters the embryo sac, the two sperm cells are released into a synergid that disintegrates and releases the sperm nuclei. One nucleus fuses with the egg cell and the other fuses with the polar nuclei. This results in the zygote ($2n$) and the endosperm ($3n$). All other cells disintegrate.
- Double fertilization resulting in a zygote and endosperm is a characteristic feature of angiosperms. See Figure 38.6.

Embryos develop within seeds.

- The success of the embryo depends on its own development, the development of the endosperm, the integuments, and the carpel. Ultimately a seed is produced that protects the dormant embryo.
- The zygote divides mitotically to produce the embryo itself and the suspensor. The suspensor pushes the embryo to the endosperm. The suspensor continues to divide and produces a filament, whereas the embryo produces a globular structure. This establishes polarity and symmetry in the embryo.
- In eudicots, the embryo progresses through the heart stage and torpedo-stage as cotyledons and organs are developed. The hypocotyl with its apical meristems is also established. See Figure 38.7.
- The endosperm also develops and holds nutrient reserves. The cotyledons may absorb the nutrients from the endosperm and increase in size.
- Once development is nearly complete, the seed loses water and the embryo ceases development. It remains in this state until germination.

- As the embryo and seed are developing, the ovary begins to form the fruit. Other parts of the flower and plant may be included in the fruit, but to be considered a fruit, only the ovary wall and the seed need be involved. The fruits aid with dispersal via various means including attaching to the coats of animals or being consumed and later deposited.

Specific environmental factors signal the onset of reproduction through flower development.

- Flowering may halt plant growth, interrupt plant growth, or have little effect on plant growth. Seed production is ultimately the purpose of the plant. Some plants go from seed to seed set and die within one growing season. These plants are referred to as annual plants. Plants that require one vegetative growing season before reproducing are called biennial plants. Plants that repeatedly flower and live for many years are called perennials.
- The first transition from vegetative growth to floral production is the alteration of the apical meristem to a floral meristem. The floral meristem differs from the apical meristem in that growth is determined. The floral meristem is programmed to produce four consecutive whorls of flower organs and to be active for a defined amount of time.
- A gene cascade leads to flower formation. Floral formation begins with the activation of a set of floral meristem identity genes. The products of the transcription and translation of these genes influence pattern formation genes which direct the organization of the floral organs. The expression of the floral organization genes in turn trigger organ identity genes which trigger additional genes leading to complete flower development.
- The gene cascade leading to flower development is controlled by photoperiod. Seasonal flowering is a result of changes in photoperiod, specifically the duration of darkness. Each plant type has a critical day length (day length corresponds to light availability that induces flowering. Short-day plants flower when the amount of light available is shorter than the critical maximum. Long-day plants flower only when the day is longer (more light available) that the critical minimum. Some plants require complex combinations of day lengths. Some plants are day-neutral and depend on other factors, such as temperature, to flower.
- Experiments have shown that plants actually respond to the length of the night rather than amount of daylight. In experiments in which daylight was interrupted, there was little effect on flowering, but in experiments in which dark was interrupted, there were significant effects on flowering. The mechanism controlling this phenomenon is far from understood, however.
- Plants, like all other organisms, have an internal mechanism for measuring time. The manifestations of this clock are referred to as circadian rhythms. Circadian rhythms have a duration referred to as a period and an amplitude, or magnitude of change over time. Circadian rhythms have specific characteristics in all organisms. They are not influenced by temperature, they are highly persistent, they can be entrained to coincide with different light–dark cycles, and light exposure causes a phase shift in the rhythm. Plants provide numerous examples of 24-hour circadian rhythms.
- Florigen is thought to be the hormone responsible for flowering even though it has yet to be isolated. A series of experiments provided evidence that once a leaf is light-induced, it sends a hormonal signal to the rest of the plant to initiate flowering.
- Vernalization is the induction of flowering by low temperatures. The discovery that plants could be induced to flower by exposure to cold has been enormously useful to agriculture.

In certain conditions, asexual reproduction is advantageous to plants' success.

- Asexual reproduction occurs through the modification of a vegetative organ. Stolons, tip layers, tubers, rhizomes, bulbs, corms, and suckers are all modifications of stems or roots that allow vegetative reproduction.
- Some plants such as dandelions reproduce asexually through seeds in a process called apomixis. Apomictic plants skip over meiosis and fertilization and produce seeds with the identical genetic makeup of the maternal plant. Sometimes apomixis does require pollination and the union of a sperm nuclei with polar nuclei to form endosperm, yet no genetic recombination occurs.
- Asexual reproduction is used in agriculture. Cuttings are commonly used in horticulture. Grafting is widely used in fruit crops where the rootstock is grafted to a scion (shoot system). This allows for a hardy rootstock to be combined with a good but often fragile fruit producer. Tissue culture (production of entire new plants from a small tissue sample) and recombinant DNA technology are leading to daily advances in agriculture.

The Big Picture

- Though plants can reproduce both asexually and sexually, maintenance of genetic variability depends on sexual reproduction. The flower is the basis of sexual reproduction in plants. The flower not only produces the necessary gametes, but it is also integrally involved in ensuring pollination. Eggs are produced through megasporogenesis, and pollen is produced through microsporogenesis. Angiosperms exhibit double fertilization resulting in a diploid embryo and a triploid endosperm.
- Embryo development takes place in the seeds of angiosperms and progresses until cotyledon(s) and the hypocotyl with its apical meristems are developed. At that point, the embryo halts development, the seed desiccates, and dormancy begins until conditions are right for germination. The seed protects the embryo, and in many cases a fruit is formed for protection until maturation, and to aid in dispersal of the mature seed.
- Flowering and seed set is triggered by the length of night. Some plants flower in response to short days, others to long days, and yet others to complex combinations of the two. Day-to-day functions of plants are triggered by photoperiod and circadian rhythms. Plants may flower, set seed, and die in one growing season or two growing seasons, or they may continue to do so for even longer periods of time.

Common Problem Areas

- Megasporogenesis is probably the most difficult concept in this chapter, but it is relatively simple if you look at the source of each cell.
- Recall that flowering, like all other plant processes, is a result of environmental signals, receptors, enzyme cascades mediated by hormones, and alterations in gene expression.

Study Strategies

- Refer to the figures in the text to understand flower structure, megasporogenesis, microsporogenesis, and fertilization.
- Spend some time thinking about the advantages and disadvantages of sexual and asexual reproduction. Related questions often show up on exams.
- Be sure you understand the particular features of double fertilization.
- Refer to the following tutorial and activity on the website:
  Tutorial 38.1 Double Fertilization
  Activity 38.1 Early Development of a Eudicot

Test Yourself

Knowledge and Synthesis Questions

1. You manage a greenhouse that produces roses for Valentine's Day. Roses normally bloom in June. Which of the following will most likely be the best lighting schedule for your roses?
   a. 16 hours of light, 8 hours of interrupted dark
   b. 16 hours of light, 8 hours of uninterrupted dark
   c. 10 hours of light, 14 hours of dark
   d. None of the above

2. After setting the correct photoperiod, you still don't have roses. Which of the following has most likely contributed to the problem?
   a. The heating system is allowing fluctuations in temperature of between 20° and 25° C.
   b. The furnace mechanic accidentally turned off the lights for an hour two days in a row.
   c. The cleaning crew turned the lights on for an hour three nights in a row.
   d. None of the above

3. You have moved into a new house. During the first summer you notice lots of plants that do not bloom. During the second summer your yard is a sea of blooms. It is now spring of the third year, and there are no plants. This can best be explained by which of the following?
   a. Your plants were annuals.
   b. Your plants were biennials.
   c. Your plants were perennials.
   d. Your plants were affected by drought.

4. You mother gives you a houseplant. You notice it sending out long stems with what looks like "little plants" attached. You allow one of these to rest in a cup of water and note that roots form. This is an example of
   a. asexual reproduction.
   b. apomixis.
   c. heterospory.
   d. parthenogenesis.

5. If plants produce pollen and mature eggs at the same time, how are self-pollination and fertilization prevented?
   a. Some plants have self-incompatibility genes.
   b. Some plants rely on physical barriers.
   c. a and b
   d. None of the above

6. Most wine grape vines are grafted onto rootstock of another species. How does this help grape yields?
   a. A hardy rootstock can replace a weak rootstock.
   b. A high-producing vine stock can replace a low-producing vine stock.
   c. The process allows vintners to select for pest resistance without losing grape quality.
   d. All of the above

7. The induction of flowering by means of exposure to low temperature is called
   a. vernalization.
   b. frigidation.
   c. apomixis.
   d. None of the above

8. The cyclic variation in an organism that corresponds to a set light–dark cycle is referred to as an organism's
   a. biorhythm.
   b. circadian rhythm.
   c. period.
   d. photoperiod.

9. The production of seeds without fertilization is called
   a. apomixis.
   b. parthenogenesis.
   c. conception.
   d. circadian rhythm.

10. Identify the following structures as sporophyte or gametophyte.
    a. Embryo sac __________
    b. Antipodal cells __________
    c. Polar nuclei __________
    d. Integument __________
    e. Receptacle __________
    f. Ovary __________
    g. Anther __________
    h. Pollen grain __________

11. In the transition from vegetative growth to floral growth the __________ must be transformed to the __________. This involves a shift from __________ growth to __________.
    a. apical meristem, floral meristem, indeterminate, determinate
    b. lateral meristem, floral meristem, indeterminate, determinate
    c. apical meristem, floral meristem, determinate, indeterminate
    d. apical cambium, floral cambium, determinate, indeterminate

12. Which of the following best describes the fate of the generative cell of the pollen grain?
    a. It coordinates growth of the pollen tube.
    b. It divides by meiosis to produce two sperm nuclei.
    c. It divides by mitosis to produce two sperm nuclei.
    d. None of the above

Application Questions

1. Compare and contrast asexual and sexual reproduction in plants. In which category does self-fertilization belong?
2. Describe egg formation in angiosperms. Begin with the sporophyte.
3. Lots of effort is spent detasseling corn (removing male flowers). Explain why corn plants cannot be sprayed with a meiosis inhibitor to halt pollen production. (Hint: male and female flowers occur on the same corn plant.)
4. Flowering is stimulated when light sets off a gene cascade. Explain how this may be hormonally controlled.
5. Describe double fertilization. What is the ploidy level of the products of double fertilization.
6. When is asexual reproduction beneficial to plants? Under what conditions is sexual reproduction beneficial?
7. Draw and label an angiosperm flower. Identify the minimum parts necessary for reproduction.
8. Define a fruit. Which of the following are fruits: tomato, pear, potato, banana, cucumber, snow pea, peanut, sunflower seed?

Answers

Knowledge and Synthesis Answers

1. **b**. Flowering is regulated by darkness. Interruptions in darkness can prevent flowering.
2. **c**. The interruptions in the dark cycle by the cleaning crew have affected the signals that tell the plant to begin flowering.
3. **b**. The plants are most likely biennials that spend one season growing vegetatively and one season producing flowers before dying.
4. **a**. The runner (stolon) is asexually propagating this plant.
5. **c**. Some plants have self-incompatibility genes that prevent the growth of the pollen tube through the style. Others have mechanical barriers to their own pollen.
6. **d**. Grafting allows selection of hardy rootstock, well-producing vine stock, and disease resistance.
7. **a**. Vernalization is noted in winter wheat and is used agriculturally to grow high yielding wheat.
8. **b**. Circadian rhythms roughly match a 24-hour cycle and can be reset by altering light and dark cycles.
9. **a**. Dandelions and other plants produce seeds without meiosis and fertilization. These seeds are genetically identical to the parent plant.
10. **a.** Embryo sac — gametophyte
    b. Antipodal cells — gametophyte
    c. Polar nuclei — gametophyte
    d. Integument — sporophyte
    e. Receptacle — sporophyte
    f. Ovary — sporophyte
    g. Anther — sporophyte
    h. Pollen grain — gametophyte
11. **a**. Apical meristems and floral meristems differ because growth from the apical meristem is indeterminate and growth from the floral meristem is determinate, leading to four whorls of floral structures.
12. **b**. The two sperm nuclei used in double fertilization are derived from the generative cell of the pollen grain.

Application Answers

1. Asexual reproduction does not involve meiosis, fertilization, or genetic recombination. The offspring from asexual reproduction are genetically identical to the parent plant. Sexual reproduction requires a meiotic event and fertilization. Self-fertilization is sexual reproduction even though genetic recombination is limited. This is because a meiotic event occurs and fertilization is necessary for reproduction to take place.
2. See Figure. 38.1 in your text.
3. Meiosis occurs in the megasporocyte as well as the microsporocyte. Inhibition of pollen formation via a meiosis inhibitor will also inhibit egg formation.
4. Light induction begins with the leaves. A single isolated leaf may stimulate floral production throughout the plant. In experiments in which leaves from one plant were grafted on other plants, an "induced" leaf caused a plant to flower even though the plant had not been exposed to the required amount of darkness. Therefore, the leaf must send a signal (most likely a hormone) that begins the gene cascade leading to flower formation.
5. Double fertilization occurs when the two sperm nuclei produced from the generative cell of the pollen grain unite with the egg nucleus and two polar nuclei respectively. The resulting zygote is diploid and the endosperm is triploid.
6. Asexual reproduction is beneficial in stable environments where many genetically identical plants are sustainable. This process can help colonize a habitat or assist a population of plants to spread. The disadvantage is lack of genetic diversity. Lack of diversity can be detrimental if the environment changes rapidly, requiring adaptability of the population.
7. See Figure 29.6. Only the ovule or the anther is required for reproduction provided that the corresponding structure (anther or ovule) is found on another plant of the same species.
8. A fruit is the seed, the ovary wall, and may include other structures of a flowering plant. With the exception of the potato, all of the listed structures are fruits.

39 Plant Responses to Environmental Challenges

Plants and their pathogens interact, each giving clues to the other.

- The action of a pathogen signals a plant to mount a defense. The plant's defense signals the pathogen. These signals go back and forth until one or the other "wins."
- The first system of plant defense is nonspecific, an attempt to prevent entry by pathogens. Cutin, suberin, and waxes cover the cork and epidermis to exclude pathogens. Plants cannot repair damaged and invaded tissues, so they seal them off to prevent systemic infection. The damaged area is replaced.
- Upon recognition of a pathogen, polysaccharides are deposited in the cell wall and seal off the plasmodesmata to form a barrier between cells. Subsequently, lignin is deposited which reinforces the barrier and acts as a pathogen toxin.
- Once invaded, plants produce chemical defenses against pathogens. Phytoalexins are nonspecific antifungal and antimicrobial compounds. Pathogenesis-related proteins (PR proteins) may function to break down the walls of pathogens or serve as signaling molecules.
- Salicylic acid, a relative of aspirin, is produced in plants due to a hypersensitive response. Pathogen-containing tissue and surrounding tissues die upon infection. As these cells are dying, they release phytoalexins and salicylic acid. These compounds "arm" the plant against subsequent invaders.
- Systemic acquired resistance protects a plant against infection. This may be due to the stimulation of PR proteins after a hypersensitive response. Plants also release methyl salicylate, which can become airborne and serve as a signal to other plant parts or to neighboring plants to mount a defense via PR proteins.
- Gene-for-gene resistance is a highly specific type of resistance and depends on the plant's having an allele for a gene that matches an allele in the pathogen. Dominant *R* genes code for resistance in plants, and dominant *Avr* genes result in less virulent strains of the pathogen by coding for a substance that elicits a defensive response in the plant. Though this mechanism is not completely understood, the gene product interactions lead to production of PR proteins.

Herbivory confers advantages and disadvantages to plants.

- Some plants depend on the grazing of herbivores to increase productivity of the plant due to coevolution of the plant and the herbivore. Removal of some leaves triggers increased photosynthesis in remaining leaves and a better partitioning of resources. Grazing also allows light to reach young leaves in plants such as grasses. Plants that are grazed have basal rather than apical growth, and continue growing after being grazed. Some produce leaves longer and flower more abundantly.
- For the majority of plants, grazing is a problem; therefore, defenses against grazing exist. Plants produce specific secondary products in defense. The effects of these chemicals are diverse, ranging from neurotoxins to hormone mimics. See Table 39.1 for classes of defensive secondary products.
- Canavanine is a defensive secondary product that plays multiple roles in the plant. It assists with nitrogen-storing in the seed, and, because it is similar to arginine, it is incorporated into herbivore proteins after consumption. The failure to distinguish between arginine and canavanine during protein synthesis alters the tertiary structure of proteins and can be lethal to the organism consuming the plant. Some organisms have the ability to distinguish between the two and are not harmed by canavanine.
- Defense is the result of a series of chemical signaling events. Figure 39.6 details the events leading from wounding to production of an insecticide. Wounding causes the release of systemin which travels to other parts of the plant and induces membrane breakdown. By-products of this breakdown are jasmonates, which enter the nucleus and activate production of a protease inhibitor that inhibits digestion in the predator.
- Some plants produce products that make them resistant to attack. These chemicals can be isolated and the genes producing them identified. Resistance can be bred into plants or transferred through biotechnology. This is not always good for crop plants because it may alter the taste or make them harmful to humans. Scientists must weigh benefits and costs of altering crop species.

Plants use a variety of strategies prevent harming themselves with their own defensive mechanisms.

- Toxic secondary products may be isolated in compartments. Water-soluble toxins are stored in vacuoles, hydrophobic poisons are stored in lactifers along with latex, and others are dissolved in the waxes covering the epidermis.

- Some toxic substances are produced only in already damaged or dying tissue. The precursors of toxic material are stored in different compartments from the enzymes that convert them to active poison, and these combine only once the cell is damaged.
- In some plants, the plant's enzyme mechanisms do not recognize the toxin.

Plants must be able to cope with water availability.

- Some plants, called xerophytes, are specifically adapted to dry areas. They survive in these areas by completing their entire life cycle in the short period that water is available. Others may have special modifications such as thickened cuticles, epidermal hairs, and sunken stomata to prevent water loss. Succulents have water-storing leaves, others drop leaves during drought, and yet others have water-storing stems and highly reduced leaves.
- Roots are also modified for drought conditions. Long taproots can reach water supplies far underground. Shallow, fibrous root systems that grow only during rainy seasons are also adapted to desert growth. Xerophytic plants are able to extract more water when it is available, by changing the osmotic and water potential of their root cells. They do this by storing proline in their vacuoles.
- Too much water can be as dangerous as too little to plant survival. Roots submerged in water do not get adequate oxygen to sustain respiration. Some plants that grow in standing water overcome this by producing pneumatophores, which are root extensions that grow above water and provide oxygen to the entire root system. Others rely on fermentation and thus grow very slowly. Yet others have leaf modifications called aerenchyma for buoyancy and oxygen storage.

Some plants are adapted to survive in saline environments.

- Saline environments have a negative water potential and pose an osmotic stress to plants. For a plant to survive, it must have a greater negative water potential to be able to extract water from the environment. Halophytes (salt-loving plants) are adapted to do this.
- Halophytes are the only plants that accumulate sodium and chloride ions which they store in vacuoles in leaves. The increased salt concentration of the halophyte means it has a more negative water potential.
- Some plants are able to excrete salt so that it does not reach toxic proportions. Salt glands move salt to the leaf surface where it can be lost to wind or rain. Salt glands on the leaf assist with water procurement from the roots, and with reduction of water loss to evaporation.
- Many of the modifications that allow survival in drought conditions allow survival in saline conditions.

Heavy metal accumulation in soils is detrimental to most plants.

- Aluminum, mercury, lead, and cadmium are poisonous to most plants. Some geographic areas are rich in these metals as a result of normal geological processes, and others have been contaminated by human activity.
- Plants living in these areas have adapted through selection to these conditions. These plants are important in bioremediation and for clean-up of these areas. The exact mechanism of survival of plants in these areas is not known.

Temperature extremes pose yet another stress to plant survival.

- High temperatures denature proteins and destabilize membranes. Cold temperatures decrease membrane fluidity, and freezing causes rupturing of membranes if ice crystals form.
- Transpiration provides evaporative cooling to plants in high temperatures, but at the expense of water loss. Adaptations to high temperatures are similar to adaptations to xeriphytic conditions. Plants produce heat shock proteins that help stabilize protein structure against denaturation.
- Plants can adjust to cold through a process of cold-hardening, which involves repeated exposure to cool nondamaging temperatures. This alters the composition of the membrane and makes it more resistant to rupture. Heat shock proteins are also produced in response to cold temperatures. Some plants have antifreeze compounds that prevent ice crystal formation.

The Big Picture

- Plants must respond to invasion by pathogens, physical damage through natural occurances and herbivory, variable water supplies, temperature fluctuations, and variable soil conditions. Natural selection has made plants uniquely suited to niches they occupy. Some thrive in very harsh environments, whereas others cannot.
- Interactions between plants and pathogens stimulate a series of chemical changes that ward off further infection. Plant strategies to isolate pathogens include sealing plasmodesmata, releasing proteins that interact with the pathogens, and signaling other parts of the plant.
- Herbivory can be advantageous for plants adapted to grazing or detrimental to those not adapted. In the latter case, the plant produces chemicals to prevent herbivory and mounts defenses around the damaged tissues.
- Specific adaptations allow plants to withstand adverse conditions. Saline environments, water loss, and high temperatures can be survived if water uptake is maximized and water loss is minimized. Sunken stomata, reduction of surface area, and increased water potential of cells are adaptations to these conditions. Extremes in temperature are survived with heat shock proteins which stabilize metabolic proteins during temperature extremes.

Common Problem Areas

- Gene-for-gene resistance can be difficult to understand, but remember that in any pathogen situation the pathogen and the host are continually interacting; therefore, over time interplay among the genomes has been selected for.
- There are many examples of the ways plants deal with pathogens and herbivory. This chapter gives examples, but these are not exhaustive.

Study Strategies

- The best strategy is to break this chapter up into components and study them one at a time so that all of the defenses/modifications do not run together. Focus first on pathogen interaction, then herbivory resistance, and so on.
- Think about how form follows function as you look at how plants are adapted to specific harsh environments.
- Review the following tutorial and activity on the website: Tutorial 39.1 Signaling Between Plants and Pathogens Activity 39.1 Concept Matching

Test Yourself

Knowledge and Synthesis Questions

1. Primary nonspecific defensive strategies in plants include which of the following?
   a. PR proteins
   b. Suberin
   c. Cutin
   d. b and c

2. Plants exhibit systemic acquired resistance much the same way people acquire resistance to pathogens; however, the mechanism is quite different. Which of the following have a role in acquired resistance in plants?
   a. Salicylic Acid
   b. PR proteins
   c. Methyl salicylate
   d. All of the above

3. Gene-for-gene resistances depends on which of the following?
   a. Compatible alleles in the plant and the pathogen
   b. Incompatible alleles in the plant and the pathogen
   c. Recessive *Avr* genes
   d. Recessive *R* genes

4. Upon infection by a pathogen, plant cells increase synthesis of polysaccharides. The function of these is to
   a. reinforce cell walls and seal off plasmodesmata.
   b. synthesize antibodies to the pathogen.
   c. isolate the pathogen in the invaded tissue.
   d. a and c

5. Canavanine is toxic to many herbivores but not to plants. This differential toxicity can best be explained by which of the following?
   a. Canavanine is confused with arginine in the predators.
   b. Canavanine is incorporated into proteins and causes them to fold improperly.
   c. Plants are able to differentiate between arginine and canavanine.
   d. All of the above

6. Which of the following best describes how plants produce their own insecticide?
   a. Wounded cells release systemin, systemin directly induces synthesis of protease inhibitors, protease inhibitors act as insecticides.
   b. Wounded cells release jasmonates, jasmonates stimulate systemin synthesis, systemin causes the production of protease inhibitors, protease inhibitors act as insecticides.
   c. Wounded cells release systemin, systemin causes membrane breakdown, membrane breakdown releases jasmonates, jasmonates induce synthesis of protease inhibitors, protease inhibitors act as insecticides.
   d. Wounded cells release systemin, systemin causes membrane breakdown, membrane breakdown releases jasmonates, jasmonates act as insecticides.

7. Which of the following are strategies for coping with drought conditions?
   a. Water storing tissues
   b. Leaf loss during drought
   c. Sunken stomata
   d. All of the above

8. Halophytes are different from all other types of plants in that they
   a. can accumulate sodium and chloride ions.
   b. have a positive water potential.
   c. are unaffected by saline conditions.
   d. All of the above

9. Which of the following conditions stimulates the production of heat shock proteins?
   a. Abnormally high temperatures only
   b. Abnormally low temperatures only
   c. Both abnormally high or abnormally low temperatures
   d. They are continually available in the plant.

10. The main function of heat shock protein is to
    a. stabilize proteins necessary to a cell's survival.
    b. reinforce membranes that lose fluidity.
    c. cause the plant to enter dormancy.
    d. act as an antifreeze compound.

Application Questions

1. Describe strategies used by plants to prevent herbivory.
2. Explain what is meant by coevolution of plants and herbivores. How can grazing actually increase the productivity of some plants?
3. It has long been a practice in rural areas to nail fencing to trees. Describe how the act of nailing a fence to a tree trunk can introduce pathogens into the tree. What measures does the tree take to ward off disease?
4. Explain how altering the osmotic and water potential of a cell can increase its ability to take up water. Under what conditions is this capability necessary?
5. At a T-intersection in a small town in the snowbelt region of Pennsylvania, a local business plants several ornamental shrubs. During the winter, snowplows push snow from the intersection around the shrubs. In an attempt to save the shrubs, the business covers the shrubs with burlap. Will this save the shrubs? What conditions in that intersection are likely to cause the greatest damage to the shrubs?
6. Bioremediation of mine and ore refinery sites is an area of intense study. How can plants help with the cleanup of sites contaminated with heavy metals?

7. Many plants produce toxins that damage eukaryotic predators. Considering that the cell structure of both the predator and the plant is similar, why is the plant not effected by its own toxins?
8. Much of crop science is focused on identifying resistance genes in agricultural crops and breeding those genes or transferring those genes to plants that may not have them naturally. What are the benefits of this technology? What are its drawbacks?
9. Pansies are planted in the winter in the southeastern United States. To assure that the plants are ready to enter cool ground, nurseries cold-harden the plants for several days. Describe this process and explain why it is done.

Answers

Knowledge and Synthesis Answers

1. **d.** Cutin and suberin, along with waxes, aid in preventing entry of pathogens into the plant by forming a barrier.
2. **d.** Plants that have been exposed to pathogens wall off the invaded tissue. That tissue releases salicylic acid to the rest of the plant which stimulates PR proteins. The tissue also releases airborne methyl salicylate which causes release of PR proteins in distant regions of the plant and even in neighboring plants.
3. **a.** The plant must have an *R* allele and the pathogen must have an *Avr* allele. Compatibility between these alleles leads to resistance.
4. **d.** Polysaccharides do seal plasmodesmata and reinforce cell walls. The purpose of this is to contain the pathogen in a minimal number of cells.
5. **d.** Canavanine's mechanism of action is that it is taken up by the predator and incorporated into its protein structures. Because the structure of canavanine is different from that of the arginine it replaces, the predator's proteins take up a structure that is lethal to the predator. The plant itself and some predators have enzyme mechanisms to distinguish canavanine from arginine and prevent its incorporation into proteins.
6. **c.** See Figure 39.6.
7. **d.** Leaf loss, when photosynthesis cannot be sustained, prevents loss of resources. Water storage tissues allow the plant to take advantage of any water that is available and hold on to it for lean times. Sunken stomata prevent excessive water loss to evaporation.
8. **a.** They can accumulate sodium and chloride ions. This makes their water potential more negative and they are able to extract more water from their surroundings than plants that do not store the ions.
9. **c.** Heat shock proteins are produced in response to any extreme in temperature.
10. **a.** Heat shock proteins are chaperonin proteins. They function to stabilize and prevent denaturation of essential proteins.

Application Answers

1. Plants prevent herbivory by producing secondary compounds that affect potential herbivores. These compounds may act as neurotoxins, inhibit digestion, or have a whole host of other consequences. See Table 39.1.
2. Coevolution of herbivores and plants occurs when the herbivore has resistance to toxins the plant may produce, and the plant develops strategies to cope with herbivory. In some cases, regular grazing increases photosynthetic productivity of the plant by reducing shading and allowing better partitioning of resources.
3. Nailed fencing on trees penetrates the nonspecific primary defenses of the tree. In response to pathogens introduced by the nail, the plant begins to isolate the affected tissues. Polysaccharides are synthesized to plug plasmodesmata, and lignin reinforces the cell walls and acts as a toxin to the invading pathogens. Phytoalexins and PR proteins are released and salicylic acid also may be released. The tissue around the nail begins to die and pathogens are isolated.
4. Altering the osmotic and water potential of a cell will affect the rate of osmosis. Water will flow to an area of more negative water potential. Anything that lowers the water potential of a cell will enhance water flow into that cell.
5. The burlap will most likely do little for the shrubs. The roots are in most danger. Because streets are heavily salted in the winter, the snow around them creates a saline environment. If the shrubs are not adapted to a saline environment, little will help them survive the salt accumulation in the soil around their roots.
6. Only plants that have evolved a strategy for taking up and coping with heavy metals can grow around these sites. By isolating and specifically growing these types of plants, the slow process of removing the metals from the soil can begin.
7. Plants may isolate the toxins into vacuoles within their cells so that the toxins do not come into contact with the active machinery of the cell. Plants may restrict toxin production to already damaged tissue. Plants may also develop enzymes that differentiate toxins and render them harmless.
8. Bred resistance to pests and pathogens means less chemical spraying of crops. This is beneficial in terms of cost and the possible health hazards of spraying. The problem is that the new or altered genes may affect the safety of the crop in other ways, or pose environmental consequences such as selecting for "super pathogens." They may also alter the production or taste of the modified crop.
9. Cold-hardening is gradual exposure to adverse temperatures. This allows for structural changes in cell membranes to cope with extreme temperatures. It also allows for the production of heat shock proteins and other compounds necessary for survival at temperature extremes.

40 Physiology, Homeostasis, and Temperature Regulation

Important Concepts

Control of the internal environment is necessary for proper cell function.

- Due to their small size, single-celled organisms can receive all of their required nutrients from the surrounding media. Small multicellular organisms, in which all the cells are only a few layers from the environment, can receive nutrients from the environment without the help of specialized cells. Larger multicellular organisms can maintain an internal environment that differs from the external environment. Specialized cells and groups of cells help to control the makeup of the internal environment and maintain the differences between the internal and external environments (see Figure 40.1).
- Homeostasis is the state of having a constant, regulated internal environment, and homeostatic mechanisms help maintain this state.

Groups of cells are organized as tissues, organs, and organ systems.

- A group of similar cells with the same form and function constitutes a tissue. There are four types of tissues: epithelial, connective, muscle, and nervous tissues.
- Epithelial tissues make up the lining of the inner and outer body surfaces, such as the skin or the lining of the various organs. There are three types of epithelial tissues, based on the layering of the cells (see Figure 40.3). A single layer of cells is a simple epithelium; a multiple layer of cells is a stratified epithelium; a single layer of cells with irregular lengths is a pseudostratified epithelium. Epithelial tissues are further categorized by their shapes: squamous, cuboidal, and columnar.
- Connective tissue is made up of an extracellular matrix that holds together a dispersed group of cells (see Figure 40.4). The protein collagen makes up the bulk of the extracellular matrix. Cartilage, bone, adipose tissue, and blood are all connective tissues.
- The muscle tissues are either skeletal, smooth, or cardiac muscles.
- Neurons and glial cells make up the two types of nervous tissues.
- Organs are groups of tissues that carry out a specific function in the body. Most organs are composed of all four of the major tissue types. The integration of a number of organs to carry out a specific function is known as an organ system, such as the digestive system (see Table 40.1).

Control systems help to regulate physiology and maintain homeostasis.

- Physiological regulatory systems require both a set point at which the parameter is to be held, and feedback that provides information about the state of the system (see Figure 40.5). Differences between the set point and the feedback information indicate to the regulatory mechanism what corrective measures are required to restore a parameter to its set point.
- Physiological control systems must take the information from the regulator systems and effect changes. Negative feedback results in a slow-down or even reversal of a process, returning the variable controlled by that process back toward its set point or regulated level.
- In contrast to negative feedback controls, positive feedback results in an amplification of a response.
- The set point maintained by an organism for a given physiological process can also be changed by feedforward information.

Temperature plays an important role in life.

- Most living cells function only between a temperature of 0°C and 45°C, with most organisms having much narrower limits in their range of survival temperatures.
- All physiological processes in organisms are sensitive to temperature. Increases in temperature usually increase the rate of a given process. The Q_{10} is a measure of change in a physiological process as the temperature increases or decreases by 10°C (see Figure 40.6). In most biological cases, Q_{10}'s range from 2 to 3.
- The metabolic rate of an organism is the total energy used by the animal and is usually measured as oxygen consumption.
- Acclimatization is the change in a physiological process that occurs over time in response to a change in the environment. Acclimatization typically results in an enhanced probability of thriving in the new environmental conditions. Metabolic compensation—acclimatization of metabolic rate—occurs in fish with changes in the seasons (see Figure 40.7). During the cold winter, fish may have a

higher metabolic rate at any given body temperature than they had during the warmer summer.

Animals can be classified as ectotherms or endotherms.

- Animals can be classified into two main groups based on the source of heat. Endotherms regulate their body temperature with internal heat production by means of a high metabolic rate and effective insulation. Ectotherms have a low metabolism and little insulation and must rely on the environment to provide body heat.
- Ectotherms are poikilotherms, with body temperatures that track the environmental temperature. An endotherm maintains a constant body temperature even as environmental temperature changes (see Figure 40.8). Endotherms are either homeotherms, which always maintain a constant temperature, or heterotherms, which do so only part of the time (such as hibernating mammals). But they both have the ability to maintain a high body temperature.
- Both endotherms and ectotherms use behavioral mechanisms to regulate body temperature. Ectotherms shuttle between warm and cool environments during the day to maintain body temperature within a given range (see Figure 40.9). Endotherms also use behavioral thermoregulation as a first line of thermoregulation (see Figure 40.10).
- There are a number of routes through which heat can be gained or lost to the environment (see Figure 40.11). Evaporation from the skin or respiratory tract cools the body. Solar radiation from the sun heats the body. Thermal radiation, convection, and conduction can all either warm or cool the body depending on the thermal environment.
- Both ectotherms and endotherms use changes in blood flow to control body temperature. Increases in blood flow result in increased heat exchange with the environment, while decreases in blood flow result in decreased heat exchange (see Figure 40.12).
- A number of ectothermic insects are able to generate heat to raise their body temperature (see Figure 40.13). This increases their physiological performance, particularly for flight.
- Some large fishes have the ability to maintain a body temperature higher than the surrounding water. They use special vascular countercurrent heat exchangers that retain metabolically produced heat in blood that remains in the core (see Figure 40.14).

Endotherms have special adaptations to help in controlling body temperature.

- The range of environmental temperatures at which the metabolic rate of endotherms does not change is known as the thermoneutral zone (see Figure 40.15). The upper and lower critical temperatures are the environmental temperatures at which metabolic rate begins to increase. Increases in heat production occur by nonshivering and shivering thermogenesis when the environmental temperature is below the lower critical temperature. Nonshivering thermogenesis involves a specialized adipose tissue, brown fat (Figure 40.16). Metabolism increases when environmental temperature is above the upper critical temperature because the animal must expend energy to cool down.
- Decreases in surface area are seen in animals that live in cold-climates. This decrease in surface area decreases the area over which heat can be lost.
- Fur and feathers provide insulation in mammals and birds experiencing cold environments. These structures help to retain heat produced in the body and help maintain a constant body temperature.
- Evaporative cooling provides an avenue for heat loss to the environment by sweating or panting.

The hypothalamus is the vertebrate thermostat.

- The hypothalamus at the bottom of the brain is involved in the control of homeostasis for many regulatory systems, including thermoregulation. The hypothalamus has a set point temperature, and changes in hypothalamus temperature result in thermoregulatory responses to offset any temperature change (see Figure 40.18). The hypothalamus also integrates thermal information received from the skin.
- The body has the ability to increase the set point temperature in response to infections, resulting in fever. Some animals are also able to decrease body temperature either on a daily scale (daily torpor), or over longer time periods (hibernation) (see Figure 40.20).

The Big Picture

- One of the most important roles of tissues, organs, and organ systems is to help maintain homeostasis within the body's cells. Only by maintaining relatively constant intracellular conditions can basic metabolic reactions continue.
- Animals can either generate their own body heat (endotherms) or rely upon the environment to determine body temperature (ectotherms). Endotherms typically have much greater insulation than ectotherms as well as a higher metabolic rate. The metabolic rates of endotherms and ectotherms respond differently to changes in environmental temperature. Endotherms increase metabolism as environmental temperature decreases. The metabolic rate of ectotherms is temperature-dependent and rises and falls with the rise and fall of environmental temperature.
- Control of body temperature can occur by either behavioral or physiological mechanisms. Both endothermic and ectothermic vertebrates use the brain's hypothalamus as the thermostat for temperature regulation.

Common Problem Areas

- Homeostasis does not necessarily mean holding every body variable absolutely constant, but rather making sure that body variables are held at the correct level for the circumstances. Even body temperature in humans, which we think of as being a constant 37°C, drops slight in the early morning hours and rises considerably during extended vigorous exercise.
- Understanding the difference between an ectotherm and an endotherm can be difficult. Think of temperature regulation in ectotherms and endotherms as the temperature

regulation of a house. A house with good insulation, furnace, and thermostat is a kind of endotherm, while a shed with poor insulation, a small wood stove, and no thermostat is an ectotherm.

Study Strategies

- Use Table 40.1 to acquaint yourself with the various organ systems and the different tissue compositions of each. This will help you when you study each of the organ systems in the later chapters.
- Refer to the following tutorial and activities on the website:
  Tutorial 40.1 Control, Regulation, and Feedback
  Activity 40.1 Thermoregulation in Endotherm
  Activity 40.2 Concept Matching: The Major Organ Systems

Test Yourself

Knowledge and Synthesis Questions

1. Which of the following contributes to the proper maintenance of homeostasis in the bodies of animals?
   a. pH buffering of the blood
   b. Control of blood flow to the skin
   c. Production of hormones
   d. All of the above

2. In an environment with an ambient temperature lower than an animal's core body temperature, which of the following would be an inappropriate physiological or behavioral response if the animal needed to eliminate excess body heat?
   a. Wallowing in a pool of water
   b. Preventing blood flow to peripheral vessels
   c. Sweating
   d. All of the above

3. A lizard lives in a desert environment where the temperature at night is low and during the day is high. What might this animal do to maintain the highest stable body temperature?
   a. Stay in a burrow during the night and shuttle between the sun and shade on the surface during the day
   b. Stay on the surface during the night and move to a burrow during the day
   c. Increase metabolism and heat production during the night and decrease it during the day
   d. None of the above

4. On a warm day your dog has been running around fetching his Frisbee and returning it to you. You notice that he is panting heavily. Which of the following does not describe the condition of your dog?
   a. Your dog is using evaporative cooling to lower core body temperature.
   b. Your dog is currently operating at a basal metabolic rate.
   c. Your dog is at the upper end of its thermoneutral zone.
   d. All of the above

5. Evaporative cooling is an effective way to increase heat loss, but carries with it the physiological drawback of
   a. increased use of ATP and substantial water loss.
   b. lowering the hypothalamic thermal set point.
   c. decreased use of ATP and substantial water gain.
   d. exhaustion of the supply of brown fat.

6. Fever
   a. is a higher than normal body temperature that is always dangerous.
   b. decreases the metabolic rate of the body to conserve energy.
   c. is regulated by chemicals that reset the thermostat of the body to a higher set point.
   d. causes the liver to release large amounts of calcium, which seems to inhibit bacterial growth.

7. Which of the following is a characteristic unique to connective tissue?
   a. Highly modified cells that show the special property of contractility
   b. Diverse anatomy with distinct specializations for information transfer
   c. Lining of the inner surfaces of the intestines and lungs in mammals
   d. Loose array of cells embedded in some kind of extracellular matrix

8. Elephants use their ears to dump heat to the environment. What mechanisms might they employ to increase heat loss from the ears?
   a. Increased convection due to flapping of the ears
   b. Moving into the sun
   c. Increased blood flow to the ears
   d. a and c

9. Which organ system is not involved in the maintenance of homeostasis in humans?
   a. Endocrine system
   b. Digestive system
   c. Muscle system
   d. All are involved in maintenance of homeostasis

10. What is the difference between a negative feedback mechanism and a positive feedback mechanism?
    a. Negative feedback mechanisms only exist in the circulatory system.
    b. Negative feedback mechanisms stabilize toward a set point and positive feedbacks move away from a set point.
    c. Negative feedback mechanisms move away from a set point and positive feedbacks stabilize toward a set point.
    d. Negative feedback mechanisms stabilize toward a set point and positive feedbacks reset the set point.

11. A number of physiological processes can undergo acclimatization. Which of the following statements about acclimatization is false?
    a. Acclimatization occurs in response to seasonal temperature changes.
    b. Acclimatization of metabolic rate occurs because enzyme expression changes.
    c. Acclimatization involves the changing of a set point.
    d. All are correct.

12. In fast-swimming cool-water fishes such as sharks and tunas, which of the following contribute to the generation and maintenance of core body temperatures in excess of ambient water temperatures?
    a. Specializations in the size and location of their blood vessels
    b. Low rates of activity in the swimming muscles
    c. An ability to acclimatize rapidly to the surrounding water
    d. Metabolic rates that are insensitive to temperature change

13. Which of the following statements about hypothalamic function is false?
    a. Circulating blood temperature is monitored by the hypothalamus.
    b. Hypothalamic thermal set points never change.
    c. The hypothalamus is a part of the central nervous system.
    d. Different thermoregulatory responses have different hypothalamic set points.

14. Some animals use brown fat as a source of heat generation. Which of the following statements about brown fat is true?
    a. Brown fat is involved in nonshivering thermogenesis.
    b. The protein thermogenin is involved in uncoupling ATP production and proton movement.
    c. Brown fat is found in endotherms.
    d. All of the above.

15. During childbirth, the pressure exerted on the mother's cervix by the emerging infant leads to increased contraction of the uterus. This interaction could be explained by
    a. negative feedback.
    b. feedforward feedback.
    c. positive feedback.
    d. metabolic compensation.

Application Questions

1. Having measured the metabolism of two similar sized animals at 15°C and 25°C, you find that the metabolic rate of the first animal was 115 and 55 ml of oxygen per hour at 15°C and 25°C, respectively. The metabolic rate of the second animal was 5.5 and 11.5 ml of oxygen per hour at 15°C and 25°C, respectively. Calculate the Q_{10} for the metabolic rate in both animals and determine if they are endotherms or ectotherms.

2. You have measured the metabolism of the first species in question 1 at both low and high environmental temperatures. You find that the metabolic rate at 40°C is higher than at 30°C. What physiological phenomenon explains why the metabolic rate of this animal rises when environmental temperature is above the upper critical limit?

3. What are the different mechanisms that animals use as insulation and what physiological controls exist to increase insulation?

4. Animals have the ability to undergo acclimatization and metabolic compensation. Describe each of these processes and how they relate to temperature regulation.

5. Why is it important for an organism to maintain homeostasis?

Answers

Knowledge and Synthesis Answers

1. **d.** All of the choices help to maintain the internal environment of an animal.

2. **b.** If the environment were cooler than the animal, the animal would want to lose heat to the environment across the skin. Increasing peripheral blood flow would move heat from the core of the animal to the surface, where it could be lost to the cooler environment.

3. **a.** In the desert, if an ectothermic lizard wants to maintain the most stable body temperature it should be in a burrow during the night, where temperatures do not drop very much, and then shuttle between the sun and shade during the day.

4. **b.** After exercise your dog will have a high metabolic rate and will employ panting in an attempt to cool down. The body temperature will most likely be above the upper thermoneutral zone.

5. **a.** Evaporative cooling is a costly means to lower body temperature. ATP must be used either during sweating or panting. Water loss can also be high during evaporative cooling.

6. **c.** Fevers are brought about by a resetting of the thermoregulatory set point in the hypothalamus in response to pyrogens.

7. **d.** Connective tissues are blood, cartilage, and bone, all of which contain a loose population of cells embedded in an extracellular matrix.

8. **d.** Elephants use both physiological mechanisms (increased blood flow to the ears), and behavioral mechanisms (flapping of ears, to dump heat to the environment).

9. **d.** All of the organ systems are involved in maintaining homeostasis.

10. **b.** Negative feedback loops use information about the state of a system to bring the internal environment back toward the set point values. Positive feedback loops amplify the response of a system, moving the system away from the set point.

11. **d.** Physiological processes acclimate in response to a seasonal change in the environment. Acclimatization may involve changes in enzyme expression and resetting of a physiological set point.

12. **a.** Large fishes such as the tuna are able to keep core body temperature above ambient temperature by using a number of mechanisms. They have a special countercurrent heat exchanger built into their blood vessels. They also have high metabolic rates and thus produce heat in their swimming muscles.

13. **b.** The hypothalamic thermal set points change daily, monthly, and seasonally.

14. **d.** Brown fat is a mechanism by which the production of ATP is uncoupled from the movement of protons by the thermogenien protein. This occurs in the endotherms, and the process of heat production is termed nonshivering thermogenesis.

15. **c.** Increased uterine contraction in response to pressure on the cervix is an example of a positive feedback.

Application Answers

1. The metabolic rate of the first animal increases with a decrease in environmental temperature, making this animal an endotherm. The metabolic rate of the second animal decreases with a decrease in environmental temperature, making this animal an ectotherm. Using the equation $Q_{10} = (R_T/R_{T-10})$, the Q_{10} for metabolic rate can be calculated. The Q_{10} of the endotherms metabolic rate is 2.1, while the Q_{10} of the ectotherm metabolic rate is 0.48 using the equation provided.

2. An endotherm in a hot environment will attempt to keep core body temperature from increasing. The heat loss mechanisms available to an endotherm, sweating and panting, require the use of ATP and result in an increased metabolic rate. Endotherms will expend energy in order to protect body temperature and lose heat to the environment.

3. Mammals use fur and fat as effective insulators, while birds use feathers. Fur and feathers are good insulators because they trap still, warm air from the body. Increases in the length of the fur or feathers will increase their effectiveness as insulators. Still air is one of the best insulators. Fat works in a similar manner by having a low thermal conductance. Humans do not have sufficient hair on their bodies for hair to provide good insulation; therefore, they wear clothes as their insulation layer. Physiologically, animals can change their blood flow patterns to increase or decrease the amount of warm blood that is received by the skin.

4. Animals undergo acclimatization in response to changes in seasonal conditions such as temperature. Often these changes are brought about by the production of enzymes that function better at the new temperature. One example of acclimatization is metabolic compensation. As seasonal temperatures change, the enzymes controlling resting metabolism adjust to compensate for the change in temperature (see Figure 40.7).

5. Homeostasis is the maintenance of a constant internal environment. The body functions with the help of proteins and enzymes. Changes in the internal environment such as pH, temperature, glucose level, and oxygen and carbon dioxide levels can all effect cellular function. Loss of homeostasis can lead to improper function of proteins and cell membranes and ultimately lead to cell death.

41 Animal Hormones

Important Concepts

Hormones are chemical signals used in physiological regulation.

- Hormones are secreted by endocrine cells, present as either individual cells or as groups of cells known as glands.
- Endocrine cells make up exocrine glands that secrete hormones through ducts to the outside of the body and endocrine glands that secrete hormones into the body. Most hormones are secreted into the blood, where they are carried to target cells and bind with a receptor to trigger its function. Not all hormones act on a global scale; some are secreted and taken up by local cells and thus act locally and are known as paracrine hormones.
- Histamine, growth factors, and neurotransmitters are all examples of paracrine hormones. Histamine acts locally in response to damage to tissue by causing dilation of blood vessels. Growth factors stimulate the growth and differentiation of cells. Neurotransmitters are paracrine hormones that carry a message from one neuron to the target cells over a very small distance.
- Hormones that act on the secreting cells are known as autocrines.
- Hormones are found in all animals and are conserved between groups. However, a given hormone may have different functions in different taxa. Prolactin is involved in reproduction in mammals, birds, amphibians, and fish, but has a different effect in each group.

Hormones are involved in the control of development in insects.

- Hormones control the molting of exoskeletons and the metamorphosis in insects.
- In an elegant set of experiments, Wigglesworth used the blood-sucking bug *Rhodnius* to examine the control of molting. Decapitation of the bugs at various times after feeding showed that there was a substance secreted in the head of these bugs that stimulates molting.
- The brains of these *Rhodnius* and other insects produce brain hormone that is stored in the *corpora cardiaca* attached to the brain. After a meal, brain hormone moves to the prothoracic gland where it stimulates the release of ecdysone, a hormone that stimulates molting.
- In molting insects, juvenile hormone results in further molts as a juvenile. In the absence of juvenile hormone, the insect molts into the adult form. In insects with complete metamorphosis, juvenile hormone results in molting into a larva, declining levels result in molting into a pupa, and in the absence of juvenile hormone the pupa molt into the adult.

The pituitary gland is the master endocrine gland in vertebrates.

- The pituitary gland is called the "master gland" because it produces hormones that control many of the endocrine glands in the body. The pituitary is made up of the anterior pituitary and posterior pituitary, each utilizing different control mechanisms and hormones.
- The posterior pituitary controls the release of antidiuretic hormone and oxytocin. Both of these hormones are produced by neurons in the hypothalamus and move down the neurons into the posterior pituitary. Antidiuretic hormone (ADH) acts on the kidneys to stimulate reabsorption of water to help maintain blood pressure and solute concentration of the blood. Pregnant females release oxytocin during birth to stimulate uterine contraction and after birth to start the flow of milk.
- The anterior pituitary gland releases four tropic hormones that control the function of the thyroid, adrenal cortex, testes, and ovaries. The anterior pituitary also produces growth hormone, prolactin, endorphins, and enkephalins. Growth hormone stimulates the liver to release hormones that stimulate bone and cartilage growth. In female mammals, prolactin stimulates milk production and secretion.
- The anterior pituitary gland is controlled by hormones produced in the hypothalamus and transported to the anterior pituitary gland by portal blood vessels. Two of the controlling hormones are thyrotropin-releasing hormone, causing the release of thyrotropin that stimulates the thyroid, and gonadotropin-releasing hormone, stimulating the release of hormones that control the gonads.
- The anterior pituitary gland is also controlled by a negative feedback loop. Hormones released from target glands can have a negative effect on the release of further hormones from the anterior pituitary gland.

The thyroid gland regulates metabolic rate and blood calcium levels.

- The thyroid gland helps regulate metabolic rate through the action of several hormones that it produces, thyroxine (T4) and triiodothyronine (T3), which increase metabolic rate in mammals. Thyrotropin from the anterior pituitary gland stimulates the thyroid gland to produce thyroxine. Poor regulation of thyroxine production can result in the condition of enlarged thyroid glands known as goiter.
- Another thyroid hormone, calcitonin, regulates calcium by stimulating osteoblasts that deposit new bone, and inhibiting osteoclasts that break down bone. The parathyroid gland produces parathyroid hormone that increases blood concentrations of calcium by stimulating osteoclasts to dissolve bone. These two hormones act antagonistically, countering the effect of each other.

The pancreas regulates blood sugar levels.

- Endocrine cells in the pancreas, called the islets of Langerhans, produce the hormones insulin and glucagon, which regulate glucose levels in the blood. Insulin binds to receptors on cells, allowing glucose to enter the cell from the blood stream. When insulin is absent or the receptors become insensitive to insulin, glucose accumulates in the blood, resulting in the disease diabetes mellitus.
- Glucagon is produced in the alpha cells of the islets of Langerhans. Glucagon stimulates the liver to convert glycogen into glucose to be used as an energy source when blood glucose levels are low.
- High blood glucose levels stimulate the pancreas to release somatostatin, which inhibits the release of insulin and glucagon.

The adrenal gland produces hormones with numerous regulatory functions.

- The adrenal gland has two regions, the adrenal medulla, making up the core of the gland, and the adrenal cortex which surrounds the medulla.
- Epinephrine and norepinephrine are produced in the adrenal medulla and act on α-adrenergic and ß-adrenergic receptors.
- The adrenal cortex produces glucocorticoids, mineralocorticoids, and sex steroids from cholesterol. The main glucocorticoid is cortisol, which stimulates cells that are not necessary for the flight-or-fight response to decrease the use of glucose. The main mineralocorticoid is aldosterone, which is involved in reabsorption of sodium and excretion of potassium at the kidney.

The gonads produce sex steroids that act as hormones.

- Male testes produce testosterone while female ovaries produce estrogens and progesterone. These sex steroids determine the sex of the developing fetus. In response to androgens, the reproductive system develops into a male; in their absence it develops into a female.
- Sex steroids increase in concentration at the time of puberty. Both luteinizing hormone (LH) and follicle-stimulating hormone (FSH) from the anterior pituitary gland control the production of sex steroids. Increased levels of LH and FSH in females and LH in males stimulate the gonads to produce more sex hormones. These hormones in turn have profound physiological effects on the development of the secondary sexual characteristics.

The pineal gland produces melatonin.

- Melatonin is involved in regulating the daily cycles in the physiology of an organism.

Hormones act on the receptors of the target cells.

- Most hormones are released in only very small quantities into the body. A hormone works only on cells that have specific receptors for a given hormone. One hormone can have various functions because there are different receptors for a hormone that have different functions.
- Hormones are either lipid-soluble or water-soluble. Lipid-soluble hormones are able to cross the plasma membrane of cells and act on receptors inside the cell. Water-soluble hormones cannot pass through the hydrophobic (water-hating) plasma membrane and act on receptors spanning the plasma membrane. Water-soluble hormones bind to a receptor on the outside of a cell and trigger the receptor on the inside of the cell to respond, usually by the release of a second messenger.
- The level of hormone receptors can be adjusted to help control the action of hormones. Downregulation can occur by a decrease in the number of receptors in response to high levels of a hormone. Upregulation occurs when a hormone stimulates a target cell to increase the number of hormone receptors.

The Big Picture

- Hormones are regulatory substances that circulate broadly throughout an animal's body, regulating functions ranging from growth to sexual maturity. In numerous ways, the endocrine system complements the nervous system as the second major regulator of body function and homeostasis.
- Every organ system in the vertebrate body is a target for hormones, which regulate their functions. Hormones are actively involved in the maintenance of homeostasis.
- Hormone action is controlled by a number of different mechanisms. In many instances there is a "hormone cascade" in which a hormone controls the release of a hormone, which controls the release of a hormone, and so on. Negative feedback occurs when a released hormone inhibits (negatively feeds back on) the tissue that may have stimulated its release in the first place.

Common Problem Areas

- Appropriate for its title "master gland," the pituitary produces numerous regulatory hormones. The functions of the anterior and posterior pituitary are quite distinct, and keeping them separate in your mind is important for your understanding of the endocrine system.

- Remember that most hormones are broadly distributed throughout the body and so all cells are exposed to the same level of hormone concentration. Hormones have specific actions because different cells have different levels of sensitivity to them, not because the hormone concentration differs throughout the body.

Study Strategies

- All of the hormones may seem to constitute a long laundry list. Try to learn the function along with the name of each hormone, as this will also help you remember the target tissues.
- Some hormones have extremely specific actions (e.g., follicle stimulating hormone), while others have broad effects on many target tissues (e.g., ephinephrine, thyroid hormones). Learning which hormones are "generalists" and which are "specialists" will help you better understand the endocrine system.
- Biologists studying the endocrine system use lots of abbreviations for the hormones. Learn these abbreviations along with the full name of the hormone.
- Refer to the following tutorial and activities on the website:
  Tutorial 41.1 Complete Metamorphosis
  Activity 41.1 Concept Matching: Vertebrate Hormones
  Activity 41.2 Endocrine Glands in Humans

Test Yourself

Knowledge and Synthesis Questions

1. A paracrine hormone is
   a. a local hormone that acts on the cell that released it.
   b. always acting on a wide variety of target tissues.
   c. a local hormone produced at one site but active at a different site in the body.
   d. None of the above

2. A target cell's response to a hormone depends on
   a. the amount of hormone reaching that target cell.
   b. the number of receptors on that target cell.
   c. the target cell's receptor-affinity for hormone.
   d. All of the above

3. "Upregulation" of hormone receptors refers to
   a. increase in hormone receptor numbers with low hormone levels.
   b. increase in hormone receptor numbers with high neurotransmitter levels.
   c. increase in hormone levels produced by increase in hormone receptor numbers.
   d. decrease in hormone levels produced by decrease in hormone receptor numbers.

4. Hormones have various regulatory functions. Which of the following statements does not describe how a hormone functions?
   a. Hormones act in very low concentration.
   b. Hormones act at sites distant from where they are produced.
   c. Hormones are transported in the blood.
   d. None of the above

5. Which of the following is a common "second messenger" substance in hormone action?
   a. Thyroid hormone
   b. ADH
   c. Cyclic AMP
   d. Epinephrine

6. A hormone circulating in the blood with target tissues 1, 2, and 3 could act selectively on only target tissue 1 if
   a. target tissue 1 had fewer hormone receptors than 2 and 3.
   b. target tissues 2 and 3 had hormone receptors with a higher hormone-affinity than 1.
   c. target tissue 1 had hormone receptors with a higher hormone-affinity than 2 and 3.
   d. the hormone was released only in the vicinity of target tissue 1.

7. Which of the following glands is considered the "master" endocrine gland in vertebrates?
   a. Adrenal glands
   b. Tthyroid gland
   c. Hypothalamus
   d. Pituitary gland

8. Which gland is involved in the function of our "biological clock"?
   a. Pineal gland
   b. Thyroid gland
   c. Adrenal glands
   d. Ovaries

9. The half-life of hormones in the plasma ranges from
   a. minutes to hours.
   b. minutes to days.
   c. days to weeks.
   d. hours to months.

10. The target tissues of hormones are those tissues that
    a. the particular hormone can actually penetrate.
    b. have specific enzymes with which hormones directly interact.
    c. have high concentrations of the "second messenger."
    d. have receptors for the particular hormone.

11. Which of the following sets of vertebrate hormones are all produced in the anterior pituitary gland?
    a. Somatostatin, vasopressin, insulin
    b. Prolactin, growth hormone, enkephalins
    c. Oxytocin, prolactin, adrenocorticotropin
    d. Estrogen, progesterone, testosterone

12. Hormones that are secreted by one endocrine gland and control the activities of another endocrine gland are called
    a. growth hormones.
    b. obstructive.
    c. tropic.
    d. selective.

13. Which of the following secretes hormones but is not considered a traditional glandular tissue?
    a. Pancreas
    b. Heart
    c. Testes
    d. Adrenal glands

Application Questions

1. Define and discuss the action of lipid-soluble and water-soluble hormones. Include how the action and timing differ.
2. The pituitary gland is made up of the anterior and posterior pituitary. What is the relationship between these two parts of the pituitary gland and the hypothalamus?
3. Hormones are regulated in numerous ways. One common method of hormonal regulation is by negative feedback. Discuss the negative feedback loop involved in the response of the thyroid to the cold.
4. Wigglesworth conducted a number of studies on the insect *Rhodnius* to examine the control of molting. Describe what would happen to a fourth-instar *Rhodnius* if Wigglesworth were to either partially or fully decapitate the insect one week after a blood meal.

Answers

Knowledge and Synthesis Answers

1. **c.** Paracrine hormones act on local cells that are different from those that secrete them. Autocrine hormones act on the very same cells that secrete them.
2. **d.** The response of target cells to a hormone depends on how much of the hormone is present and acting on the target, how many receptors are present on the surface of the cell that the hormone can act on, and how well the hormone binds with the receptor.
3. **a.** "Upregulation" of hormone receptors on a cell is the production of more receptors when the hormone is present over time in the blood or other fluids surrounding the cell.
4. **d.** All of the statements are correct. Hormones are found and act at low concentrations in the body. They are transported by the blood or diffuse through interstitial space to reach target cells that may be some distance from the secreting gland.
5. **c.** In many cases, hormones trigger cyclic AMP when they bind with the target cells. Cyclic AMP then carries out the action of the hormone within the cell as a second messenger (see Chapter 15).
6. **c.** Because hormones are typically distributed throughout the blood, all potential target tissues will experience the same hormone concentration. Only by having either higher numbers of receptors, or receptors with greater hormone affinity, can one tissue be stimulated preferentially over other tissues by the same hormone.
7. **d.** The pituitary gland is considered the master gland because it produces hormones that control endocrine glands throughout the body.
8. **a.** The pineal gland secretes melatonin, a hormone that controls the daily cycles and acts as the "biological clock."
9. **b.** Hormones tend to act over time periods of minutes to days. Epinephrine has a half-life of up to three minutes, while cortisol has a half-life of many days.
10. **d.** For a hormone to act on a target cell, the target cell must have the receptors for the specific hormone to bind and trigger the hormonal action.
11. **b.** All of these hormones are produced in the anterior pituitary.
12. **c.** Hormones that control endocrine gland function are known as tropic hormones.
13. **b.** The heart is not typically considered a gland, but it secrets atrial natriuretic hormone that acts on the kidneys to promote sodium excretion.

Application Answers

1. The plasma membrane of cells is hydrophobic. Lipid-soluble hormones, which as their name suggests dissolve in lipids (fats), cross the plasma membrane and act on receptors inside the target cells. Water-soluble hormones cannot cross the lipid-based plasma membrane and act on receptors on the outside of the target cells through a second messenger such as cyclic AMP. Because lipid-soluble hormones cross the plasma membrane and enter the cell, they tend to take longer to act, but they also remain active for a longer time period than water-soluble hormones.
2. The hormones secreted by the posterior pituitary gland are produced by neurons in the hypothalamus. The neurohormones move down the neuron and into the posterior pituitary gland where they are stored and secreted. The anterior pituitary produces the hormones it secretes. However, neurohormones from the hypothalamus control the release of these hormones.
3. In response to cold the body tends to increase metabolic rate mediated by increased levels of thyroxine. The hypothalamus is stimulated by the cold to release thyrotropin-releasing hormone. Thyrotropin-releasing hormone stimulates the anterior pituitary gland to secrete thyrotropin that acts on the thyroid. The tropic hormone thyrotropin stimulates the thyroid to release thyroxine, resulting in increased metabolic rate. The increase of thyroxine in the system acts as a negative feedback to inhibit the secretion of thyrotropin by the anterior pituitary gland.
4. The insect *Rhodnius* can live for long periods of time after it has been decapitated. If the insect is partially decapitated, leaving the *corpora allata*, the insect will molt into a fifth-instar juvenile. This is because the *corpora allata* which produces juvenile hormone that stimulates molting into another juvenile instar is still present. The animal that was fully decapitated would molt into an adult rather than another juvenile instar. The fully decapitated animal had enough brain hormone circulating one week after the meal that it would molt; however, there would be no juvenile hormone present and the animal would molt into an adult.

42 Animal Reproduction

Important Concepts

Some animals reproduce asexually.

- Many invertebrates employ asexual reproduction, producing genetically identical offspring from the parent.
- New offspring can be produced either by budding off outgrowths to produce a new individual or by regenerating a new individual from pieces of an animal (see Figure 42.1).
- Animals can produce new individuals asexually through parthenogenesis—the development of unfertilized eggs into new offspring.

Sexual reproduction produces genetic diversity.

- Sexual reproduction in animals produces genetic diversity when two haploid gametes join to form a diploid cell.
- Sexual reproduction has three stages: gametogenesis, mating, and fertilization.
- Gametogenesis produces haploid gametes using meiotic cell division. Gametogenesis occurs in the animal's gonads. Male gametes—sperm—are produced in the testes, and female gametes—eggs or ova—are produced in the ovaries. Gametes are derived from germ cells to produce diploid oogonia and spermatogonia. They then turn into primary oocytes and primary spermatocytes.

Males produce sperm by spermatogenesis.

- Spermatogenesis begins with mitosis of a male germ ($2n$) cell into a spermatogonium (see Figure 42.4*a*). The first meiotic division produces secondary spermatocytes (n) and a second meiotic division produces the spermatids (n).
- Mammalian spermatids share gene products of the X chromosomes at this stage.
- The spermatids differentiate and mature into the sperm cell gametes.

Females produce eggs by oogenesis.

- Oogenesis begins with differentiation of female germ cell ($2n$) into an oogonium ($2n$) and then into primary oocytes ($2n$) (see Figure 42.4*b*).
- The primary oocytes enter prophase and in human females remain in prophase for up to 50 years. No new primary oocytes will be produced during the life of the female.
- Upon exiting prophase, the primary oocyte goes through a meiotic division to produce a secondary oocyte and a polar body. The second meiotic division produces the ootid and another polar body. This second meiotic division may not be completed until fertilization in some species.

Some animals have both male and female reproductive systems.

- Dioecious species have separate male and female individuals.
- Hermaphroditic species have both the male and female members on the same individual. They can be either simultaneous hermaphrodites (both male and female at the same time), or sequential hermaphrodites (changing sexes at some point during their life span). Sequential hermaphrodites have less of a chance of inbreeding than simultaneous hermaphrodites.

Fertilization occurs differently in the water and on land.

- Fertilization is the joining of a sperm with an egg to form a diploid zygote. Many aquatic animals have external fertilization of their eggs. The eggs and sperm are released into the water, leaving possible contact and fertilization to chance.
- Many fish and amphibians ensure fertilization by using special behaviors of the male and female to bring the gametes closer to each other.
- Some mite and scorpion males produce a sperm-filled spermatophore that they deposit in the environment. When a female comes across it she straddles it and takes the sperm into her reproductive tract.
- Internal fertilization requires accessory sex organs that pass the sperm on to the female during copulation. The penis is a male accessory sex organ and the vagina in some species or the cloaca in other species is the female accessory sex organ.

The move to land required a reproductive system that functions in a dry environment.

- Amphibians were the first vertebrates to live on land, but they require water to reproduce.
- Many reptiles and all birds have solved the problem of a desiccating environment by being oviparous, laying pro-

tective shelled eggs. Internal fertilization must occur prior to shell formation around the egg. Some reptiles are ovoviviparious, with the embryo developing in a shelled egg that is retained in the female.

- All mammals except the monotremes are viviparious, retaining their embryo within the female uterus. The marsupials have a simple uterus and a pouch, or marsupium, into which the early embryo crawls after leaving the uterus.

In humans, males produce sperm and deliver it to the female.

- Males produce semen containing sperm (n) and fluids that facilitate fertilization. In the testis, sperm cells are produced by spermatogenesis in the seminiferous tubules and then surrounded by Sertoli cells to provide protection and nourishment (see Figure 42.8).
- Sperm develop a flagellum at the back and an acrosome cap at the front containing digestive enzymes. Mitochondria provide energy for the flagellum (see Figure 42.9). Sperm are stored in the epididymis, which connects to the vas deferens and then the urethra and opens to the outside of the body through the penis.
- The other fluids in semen come from various glands. The bulbourethral glands produce an alkaline and mucoid secretion that helps control pH. The seminal vesicles produce seminal fluid containing energy for sperm in the form of fructose, mucus, and proteins. The prostate gland produces the milky prostate fluid and a clotting enzyme that works on the protein in seminal fluid to turn semen into a gelatinous mass.
- During sexual arousal, the penis becomes engorged with venous blood and contraction of smooth muscle in the ducts and at the base of the penis results in emission and ejaculation of semen. After sexual stimulation the autonomic nervous system causes the vessels at the base of the penis to constrict, leading to the loss of erection.
- Testosterone, produced in the Leydig cells of the testes, is the male sex hormone that controls sexual function and sperm production (see Figure 42.10).

Females produce eggs, receive sperm, and incubate the embryo.

- The ovaries release eggs into the body, where they enter the fallopian tube or oviduct and move toward the uterus (see Figure 42.11). Sperm swim up the vagina, through the opening to the uterus (cervix), through the uterus, and into the oviduct. Embryos develop in the uterus after fertilization of the eggs by sperm in the oviducts.
- The initial division of the zygote produces a blastocyst that moves into the uterus, where it imbeds into the uterus wall (the endometrium) to form the placenta. The placenta provides nutrients and removes wastes from the developing embryo.
- Mature eggs are produced during a 28-day ovarian cycle (see Figure 42.12). Initially, six to twelve primary oocytes begin to mature and are surrounded by follicle cells to produce individual follicles. After one week, one primary oocyte continues to grow while the others shrink. Prior to ovulation, the primary oocyte matures into a secondary oocyte and is expelled from the ovary. The remaining follicle cells become the corpus luteum, which produces progesterone and estrogen.
- The uterine cycle, in concert with the ovarian cycle, prepares the endometrium to receive the blastocyst. If a blastocyst does not implant by around day 14 of the uterine cycle, the endometrium is broken down and expelled in the process of menstruation. The human is the only mammal that goes thorough menstruation. Other mammals reabsorb the endometrium at the end of the uterine cycle.
- Luteinizing hormone (LH) and follicle-stimulating hormone (FSH) control the timing of the ovarian and uterine cycle (see Figure 42.13). The first step of the uterine and ovarian cycle is menstruation. Prior to day 10 of the uterine cycle, FSH and LH levels are kept low by gonadotropins, estrogen, and progesterone. Around day 12–14, estrogen stimulates a rise in FSH levels, which in turn stimulate the follicles to begin to mature. LH levels also increase, triggering the release of the egg from the mature follicle and the growth of the corpus luteum. Estrogen and progesterone produced by the corpus luteum stimulate the growth and maintenance of the endometrium. In the absence of fertilization, the corpus luteum degenerates, estrogen and progesterone levels decrease, and menstruation occurs to start the cycle again.
- When the egg is fertilized and implants in the uterus, human chorionic gonadotropin is produced, which keeps the corpus luteum functional. Estrogen and progesterone stimulate development of the endometrium, preparing the uterus for pregnancy.

Humans use several methods to prevent pregnancy.

- The rhythm method involves avoiding sex during the time period that the female is most likely to be fertile and has a 15 to 35 percent failure rate.
- Condoms, diaphragms, and cervical caps work by creating a barrier to sperm. They have a failure rate of 15 percent.
- A number of methods employ hormones to prevent ovulation by the female. Oral contraceptives contain estrogens and progesterone that interfere with the development of the ova and suspend the ovarian cycle. The failure rate is less than 1 percent.
- Intrauterine devices work by blocking the implantation of the fertilized egg in the uterus.
- Both males and females can be sterilized. Males undergo a vasectomy in which the vas deferens is cut, while females have their oviducts cut in tubal ligation (see Figure 42.15).
- The termination of pregnancy is known as abortion. A spontaneous abortion occurring early in pregnancy is commonly called a miscarriage.
- Controlling female fertility is much easier than controlling male fertility. Males continuously produce millions of sperm. Continuous chemical intervention would be needed to block spermatogenesis completely.

Medical science has found ways to help overcome infertility.

- A number of problems may prevent couples from conceiving. The male may not be producing adequate amounts of sperm, or the environment of the uterus may be poor for hosting sperm or the fertilized egg.
- Artificial insemination places sperm in the oviduct where fertilization can occur to increase the chance of fertilization.
- Assisted reproductive technologies take eggs from a female and then either fertilize them outside of the body or place a mixture of sperm and egg in the female oviduct.

Some diseases are sexually transmitted.

- Sexually transmitted diseases have been around since ancient times.
- Over 10 million new cases of STD's occur in the United States each year.
- STD's can be caused by viruses, bacteria, yeasts, or protozoans

The Big Picture

- Animals reproduce both asexually and sexually. Asexual reproduction includes budding, regeneration, and parthenogenesis. Sexual reproduction has the advantage over asexual reproduction because it helps to increase genetic diversity. Sexual reproduction has three main stages: gametogenesis, mating, and fertilization.
- Gametogenesis is the process by which the sex cells are produced. Haploid sperm develop from diploid spermatogonia by the process of spermatogenesis. The haploid egg of the female develops from diploid oogonia by the process of oogenesis.
- The female has two cycles, the uterine and the ovarian cycles, during which an egg is produced and the uterus is prepared for pregnancy. These two cycles occur in parallel and are coordinated by changes in hormone levels. Luteinizing hormone, follicle-stimulating hormone, gonadotropins, estrogen, and progestrogen all play a role in the control of these cycles. The uterine cycle gets the uterus ready for implantation and the ovarian cycle produces and releases the egg.
- Mating in aquatic animals involves external fertilization of the egg by the sperm. On land, animals use internal fertilization with the help of secondary sex organs to ensure fertilization.
- A number of methods are available both for preventing contraception and for overcoming infertility.

Common Problem Areas

- Understanding the ploidy level of the developing egg and sperm can be very confusing, but it is essential in determining how genetic inheritance works.
- Following the hormone changes and hormone cues associated with the human menstrual cycle can be confusing. Create a chronological sequence of hormones and the events they trigger during the 28-day uterine cycle.

Study Strategies

- Recognize that while sperm and egg formation are completely separate processes, each with their own characteristics, the same general events are happening in both. This will help you to recognize and remember the stages of spermatogenesis and oogenesis.
- To better understand the changes in ploidy in gametes and the products of their combination, create a "flow chart" that shows how the haploid state changes over one or two cycles.
- Review the following tutorial and activities on the website:
  Tutorial 42.1 The Ovarian and Uterine Cycles
  Activity 42.1 Human Female Reproductive Tract
  Activity 42.2 Human Male Reproductive Tract
  Activity 42.3 Spermatogenesis

Test Yourself

Knowledge and Synthesis Questions

1. Asexual reproduction is an effective strategy in stable environments because
   a. gametogenesis is most efficient under these conditions.
   b. the offspring, genetically identical to their parents, are preadapted to the environment.
   c. asexual parthenogenesis produces a large amount of genetic diversity.
   d. animal cells tend to be more totipotent under stable conditions.

2. An important difference between a sperm and an egg is
   a. their size.
   b. the amount of cytoplasm they contain.
   c. whether or not they are motile.
   d. All of the above

3. Which of the following would not be present in the body or lifestyle of a terrestrial male animal that reproduces sexually?
   a. Internal fertilization
   b. A secondary sexual organ as part of the genitalia
   c. Synchrony of reproductive physiology with the female
   d. Oogenesis

4. Which of the following best represents the normal path of a sperm cell as it makes its way from the point of entry into the female's reproductive tract to the place where fertilization typically occurs?
   a. Cervix, vagina, ovary, oviduct
   b. Vagina, cervix, uterus, oviduct
   c. Uterus, cervix, vagina, oviduct
   d. Vagina, uterus, cervix, oviduct

5. The function of the seminal vesicle is to
   a. produce a solution of fructose to provide energy for the mitochondria of the sperm.
   b. secrete alkaline fluids that neutralize the acidity of the female's reproductive tract.
   c. initiate the muscular contractions leading to emission.
   d. produce prostaglandins that stimulate contractions of the male reproductive organs.

6. Which of the following is an example of positive feedback control in the reproductive cycle of males or females?
   a. The increased response of the hypothalamus and anterior pituitary gland in response to estrogen
   b. The decreased response of the hypothalamus and anterior pituitary gland in response to estrogen
   c. The inhibition of luteinizing hormone by high levels of testosterone
   d. The stimulation of luteinizing hormone by low levels of testosterone

7. Which of the following statements about oogenesis is false?
   a. The polar bodies degenerate after the second meiotic division.
   b. The ovum produced is haploid.
   c. The major growth phase of the primary oocyte occurs in prophase I.
   d. The primary oocyte is haploid.

8. Which of the following is not an accessory sex organ?
   a. Penis
   b. Uterus
   c. Gonads
   d. Cloaca

9. For approximately how long during the human female's menstrual cycle are progesterone concentrations high enough to maintain the uterus in a proper condition for pregnancy?
   a. All of the cycle
   b. None of the cycle
   c. During the first half of the cycle
   d. During the second half of the cycle

10. Which of the following is not true about the birth control pill?
    a. The pill works by preventing ovulation.
    b. The pill works by preventing implantation.
    c. The ovarian cycle is suspended by the birth control pill.
    d. The birth control pill contains low doses of estrogen and progesterone.

11. Which one of the following is not a biological or medical reality for human reproduction?
    a. Negative feedback control of the hypothalamus by sex hormones
    b. Contraception through complete chemical blockage of sperm production
    c. Surgical transfer of gametes from one area of the reproductive tract to another
    d. Susceptibility of reproductive organs to cancer

12. If you compared the genetic makeup of an animal produced by parthenogenesis with that of its mother, which of the following would you expect?
    a. About 100 percent genetic similarity
    b. About 50 percent genetic similarity
    c. No genetic similarity
    d. Parthenogenetic animals have no mother.

13. Which of the following definitions about oviparity is false?
    a. Only birds and reptiles are oviparous.
    b. The large amount of yolk provides the nutrients for the developing embryo.
    c. The shell protects the egg from dehydrating.
    d. Both oxygen and carbon dioxide can diffuse through the shell.

14. During spermatogenesis a single male germ cell produces ______ sperm cell(s).
    a. 1
    b. 2
    c. 3
    d. 4

Application Questions

1. *Daphnia* (water fleas) can reproduce sexually or asexually. When and why might these animals switch between sexual and asexual reproduction?

2. While visiting the shore of a small pond, you notice two salamanders mating in the water. Based on what you know about external fertilization, what will ensure that the female's eggs will be fertilized?

3. How is it possible for a sea star to reproduce both sexually and by regeneration? How will the genetic makeup of offspring from each method differ?

4. The following graph depicts the average blood concentrations of three circulating hormones collected from 100 healthy adult women. Each hormone plays a role in regulating some aspect of the menstrual cycle. Identify the hormone that each curve most likely represents, and explain the criteria used in making your choice.

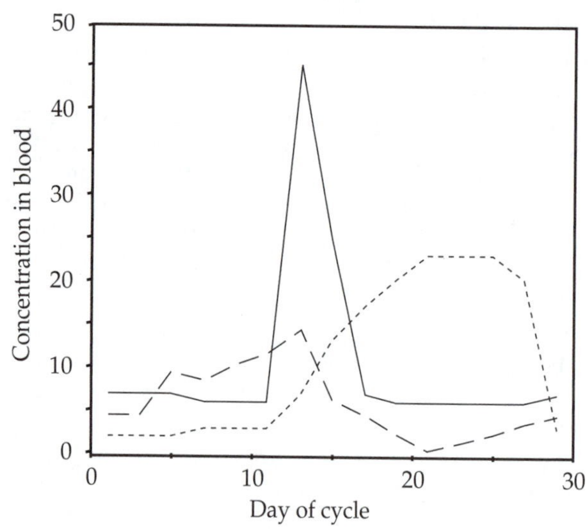

Answers

Knowledge and Synthesis Answers

1. **b.** The parents that have survived to reproduce asexually are able to survive in the current stable environment. Therefore, the offspring should be preadapted for this stable environment.

2. **d.** There are many differences between the egg and sperm. They are produced by the different sexes, sperm is motile and the egg is not, and the egg is much larger than sperm.
3. **d.** In males, sperm is produced during spermatogenesis. The egg is produced during oogenesis. Males have a secondary sexual organ such as the penis that aids in internal fertilization. External fertilization does not work well on land.
4. **b.** A sperm is ejected by the male into the vagina. From the vagina the sperm move through the cervix into the uterus and finally the oviduct where fertilization occurs.
5. **a.** The seminal vesicles are involved in producing the seminal fluid. One of the components of the seminal fluid is an energy source for the sperm in the form of fructose.
6. **a.** During days 12 through 14, a positive feedback occurs in response to estrogen. The hypothalamus and anterior pituitary gland are stimulated to release LH and FSH.
7. **d.** During oogenesis, the primary oocyte is diploid; after the first meiotic division into the secondary oocyte the cell becomes haploid.
8. **c.** The gonads are not secondary sex organs; they are the primary sex organs producing the sperm and the egg.
9. **d.** High levels of progesterone are needed to maintain the uterus in the proper condition for pregnancy. The levels of progesterone are high only during the second half of the uterine cycle.
10. **b.** The birth control pill interferes with the maturation of the follicles and the ova, inhibiting the release of an egg.
11. **b.** Males are constantly make sperm so that any chemical block of sperm production would have to be constant. Chemical blocks of sperm would also have serious medical side effects.
12. **a.** Species that exhibit parthenogenesis develop from unfertilized eggs produced by the mother. Therefore, the genetic make-up should be 100 percent the same as the mother.
13. **a.** The birds and reptiles are not the only oviparous species. A group of mammals, the monotremes, are also egg layers.
14. **d.** During spermatogenesis a single male germ cell will under go two meiotic divisions. This results in the production of 4 haploid sperm cells.

Application Answers

1. Female *Daphnia* tend to produce new females parthenogenically when conditions are good. When conditions turn unfavorable, they produce males instead of females. After the males are produced, the females reproduce sexually with the males, producing a resting egg. The mating with the males allows for genetic mixing of the population and the production of a resting egg that will produce a new parthenogenic female when conditions turn favorable again.
2. For successful external fertilization in water, the two salamanders must engage in a behavior that will bring the male's sperm in close proximity to the female's eggs. The second important thing that must occur is that the male and female must both produce large numbers of gametes to ensure that they are fertilized. Finally, the male and female must be at the same reproductive stage for successful mating to occur.
3. If a sea star loses one of its arms, it can regenerate a new one to replace it. If the detached arm remains alive, it can regenerate an entire new sea star. The new sea star will be genetically identical to the original one. Sea stars also produce eggs and sperm that join during sexual reproduction to produce new individuals. The new sea stars will share a genetic makeup with both parents, increasing genetic diversity.
4. The corpus luteum produces progesterone during the second half of the uterine cycle to maintain the endometrial lining of the uterus. The levels drop off by day 28 if implantation has not occurred. As the follicle grows during the first half of the ovarian cycle, follicle stimulating hormone (FSH) slowly increases. When the follicle ruptures and releases the egg, FSH is inhibited and the levels decrease. Luteinizing hormone peaks between day 12 and 14 to stimulate ovulation.

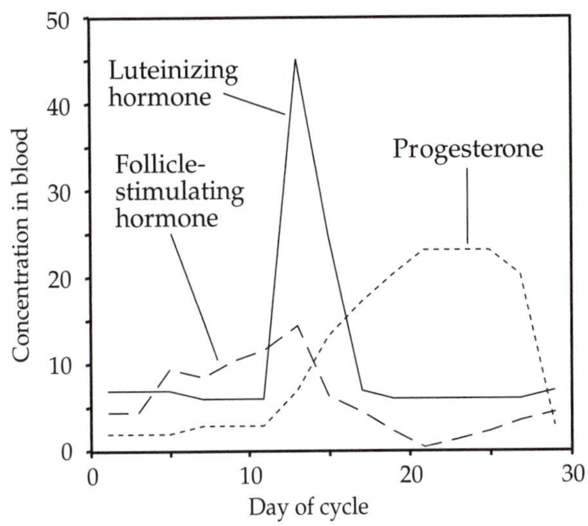

43 Animal Development

Important Concepts

Fertilization is the joining of the sperm and the egg.

- Specific mechanisms control recognition between the sperm and the egg to ensure that cross-species fertilization does not occur.
- In externally fertilizing species, eggs are surrounded by a protective vitelline envelope and jelly coat. At the tip of the sperm is an acrosome that contains enzymes to help trigger the digestion of the jelly coat of the egg. An acrosomal process containing the protein bindin extends from the sperm down to the vitelline envelope. Bindin is species-specific and helps initiate entrance of the sperm head into the egg via the egg's fertilization cone (see Figure 43.1).
- The egg of internally fertilizing animals is covered by a gelatinous matrix, the cumulus, and the zona pellucida which provides protection similar to the vitelline envelope above (see Figure 43.2).
- After entering the vagina, sperm undergo capacitation during which surface proteins are altered and the sperm become receptive to an egg.
- Enzymes on the tip of the sperm help digest the cumulus, the sperm binds to species-specific glycoproteins on the egg, and an enzyme for digesting the zona pellucida is released.

Mechanisms exist to block polyspermy.

- The first order of business after sperm entry is to initiate a mechanism that prevents any other sperm from fertilizing the egg.
- In sea urchins, the plasma membrane of the egg changes electrical potential in a fast block after which no other sperm can fuse with the egg plasma membrane (see Figure 43.3).
- A fertilization envelope around the egg acts as a slow block that is triggered by the release of calcium into the cytoplasm. The calcium triggers the cortical granules to fuse with the plasma membrane and allow water to flow between the vitelline envelope and the plasma membrane, which becomes the fertilization envelope (see Figure 43.4).
- The fast response in mammals does not trigger a rapid change in membrane potential, but rather a phosphatidylinositol (PTI) signaling system. The metabolism of the egg, cytoplasm pH, and protein synthesis all increase.
- The egg and sperm of mammals do not fuse until 12 hours after initial fertilization because the egg must first complete its second meiotic division.

The egg and sperm contribute differently to zygote formation.

- The egg contributes the majority of the cytoplasm and the mitochondria to the zygote.
- The sperm contributes its haploid nucleus and a centriole that becomes the centrosome and eventually forms the mitotic spindles.
- Upon fertilization, the cytoplasm of the egg undergoes rearrangement. The nutrients concentrate in the vegetal hemisphere of the egg in the lower half, while the nucleus is found in the upper animal hemisphere (see Figure 43.5).
- Sperm enters in the animal hemisphere at what will become the ventral side in frogs and causes a rotation of the cortical cytoplasm to create a gray crescent, which will become the dorsal region of the frog. This helps to set up the left-to-right axis and the anterior-to-posterior axis of the developing animal (see Figure 43.5).
- During the movement of the cytoplasm, the transcription factor β-catenin redistributes in the egg so that it is concentrated in the dorsal side of the embryo (see Figure 43.6). The protein GSK-3 moves to the ventral side of the zygote where it degrades β-catenin. On the dorsal side, proteins inhibit the action of GSK-3.

Cleavage is the first series of cell divisions in development.

- The first divisions of the cells occur rapidly with little growth or differentiation in a process known as cleavage.
- The mass of cells that is generated, the blastula, has a central cavity, the blastocoel. Individual cells in the blastula are known as blastomeres.
- The amount of yolk in the egg determines the plane along which cell division occurs. Cleavage furrows make up the site of cell division. Cleavage furrows are impeded within the egg by the yolk. Animals with small amounts of yolk have complete cleavage (see Figure 43.7). Fish, reptile, and bird eggs with large amounts of yolk have incomplete cleavage, with the embryo forming a disc of cells sitting on top of the yolk known as the blastodisc. Some insects undergo superficial cleavage.

- The mitotic spindles help to orient the cleavage planes into a radial cleavage in the sea urchin, rotational cleavage in the mammals, or spiral cleavage in the mollusks.

In mammals, cells separate into two groups between the 16-cell and 32-cell stage.

- Mammals have unique cleavage. The zygote produces both the embryo and extraembryonic structures that interact with the mother. Unlike in other species where cleavage is directed by molecules already present in the egg, in mammals genes are expressed during cleavage and play a large role.
- During cleavage from the 16-cell to the 32-cell stage, the cells separate into an inner cell mass that will become the embryo and a trophoblast that will become the placenta and attach to the uterus during implantation. At the 32-cell stage the mass of mammalian cells is called a blastocyst (see Figure 43.8).

Blastomeres become determined during late cleavage.

- During cleavage, cytoplasm is distributed to the cells of the blastula such that cells in different regions have different levels of nutrients and informational molecules. Specific blastomeres will become specific tissues and organs, and this can be mapped out in the developing cells as a fate map of the blastula (see Figure 43.9). The blastomeres become committed to specific fates at different times for different species.
- Animals with mosaic development have blastomeres that are set to contribute to specific parts of the embryo, while in animals with regulative development the blastomeres can be removed and other cells will compensate for the loss.

Gastrulation involves movement and differentiation of cells.

- During gastrulation, the germ layers of the tissues form and position themselves in the embryo. Inductive interactions occur during gastrulation due to the spatial relationships between cells, allowing for differentiation and organ formation.
- Three germ layers exist. The endoderm inside the embryo will become the gut tissues, the ectoderm outside the embryo will become the skin and nervous system, and the mesoderm between the two layers will become the tissues of the organs.

Gastrulation in sea urchins

- The vegetal pole in sea urchins moves into the cavity of the blastocoel to become the primary mesenchyme cells that make up the mesoderm (see Figure 43.11). The flat portion of the vegetal pole moves inward as an involution to form the endoderm and primitive gut, known as the archenteron.
- More mesenchyme cells are produced from cells that break off at the tip of the archenteron and enter the blastocoel.
- The opening of the archenteron or blastopore will become the anus, and the place where the tip of the archenteron meets the ectoderm will become the mouth.

Gastrulation in the frog

- The frog embryo has more yolk than the sea urchin and gastrulation is more complex (see Figure 43.13). Initially, cells near the gray crescent begin to bulge into the blastocoel and invaginate. These initial cells (bottle cells) move along the interior of the blastula and pull the outer surface of cells along with them, creating a dorsal lip. The first cells moving in will become endoderm and the archenteron.
- This process of the movement of cells from the surface into the blastopore is known as epiboly. As epiboly continues, the dorsal lip forms a complete circle around a plug of yolk-rich cells. At the end of gastrulation, the specific cells have become fate-determined and are layered as the ectoderm, endoderm, and mesoderm.
- Hans Spemann and Hilde Mangold examined the timing and fate of cells during gastrulation to determine if they were totipotent or able to direct development (see Figure 43.14). They found that the dorsal lip controlled the ability to form an embryo when it was transplanted to another gastrula, resulting in two embryos attached at the belly, thus acting as the primary embryonic organizer (see Figure 43.15). β-catenin is localized in the dorsal lip and is necessary for gastrulation to occur.

Gastrulation in reptiles and birds

- Cleavage produces a bastodisc on top of the large amount of yolk in the egg.
- The blastula is composed of an outer epiblast layer of cells and an inner hypoblast layer of cells. The fluid-filled space between the two layers is the blastocoel (see Figure 43.16).
- The blastopore is formed by the movement of cells in the epiblast toward the midline, creating the primitive streak. Along the primitive streak a primitive groove forms and cells move through this blastopore, becoming endoderm and mesoderm in the blastocoel.
- In the chicken there is no archenteron. A group of cells known as Hensen's node act in the same manner as the dorsal lip of the frog blastopore.

Gastrulation in mammals

- In mammals, the cells of the embryo segregate into an epiblast and a hypoblast with a blastocoel in between, just as in birds and reptiles. The embryo develops from the epiblast and the extraembryonic membranes develop from the hypoblast.
- Gastrulation in mammals occurs just as in birds, with the formation of a primitive streak.

The nervous system begins to develop during neurulation.

- After gastrulation, the organs begin to develop in the process of organogenesis. The nervous system develops early in organogenesis during neurulation.
- During neurulation, the notochord develops in the blastocoel from chordomesoderm cells derived from the dorsal mesoderm (see Figure 43.18).

- A neural plate forms above the notocord from the ectoderm, which will begin to fold into a cylinder forming the neural tube. The anterior end of the neural tube will become the major parts of the brain, with the rest forming the spinal cord.
- The repeating pattern of the vertebrate body plan forms from somite tissue derived from mesoderm located on both sides of the neural tube (see Figure 43.19). The somites will become the skin, muscles, and cartilage of the vertebrae and ribs.
- Neural crest cells break from the neural tube and migrate inward to become peripheral nerves.

Homeotic genes control differentiation in the anterior–posterior directions during development.

- The *Hox* genes in mice control the process of differentiation in the anterior–posterior directions during development (see Figure 43.20).
- *Hox* genes are found on different chromosomes and are expressed during development in their order on the chromosomes.
- Dorsal–ventral differentiation is controlled by a separate set of genes. The *sonic hedgehog* gene controls the development of the cells over the neural tube that becomes part of the ventral spinal cord.
- *Pax* genes are also important in development of the nervous system and somite development.
- After segmentation occurs, organs and organ systems develop rapidly.

Birds and reptiles have four extraembryonic membranes.

- Extraembryonic membranes surround the embryos of reptiles, birds, and mammals (see Figure 43.21).
- In birds, the yolk sac is derived from the hypoblast and surrounds the entire yolk to help retrieve nutrients from the yolk and deliver them to the embryo.
- Cells from the ectoderm and mesoderm form the amnion, which helps provide an aqueous environment, and the chorion, which develops just under the shell. The chorion regulates water, oxygen, and carbon dioxide exchanges across the shell.
- The allantois is derived from endoderm and is a membrane that forms a sac to store metabolic waste products.

The placenta of mammals provides exchanges between the mother and the embryo.

- In mammals, the first extraembryonic membrane, the trophoblast, interacts with the endometrium to attach to the uterine wall and begin implantation.
- The hypoblast cells interact with the trophoblast to form the chorion, which along with tissues from the uterine wall forms the placenta (see Figure 43.22).
- The amnion of the developing mammal surrounds the embryo to produce a closed fluid-filled environment.
- The allantois of mammals has the function of removing nitrogenous wastes, but its function is relatively minor in many mammals. The allantoic tissues help to form the umbilical cord, which carries major blood vessels that provide a route for exchanges of nutrients, carbon dioxide, and oxygen between the mother and fetus.
- Amniocentesis and chorionic villus sampling are two methods used to detect genetic markers for diseases (see Figure 43.23).

Human pregnancy is divided into trimesters.

- In humans pregnancy lasts about 9 months, compared with 21 days in mice and 600 days in elephants.
- During the first trimester of human development, the embryo develops into a miniature version of the adult. At this stage the developing fetus is most sensitive to environmental disruptors. The corpus luteum continues to produce estrogen and progesterone during this trimester.
- During the second trimester of development, the first fetal movements are felt, the fetus reaches a size of 600 g, and the woman begins to show physical evidence of her pregnancy.
- The third trimester is marked by rapid growth of the fetus and the maturation of the internal organs.
- Parturition or childbirth is brought on by both hormonal and mechanical stimuli. Estrogen and oxytocin both stimulate the uterine muscles to contract. Stretch receptors in the uterus stimulate the release of oxytocin, resulting in continued muscle contraction.
- Initial contractions cause the cervix to dilate.
- The mother continues to push with her abdominal muscles to deliver the baby, which typically arrives headfirst.

The Big Picture

- Animal development involves a number of steps: fertilization, cleavage, gastrulation, and neurulation. Fertilization involves the joining of the egg and sperm to form a zygote. A number of mechanisms exist to block additional sperm from entering the egg once fertilization has begun.
- Following fertilization, the cells of the embryo begin to divide in a process known as cleavage. Eggs undergo either complete cleavage or incomplete cleavage depending on the amount of yolk present. Those that undergo complete cleavage form a blastula, and those that undergo incomplete cleavage form a flat blastodisc. Cleavage in mammals is different from other groups. It is much slower and involves the production of proteins. The mammalian egg produces cells that will become the embryo and extraembryonic structures.
- During gastrulation, the germ tissues—endoderm, ectoderm, and mesoderm—form and move into specific positions. In the sea urchin, an archenteron that will become the gut forms and specific cells form the three germ layers. In frogs, a dorsal lip is formed and the germ cells move

into place. The dorsal lip is considered the primary embryonic organizer. The blastodisc of reptiles, birds, and mammals goes through a much different pattern of gastrulation.

- During neurulation, much of the organogenesis occurs. The nervous system begins to form a notochord and neural plates. The neural plates form into a neural tube that will become the brain and spinal cord. Somites produce the repeating segments of the vertebrate body plan. A number of genes play a large role in the differentiation of tissues.
- Reptiles and birds have four major extraembryonic membranes: yolk sac, amnion, chorion, and allantois. Mammals produce a special placenta that provides nutrients between the mother and the developing fetus.

Common Problem Areas

- All fertilized cells must divide to grow into adult animals. Sorting out the features of division that are common to all animals from those specific to certain taxonomic groups can be difficult and confusing until you begin to recognize the basic patterns.
- Students are often confused about the role of the placenta and its various membranes and structures. Is it embryonic? Is it maternal? And how do nutrients, gases, and wastes travel between the mother and embryo or fetus?

Study Strategies

- There is a great deal of terminology in this chapter, particularly with respect to the reproductive anatomy. Yet there are also many terms that are common to worms and wolves alike. Create a list of new terms from this chapter. Then determine which are the more general terms applicable to numerous animals (e.g. "somites," "gastrulation") and learn these first. Then tackle those referring to reproduction in specific animals.
- A key point in understanding animal development is that all specialized tissues arise from the three germ tissues: endoderm, ectoderm, and mesoderm. More than just committing these three tissues to memory, be sure to learn the specialized tissues that derive from the germ tissues. This will better enable you to make predictions about tissue functions and to compare body plans among animals.
- Review the following tutorials and activity on the website:
  Tutorial 43.1 Fertilization in the Sea Urchin
  Tutorial 43.2 Gastrulation
  Activity 43.1 Extraembryonic Membranes

Test Yourself

Knowledge and Synthesis Questions

1. Which of the following is not a mechanism used to block polyspermy?
   a. The formation of a fertilization envelope
   b. Changes in membrane potential upon fertilization
   c. The phosphatidylinositol signaling system
   d. The process of genomic imprinting
2. The sperm and the egg make different contributions to the zygote. Which statement about their contributions is false?
   a. The sperm contributes most of the mitochondria.
   b. The egg contributes most of the cytoplasm.
   c. Both the sperm and the egg contribute a haploid nucleus.
   d. All of the above are true.
3. What must a mammalian sperm penetrate before it can fertilize the egg?
   a. The cumulus
   b. The cumulus and the zona pellucida
   c. The jelly coat and the zona pellucida
   d. The cumulus and the vitelline envelope
4. Why do some species have complete cleavage and others incomplete cleavage?
   a. Incomplete cleavage occurs in species with small volumes of cytoplasm.
   b. Complete cleavage, found in mammals, is a more evolved characteristic.
   c. Incomplete cleavage occurs in species with large amounts of yolk.
   d. Complete cleavage occurs only in eggs that have been fertilized by two sperm.
5. A fate map can be used to map out the tissues and organs that the germ layers will become in a frog blastula. What is the fate of the dorsal lip of the blastopore?
   a. The lining of the gut
   b. The notochord
   c. The epidermal layer of skin
   d. The heart
6. Whereas the location of the _________ determines the anterior–posterior axis of the embryo, the _________ determines the embryo's bilateral symmetry.
   a. primitive streak, epiblast
   b. blastopore, neural plate
   c. vegetal hemisphere, notochord
   d. hypoblast, blastopore
7. What is the order of the germ layers from the inside to the outside?
   a. Mesoderm, ectoderm, endoderm
   b. Endoderm, ectoderm, mesoderm
   c. Ectoderm, mesoderm, endoderm
   d. Endoderm, mesoderm, ectoderm
8. Birds develop from a blastodisc while frogs develop from a blastula. During gastrulation in frogs, a dorsal lip forms that will eventually form the notochord. What is the equivalent in the chicken blastodisc?
   a. Epiblast
   b. Hypoblast
   c. Hensen's node
   d. Bottle cells
9. Birds develop extraembryonic membranes during development. Which statement about avian extraembryonic membranes is *false*?

a. The yolk sac surrounds the yolk and provides nutrients.
b. The amnion and chorion are derived from ectoderm and mesoderm.
c. The allantois stores nutrients.
d. The chorion exchanges gases and water between the embryo and the environment.

10. Which of the following statements about the mammalian blastocyst is false?
a. The trophoblast gives rise to the embryo proper.
b. Maternal genes are expressed during cleavage.
c. The blastocyst implants in the mother's uterus.
d. Early mammalian development is slow.

11. The role of the corpus luteum and the hormones it produces ends
a. at the time of embryonic implantation in the uterus.
b. with the formation of the placenta.
c. during the second trimester of pregnancy.
d. at birth.

12. In which of the following animal eggs would you expect to find a zona pellucida covered by a cumulus?
a. Human
b. Frog
c. Sea urchin
d. Chicken

13. Why did Hans Spemann call the dorsal lip of the blastopore the embryonic organizer?
a. It is the point where gastrulation begins.
b. It becomes part of the nervous system.
c. It becomes part of the notochord.
d. It leads to the establishment of the embryonic axes.

14. During its development, the human embryo is contained within a fluid-filled chamber bounded by the membranous
a. yolk sac.
b. amnion.
c. chorion.
d. allantois.

Application Questions

1. The development of the shelled egg had a profound effect on the evolution of reptiles and birds. Why was the development of this structure so important to the evolution of terrestrial vertebrates?

2. In the following diagram of a sea urchin gastrula, label the ectoderm, endoderm, primary mesenchyme, secondary mesenchyme, blastopore, blastocoel, and archenteron.

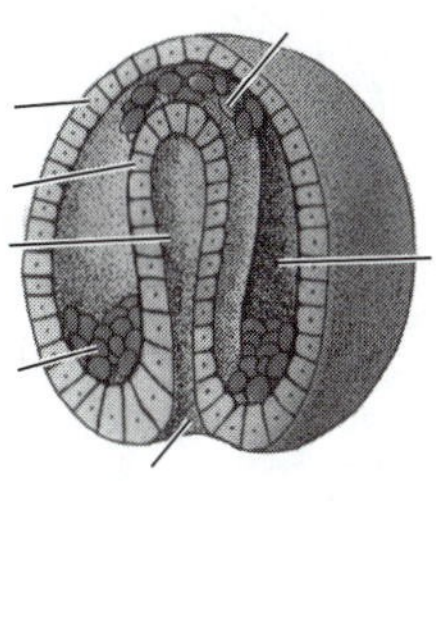

3. Label the listed structures in the following longitudinal diagram of a late frog gastrula: blastopore, area where brain will form, yolk plug, notochord, neural plate, archenteron, ectoderm, mesoderm, and endoderm. Also indicate the (A) anterior, (P) posterior, (D) dorsal, and (V) ventral ends of the embryo.

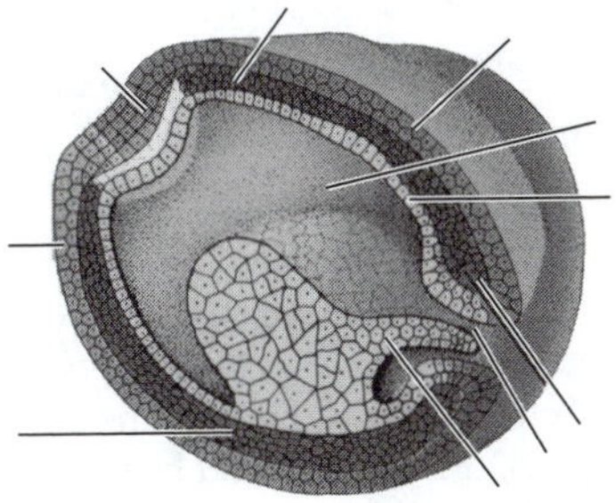

4. Describe the blastodisc and blastula of the birds and mammals.

5. How does the development of the extraembryonic membranes of the birds and mammals differ and how are they similar? What is the main reason for the differences in the development of the extraembryonic membranes in birds and mammals?

Answers

Knowledge and Synthesis Answers

1. **d.** The process of genomic imprinting is the phenomenon by which certain genes function only if they are derived from the sperm and other genes function only if they are derived from the egg. This has been established for mammals.

2. **a.** The mitochondria of the sperm degenerate. The egg donates all of the mitochondria to the zygote. This fact is used by some scientists to determine how species are related.

3. **b.** The two outer layers of the mammalian egg are the cumulus and the zona pellucida. In order to reach the plasma membrane of the egg, the sperm must travel through these two membranes.

4. **c.** Incomplete cleavage occurs because the cleavage furrows cannot completely penetrate the yolk. Incomplete cleavage occurs in bird and reptile species with large amounts of yolk.

5. **b.** The dorsal lip of the blastopore in a frog will form the notochord.

6. **b.** The blastopore, which will eventually become the anus, marks the posterior of the embryo; the embryo is bilaterally symmetrical with respect to the neural plate.

7. **d.** The germ layers that are formed during gastrulation are the inner layer of endoderm, the middle layer of mesoderm, and the outer layer of ectoderm.

8. **c.** Chickens develop a primitive streak and groove along the top of the blastodisc. At one end of the primitive streak is the Hensen's node that is the equivalent of the dorsal lip in the developing frog.
9. **c.** During development, an avian embryo produces wastes, but because they develop in a shell the wastes must be stored in the egg. The allantois forms a sac that is used for the storage of metabolic wastes.
10. **a.** In the mammalian blastocyst, the trophoblast forms the fetal part of the placenta. A disc-shaped portion of the inner cell mass surrounded by the trophoblast becomes the embryo.
11. **c.** The corpus luteum produces estrogen and progesterone. During early pregnancy, the hormones produced by the corpus luteum are required for the maintenance of early pregnancy. During the second trimester, the placenta is fully formed and takes over hormone production.
12. **a.** The eggs of mammals are surrounded by the cumulus and the zona pellucida; therefore, a human egg would fit the criteria.
13. **d.** The dorsal lip leads to the establishment of the embryonic axes.
14. **b.** The amnion makes up the membrane that surrounds the developing mammalian fetus.

Application Answers

1. The development of the shelled egg in ancestral reptiles and birds allowed reproduction to occur outside of an aquatic environment. Rather than laying eggs in water, amphibians, reptiles, and then birds could lay eggs even in quite dry terrestrial environments, with the egg shell and its low water permeability preventing desiccation of the eggs during the incubation period.
2.

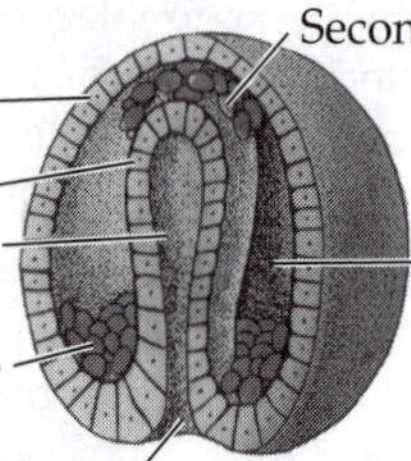

3.

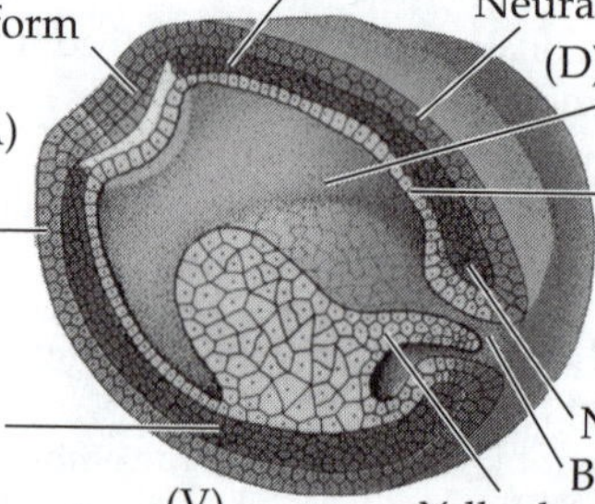

4. The blastodisc is the product of cleavage in both the birds and mammals. Due to the large amount of yolk in the avian egg, cleavage is incomplete and the result is a blastodisc that sits on top of the yolk. The blastula forms from the blastodisc. It is formed from a continuous layer of cells called the hypoblast, which is covered by an upper layer of cells called the epiblast. In between these two layers lies a fluid-filled space, or blastocoel. Mammals undergo the same incomplete cleavage, even though they lack the large amount of yolk. The blastodisc and blastula are similar in the mammals and birds.
5. Both the birds and mammals develop extraembryonic membranes. The extraembryonic membranes of the birds are used to obtain food from the large reserve of yolk (yolk sac), to exchange water and gases with the environment across the shell (amnion and chorion), and to store nitrogenous wastes (allantois). The mammals have extraembryonic membranes that form a placenta involved in exchanging nutrients, oxygen, and carbon dioxide with the mother. The hypoblast cells that form the yolk sac in birds along with the trophoblast cells and the chorion all contribute to the placenta. An allantois also develops, but has limited function in humans.

44 Neurons and Nervous Systems

Important Concepts

The nervous system consists of neurons and glial cells.

- The nervous system is made up of two major types of cells: neurons and glial cells.
- Neurons transmit information as electric signals from sensory cells to effectors such as muscles.
- Simple animals such as the sea anemone posses a nerve net (see Figure 44.1).
- The nervous system of invertebrates is made up of groups of neurons distributed throughout the body, known as ganglia.
- Vertebrates have a central nervous system—the brain and spinal cord—and a peripheral nervous system—the neurons in the rest of the body.
- A neuron is composed of four parts: the cell body, dendrites, axons, and axon terminals (see Figure 44.3). The cell body contains the nucleus, with dendrites coming off the cell body and receiving information from other neurons. The plasma membrane of neurons has an electrical charge or action potential that helps transmit information away from the cell body along an axon. The axon terminals interact with the neuron's target cells to form a synapse.
- Neurons work together in the nervous system. The simplest neuronal network is a sensory neuron connected to a motor neuron connected to a muscle cell.
- Glial cells serve many functions in the nervous system, such as supplying nutrients to neurons, maintaining proper ionic balance, and removing waste particles. Glial cells that insulate axons are Schwann cells and oligodendrocytes. Astrocytes found around blood vessels in the brain form a barrier to some toxins as well as certain drugs.

Neurons have an electrically charged cell membrane at rest.

- Resting neurons have a negative charge inside and a positive charge outside, resulting in a difference in electrical charge across the membrane known as the membrane potential (see Figure 44.5).
- The electrical charge across the membrane at rest is due to differences in concentrations of the charged ions sodium (Na^+), chloride (Cl^-), potassium (K^+), and calcium (Ca^{2+}).
- The lipid phase of the plasma membrane is impermeable to ions. Ions move across the plasma membrane through channels or by ion pumps.
- The sodium–potassium pump maintains higher concentrations of Na^+ ions outside the cell and higher concentrations of K^+ ions inside the cell. This pump transports Na^+ out of the cell and K^+ into the cell (see Figure 44.6).
- Ion channels are selective pores in the plasma membrane that allow specific ions to diffuse through the membrane down their concentration gradient. Ion channels can be voltage-gated, responding to changes in the voltage across the membrane, or chemically gated, responding to the presence of a specific chemical.
- At rest, neurons have a specific charge due to K^+ movement to the outside of the cell, resulting in a resting potential. K^+ channels are the most common open channels, allowing K^+ to diffuse out of the cell down the concentration gradient that has been set up by the Na^+–K^+ pump (see Figure 44.7).

Neurons generate action potentials to conduct nerve impulses.

- Changes in resting membrane potential ("depolarization") produce action potentials or nerve impulses.
- Depolarization occurs when the inside of a neuron becomes less negative compared to the resting potential (see Figure 44.9). Hyperpolarization occurs when the inside of a neuron becomes more negative compared to the resting potential (see Figure 44.9).
- Changes in polarity of the plasma membrane due to opening and closing of ion channels are passed down along an axon to transmit a signal as an action potential.
- Action potentials are very short-lived changes in membrane potential (see Figure 44.10). Action potentials are generated when the membrane reaches a threshold potential, Na^+ voltage-gated channels open, and Na^+ enters the cell to make the inside of the axon positive. Voltage-gated K^+ channels then open, allowing positive charged K^+ to leave the axons to help return the membrane potential back to the resting level. As the K^+ channels open, the Na^+ channels close and cannot be opened for about 1 to 2 milliseconds, which is the refractory period. The Na^+–K^+ pump helps return the concentration of ions back to the resting levels.

- Action potentials travel down an axon by a positive feedback mechanism that stimulates adjacent regions of an axon to generate an action potential. The Na^+ ions that enter during an action potential flow to adjoining regions of the axon, stimulating depolarization and the movement of the action potential along the axon (see Figure 44.11).
- The refractory period during which the Na^+ channels cannot act keeps an action potential moving in one direction.
- Action potentials are an all-or-nothing response. The depolarization must reach a threshold level for an action potential to occur (see Figure 44.10).

Action potentials can jump down an axon.

- Some axons are myelinated. The protein myelin wrapped around some axons insulates portions of them from depolarization (see Figure 44.13).
- Nodes of Ranvier are gaps in the myelin wrapping at which depolarization can occur. Depolarization jumps from node to node, increasing the speed of transmission of an action potential along the axon in a process known as saltatory conduction.

Neurons communicate across synapses.

- Transfer of information from one cell to another by either a chemical or electrical message occurs in a small gap junction called a synapse. Presynaptic nerve cells send a message across the synapse to postsynaptic cells.
- A chemical synapse uses a chemical messenger, or neurotransmitter, to communicate between the presynaptic and postsynaptic cells (see Figure 44.14). The neurotransmitter is produced in the axon terminal and packaged into vesicles. The neurotransmitter is released into the synaptic cleft when the vesicle binds with the presynaptic membrane. The neurotransmitter crosses the synaptic cleft and binds with receptors on the surface of the postsynaptic cell.

Acetylcholine is a common synaptic neurotransmitter.

- The neurotransmitter acetylcholine is the chemical messenger carrying information between motor neurons and muscle cells at neuromuscular junctions (see Figure 44.16). Acetylcholine is enclosed in a vesicle at the presynaptic synapse which fuses with the membrane to release the acetylcholine into the 20–40 nm wide gap between the two cells, the synaptic cleft.
- Ca^{2+} channels open when the action potential reaches the axon terminal, causing Ca^{2+} to rush in and regulate the fusing of the acetylcholine-containing vesicles to the presynaptic membrane.
- Acetylcholine released into the synaptic cleft binds with receptors in the postsynaptic membrane, the motor end plate, which opens Na^+ channels, resulting in depolarization of the motor end plate (see Figure 44.15).
- Acetylcholinesterase breaks down acetylcholine in the synapse junction to halt the action of the released acetylcholine.

Synapses can either depolarize or hyperpolarize postsynaptic membranes.

- Synapses in vertebrates can be excitatory and depolarize or inhibitory and hyperpolarize the postsynaptic membrane. Hyperpolarization occurs by the neurotransmitters γ-amino butyric acid (GABA) or glycine, which activate Cl^- channels.
- Neurons many receive synaptic inputs from many neurons. Excitatory and inhibitory postsynaptic potentials are summed up by adding simultaneous potentials (spatial summation) or by summing up rapid firing of one postsynaptic potential (temporal summation) (see Figure 44.17).
- The region of the cell body at the base of the axon is the "decision-making" area of a neuron. If this axon hillock is depolarized, the axon will fire an action potential.

Two main types of neurotransmitter receptors exist.

- Ionotropic receptors found on the postsynaptic membrane are ion channels that are activated by the binding of the neurotransmitter.
- Metabotropic receptors found on the postsynaptic membrane act by initiating a second messenger, such as G proteins, to activate ion channels (see Figure 44.18).
- Metabotropic receptors are categorized depending on the neurotransmitter that activates them (see Figure 44.19). α-adrenergic receptors are stimulated by norepinephrine, while muscanaric receptors are stimulated by acetylcholine.
- There are over 25 neurotransmitters, including amino acids, dopamine, norepinephrine, serotonin, peptides, carbon monoxide, and nitric oxide (see Table 44.1). One neurotransmitter can act on a number of different receptors which dictate their action. The action of a neurotransmitter depends on the receptor to which it binds.
- In the CNS the neurotransmitters are glutamate, glycine, and GABA.
- The action of neurotransmitters is stopped by enzymes, diffusion away from the synaptic cleft, or by active uptake.

NMDA and AMPA receptors may have a role in memory.

- Two types of receptors—NMDA and AMPA—may play a role in learning. NMDA receptors allow for a slow, long-lasting influx of Na^+, while the AMPA receptors allow for rapid influx. The NMDA receptors also let Ca^{2+} into the postsynaptic cell, resulting in long-term cellular changes.
- Both of these receptors are excited by glutamate.
- Potentiation is an enhanced response in the postsynapse with stimulation (see Figure 44.21). Long-term potentiation may be involved in long-term memory.
- Mice with NMDA receptors that stayed open longer than usual tended to be smarter than regular mice.

Electrical synapses connect some neurons.

- Electrical synapses, or gap junctions, are formed by direct contact between adjacent neurons.

- Two neurons forming an electrical synapse are joined by connexons, which are tunnels between the two neurons allowing ions to pass between the two cells.
- Electrical synapses are very fast connections and can transmit an action potential in either direction between cells, making them good for rapid communication.
- Gap junctions are less common than chemical synapses. With electrical synapses, there can be no temporal summation of synaptic inputs. They also require large areas of contact and cannot be inhibitory.

The Big Picture

- The neuron, with support from surrounding glial cells, is the functional unit of the nervous system. The neuron is composed of a cell body, dendrites, and an axon. The membrane of a neuron has a difference in voltage across it. Nerve impulses are passed down the axon of a neuron as action potentials. An action potential is a temporary disruption of the "battery-like" state of the resting neuron membrane due to the opening and closing of sodium and potassium voltage-gated channels. The action potential is an all-or-nothing response that occurs when the depolarization of an axon reaches a threshold level.
- At the end of the axon synapse, information flow is controlled through excitation or inhibition of synapses. There are many different neurotransmitters that transmit a nerve impulse from the presynaptic cell to the postsynaptic cell. The action of a specific neurotransmitter will depend on the receptor with which it binds.

Common Problem Areas

- One of the most difficult challenges with the nervous system is understanding how the resting membrane potential and action potential are produced. Remember that the neuron at rest is like a battery and that the charge of the membrane is dependent on the ions that are on either side and moving across the membrane.
- Students often have difficulty with the function of a synapse and how the action potential is transmitted across the synapse. Remember that in many cases the signal goes from an electrical signal, to a chemical signal, and back to an electrical signal.

Study Strategies

- Students need to nail the concept of "pre-" and "postsynaptic" neurons. Try and remember the sequence of events and relate them to the function of the neuron, transmitting a stimulus from one cell to another.
- One of the major problems that students have is understanding how the membrane potential is set up. Students should understand what each important anion and cation does at rest and during the action potential. Then piece by piece put together a sense of how the neuron membrane is charged at rest, and what happens during an action potential.
- Review the following tutorials and activity on the website:
  Tutorial 44.1 The Resting Membrane Potential and the Action Potential
  Tutorial 44.2 Synaptic Transmission
  Activity 44.1 Concept Matching: Neurotransmitters

Test Yourself

Knowledge and Synthesis Questions

1. The extensions of postsynaptic neurons that provide the main receptive surface for presynaptic neurons are
   a. nuclei.
   b. somas.
   c. axons.
   d. dendrites.

2. The substance that wraps the axon of many neurons and provides for increased conduction speed is
   a. dendrase.
   b. histamine.
   c. acetylcholine.
   d. myelin.

3. The long extensions from the neurons that provide the pathway for action potentials to synapse are
   a. dendrites.
   b. cell bodies.
   c. axons.
   d. presynaptic membranes.

4. The threshold of a neuron is
   a. the amount of inhibitory neurotransmitter required to inhibit an action potential.
   b. the membrane voltage at which an axon potential will be suppressed.
   c. the amount of excitatory neurotransmitter required to elicit an action potential.
   d. the membrane voltage at which the membrane potential develops into an action potential.

5. When a membrane is at the resting potential, the concentration of
   a. sodium and potassium ions is higher on the inside of its membrane.
   b. sodium and potassium ions is higher on the outside of its membrane.
   c. sodium ions is higher on the inside of its membrane and of potassium ions is higher on the outside.
   d. sodium ions is higher on the outside of its membrane and of potassium ions is higher on the inside.

6. Glial cells are specialized to do all of the following except
   a. receive neural impulses.
   b. insulate the axons.
   c. supply neurons with nutrients.
   d. help maintain proper ionic environment for the neuron.

7. The cells that create the blood–brain barrier keeping toxic substances from entering the brain are ________ and belong to a type of neural tissue called ________.
   a. endothelial cells, Schwann cells

b. astrocytes, glial cells
c. glial fibers, axons
d. dendrites, synapses

8. A particular disease of the nervous system specifically involves the Ca^{2+} ion channels at the chemical synapses of motor neurons where neurotransmitter is stored and released. In other words, this disease affects the
   a. axon terminals of the presynaptic cell and the release of acetylcholine.
   b. axon terminals of the postsynaptic cell and the release of K^+ ions.
   c. electrical synapses.
   d. axon terminals of the presynaptic cell and the release of K^+ ions.

9. Which of the following statements about gap junctions or electrical synapses is *false*?
   a. Connexons form molecular tunnels between two cells.
   b. Electrical synapses cannot be inhibitory.
   c. Electrical synapses do not allow for temporal summation.
   d. Electrical transmission is very slow and is bidirectional.

10. One group of neurotransmitter receptors is the metabotropic receptors. Which one of the following statements about metabotropic receptors is false?
    a. Both muscanaric and β-adrenergic receptors are metabotropic receptors.
    b. Metabotropic receptors are coupled with A proteins.
    c. Metabotropic receptors can stimulate gene expression.
    d. Muscanaric receptors bind with acetylcholine.

11. The rapid depolarization of a neuron during the first half of an action potential is due to the
    a. exit of K^+ ions from the cell through gated potassium channels.
    b. rapid reversal of ion concentration caused by the action of the sodium–potassium pump.
    c. entry of Na^+ ions into the cell through gated sodium channels.
    d. movement of both Na^+ and K^+ ions through appropriate open channels.

12. The refractory period of a neuron results from
    a. the period when the sodium–potassium pump is nonfunctional.
    b. activation of voltage-gated chloride channels.
    c. closing of inactivated voltage-gated sodium channels.
    d. the action potential reaching the synapse.

13. Which of the following statements about the process of summation in a neuron is false?
    a. Slight perturbations of the membrane potential spread across the postsynaptic cell body.
    b. Axons that terminate closer to the axon hillock have more influence on the summation process.
    c. Summation consists essentially of comparing the total of excitatory and inhibitory postsynaptic inputs.
    d. The concentration of voltage-gated sodium channels is highest in the dendrites of the postsynaptic cell.

14. The amount of neurotransmitter released from a presynaptic cell is a function of
    a. the voltage of the individual action potential in the presynaptic neuron.
    b. the frequency of arriving action potentials in the presynaptic neuron.
    c. whether the action potential is excitatory or inhibitory.
    d. the receptors in the postsynaptic cell.

15. The electrical events labeled as EPSPs are the result of
    a. hyperpolarization of the postsynaptic membrane.
    b. depolarization of the postsynaptic membrane.
    c. hyperpolarization of the presynaptic membrane.
    d. depolarization of the presynaptic membrane.

Application Questions

1. You are testing to determine the role of the potassium channels in a neuron. To do this you have knocked out all of the functional potassium channels and you depolarize the membrane potential. What will happen to the membrane potential after depolarization?

2. The active ingredients in many nerve gases belong to a class of chemicals called anticholinesterases. Suggest a possible synaptic mechanism to explain how these chemicals can do such damage to an animal's nervous system.

3. The following diagram represents a monosynaptic reflex loop. Label the sensory neuron, motor neuron, and effector in the simple neural circuit. Also indicate which part of the circuit is the sensory side and which is the motor side.

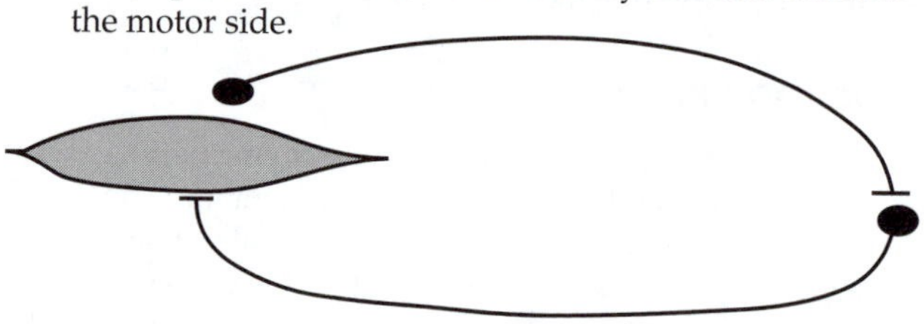

4. Explain how an action potential travels faster down an axon wrapped in myelin than one without the insulating myelin.

Answers

Knowledge and Synthesis Answers

1. **d.** The neuron is composed of a cell body, an axon, and dendrites. The dendrites form synapses with presynaptic cells to form the junction where information from one neuron is transferred to another neuron.

2. **d.** The cells that coat the axon of some neurons are myelin. Myelin is a specific type of protein found only in glial cells.

3. **c.** The neuron is composed of the cell body, the dendrite, and the axon. The axon carries action potentials away from the cell body to the synapses.

4. **d.** For an action potential to occur in an axon, the membrane must be depolarized above a certain level. This level is known as the threshold.

5. **d.** The resting potential of a neuron membrane occurs when the sodium ion concentration is higher on the outside and the potassium ion concentration is higher on the inside.

6. **a.** Glial cells perform many functions in the nervous system, but they do not receive neural impulses.

7. **b.** The blood–brain barrier is formed by astrocytes that wrap around the blood vessels traveling through the brain. The astrocytes are a special kind of glial cell.

8. **a.** If the disease acts on a chemical synapse where the neurotransmitter is stored and released, it is affecting the axon terminals of the presynaptic cell. Ca^{2+} channels are involved in regulating the release of acetylcholine by allowing Ca^{2+} to enter the presynaptic cell and promoting the fusing of acetylcholine containing vesicles to the membrane.

9. **d.** Electrical synapses join two cells with protein tunnels known as connexons. The junctions that are formed provide for very fast transmission between cells.

10. **b.** Metabotropic receptors are coupled with G proteins that bind GDP. When the receptor binds a neurotransmitter, the GDP is turned into GTP and a subunit of the G protein separates and activates effector proteins.

11. **c.** The first step in an action potential is the influx of Na^+ ions leading to a depolarization of the axon membrane. Na^+ ions rush into the cell due to the higher concentration outside of the cell and the negative membrane potential.

12. **c.** After the spike of the depolarization, the sodium voltage-gated channels close. One of the properties of these channels is that they will open again only after a short delay. This short delay is known as the refractory period when the sodium voltage-gated channels are inactive.

13. **d.** Dendrites and most of the cell body have few gated sodium channels. These channels mediate the action potentials that travel down the axon, where their levels are high.

14. **b.** Neurotransmitter is released from the presynaptic cell when an action potential reaches the synapse. Increases in the frequency of arriving action potentials in the presynaptic neuron will increase the release of neurotransmitter into the synapse.

15. **b.** Excitatory postsynaptic potentials (EPSP) make it easier for an action potential to occur, so they depolarize the postsynaptic membrane.

Application Answers

1. The potassium voltage-gated channels are responsible for setting up the resting potential of a membrane. Potassium ions have a tendency to diffuse out of the cell, leaving negative charge inside. Knocking out the function of the potassium voltage-gated channels would result in the cell's being unable to maintain resting potential. If the cell is depolarized by the opening of sodium voltage-gated channels, it might not repolarize because the potassium channels that help to repolarize the membrane would not be functioning.

2. The neurotransmitter at many synapses is acetylcholine. Acetylcholine transmits the action potential from a presynaptic cell to a postsynaptic cell. Acetylcholinesterase is found in the synaptic cleft and it cleaves acetylcholine to help remove it from the synaptic cleft after an action potential. A nerve gas that was composed of anticholinesterases would block the action of acetylcholinesterase, causing acetylcholine to build up in the synaptic cleft. This build-up would mean that the receptors on the postsynaptic cell would remain bound with acetylcholine.

3.

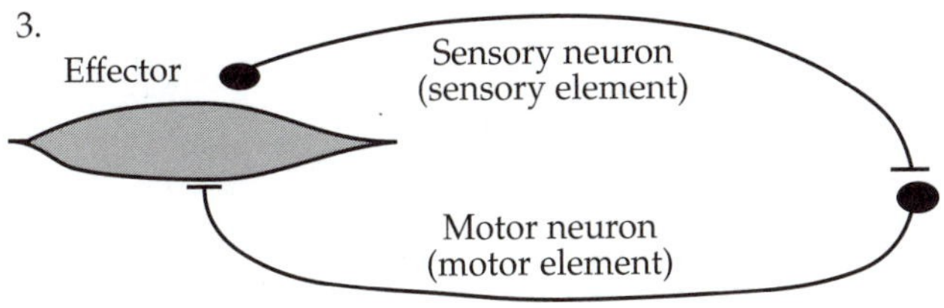

4. The conduction of an action potential down a myelinated axon is known as saltatory conduction. The myelin acts to insulate areas of the axon, preventing depolarization. The areas of the axon between the myelin are known as nodes of Ranvier. Depolarization can only occur at these nodes. As the action potential moves down a myelinated axon, the influx of sodium ions at one node diffuses down the axon. This diffusion of sodium ions results in the depolarization of the next node of Ranvier. Depolarization can occur only in the downstream nodes because the upstream nodes are in a refractory period. In this way the action potential moves quickly down the axon to the synapse.

45 Sensory Systems

Important Concepts

Sensory cells detect stimuli and transmit the information to the CNS.

- Sensory cells are modified neurons specialized to detect internal and external stimuli, and to transmit information on these stimuli as action potentials to different sites in the central nervous system (CNS).
- Sensory cells are involved in maintaining homeostasis by transmitting information about the status of the internal environment to the CNS.
- Sensory organs are groups of sensory cells along with other cells which collect, filter, and amplify stimuli.
- Sensory transduction is the process by which a mechanical, thermal, or chemical stimulus to a sensory cell is transformed ("transduced") into an action potential. Sensory transduction can occur by direct activation of receptor proteins that open or close ion channels to change membrane potential. Alternatively, the stimulus can activate a second messenger within the cell that couples with a G protein to eventually open ion channels.
- Receptor potentials are transmitted into action potentials by the release of a neurotransmitter or by generation of action potentials within the sensory cell though the opening of Na^+ voltage-gated channels.
- Many sensory cells have diminished responses to a stimulus over time in a process known as adaptation. Adaptation allows organisms to ignore background conditions and focus on new information.

Chemoreceptors produce action potentials in response to a chemical stimulus.

- Smell, taste, and chemicals circulating within the body stimulate chemoreceptors.
- Insects use pheromones or chemical signals secreted into the environment to attract mates by remotely stimulating their target's chemoreceptors (see Figure 45.4). The concentration of the pheromone released by a female provides a gradient that provides information on the insect's specific location.
- Olfaction is the sense of smell. The olfactory sensors of vertebrates are neurons with the axon extending to the olfactory bulb of the brain and the dendrites exposed as hairs to the environment (see Figure 45.5). Olfactory receptor proteins are found on the hairs and each receptor binds with specific odorants. The ability to discriminate many different odors is due to the high number of specific receptors. The strength of a smell is related to the number of receptors that are bound.
- Binding of an odorant results in depolarization of the cell through a G protein that activates a second messenger that opens sodium channels.
- Gustation is the sense of taste. Taste buds contain sensory cells that are not neurons (see Figure 45.6). Binding of the stimulus to receptor proteins causes a change in the membrane potential and the release of neurotransmitters that stimulate the underlying sensory neurons.

Mechanoreceptors detect mechanical force.

- Mechanical force causes distortion of the membranes of mechanoreceptors, causing ion channels to open and an action potential to be generated.
- The skin has a number of different types of mechanoreceptors (see Figure 45.7). Meissner's corpuscles are nerve endings located in the outer layer of skin and are sensitive to light touches of the skin. Expanded-tip tactile receptors are also located in the outer layer of skin and provide steady-state information about objects in continual contact with the skin. The density of these receptors varies throughout the body. Pacinian corpuscles are deep in the skin and respond to vibrations.
- Stretch receptors are mechanoreceptors that sense position and stress on limbs. Muscle spindles are stretch receptors in the connective tissue of muscles that perceive muscle stretch. The Golgi tendon organs are found in the tendons and ligaments and provide information about forces in contracting muscles. Collectively, these mechanoreceptors provide information on body posture and limb position, as well as muscle and tendon tension.
- Hair cells are mechanoreceptors that release neurotransmitters in response to the bending of stereocilia on the hairs. Invertebrates have hair cells in equilibrium chambers called statocysts that help maintain equilibrium in the presence of gravity (see Figure 45.10). In mammals, the inner ear uses hair cells in the semicircular canals and the vestibular apparatus to maintain equilibrium. The lateral line system of fish is composed of hair cells located under

the scales that detect movements of water around the fish (see Figure 45.9).

- The inner ear of vertebrates has hair cells that detect a body's position relative to gravity (see Figure 45.11). In the semicircular canal of the ear, flow of semicircular fluid shifts the cupulae containing the hair cells when the head changes position. In the vestibular apparatus, layers of calcium carbonate resting on hair cells are moved by gravity when the head changes position. Both of these stimuli to the hair cells inform the organism of its position in relation to gravity.

The auditory system contains mechanoreceptors specialized for sound reception.

- The auditory system takes in sound as pressure waves and transforms them into action potentials (see Figure 45.12).
- The ear pinnae are the outer portion of the mammalian ear that directs sound waves into the auditory canal. The end of the auditory canal is covered with a tympanic membrane which vibrates and transmits sound waves to ear ossicle bones in the middle ear. The ossicle bones are the malleus (hammer), incus (anvil), and stapes (stirrup), and together they transmit sound waves to the oval window membrane.
- Sound travels through the oval window into the fluid-filled cochlea where the pressure waves are turned into action potentials (see Figure 45.13). The cochlea, or inner ear, of a mammal is a three-canal chamber with two membranes: Resissner's membrane and the basilar membrane divide the chamber. The organ of Corti contains hair cells with stereocilia that are in contact with a tectorial membrane that transduces movement of the basilar membrane into action potentials to the cochlea nerve.
- The upper and lower chambers of the cochlea are separated at the distal end by the basilar membrane. Sound waves enter through the oval window as pressure waves. The waves move down the cochlea, reach the end, and move back down where they are absorbed by the flexible membrane of the round window.
- If the oval window vibrates at faster frequencies, the wave cannot make it all the way down the cochlea. It crosses the basilar membrane, causing it to flex. The flexing of the membrane stimulates hairs and the organ of Corti. Different pitches of sound cause the basilar membrane to flex at different locations, stimulating different hairs.

Light-sensitive pigments are used in photoreceptors.

- Rhodopsin is a family of pigments made up of the protein opsin and the light-absorbing prosthetic group 11-*cis*-retinal (see Figure 45.14). The 11-*cis*-retinal absorbs photons of light and changes conformation to all-*trans*-retinal, causing opsin to change conformation and become photoexcited rhodopsin.
- Photoexcited rhodopsin triggers a cascade that ultimately leads to changes in membrane potential and the photoreceptors response to light. Excited rhodopsin activates the G protein transducin. Transducin converts cyclic GMP (which opens Na^+ channels), to 5'-GMP, allowing Na^+ channels to close and the cell to hyperpolarize.
- In vertebrate eyes, rod cells are photoreceptor cells that contain an inner segment, a synaptic terminal, and an outer segment with many rhodopsin molecules (see Figure 45.15). Rods are found in the retina, along with a layer that transduces visual information into action potentials.
- Rod cells become hyperpolarized in response to light and transmit this by decreasing the levels of neurotransmitter released (see Figure 45.16).

Invertebrates use rhodopsin in their visual systems.

- Light-sensitive photocups containing rhodopsin act as simple eyes in flatworms (see Figure 45.18). They are used to orient away from light sources.
- Arthropods have compound eyes with 800 to 10,000 ommatidia per eye. The ommatidia contain light-sensitive photoreceptors called retinula cells (see Figure 45.19). The inner border of the retinula cells contains rhodopsin containing microvilli. The microvilli of adjoining retinula cells overlap to form a central rod, the rhabdom.

Components of the image-forming eyes of vertebrates

- The vertebrate eye is surrounded by the sclera, formed from connective tissue (see Figure 45.21). The cornea is the transparent sclera through which light passes.
- The iris controls the amount of light entering the eye through the pupil.
- The lens focuses images onto the retina at the back of the eye by changing from a spherical to a flattened shape. We focus on near and far objects by contracting and relaxing ciliary muscles that round up or flatten the lens to bring things into focus (see Figure 45.21).
- The center of the retina is an area with the highest density of photoreceptor cells, known as the fovea. A blind spot with no photoreceptors in the retina is located where the blood vessels and optic nerve exit the back of the eye.
- The human retina contains both rods (which are more light-sensitive) and cones (which absorb light of various wavelengths, allowing for color vision). Cones have different opsin molecules which absorb blue, green, yellow, or red and are the only cells found in the fovea (see Figure 45.23).

The vertebrate eye focuses light onto photoreceptors.

- There are five layers of neurons in the retina, with the photosensor rods and cones in the last layer (see Figure 45.24).
- The first layer of cells consists of ganglion cells (which create the action potential) and the axons of the optic nerve. Bipolar cells are stimulated by neurotransmitters from the photoreceptors to transmit the signal from the photoreceptor to the ganglion cells by the release of a neurotransmitter.
- The two other layers are the horizontal cells (which connect adjoining groups of photoreceptors and bipolar cells) and amacrine cells (which connect adjoining groups of bipolar cells and ganglion cells).

- One ganglion cell receives and integrates the information from many groups of photoreceptors that make up a circular receptive field with a center and a surround (see Figure 45.25). On-center receptive fields excite ganglion cells and are connected by the bipolar cells. The off-center receptive fields inhibit ganglion cells and connect though horizontal cells. Receptive fields of ganglion cells can overlap.

Some animals can sense another world.

- Humans can see light from only a narrow region of the electromagnetic radiation of the environment.
- Insects are sensitive to ultraviolet light, allowing them to see ultraviolet light patterns on many flowers.
- Some snakes are sensitive to infrared radiation, detecting heat differences in the environment.
- Bats and marine mammals use echoes of sound to visualize the environment.
- The lateral lines of some fishes are able to pick up electrical fields that help locate prey.

The Big Picture

- Sensory structures work by converting some form of stimulus—mechanical, chemical, light, sound—into action potentials in the nervous system, which are then interpreted by the central nervous system as a perceived sense.
- Receptors are named on the basis of their sensitivity. Chemoreceptors respond to chemical stimulation, mechanoreceptors respond to mechanical stimulation, auditory receptors respond to sound, and visual receptors respond to light.
- Different animals have different sensitivities of senses, as well as different types of senses.

Common Problem Areas

- The action potentials produced in the ear, eye, knee, or stomach are identical. What makes action potentials coming from the eye, for example, to be interpreted as light is the region of the brain that receives and analyzes these action potentials.
- Many of the more complexly structured receptors have both neural and non-neural components. The non-neural components (e.g., the ear pinnae) help channel or otherwise alter or filter the stimulus that will arrive at the neural component of the receptor.

Study Strategies

- The travel of sound in the ear can be very confusing. View the cochlea in the uncoiled form as in Figure 45.13. This will help you visualize how pressure waves of different wavelengths produce different sounds.
- All of the senses have what appear to be very different mechanisms for the transmission of information to the brain. To help sort all of this out, remember that there are only a few types of receptors that respond to stimuli and that they all generate action potentials. The steps in sensory transduction are also similar between all the different sensory systems.
- Review the following tutorial and activities on the website:
  Tutorial 45.1 Receptive Fields in the Retina
  Activity 45.1 Structure of the Retina
  Activity 45.2 Structure of the Eye
  Activity 45.3 Sound Transduction

Test Yourself

Knowledge and Synthesis Questions

1. An electrode is inserted into a chemosensory nerve leading away from a taste bud in the mouth of a dog. A mild acid solution is then flushed continuously over the sensors associated with this nerve. Initially the nerve responds to this stimulation but over time ceases to carry action potentials. This observation would best be explained by
   a. translocation.
   b. adaptation of the sensory cells.
   c. depletion of neurotransmitter in the sensory nerve.
   d. second messenger influences that increase cell membrane potentials.

2. There are five different types of sensory cells that contribute to the sensory system. Which of the following statements about sensory cells is false?
   a. Mechanosensors open ion channels in response to a pressure change.
   b. Electrosensors control ion channels in response to changes in electric charge.
   c. Thermosensors involve an intracellular messenger cascade.
   d. Photoreceptors exhibit a conformational change when stimulated by light.

3. Silkworms use chemosensory molecules for mate attraction that are know as
   a. rhodopsin.
   b. hormones.
   c. pheromones.
   d. G proteins.

4. Stretch receptors in the aorta and carotid artery sense changes in arterial pressure. These sensors are
   a. chemosensors.
   b. thermosensors.
   c. electrosensors.
   d. mechanosensors.

5. All but which one of the following mechanosensor systems employs hair cells as its transducer?
   a. Meissner's corpuscle
   b. Statocyst
   c. Organ of Corti
   d. Vestibular apparatus

6. Which of the following statements about human gustation is false?
   a. Taste sensors are relatively short-lived cells because of the high degree of abrasion they encounter.
   b. Lemon juice is tasted most clearly on the sides of the tongue.
   c. Receptors on the microvilli of taste sensors bind to food molecules.
   d. Humans perceive only three categories of tastes: sweet, sour, and bitter.

7. Which of the following statements about the photosensitive molecule rhodopsin is *false*?
   a. Opsin is converted from the 11-*cis* to the all-*trans* form upon light-induction.
   b. The retinal is the light-absorbing group.
   c. Photoexcited rhodopsin binds with transducin.
   d. None of the above.

8. Which of the following cells in the human visual system send information directly to the brain?
   a. Amacrine cells
   b. Bipolar cells
   c. Ganglion cells
   d. Rods and cones

9. Through which of the following cell layers must a photon of light pass before striking a cone cell in the eye of a hawk?
   a. Amacrine
   b. Bipolar
   c. Ganglion
   d. All of the above

10. Which of the following statements about animal vision is *false*?
    a. Photoreceptors in simple invertebrates produce image vision.
    b. There are several types of cones in vertebrate retinas, each of which detects different wavelengths.
    c. An ommatidium is only one visual element within the compound eye of an arthropod.
    d. Although retinal actually absorbs the photon, the wavelength it absorbs best is determined by opsin.

11. Which of the following is true about stimulation of an off-center receptive field?
    a. Stimulating the surround inhibits it.
    b. Stimulating the surround excites it.
    c. Stimulating the center excites it.
    d. a and c

12. Which of the following is not a membrane found in the cochlea?
    a. Reissner's membrane
    b. Tectorial membrane
    c. Tympanic membrane
    d. Basilar membrane

13. Which of the following about generator potentials is *false*?
    a. They create an action potential by opening sodium voltage-gated channels.
    b. The receptor potential of a sensory cell becomes a generator potential at the axon hillock.
    c. Generator potentials help transmit sensory stimuli.
    d. A generator potential is created by neurotransmitter release at the axon hillock.

Application Questions

1. This past winter was an awful year for colds. You had a cold for a number of weeks, during which time your sense of smell was diminished. What was the cause of your loss of smell?
2. After a loud rock concert your hearing appears to be dampened. What portion of the ear has been altered to dampen your hearing?
3. You have just given a presentation in your biology class that had many elaborate red- and green-colored slides. Afterward, a male friend tells you that he could not see any of the differences you were reporting. Why could your friend not see the differences?
4. Label each of the following structures on the diagram of the human eye shown below: sclera, cornea, iris, pupil, lens, retina, fovea, and blind spot.

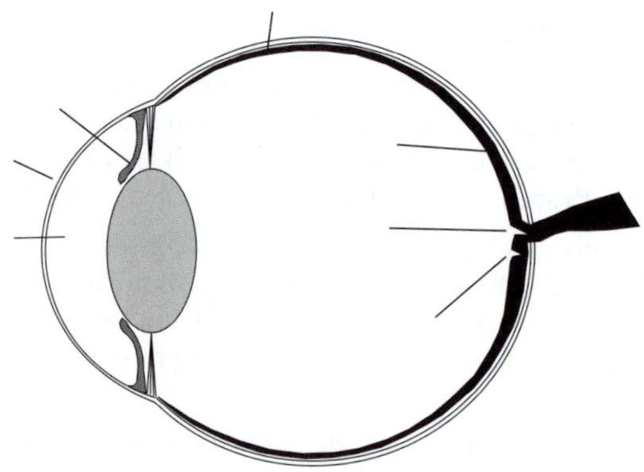

Answers

Knowledge and Synthesis Answers

1. **b.** When a sensor cell is stimulated by an unchanging, steady-state stimulus, it will become adapted. This allows the sensory system to ignore the unchanging stimulus while still being able to respond to new information.

2. **c.** Thermosensors involve a membrane enzyme that controls an ion channel.

3. **c.** Pheromones are chemical molecules that a female releases into the environment. The male uses chemoreceptors to follow the pheromone to the source, the female.

4. **d.** The stretch receptors of the aorta which detect changes in pressure are examples of mechanosensors.

5. **a.** The Meissner's corpuscle of the skin does not use hair cells to sense a stimulus. The cell membranes of the Meissner's corpuscle deform in response to light touching of the skin.

6. **d.** Humans can actually perceive four tastes: sweet, salty, sour, and bitter. The combination of taste and smell help to provide the complex subtle flavors of the food we eat.

7. **a.** Rhodopsin contains two groups: the protein opsin and the light-sensitive group retinal. Retinal is converted from the 11-*cis* to the all-*trans* form upon light-induction, not opsin. Opsin does change conformation in response to a change in the rhodopsin to signal the detection of light.
8. **c.** The ganglion cells transmit information from the bipolar cells to the brain. The axons of the ganglion cells connect with the optic nerve.
9. **d.** The photoreceptive cells are located at the back of the retina. Light must pass through a layer of ganglion cells, a layer of amacrine and bipolar cells, and a horizontal cell layer.
10. **a.** The photosensors in simple invertebrates such as the flatworm only detect light-dark contrasts.
11. **b.** The stimulation of the surround of an off-center receptive field excites it, while the stimulation of the center will inhibit it. The reverse is true of an on-center receptive field.
12. **c.** Although the tympanic membrane is found in the human ear, it is not found in the cochlea. The tympanic membrane is the membrane that transmits sounds from the auditory canal and the middle ear.
13. **d.** The receptor potential is transformed to a generator potential at the axon hillock of the sensory cell.

Application Answers

1. Your sense of smell depends on olfactory cilia receptors binding with odorant molecules. This triggers an action potential that is sent to the olfactory bulb. The olfactory cilia line the surface of the nasal epithelium. Usually there is a thin layer of protective mucus covering this epithelium. However, when you have a cold the production of mucus increases. This increased mucus covers the epithelium and the olfactory cilia, making it more difficult for odorant molecules to penetrate the mucus and reach the cilia. Thus you are not receiving as much stimulation and your sense of smell is decreased.
2. Sound is transmitted into the ear and into the fluid-filled cochlea. Sounds that are too loud will eventually damage the hair cells of the organ of Corti. Since these hair cells transmit the sound to the organ of Corti, any damage will decrease the response of the organ of Corti to sounds. This damage to the hair cells is cumulative and permanent.
3. Your friend is red-green color blind. The cones in our eyes allow us to see color. We have cones for red, green, and blue. Your friend either has red and green cones that do not function properly, or they are missing altogether.
4.

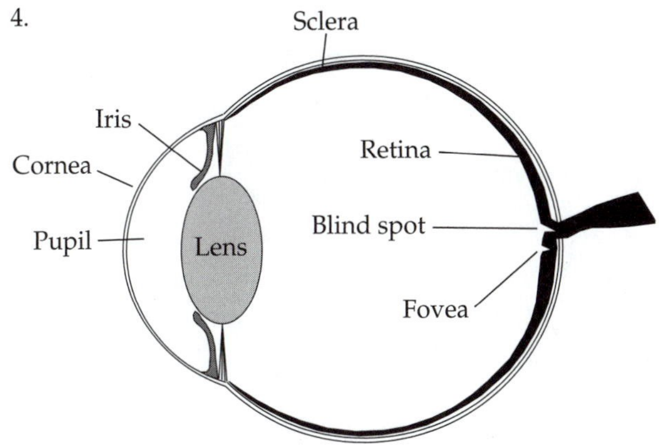

46 The Mammalian Nervous System: Structure and Higher Functions

Important Concepts

The mammalian nervous system consists of the central nervous system and the peripheral nervous system.

- The central nervous system (CNS) comprises the brain and the spinal cord.
- The nerves of the peripheral nervous system transport information between every tissue in the body and the CNS.
- A nerve is a bundle of axons that carries information in both directions between the body and the CNS (see Figure 46.1).
- The peripheral nervous system is composed of afferent nerves that carry information such as hearing and vision to the CNS and efferent nerves that carry information from the CNS to muscles and glands (see Figure 46.2).

The CNS develops from the embryonic neural tube.

- The neural tube is a tube of neural tissue running down the length of a vertebrate embryo in its early developmental stages (see Figure 46.3). The anterior end of the neural tube will develop into the main parts of the brain—the hindbrain, the midbrain, and the forebrain—with the spinal cord developing from the rest of the neural tube.
- The hindbrain is made up of the medulla and pons, which help control physiological functions, and the cerebellum, which helps coordinate commands leaving the brain with the actual actions of the muscles.
- The midbrain receives and processes auditory and visual information.
- The forebrain consists of a central diencephalon region and a surrounding telencephalon region. The diencephalon contains the thalamus, a relay station for sensory information, and the hypothalamus, which is involved in maintenance of homeostasis. The cerebral hemispheres (cerebrum) make up the telencephalon.

The spinal cord is the main neural highway between the body and the brain.

- The spinal cord is a bidirectional neural pathway for information flow between the peripheral nervous system and the brain.
- The spinal cord also converts some afferent information from the peripheral nervous system into efferent information sent back to the peripheral nervous system in a process known as spinal reflex (see Figure 46.4).
- A monosynaptic reflex, such as the knee-jerk reflex, involves only an afferent neuron, an efferent neuron, and one synapse. Leg muscle stretch receptors trigger the sensory neuron to conduct action potentials to the spinal cord upon stretching. The action potential is passed through the synapse of the sensory and motor neuron located in the central gray matter of the spinal cord.
- Flexor muscles, which flex the limb, and extensor muscles which straighten the limb, control limb movement. The motor neurons that stimulate these muscles are antagonistic and polysynaptic, with each sensory neuron stimulating one motor neuron while inhibiting the other motor neuron at two distinct synapses.

The reticular system and limbic system interact with the forebrain.

- The reticular system contains groups of neurons forming a nucleus (see Figure 46.5). Such nuceli are located through the medulla, pons, and midbrain.
- Many of the connections involved in functional control of the body occur in the reticular system, along with the control of sleep and waking.
- The limbic system is involved in instincts and emotions (see Figure 46.6). A portion of the limbic system known as the hippocampus helps to transfer short-term memory to long-term memory.

The cerebrum is the largest portion of the mammalian brain.

- The cerebral hemispheres are covered by a convoluted cerebral cortex, a sheet of gray matter that processes sensory information and higher-order information in the association areas (see Figure 46.7). The convolutions are ridges known as gyri and valleys known as sulci. Underneath the gray matter is white matter containing axons that connect cell bodies in the cortex and other areas of the brain. Each cerebral hemisphere is broken down into four main regions: the temporal lobe, the frontal lobe, the occipital lobe, and the parietal lobe. Each region has an association area that is involved in various functions.
- The size and complexity of the cerebral cortex is larger and greater in animals that exhibit complex behaviors.

- The temporal lobe processes auditory information and the association areas are involved with recognition, identification, and naming. Damage to the temporal lobe results in a condition in which individuals are unable to identify a stimulus even though they can perceive it (see Figure 46.8). Integration of spoken language can also be diminished with damage to the temporal lobe.
- The frontal lobe contains the primary motor cortex made up of axons that project to muscles in the body. The primary motor cortex can be mapped according to the locations that control movements of various body parts (see Figure 46.9). The association areas of the frontal lobe involve planning.
- The central sulcus separates the parietal lobe from the frontal lobe. The primary somatosensory cortex is located in the parietal lobe. As with the motor cortex, the location of the body that is sensed by the somatosensory cortex can be mapped on the brain (see Figure 46.9). Contralateral neglect syndrome results from damage to the right parietal lobe and is characterized by the inability to detect stimuli from the left side of the body.
- Visual information is processed by the occipital lobe of the cerebrum. The association areas for the occipital lobe involve translation of the visual stimuli into language.

The sympathetic and parasympathetic systems make up the autonomic nervous system.

- The sympathetic and parasympathetic divisions of the autonomic nervous system have antagonistic effects on the organ systems they innervate and play a large role in maintenance of homeostasis.
- The sympathetic division is involved in the fight-or-flight response of increased heart rate, blood pressure, and cardiac output. The parasympathetic division slows down heart rate and decreases blood pressure and cardiac output.
- Preganglionic neurons of the autonomic efferent pathway have cell bodies in the brain stem or spinal cord. They use the neurotransmitter acetylcholine at synapses where the ganglion lies outside the CNS. The axon that runs from the ganglion to the target cells is known as the postganglionic neuron.
- Norepinephrine is the neurotransmitter in postganglionic neurons of the sympathetic system, while acetylcholine is the neurotransmitter in postganglionic neurons of the parasympathetic system.
- Both acetylcholine and norepinephrine stimulate the pacemaker cells in the heart to control heart rate.
- The parasympathetic system has preganglionic neurons that come from the brain stem and last segment of the spinal cord. In the sympathetic system they come from upper regions of the spinal cord (see Figure 46.11).

Visual images are constructed in the occipital cortex.

- Information from the retina is transferred by the optical nerve to the thalamus and then to the visual cortex located in the occipital cortex.
- The visual cortex is composed of cells with specific receptive fields associated with specific areas of the retina that respond to specific light patterns. Simple cells are stimulated by static bars of light with specific orientations (see Figure 46.12). Complex cells receive input from several simple cells that have receptive fields in different locations on the retina. The mental image of the world is determined by analyzing the edges of the patterns of light falling on the retina.
- The optic nerve from the two eyes joins at the optic chiasm where half of the axons from one eye cross over to the opposite side of the brain (see Figure 46.13). Binocular vision occurs because the overlap in the field of view for each eye is transmitted to the same location in the visual cortex, the binocular cells. Binocular cells receive overlapping visual information from both eyes and interpret the disparity between the overlapping information from the two eyes to produce a three-dimensional image.

Sleep involves complex cerebral function.

- An electroencephalogram (EEG) measures the electrical activity of neurons in the cerebral cortex. An electromyogram (EMG) records electrical activity of muscles.
- There are two main states of sleep in humans: rapid-eye movement (REM) sleep and non-REM sleep (see Figure 46.14).
- Non-REM sleep has four states progressing from state 1 to the restorative states 3 and 4. During non-REM sleep the neurons in the thalamus and cerebral cortex become hyperpolarized due to the opening of K^+ channels. During this hyperpolarization the cells fire off in synchronized burst due to Ca^{2+} channels deactivating, which produces a slow-wave pattern. Increases in adenosine due to depletion of brain energy reserves may also play a role in the hyperpolarization, with non-REM sleep restoring the energy reserve.
- During REM sleep dreams and nightmares occur along with complete paralysis of skeletal muscles. Neurons that were hyperpolarized during non-REM sleep return to waking levels, allowing information to be processed. Afferent and efferent pathways are inhibited during REM sleep.
- During a night of sleep, the brain cycles between REM sleep and non-REM sleep, with 80 percent of sleep being non-REM sleep.

Learning and memory involve long-lasting synaptic changes.

- Long-lasting synaptic changes must occur in order for learning to take place. Long-term potentiation (LTP) occurs when high-frequency electrical stimulation makes circuits of the hippocampus more sensitive to subsequent stimulation. Ion channels controlled by glutamate allow both Na^+ and Ca^{2+} ions to enter neurons. This acts to enhance the responsiveness of the neuron.
- Long-term depression (LTD) results when hippocampus circuits become less sensitive to low-level stimulation.
- Associative learning in animals involves linking two unrelated stimuli, as in the conditioned reflex of Pavlov's dogs.

The dogs were conditioned to associate eating with the ringing of a bell, so that even in the absence of food the mere ringing of a bell stimulated salivation.

- Immediate memories are very short-term vivid memories of what has just happened. Short-term memories last between 10 and 15 minutes and contain less information than long-term memories. Repetition enhances the transfer of short-term memory to long-term memory.
- Declarative memory is memory of people, places, events, and things. Procedural memory is memory of how to perform motor tasks.

The language center is located in the left cerebral hemisphere.

- In most people the ability to produce and interpret language occurs in the left hemisphere of the cerebrum.
- The two cerebral hemispheres are connected by the corpus callosum. This connection allows the two hemispheres to communicate. Severing the corpus callosum results in an individual being unable to express in language the knowledge that is in the right hemisphere.
- A number of areas have been located that are important for language (see Figure 46.17). The frontal lobe contains Broca's area, which is involved in speech. The Wernicke's area is located in the temporal lobe and involves sensory aspects of language. The angular gyrus is thought to be involved in integrating spoken and written language.
- Language involves the flow of information between the various regions mentioned above. Damage to any of the areas involved in language will break the chain of information and result in aphasia, the condition of not being able to use or understand written or spoken words.

The Big Picture

- The mammalian nervous system can be divided both anatomically and functionally. Anatomically, it consists grossly of the central and peripheral nervous system. Functionally, it consists of the sympathetic and parasympathetic nervous system with afferent nerves carrying information toward the central nervous system from the peripheral nervous system, and efferent nerves carrying information from the central to the peripheral nervous system.
- The brain is divided up both anatomically and functionally into many different regions, each responsible for the many vital actions of the brain. The "higher" brain centers of the cerebrum are responsible for conscious thought and deliberate (voluntary) movements. The "lower" brain centers such as the cerebellum, pons, and medulla regulate involuntary movements and are involved primarily in maintaining homeostasis throughout the body.

Common Problem Areas

- It is easy to confuse "afferent" and "efferent", and it is important in your understanding of the nervous system to differentiate them. Think of *a*fferent as *a*rriving, and *e*fferent as *e*xiting the reference point—usually the CNS.
- It is important to recognize that while the neurotransmitters acteylcholine and norepinephrine always have antagonistic effects on each other, there is no universal pattern as to which stimulates tissue and which inhibits tissue. For example, acetylcholine causes the smooth muscle in blood vessels in many regions of the body to relax, while it causes the smooth muscle in the stomach and intestines to contract.
- It is often difficult for students to appreciate that, when dealing with sensory input to the brain, specific regions of the body "map onto" specific regions of the cerebrum. Similarly, with regard to control of movements, specific regions of the brain "map onto" specific regions of the body. Moreover, the amount of brain matter devoted to a particular body region depends on the amount of muscle control and sensors contained in that body region. Thus, a relatively small area of the cerebral hemispheres is devoted to the upper leg (which has relatively limited ranges of movement and sensation), while the tongue, with its many sensory receptors and high degree of mobility, commands more of the tissue of the cerebrum.

Study Strategies

- There are several key terms in this chapter that occur repeatedly: parasympathetic, sympathetic, afferent, efferent, agonist, antagonist, preganglionic, postganglionic. Until you master this vocabulary, it will be difficult to put together a comprehensive picture of the nervous system. Create your own list of the crucial terms you see repeatedly, and make sure that you understand their meanings.
- Brain anatomy is complex. Start by dividing up the structures into forebrain, midbrain, and hindbrain regions. Then identify each of the subcomponents. Your understanding will be more complete if you learn the general functions of each brain section as you go along. That is, rather than learning the anatomy of the brain and then starting over to learn the functions, learn the two at the same time.
- Review the following tutorial and activities on the website:
  Tutorial 46.1 Information Processing in the Spinal Cord
  Activity 46.1 The Human Cerebrum
  Activity 46.2 Brain Structures
  Activity 46.3 Brain Language Areas

Test Yourself

Knowledge and Synthesis Questions

1. A man has damage to his brain that affects his ability to recognize the faces of people he knows. Where did the damage occur?
   a. Hypothalamus
   b. Temporal lobe
   c. Parietal lobe
   d. Frontal lobe

2. The secretion of hormones from an endocrine gland is most likely under the control of which of the following components of the nervous system?
   a. Autonomic
   b. Voluntary
   c. Dendritic
   d. Limbic

3. The motor cortex and somatosensory cortex of the cerebrum are
   a. mapped from the head region on the upper side of the cortex to the lower part of the body on the lower side of the cortex.
   b. areas of the brain where the most sensitive body areas are represented by the largest areas of the cortex.
   c. located in the parietal lobe.
   d. None of the above.

4. Which of the following statements about the sympathetic division of the autonomic nervous system is *false*?
   a. It increases heart rate.
   b. It relaxes the urinary bladder.
   c. It dilates blood vessels.
   d. It stimulates secretion by the sweat glands.

5. Which of the following is *not* part of the central nervous system?
   a. Brain stem
   b. Spinal gray matter
   c. Cerebellum
   d. Neuronal cell body of a sensory afferent

6. Observations of patients with aphasias indicate that
   a. only Broca's area is essential for normal language skills.
   b. language skills depend on proper neural flow between the temporal lobes and motor cortex.
   c. the right hemisphere dominates the production and use of language in humans.
   d. most people can control language functions from either cerebral hemisphere.

7. A friend wakes you from sleep, and you have the sensation of having just experienced a vivid dream. Which of the following would be *false*?
   a. Your hands and feet have been twitching slightly.
   b. Your eyes have been twitching.
   c. Most of your voluntary body muscles have been inactive.
   d. Your cerebral cortex was not as active as when you were awake.

8. If you want to find a complex cell in a cat's visual system, you need to look in
   a. Broca's area.
   b. the spinal cord.
   c. the occipital cortex.
   d. the reticular activating system.

9. Which one of the following statements about the peripheral nervous system is *false*?
   a. Afferent portions of the peripheral nervous system carry information to the CNS.
   b. Efferent portions of the peripheral nervous system carry information from the CNS.
   c. The peripheral nervous system communicates only with the circulatory and digestive systems.
   d. The efferent portion is broken down into voluntary and involuntary divisions.

10. Development of the vertebrate brain begins early in development. Which one of the following statements about the developing CNS is false?
    a. The CNS develops from a solid neural cylinder.
    b. The midbrain becomes part of the brain stem.
    c. The forebrain develops into both the diencephalon and the telencephalon.
    d. The hindbrain develops into the medulla, the pons, and the cerebellum.

11. In which of the following cortical lobes can association areas and long-term memory be found?
    a. Temporal
    b. Parietal
    c. Occipital
    d. All of the above

12. Which of the following animals would likely have the smallest ratio of telencephalon size to body size?
    a. Fishes
    b. Amphibians
    c. Mammals
    d. Reptiles

13. Which of the following statement about the knee-jerk reflex is *false*?
    a. It is a monosynaptic reflex.
    b. The leg extends in response to a hammer tap.
    c. Chemoreceptors sense the hammer tap.
    d. The afferent nerve travels from the receptor to the spinal cord.

14. The ridges of the cerebral cortex are called ___________ and the valleys on the cerebral cortex are the _________.
    a. gyri, interneuron
    b. gyri, sulci
    c. sulci, gyri
    d. interneuron, gyri

Application Questions

1. What are the differences between the brain functions involved in reading a written sentence out loud and repeating a sentence one has just heard?
2. You have been sitting up all night watching movies in the living room of your apartment and your roommate enters from a bedroom, apparently sleepwalking. What type of sleep is your roommate in at the moment, and how can you tell?
3. You have been eyeing a piece of hard candy in a dish, and finally decide to unwrap a piece and eat it. As you begin to suck on the candy, your salivary glands begin to secrete saliva. What parts of the nervous system are involved in the sequence of events?
4. Humans have the ability to see things in three dimensions. What allows us to have binocular vision?

Answers

Knowledge and Synthesis Answers

1. **b.** The temporal lobe is involved in recognition of people and objects. A person who has had damage to the temporal lobe will not be able to identify someone by the face, and must use other cues.
2. **a.** Secretion of hormones by endocrine glands is involuntary. The autonomic nervous system consists of efferent pathways that link the CNS with many physiological functions.
3. **a.** The motor and sensory axons for the head region are mapped on the lower side of the cortex and the lower part of the body maps on the upper side of the cortex.
4. **c.** The sympathetic division of the autonomic nervous system constricts blood vessels rather than dilating them.
5. **d.** The cell bodies of the sensory neurons are located in the periphery and send their axons to the CNS.
6. **b.** Speaking a written or heard word requires neural flow from Wernicke's area in the temporal lobe to Broca's area in the frontal lobe. Therefore, language skills depend on proper neural flow between the temporal lobes and the motor cortex.
7. **d.** During REM sleep the body is paralyzed except for twitching of muscles. During this stage of sleep the brain is as active as when you are awake.
8. c. The occipital cortex is the higher brain region controlling vision.
9. **c.** The peripheral nervous system is in contact with every tissue in the body sending and receiving information to the CNS.
10. **a.** The CNS develops from a hollow tube composed of neural tissues.
11. **d.** Association areas and long-term memory are present in all of these regions of the cerebral cortex.
12. **a.** Fishes have the most undeveloped telencephalon of the group listed, resulting in the smallest ratio.
13. **c.** The knee-jerk reflex is an example of a monosynaptic reflex. Stretch receptors sense the hammer tap on the tendon.
14. **b.** The ridges on the surface of the cerebral cortex are the gyri and the valleys are the sulci.

Application Answers

1. Both speaking written language and repeating heard language follow similar paths in the brain. The main difference has to do with the initial region of the brain perceiving the word. In reading a word, the area at the back of the cerebrum is used to visualize it. The spoken word uses an area of the cerebrum just behind the area used for speech. Once the word has been processed by the initial centers, the path used for speaking the word is the same. Wernicke's area is stimulated, followed by the Broca's area, and then the motor area.
2. Your roommate is sleepwalking during non-REM sleep. During REM sleep the skeletal muscles of the body become paralyzed and the sleeper is not able to move. You can tell that your roommate is not in REM sleep because you will not see the eye movement typical of this stage of sleep. So just guide your roommate back to bed and continue watching your movie.
3. The peripheral system contributes to both seeing the candy and the movements of the arms and legs that you used to pick it up. The parasympathetic branch of the autonomic nervous system stimulated salivation. The central nervous system was involved in the recognition and decision to unwrap and eat the candy.
4. The right side of your brain receives visual information from the left visual field and the left side receives information from the right visual field. In the visual cortex, the cells are organized in columns that alternate between receiving information from the right and left eyes. At the borders of the columns, the inputs from the right and left eyes overlap. The cells that receive the overlap are called binocular cells, and they interpret the disparity between what the two eyes sense. This provides a three-dimensional image.

47 Effectors: Making Animals Move

Important Concepts

Microtubules and microfilaments generate movement at the cellular level.

- Flagella and cilia have a core, the axoneme, made of nine pairs of microtubules that slide past one another resulting in bending of the flagella or cilia (see Figure 47.4). Movement of the axoneme translates into movement of the entire flagellum or cilium. The force generated by flagella or cilia is powered by the hydrolysis of ATP by the enzyme dynein. Dynein molecules are attached to the microtubules and form cross-links; upon ATP hydrolysis the dynein changes shape, pushing the microtubule pair. Dynein then detaches from the microtubule pair, changes back to the resting state, and may repeat the cycle for more movement.
- Microtubules are active in movement of organelles and in helping to change the shape of cells.
- Small multicellular animals use cilia to move through liquids. Cilia also move fluids over the body of the animals, helping to deliver nutrients and oxygen. Cilia move with two strokes, a fast power stroke and a slow recovery stroke (see Figure 47.3).
- Flagella provide movement of a body or a fluid over a body's surface.
- Actin is a microfilament that is involved, along with the protein myosin, in movement of muscles in vertebrates. Amoeboid movements are produced by the interaction of actin and myosin.

Smooth, skeletal, and cardiac muscle types are found in the human body.

- Actin and myosin are arranged in filaments in muscle cells.
- Smooth muscles are involuntary muscles comprised of long, simple cells, each with a single nucleus, and are controlled by the acetylcholine and norepinephrine of the autonomic nervous system (see Figure 47.6). When smooth muscle is stretched it contracts with strength proportional to the stretch of the muscle.
- Cardiac muscle is found in the heart and is composed of branched muscle cells that form a strong meshwork. Intercalated discs add additional strength by holding the cells together. Heartbeats originate at the pacemaker cardiac muscle cells and spread rapidly through gap junctions in the muscle. The heartbeat is said to be myogenic because it is determined by the pacemaker.
- Skeletal muscles are striated voluntary muscles made of large muscle fibers. Muscle fibers have multinuclei and are composed of groups of actin and myocin arranged into myofibrils.

Muscle contraction is described by the sliding filament theory.

- The contracting unit of the myofibrils is the sarcomere, which contains actin and myocin filaments and has very distinct repeating patterns (see Figure 47.7).
- In a sarcomere, the actin filaments are anchored by the Z lines and the myosin filaments are found at the center in the A band.
- In relaxed muscle, the H zone and I band are the regions where there is no overlap of actin and myosin. During muscle contraction, the Z lines move toward each other and the H zone and I band shrink in size due to the sliding of actin filaments along the myosin filaments.

The proteins myosin and actin are the key to muscle contraction.

- Myosin molecules are made of two polypeptide chains wrapped around each other, each with a globular head at one end, much like two twisted golf clubs. A myosin filament is composed of many myosin molecules (see Figure 47.8).
- Actin filaments are composed of two monomer chains in a helical arrangement, appearing as two linear strings of pearls wrapped around each other. The proteins tropomyosin and troponin are associated with actin (See Figure 47.8).
- The myosin heads change confirmation when they bind to actin filaments at myosin binding sites, forming a cross-bridge connection. The conformational change in the myosin pulls the actin in toward the middle of the sarcomere. ATP then binds to an ATP binding site on myosin, resulting in the bound actin's release from the myosin and the myosin's return to the original conformation. Many myosin molecules cycle through binding with actin to shorten the sarcomere.

Calcium ions regulate myosin and actin filament movements.

- In relaxed muscle, tropomyosin and troponin cover the myosin binding sites on actin filaments, inhibiting muscle contraction. Calcium regulates contraction by binding with troponin, causing the tropomyosin to change conformation and expose the actin–myosin binding sites on the actin filaments (see Figure 47.10).
- Action potentials form at motor neuron synapses with hundreds of muscle fibers all contracting at the same time. All the fibers activated by a single motor neuron constitute a motor unit. Action potentials spread deep into the cytoplasm or sarcoplasm of the muscle through transverse tubules or T tubules that are in contact with sarcoplasmic reticulum throughout the sarcoplasm (see Figure 47.9). Sarcoplasmic reticulum take up and release Ca^{2+} ions into the sarcoplasm, thereby controlling relaxation and contraction of the myofibrils.
- In smooth muscle, contraction is controlled by a calmodulin–Ca^{2+} complex. This complex activates myosin kinase, an enzyme that phosphorylates the myosin head to cause contraction. Myosin phosphatase works in the opposite direction by dephosphorylating myosin.

Skeletal muscles show graded contractions, ranging from twitches to tetanany.

- An action potential in a skeletal muscle fiber causes twitch or contraction of the muscle. Twitches can occur as discrete contractions, or if they occur frequently enough, the twitches are summed together (see Figure 47.11).
- Maximum muscle tension or tetanus occurs when there is a high rate of stimulation by action potentials. Temporal summation occurs when there is fast twitching of individual fibers, and spatial summation occurs when more than one motor unit stimulates contraction.
- During tetanic contraction, actin and myosin bonds cycle to help keep the muscle fiber from stretching. ATP levels control the length of a tetanic contraction because it is the molecule that provides the energy for myosin to break the bond with actin.
- Slow-twitch muscle fibers are highly resistant to fatigue because they have lots of myoglobin, mitochondria, and blood vessels, giving them the name "red muscle" (see Figure 47.12).
- Fast-twitch muscle fibers fatigue quickly but rapidly develop maximum tension because they have lower levels of mitochondria, myoglobin, and blood vessels, and are called "white muscle" (see Figure 47.12).

Invertebrates use hydrostatic skeletons and exoskeletons for support and movement.

- Many soft-bodied invertebrates have a fluid-filled body cavity that acts as a hydrostatic skeleton. Earthworms have circular muscles and longitudinal muscles that oppose each other to act upon the hydrostatic skeleton to control elongation and shortening of body segments (see Figure 47.13).
- Arthropods have an exoskeleton or hard outer shell composed of chitin that offers protection and provides sites for muscle attachment. Each flexible joint has muscles that extend the body parts (extensor muscles), and muscles that flex body parts (flexor muscles) (see Figure 47.14).
- As an animal with an exoskeleton grows, it undergoes the process of molting. With each molt the old exoskeleton is shed, revealing a new one that has developed underneath.

The vertebrate endoskeleton provides a frame for support and movement.

- Endoskeletons are growing, living tissue that provide sites for muscle attachment and support for the body.
- The human skeleton is composed of a central axial skeleton and an appendicular skeleton (see Figure 47.15).
- The pliable portions of the endoskeleton, such as the framework of the nose, and the surface of the joints of the endoskeleton are composed of cartilage containing the protein collagen.
- The bone in the endoskeleton is strong and solid and is composed mainly of collagen and calcium phosphate. Bone is constantly being reworked by two types of cells, osteoblasts which lay down new bone and osteoclasts which break down bone (see Figure 47.16).
- Developing bone is created as membranous bone growing on a scaffolding of connective tissue, or as cartilage bone hardening from an initial cartilage state. Compact bone is solid, while cancellous bone is lightweight with many cavities. Compact bone is known as Haversian bone because it is composed of concentric rings with blood vessels running through the middle known as Haversian systems (see Figure 47.18).

Movable bones join at joints.

- There are six types of joints where bones meet, allowing for a variety of bone movement (see Figure 47.20).
- The muscles attached to bones at joints work antagonistically. The flexor muscles bend the joints while the extensor muscles straighten the joints.
- Two types of connective tissues hold joints and bones together. Ligaments hold bone to bone, and tendons hold muscle to bone.

Effectors can be cells other than muscle cells.

- Nematocysts are specialized cells found in hydras, jellyfish, and cnidarians that are fired at prey either to entangle or poison the prey.
- Chromatophores are specialized cells that contain pigment and allow an animal to change color.
- Glands also act as effector organs.
- Electric organs in some animals are able to produce large electrical currents.

The Big Picture

- Actin and myosin are the "universal" proteins for motion. Whether they are located in a cilium, in a unicellular animal or the leg muscle of a human, the molecular interactions of actin and myosin produce movement in the structures in which they reside.

- The sliding filament hypothesis provides a model for how actin and myosin filaments slide past each other, resulting in the shortening (contraction) or lengthening (relaxation) of muscle cells. The whole process of actin and myosin interaction is tightly regulated by the movement of calcium ions into and out of the intracellular spaces of muscle cells, which in turn is activated by the arrival of action potentials in motor neurons. Muscle contraction requires the use of energy in the form of ATP.
- Muscles act in concert with skeletons that are either external (invertebrates) or internal (vertebrates). In vertebrates, bones are articulated, forming joints that provide for specialized directional movements of the tissues supported by the bones (e.g., limbs). Muscles controlling joint movement are often located in pairs acting antagonistically, with one set of muscles causing bending or flexion of the joint and the other causing straightening or extension of the joint.

Common Problem Areas

- The sarcomere is the functional unit of the muscle cell, and until you understand its fine structure—Z lines, H lines, etc.—it will be difficult for you to appreciate how the sarcomere shortens through the actions of actin and myosin.
- The way that movements of multiple muscle groups cause both flexion and extension of a joint can be confusing. Thinking of joints in terms of levers and pulleys may help you understand their actions.

Study Strategies

- The sliding filament hypothesis can hardly be called a hypothesis anymore, with so much supporting evidence having accumulated. The interactions of myosin, actin, troponin, troponomyosin, and calcium, and the role of action potentials in stimulating muscle contraction make up a complex, multistep process.
  - First, break the process down into the constituents and learn their location and general structure.
  - Second, determine how actin and myosin move relative to each other through a series of power strokes.
  - Finally, ascertain how calcium ions released from the sarcomeres by action potentials initiate and maintain the whole process of muscle contraction.
- Review the following tutorial and activities on the website:
  Tutorial 47.1 Molecular Mechanism of Muscle Contraction
  Activity 47.1 The Neuromuscular Junction
  Activity 47.2 The Structure of a Sarcomere
  Activity 47.3 Joints

Test Yourself

Knowledge and Synthesis Questions

1. Both cilia and flagella are used for movement in many organisms. What protein uses the energy from ATP hydrolysis to cause the cilia or flagella to move?
   a. Calmodulin
   b. Dynein
   c. Axoneme
   d. None of the above

2. A motor unit is best described as:
   a. all the nerve fibers and muscle fibers in a single muscle bundle.
   b. one muscle fiber and its single nerve fiber.
   c. a single motor neuron and all the muscle fibers that it innervates.
   d. the neuron that provides the CNS with information on the state of contraction of the muscle.

3. Which of the following statements is *false*?
   a. Cardiac muscle is striated.
   b. Smooth muscle does not contain actin.
   c. Skeletal muscle is considered voluntary.
   d. Smooth muscle is found in the iris of the eye and the walls of the bladder.

4. The compound acting as an oxygen store in skeletal muscle is
   a. myoglobin.
   b. hemoglobin.
   c. ATP.
   d. myokinase.

5. The action potential that triggers a muscle contraction travels deep within the muscle cell by means of
   a. sarcoplasmic reticulum.
   b. transverse tubules.
   c. synapses.
   d. motor end plates.

6. A sarcomere is best described as a
   a. moveable structural unit within a myofibril bounded by H zones.
   b. fixed structural unit within a myofibril bounded by Z lines.
   c. fixed structural unit within a myofibril bounded by A bands.
   d. moveable structural unit within a myofibril bounded by Z lines.

7. ATP provides the energy for muscle contraction by allowing for
   a. an action potential formation in the muscle cell.
   b. cross-bridge detachment of myosin from actin.
   c. cross-bridge attachment of myosin to actin.
   d. release of calcium by sarcoplasmic reticulum.

8. Ca^{2+} binds to _______ in skeletal muscle and leads to exposure of the binding site for _______ on the ______ filament.
   a. troponin; myosin; actin
   b. troponin; actin; myosin
   c. actin; myosin; troponin
   d. tropomyosin; myosin; actin

9. Tropmyosin is moved by which of the following proteins?
   a. Calmodulin
   b. Acetylcholine
   c. Actin
   d. Troponin

10. Summation of frequent muscle twitches to give a smooth sustained contraction is called
   a. motor unit summation.
   b. treppe.
   c. facilitation.
   d. tetanus.

11. _______ are responsible for regulating extracellular calcium balance by releasing or sequestering Ca^{2+} ions in the bone matrix.
   a. Osteoblasts
   b. Osteoblasts and osteoclasts
   c. Osteoclasts and osteocytes
   d. Osteoblasts, osteoclasts, and osteocytes

12. The primary difference between an endoskeleton and an exoskeleton has to do with
   a. the presence of both circular and longitudinal muscles.
   b. whether or not the skeleton is on the inside of the body.
   c. the presence or absence of joints.
   d. the amount of fluid in the body.

13. A soccer player suffers a knee injury that damages the tissue holding his upper and lower leg bones together. The damaged tissue is probably a kind of
   a. muscle.
   b. tendon.
   c. ligament.
   d. cartilage.

Application Questions

1. Smooth muscle contracts involuntarily, while skeletal muscle contraction is under voluntary control. What is the difference between the mechanism that controls smooth muscle and skeletal muscle contractions?
2. White muscle and red muscle are found in different parts of the body and are used for different types of movement. What are the physiological and morphological characteristics that separate the two types of muscle?
3. Smooth muscle contracts involuntarily in the digestive tract, blood vessels, and urinary bladder. Describe the two main ways that smooth muscle contraction and the membrane potential of smooth muscle are controlled?
4. Earthworms have a hydrostatic skeleton. Describe how the hydrostatic skeleton is used to propel them through the soil.

Answers

Knowledge and Synthesis Answers

1. **b.** The core of cilia and flagella is composed of units known as axonemes. The protein dynein uses the energy from ATP to slide microtubules by each other, causing the cilia and flagella to move.
2. **c.** A motor unit is a single motor neuron and all the muscle fibers that it innervates.
3. **b.** Although contraction of smooth muscle is controlled differently from that of skeletal muscle, smooth muscle does contain actin and myosin.
4. **a.** Myoglobin is the main oxygen-carrying molecule in muscle.
5. **b.** The action potential arriving to the muscle travels into the muscle through the transverse tubules.
6. **d.** A sarcomere is a structural unit within a myofibril bounded by Z lines that contain actin and myosin.
7. **b.** ATP provides energy that is used to detach myosin from actin.
8. **a.** Calcium is released from the sarcoplasmic reticulum and binds with troponin, resulting in the myosin binding site on actin to be exposed.
9. **d.** The binding of calcium with troponin causes a conformational change in tropomyosin.
10. **d.** Tetanus is the summation of twitches for a constant muscle contraction.
11. **d.** Osteoblasts, osteoclasts, and osteocytes are all involved in extracellular calcium balance.
12. **b.** Endoskeletons (such as those of mammals) are found inside the body, and exoskeletons (such as those of insects) are found outside the body.
13. **c.** Ligaments hold bones together.

Application Answers

1. Both smooth muscle and skeletal muscle contraction are regulated by the presence of Ca^{2+}. In smooth muscle, Ca^{2+} joins with the protein calmodulin to activate myosin kinase in the sarcoplasm. Myosin kinase then phosphorylates the myosin head, allowing the myosin head to bind with actin. In skeletal muscle Ca^{2+} binds with troponin on the actin filaments. This binding of Ca^{2+} causes a conformational change in the tropomysin and uncovers the myosin binding sites on the actin, allowing myosin to bind with actin.
2. Fast-twitch fibers are known as white muscle, which has few mitochondria, small amounts of myoglobin, and few blood vessels. Slow-twitch fibers are known as red muscle. Red muscle has lots of mitochondria, large amounts of myoglobin, and many blood vessels.
3. Smooth muscle is sensitive to stretching. It will depolarize in response to any stretching and contract, with the strength of the contraction proportional to the amount of stretch in the muscle. Parasympathetic inputs of acetylcholine also cause depolarization of the muscle membrane resulting in contraction, while norepinephrine hyperpolarizes the muscle membranes.
4. The hydrostatic skeleton of the earthworm is an incompressible fluid-filled cavity surrounded by longitudinal and circular muscles. Contraction of the longitudinal muscles causes the segments to contract and the body to shorten. Contraction of the circular muscles causes the body segments to elongate and the body to lengthen. Alternating contractions between the longitudinal and the circular muscles moves the animal in a push and pull manner. Bristles on the body help hold the animal in place after elongation, and the body is pulled forward during shortening.

48 Gas Exchange in Animals

Important Concepts

Respiratory gas exchange occurs by diffusion.

- The respiratory gas oxygen (O_2) is required by cells to produce energy in the form of ATP.
- Carbon dioxide (CO_2) is one of the waste by-products of ATP production and must be eliminated from an animal's body.
- The transfer of these gases occurs by simple diffusion in the respiratory systems of animals.

Several physical parameters affect respiratory gas exchange.

- There are major differences in the O_2 capacity of air and water.
- Water, because of its density, is more expensive to move than air. Water also contains far less O_2 than air, and O_2 also diffuses far more slowly in water then in air.
- Temperature affects respiration of aquatic animals because there is less O_2 in warm water than in cold and an animal's need for oxygen increases with temperature (see Figure 48.2).
- The tendency for a gas to move by diffusion across gills, skin, or lungs depends on its partial pressure. The sum of all the gases' partial pressures in a gas mixture is the total partial pressure, which in the atmosphere equals the atmospheric pressure.
- Oxygen makes up 20.9 percent of the atmospheric pressure. As elevation increases and atmospheric pressure decreases, the total amount of oxygen in air decreases (see Figure 48.2).
- CO_2 diffuses across the respiratory organs in both air and water. In air, diffusion is rapid because of the large partial pressure gradient between the blood and the atmosphere. In aerated water, CO_2 diffuses easily, but if the water is stagnant a large partial pressure gradient may not exist and CO_2 diffusion will be slower.

Fick's law of diffusion describes the exchange of gas in the respiratory system.

- The partial pressure gradient is the difference between the partial pressure of a gas at two locations.
- The partial pressure gradients of O_2 and CO_2 drive the movement of O_2 into the body and CO_2 out of the body.
- Diffusion, described by Fick's law of diffusion, depends on a diffusion coefficient that varies according to temperature, the medium, the diffusing molecules, the total surface area over which diffusion occurs, and the distance of diffusion between the two locations.

Animals have adaptations to maximize gas exchange.

- Animals maximize respiration by decreasing the distance that gases must diffuse between the blood and the external environment. They achieve this with very thin respiratory surface membranes.
- Animals also actively move the environmental respiratory fluid over the gills and through the lungs, and perfuse the internal side of the respiratory organ to carry the respiratory gases.
- Some aquatic amphibians and insects have external gills with large surface areas for exchange of respiratory gases with water (see Figure 48.4).
- Internal gills have a large surface area and are protected from the environment by an animal's body cavity. They must be actively ventilated with water to achieve gas exchange.

Diverse types of respiratory organs have evolved in animals.

- Insects have tracheae—air tubes that end at the tissue cells as air capillaries and open to the environment through spiracles. Gases diffuse through the tracheae into the air capillaries, but are also moved by movement of the animal's body parts (see Figure 48.5).
- Some aquatic insects carry a bubble of air with them underwater. As they use the O_2 in the air bubble, the PO_2 in the bubble decreases resulting in a PO_2 gradient between the water and the air bubble. Oxygen diffuses into the air bubble from the water down the PO_2 gradient, restoring oxygen in the bubble.
- Fish have internal gills with a large surface area that they ventilate with unidirectional flowing water to maximize the external PO_2 levels. Each gill has hundreds of gill filaments with folds, or lamellae, that act as the respiratory gas exchange surfaces. Countercurrent flow of blood in the lamellae and water over the lamellae maximize the PO_2 gradient between the water and the blood (see Figures 48.6 and 48.7). As blood flows through the lamellae it is always in contact with water that has a higher oxygen

level, resulting in a continuous oxygen gradient to maximize the uptake of oxygen into the blood.

- Birds create a unidirectional flow of air through air sacs and lungs, the site of gas exchange, which make up the avian respiratory system. Air requires two cycles of inhalation and exhalation to move from the posterior air sac, to the lung, to the anterior air sac, and out of the respiratory tract. Gas exchange occurs in the parabronchi and air capillaries of the lung (see Figure 48.9).

The lung is the respiratory organ in mammals.

- Mammalian lungs have a very large surface area and a very short gas diffusion pathway.
- Air enters at the oral cavity or nasal passage, travels down the pharynx through the larynx, and then through the trachea, which branch into two bronchi leading to each lung. In the lung, more branching occurs to produce bronchioles and finally the site of gas exchange—small air sacs, the alveoli (see Figure 48.12).
- Oxygen diffuses across the thin walled alveoli (less than 2 μm) into many surrounding capillaries.
- The alveoli have a surface tension due to an aqueous layer that makes it difficult to inflate the alveoli. This is overcome by a surfactant produced by the cells forming the alveolar walls. The aqueous layer is needed to ensure the diffusion of gases across the membrane.
- Mucus, produced along the lung's larger airways, catches and removes inhaled dust and microorganisms. Cilia along the airways move the mucus up the airways and into the throat, acting as a mucus escalator.

Mammalian lungs are ventilated by tidal ventilation.

- The tidal breathing of mammals can be described by a series of lung volumes (see Figure 48.11). The normal volume of air moved during one cycle is known as the tidal volume. The total lung capacity is made up of the resting tidal volume, plus the additional volume of air that can be taken in with a large breath and the additional volume of air that can be force out with a large exhale. Not all of the air can be forced out of the lungs. Air remains in the bronchi and trachea, making up the dead space.
- Lungs are inflated by negative pressure created by contraction of the diaphragm muscle in the thoracic cavity. Closed pleural membranes line the thoracic cavity and allow for the development of negative pressure. During inhalation, external intercostal muscles on the ribs pull up the ribs and help expand the pleural cavities. During exhalation, internal intercostal muscles pull the ribs down (see Figure 48.13).

Oxygen is transported in the blood by hemoglobin in vertebrates.

- Gases diffuse between the respiratory organ and the circulatory system, where they are transported in blood. The liquid blood plasma transports only a small amount of dissolved O_2, with the vast majority of O_2 transported by the oxygen-binding pigment hemoglobin found in the red blood cells of vertebrates.
- Hemoglobin contains two alpha subunits and two beta subunits, each with a heme group that reversibly binds O_2. The amount of O_2 bound to hemoglobin depends on the partial pressure of O_2. At the high blood PO_2, characteristic of blood leaving the lungs, hemoglobin is almost 100 percent saturated with O_2. At the tissues, the PO_2 is lower and hemoglobin releases approximately 25 percent of its bound O_2. The relationship between the PO_2 of the environment and the percentage of hemoglobin bound with oxygen is known as the S-shaped oxygen dissociation curve (see Figure 48.14). At low levels of PO_2, only one of the hemoglobin subunits can bind O_2. Once it binds a single O_2 molecule, the conformation of the hemoglobin changes, making it easier for the other three sites on the hemoglobin to bind oxygen in a process known as positive cooperativity.
- Fetal hemoglobin has a different structure from the adult form, with a higher affinity for O_2.
- The Bohr effect is the shifting of the hemoglobin binding curve to the right, in response to a decrease in pH of the blood, resulting in more oxygen being released at a given environmental PO_2 at the lower pH (see Figure 48.16).
- Hemoglobin's affinity for O_2 is lowered by 2,3 diphosphoglyceric acid (DPG), which binds with hemoglobin to cause a conformational change in the hemoglobin, resulting in a right shift in the oxygen-binding curve.
- In muscle, myoglobin is the O_2 carrying molecule, and functions primarily to store O_2 intracellularly. The O_2 affinity is higher in myoglobin than in hemoglobin, but myoglobin has only one unit to bind one O_2 molecule.

Carbon dioxide is transported as both dissolved CO_2 and bicarbonate ions.

- Tissues produce CO_2 as a by-product of metabolism. It must be moved from the tissues and excreted into the environment.
- Significant amounts of CO_2 dissolve in the plasma and are carried in this form to the lungs.
- Most of the CO_2 is transported as bicarbonate ions. CO_2 reacts with water to make bicarbonate ions and a proton. Carbonic anhydrase speeds up the conversion of CO_2 to carbonic acid which then dissociates into bicarbonate ions. CO_2 can also bind with hemoglobin to form carboxyhemoglobin. The reverse reaction occurs at the lungs (see Figure 48.17).

Breathing is controlled by the brain and blood pH levels.

- Breathing is controlled by the autonomic nervous system from neurons in the medulla.
- Receptors on the medulla are sensitive to CO_2 levels mediated by changes in pH in the blood. They increase breathing when CO_2 levels increase and pH decreases.
- Control of breathing is relatively insensitive to blood O_2 levels, but chemoreceptors in the arteries leaving the heart can signal the medulla in response to low levels of O_2.

The Big Picture

- Cellular metabolism requires O_2 and produces CO_2 as a waste product that must be eliminated. Animals have evolved a variety of structures and mechanisms for exchanging these gases with the environment. Tracheal systems in insects, gills in fishes, and lungs in terrestrial vertebrates are three examples of gas-exchange organs. All are characterized by very short diffusion distances and very large surface areas—adaptations designed to maximize gas exchange.
- Oxygen is transported by the protein hemoglobin in vertebrates. Hemoglobin–oxygen binding is determined by the PO_2 levels surrounding hemoglobin. If PO_2 is high, hemoglobin accepts O_2, but it readily gives up its O_2 when the PO_2 falls. These properties allow hemoglobin to bind O_2 in the gas-exchange organ, transport it to the metabolizing tissues, and then release its oxygen for consumption in cellular metabolism.

Common Problem Areas

- It is easy to think, incorrectly, that the exchange of O_2 and CO_2 is a "two way street," with O_2 taking the immediate place of CO_2 in the lungs and CO_2 then taking the place of O_2 in the tissues. In fact, the mechanisms for exchange of these two gases are completely different.
- Understanding how O_2 is picked up, transported by, and released from hemoglobin can be quite confusing. Think of the oxygen dissociation curve as a "tool." Think first of the situation at the top of the curve where PO_2 is high (the lungs or gills, for example). If the PO_2 is high, hemoglobin will maximally bind oxygen. Think then of transporting that blood to a region with a given PO_2, and read from the curve what must happen to O_2 saturation under the new PO_2 level. The O_2 given from hemoglobin is now available for tissue respiration.

Study Strategies

- Use Fick's law of diffusion as a guideline for understanding gas exchange. The various components of the law—distance for diffusion, partial pressure gradient, surface area over which gas exchange occurs—provide a good framework for understanding why gas-exchange organs have evolved with certain characteristics in common. For example, gas-exchange organs tend to have large surface areas, very short diffusion distances, and to experience large partial gradients for O_2 across their surfaces. All of this makes sense in the context of Fick's law of diffusion. Fick's law of diffusion also governs the movement of O_2 and CO_2 in the body, and it can help you remember where and why these gases are picked up and dumped in the body.
- Review the following activities and tutorials on the website:
  Tutorial 48.1 Airflow in Avians
  Tutorial 48.2 Airflow in Mammals
  Activity 48.1 Oxygen-Binding Curves
  Activity 48.2 Concept Matching
  Activity 48.3 Human Repiratory System

Test Yourself

Knowledge and Synthesis Questions

1. Which of the following is *not* a reason that breathing air is easier than breathing water?
   a. A given volume of air is easier to move across the respiratory organs than the same volume of water.
   b. Air holds more oxygen per unit volume than water.
   c. Water breathers have a difficult time ridding themselves of CO_2 because CO_2 does not dissolve well in water.
   d. Temperature increases affect the O_2 content of water more than they do that of air.

2. External gills, tracheae, and lungs all share which of the following sets of characteristics?
   a. Part of gas-exchange system; exchange both CO_2 and O_2; increase surface area for diffusion
   b. Used by water breathers; based on countercurrent exchange; use negative pressure breathing
   c. Exchange only O_2; are associated with a circulatory system; found in vertebrates
   d. Found in insects; employ positive-pressure pumping; based on crosscurrent flow

3. Which of the following represents a larger volume of air than is normally found in the resting tidal volume of a human lung?
   a. Residual volume
   b. Inspiratory reserve volume
   c. Expiratory reserve volume
   d. All of the above

4. Which of the following is shared by bird and mammal lungs?
   a. Both need two cycles of inhalation and exhalation for air to move through the lungs.
   b. Both contain alveoli as the terminal end of the lung.
   c. Both have an anatomical dead space.
   d. Both exchange O_2 and CO_2 with blood in capillaries.

5. According to Fick's law of diffusion, what does not play a role in the diffusion of oxygen across a membrane?
   a. Surface area
   b. Volume
   c. Difference in concentration or partial pressure
   d. Diffusion distance

6. Because of the relatively high altitude of Antonito, Colorado, the town has a normal barometric pressure of about 600 mm Hg rather than 760 mm Hg as at sea level. The partial pressure of oxygen in Antonito's air is approximately
   a. 75 mm Hg.
   b. 126 mm Hg.
   c. 160 mm Hg.
   d. 760 mm Hg.

7. Gas flows into the lungs of mammals during inspiration because
   a. the pressure in the lungs falls below atmospheric pressure.
   b. the volume of the lungs decreases.

c. the pressure in the lungs rises above atmospheric pressure.
d. the diaphragm moves upward toward the lungs.

8. The movement of O_2 and CO_2 between the blood in the tissue capillaries and the cells in tissues depends most directly upon
 a. active transport of O_2 and CO_2.
 b. total atmospheric (barometric) pressure differences across the cell membranes.
 c. diffusion of O_2 and CO_2 down a concentration gradient.
 d. diffusion of O_2 and CO_2 down a partial pressure gradient.

9. The alveoli of the lungs do not contain "air" because
 a. we normally do not ventilate our lungs at a high enough rate.
 b. the lungs have too many alveoli to ventilate.
 c. there is "dead space" in the trachea and bronchi.
 d. the trachea and bronchi are too small in volume.

10. Which of the statements about hemoglobin is false?
 a. Hemoglobin allows the blood to carry a large amount of O_2.
 b. Hemoglogin contains a single polypeptide chain with very high affinity for O_2.
 c. Hemoglogin is packaged inside red blood cells
 d. Fetal hemoglobin is structurally different from adult hemoglobin.

11. As blood becomes fully O_2-saturated, hemoglobin is combining with _________ molecule(s) of oxygen.
 a. 1
 b. 2
 c. 4
 d. 8

12. The Bohr shift describes
 a. the outward movement of Cl^- from the blood cell in exchange for HCO_3^- moving into the cell.
 b. the leftward shift of the entire oxygen equilibrium curve when temperature rises.
 c. the rightward shift of the entire oxygen equilibrium curve when pH rises.
 d. the rightward shift of the entire oxygen equilibrium curve when pH falls.

13. The presence of CO_2 in blood will lower pH because CO_2 combines with _________, with the rate of reaction increased by _________.
 a. H_2O to form H+ and HCO_3^-, carbonic anhydrase
 b. H_2O to form only HCO_3^-, carbonic anhydrase
 c. H_2O to form only H^+, carbonic ions
 d. H^+ to form HCO_3^-, oxyhemoglobin

14. The largest proportion of CO_2 carried by the blood is in the form of
 a. bicarbonate ions (HCO_3^-) carried in the plasma.
 b. molecular CO_2 dissolved in the plasma.
 c. bicarbonate ions (HCO_3^-) carried within the red blood cells.
 d. molecular CO_2 chemically bound to hemoglobin.

15. Which of the following is not involved in the neural control of ventilation?
 a. Neurons in the medulla
 b. The vagus nerve
 c. The contraction state of the diaphragm
 d. Chemosensors on the surface of the medulla

Application Questions

1. You have just captured in a dredge an aquatic animal that has no specialized respiratory surfaces. Describe how the organism can get enough oxygen to all of its cells to survive.
2. The gills of fish employ a countercurrent exchange mechanism to insure that a large amount of oxygen is extracted from the water. Describe how the countercurrent exchange works.
3. What is the difference between the myoglobin oxygen-binding curve and the hemoglobin oxygen-binding curve?

Answers

Knowledge and Synthesis Answers

1. **c.** Water breathing is more difficult than air breathing because the higher density of water makes it more expensive to move across the respiratory surfaces than air, it has less oxygen than air, and the oxygen content is very dependent on the temperature of the water. CO_2 however, dissolves easily into water and is easy to get rid of.
2. **a.** The external gill, tracheae, and lungs are all examples of gas-exchange systems used by various animals to exchange both CO_2 and O_2. One of the special properties of all respiratory systems is that they increase surface area for diffusion.
3. **d.** All of the listed choices have a larger volume than the resting tidal volume.
4. **d.** The lungs of birds and mammals are both the sites for the exchange O_2 and CO_2 with blood in capillaries. The lungs of birds take two cycles for air to move through them, while the mammalian lung only takes one cycle. The mammalian lung is the only one to end in alveoli and have an anatomical dead space.
5. **b.** The volume of the gas is not included in Fick's law of diffusion. All of the other parameters are important in determining the diffusion of gases in the respiratory system of animals.
6. **b.** Air at sea level with an atmospheric pressure of 700 mm Hg has a partial pressure of oxygen of 160 mm Hg (700×21). Therefore, the partial pressure of O_2 at Antonito, Colorado, is $600 \times 21 = 126$ mm Hg.
7. **a.** Inspiration in the mammalian lung occurs by means of negative pressure produced by contraction of the diaphragm. Thus the pressure in the lungs falls below atmospheric pressure.

8. **d.** Movement of O_2 and CO_2 from the blood to the tissues always occurs by diffusion of O_2 and CO_2 down their partial pressure gradients.
9. **c.** The alveoli do not contain "air" with 20.9 percent O_2 because incoming air is mixed with air left in the "dead space" of the trachea and bronchi that has had some of the O_2 removed by the lungs.
10. **b.** Hemoglobin is composed of four subunits which act with positive cooperativity to bind O_2. Oxygen-binding myoglobin found in muscle has only one unit.
11. **c.** Hemoglobin has 4 subunits, each of which binds to 1 molecule of O_2 for a total of 4 molecules of O_2 bound to 1 hemoglobin molecule.
12. **d.** The Bohr effect describes the action of pH on the oxygen-binding curve. Decreases in pH result in a net right shift of the entire oxygen equilibrium curve.
13. **a.** Carbon dioxide combines with H_2O in the plasma to form H^+ and HCO_3^-. The enzyme carbonic anhydrase catalyzes the reaction.
14. **a.** The majority of CO_2 is carried as bicarbonate ions (HCO_3^-) in the plasma.
15. **d.** The contraction state of the diaphragm does not influence the neural control of ventilation.

Application Answers

1. Having no respiratory surfaces, the animal must rely on simple diffusion from the environment into the cells. In order for simple diffusion to be an effective means of oxygen transport, the animal must be very small, with all of its cells only a few cell layers from the environment. The metabolic rate of this animal is most likely very low in order to accommodate the lack of any special respiratory surfaces.
2. In the countercurrent exchange in the gills of fish, water and blood move in opposite directions. This results in blood with low oxygen content entering the gill and coming into water that has a higher oxygen content. Oxygen diffuses down the partial pressure gradient into the blood. As blood moves along the lamella, it picks up oxygen, but it always has a lower oxygen content than the water moving in the opposite direction. In this way the blood is able to optimize oxygen uptake during the entire trip through the lamella.
3. The oxygen-binding curve of myoglobin is shifted to the left in relationship to the oxygen-binding curve of hemoglobin. This allows myoglobin to pick up O_2 and hold it at lower partial pressures of O_2.

STUDY GUIDE

49 Circulatory Systems

Important Concepts

Small organisms do not need circulatory systems.

- An animal's circulatory system transports nutrients and wastes to and from tissues and cells.
- In the smallest multicellular animals, every cell within the body is no more than one to two cells from the environment. Such small animals do not require a circulatory system because materials can move efficiently into and out of cells by diffusion alone.
- Some aquatic invertebrates have a gastrovascular cavity in which exchange of gases and nutrients occurs (see Figure 49.1). Other organisms are flattened to increase the surface-to-volume ratio, ensuring that cells are close to the outside.
- Large animals need a circulatory system to carry wastes, gases, and nutrients to and from all the cells in the body.

Circulatory systems may be classified structurally as open or closed.

- In an open circulatory system, blood and other tissue fluids are not separated (see Figure 49.2). A pump may be present to help move body fluids through the various large compartments, assisted by movements of the animal.
- In a closed circulatory system, blood is pumped through a series of interconnected closed vessels (see Figure 49.3). There are numerous advantages to closed systems compared to open ones: rapid transport of fluid in the closed vessels, the ability to divert the flow of blood between different tissues, and direct hormone and nutrient transport to the tissues.
- The closed system typical of vertebrates is composed of arteries and arterioles which carry blood from the heart to the body, veins and venules which carry blood from the body to the heart, and an intervening capillary bed. In birds and mammals, the systemic circuit flows from the heart to the body tissues and the pulmonary circuit flows from the heart to the gas exchange organ.

The gills (gas exchanger) and systemic tissues play an important role in the fish circulatory system.

- Fishes pump blood with a heart with one atrium and one ventricle. Blood leaves the heart and enters the gills where gas exchange occurs. Some arterial blood pressure is lost in passing through the gills.
- Blood leaving the gills collects in the aorta and travels to the systemic organs and tissues, eventually returning to the heart.
- African lungfish have a well vascularized lung for gas exchange that is formed embryonically from an outpocketing of the gut. A heart with a partially divided atrium helps to separate oxygenated blood returning from the lung and deoxygenated blood returning from the tissues as they enter the single ventricle. A large proportion of oxygenated blood bypasses the gills and goes directly to the tissues.

Amphibians have partial separation of pulmonary and systemic blood.

- Amphibians have two atria, one receiving deoxygenated blood from the tissues and one receiving oxygenated blood from the lungs.
- The outflow tract of the ventricle has a long spiral valve that helps directs deoxygenated blood preferentially to the lungs and skin (which also acts as a gas exchange organ in amphibians) and oxygenated blood to the systemic body tissues. Even though the two circuits are not fully divided, blood tends to flow in parallel.

Reptiles control their intracardiac shunt to regulate the distribution of oxygenated and deoxygenated blood leaving the heart.

- Turtles, snakes, and lizards have two atria and a ventricle that is internally divided into subchambers that allow for control of the distribution of blood to the lungs and body.
- During air breathing, blood from the right side of the ventricle is preferentially pumped to the lungs because there is a higher resistance in the systemic circuit than in the pulmonary circuit. Also, there is an asynchronous ejection from the heart as pumping occurs first from the right and then left side of the ventricle. By the stage in the cardiac cycle that the blood from the left side leaves the heart, the resistance is higher in the pulmonary circuit. Thus the last blood leaving the heart tends to be deoxygenated blood and is preferentially directed to the pulmonary circulation.
- When the reptile stops breathing, vessel constriction in the lungs causes pulmonary circuit resistance to increase, causing blood from the right side to bypass the pulmonary circulation and flow to the systemic circuit.

- Alligators and crocodiles are unique in that they have a totally divided atrium and ventricle, resembling that of a bird or mammal. However, alligators and crocodiles retain a left aorta that arises from the right side of the heart. There is a small passageway between the left and right aorta through which blood can bypass the circulation. Thus, crocodiles and alligators can operate their heart like that of a mammal, distributing blood equally to the pulmonary and systemic circulations, or they can partially bypass the lungs by directing blood from the left aorta into the right aorta and on to the systemic circulation.

Birds and mammals have a completely parallel circulatory system.

- Birds and mammals have a four-chambered heart that allows for complete separation and no mixing of systemic and pulmonary blood.
- This cardiovascular arrangement maximizes O_2 transport by eliminating mixing between pulmonary and systemic circulations.
- The pulmonary circulation operates at low pressure, protecting the delicate lung membranes, while the systemic circuit operates at high pressures, allowing the perfusion of tissues and organs distant from the heart.

The human heart contains separate left and right sides pumping to the lungs and tissues, respectively.

- The human heart has two atria and two ventricles (see Figure 49.4). The atriventricular valve between the atrium and ventricle stops backflow of blood from the ventricle into the atrium. The pulmonary and aortic valves reside between the ventricles and arteries, and help maintain one-way flow through the heart.
- Deoxygenated blood flows from the body through the superior and inferior vena cava into the right atrium and into the right ventricle by passive filling during heart relaxation. During heart contraction, blood flows from the right ventricle to the pulmonary artery and the lungs, where blood becomes oxygenated.
- Oxygenated blood from the lungs flows back to the heart and into the left atrium, then into the left ventricle, and finally out through the aorta to the systemic tissues.
- Resistance in the blood vessels of the body's systemic tissues is higher than in the lungs, so the left ventricle must be larger and more muscular than the right ventricle in order to generate more force while pumping the same volume of blood.
- Systole is the contraction of the ventricle and diastole is the relaxation of the ventricle and contraction of the atria. Together these two phases make up the cardiac cycle.
- During the cardiac cycle the pressure is highest during ventricle contraction (systole) and lowest during relaxation (diastole). see Figure 49.5.

The human heartbeat is generated in the heart's pacemaker.

- The pacemaker, or sinoatrial node, is a small section of highly modified muscle tissue located at the junction of the superior vena cava and right atrium.
- The pacemaker rhythmically produces action potentials that pass on to the rest of the heart and stimulate a heartbeat (see Figure 49.8).
- Gap junctions connect cardiac muscle cells though which the action potentials spread rapidly throughout the atrium fibers, providing a coordinated atrial contraction.
- The action potential moves from the atrium to the ventricles through the atrioventricular node (AV node). The AV node slows down the transmission of the action potential from the atrium to the ventricle. The action potential passes rapidly down the bundle of His and to the Purkinje fibers fanning out in the ventricular tissue. From the Purkinje fibers the action potential spreads through both ventricles, causing contraction to occur slightly after the contraction of the atrium.
- The autonomic nervous system also plays a role in the control of the heartbeat. Sympathetic activity increases heart rate and parasympathetic activity decreases heart rate (see Figure 49.7).
- The electrocardiogram (EKG) measures the electrical activity of the heart as a complex waveform (Figure 49.9). The wave has five parts: P (depolarization and contraction of the atrium), Q, R, and S (depolarization of the ventricles), and T (relaxation of the ventricles).
- The strengthening of a heartbeat is due to the Frank–Starling law which states that stretching of cardiac cells due to increased blood return to the ventricle results in increase contraction force.

The vascular system is composed of arteries, capillaries, and veins.

- Arteries carry blood away from the heart to the body under relatively high pressure. Veins carry blood toward the heart from the body under low pressure (see Figure 49.10). The very fine capillaries interconnect the arteries with the veins in the tissues, dissipate pressure, and are the site for all exchanges of gases, nutrients, and wastes.
- Arteries and the smaller arterioles are elastic, allowing them to stretch during systole and rebound during diastole. In vertebrates the walls of vessels contain smooth muscle that controls the vessel's resistance by effecting changes in diameter.
- Blood returns to the heart at low pressure in expandable veins. Because of the low pressure, veins that act against gravity have valves, and blood flowing toward the heart is helped by the movement of the surrounding skeletal muscle (see Figure 49.13). Gravity can cause pooling of blood in the veins if a person is inactive for a long period of time.

Capillaries are the site of gas, nutrient, and transport exchange, as well as the site of considerable fluid loss and recovery.

- Capillaries connect the arteries with the veins in the tissues. Capillaries have a large cross-sectional area compared with arteries and veins, and have extremely thin walls to allow the diffusion of gases, nutrients, and wastes (see Figure 49.10).

- As blood enters the capillaries it is moving slowly. Blood pressure from the arterial side produces a hydrostatic pressure that forces fluids, ions, and small molecules out of the capillaries through small holes called fenestrations in a process known as filtration (see Figure 49.12). Near the venule end of the capillaries, the difference in osmotic potential between the plasma inside the capillaries and the surrounding fluids creates an osmotic pressure resulting in fluid recovery back into the capillaries. Thus the net flow of water between the plasma and tissue fluids is determined by the difference in hydrostatic and osmotic pressure.
- Some capillaries are selective about which ions and molecules can pass through them, while others are less selective.

The lymphatic system is a system of vessels with one-way valves that returns tissue fluid not reclaimed by the capillaries to the blood.

- In humans, lymph is moved through the lymphatic vessels up the body by surrounding skeletal muscle contractions. Lymph is returned to the blood through the thoracic duct at the superior vena cava.
- Lymph nodes exist in mammals and birds and contribute to the defenses of the body. Foreign materials and microorganisms are removed by the phagocytic action of the lymph nodes.

Cardiovascular disease affects the arteries and veins.

- Hardening of the arteries is known as atherosclerosis.
- Plaque deposits begin to form at sites of endothelial damage and lipids tend to accumulate on the deposits. Connective tissue invades the deposits and makes the wall of the artery less elastic and narrow. This can lead to the formation of a blood clot, or thrombus.
- The coronary arteries that supply the heart with blood are highly susceptible to blockage, resulting in a heart attack.
- An embolus is a piece of a thrombus that breaks off, travels in the vessels, and then lodges in a vessel. When this happens in the brain it is known as a stroke and results in denial of blood flow to a particular region of the brain, causing paralysis, loss of cognitive abilities, or death.

Blood is a tissue composed of a fluid plasma and many different cell types.

- The fluid matrix of blood is known as plasma and contains water, salts, and proteins. The hematocrit is the percentage of blood volume composed of cells.
- A number of different cell types in blood carry out a wide range of functions (see Figure 49.15). The cell types include the erythrocytes, basophils, eosinophils, neutrophils, lymphocytes, monocytes, and platelets.
- Erythrocytes, or red blood cells, are shaped like discs with high surface areas carrying hemoglobin and the oxygen in hemoglobin from the lungs to the tissues.
- Erythrocytes are produced in stem cells located in the marrow of bone. Once they are released into the blood stream they survive for about 120 days. The hormone erythropoietin is produced in the kidneys and regulates the production of red blood cells (see Figure 49.16).
- The spleen serves as a reservoir for red blood cells.
- Platelets that help to start blood clotting are also found in blood and are produced by the breaking off of cells from megakaryocytes in the bone marrow. Platelets bind to the edges of a break in a vessel wall, not only providing a mechanical plug but also starting a chemical chain reaction that leads to the conversion of the circulating protein prothrombrin into thrombin. Thrombin stimulates the formation of sticky fibrin threads which entrap red blood cells and other fibrin threads into a meshwork that covers the hole in the vessel (see Figure 49.17).
- Plasma contains many gases, ions, nutrients, proteins, and other molecules. Albumin in the plasma helps to maintain the osmotic pressure in the capillaries. The plasma also contains the hormones that are circulated by the blood to target cells.

The circulatory system is under hormonal, neural, and local control.

- Blood flow in capillary beds in many tissues is controlled locally by autoregulatory mechanisms. Precapillary sphincters at the arterial end of the capillaries control the flow of blood, with contraction of these sphincters limiting or stopping blood flow through the capillary bed (see Figure 49.18). Low levels of O_2 and high levels of CO_2 can open the precapillary sphincters.
- Norepinephrine from sympathetic neurons and epinephrine released from the adrenal medulla cause arteries and arterioles to contract, reducing blood flow and increasing central blood pressure.
- Autonomic control of heart rate and constriction of blood vessels occurs from the medulla. Stretch receptors in the aorta and carotid arteries help regulate blood pressure. With rising blood pressure they inhibit sympathetic output and cause the heart to slow and arterioles to dilate. When pressure is low, sympathetic output is stimulated, increasing heart rate and constriction of arterioles.
- The hypothalamus also helps in control by releasing vasopressin which stimulates arterioles to constrict, leading to increased blood pressure.

The Big Picture

- Cardiovascular systems are needed when animals grow too big to acquire nutrients and eliminate waste by diffusion alone. Animals have evolved a wide variety of cardiovascular systems, which can be classified generally as open or closed. Open systems have relatively low pressures and have large sinuses that bath tissues and organs in circulating fluid. Closed systems have an interconnected series of vessels that keeps blood separated from tissue but allows gases, nutrients, and wastes to diffuse through their walls.
- The evolution of vertebrates has seen the heart change from a simple two-chambered heart in fish to the complex four-chambered heart of birds and mammals. However,

the hearts of amphibians and reptiles do provide for complex intracardiac shunting and bypass of blood flow to the lungs, which is an adaptation that allows them to stop breathing or breath intermittently.

- The human heart is really two pumps in one, with a left pump providing oxygenated blood under high pressure to the systemic tissues, and a right pump sending deoxygenated blood under low pressure to the lungs. A cardiac pacemaker sets the cardiac rhythm.
- The flow of blood through arteries, capillaries, and veins is highly regulated, either at the local level in capillaries because of changes in their local surroundings, or at the level of arteries and veins by the combined action of neural activity and circulating hormones.

Common Problem Areas

- It is frequently thought that the hearts of amphibians and reptiles are primitive structures, awaiting "repair through evolution" to the more highly evolved bird and mammal heart. In fact, the hearts of amphibians and reptiles are highly adapted to suit their own particular lifestyle, particularly their intermittent breathing and relatively low metabolic rate. By being able to shunt blood within the heart, these animals can save the energy that would otherwise be wasted in sending blood to the temporarily nonventilated lungs.
- Students often fail to appreciate that the human heart is really two completely distinct pumps—the right heart pumping blood to the lungs, and the left heart pumping blood to the body. The two parts just happen to be packaged in the same structure—the heart. In fact, cardiologists refer to the "left heart" and "right heart," emphasizing the separate nature of these two pumps. Figuring out the flow of blood through the human heart becomes easier when it is considered from this perspective.

Study Strategies

- Determining the pattern of blood flow through the human heart will be easier if you mentally "unfold" the circulation into a great circle, in which there are two pumps at opposite sides of the circle with a vascular bed intervening between each pump. Once you have mastered this concept, you will see that the only way to place the two pumps in proximity in the same structure (the heart) is to twist the circle into a folded figure eight—the typical diagram of the circulation in the textbook. It is as much a problem in typography as anatomy!
- Following from the above, learn the various chambers and valves in the context of a sequence of structures in each of the right and left pumps of the heart, rather than as a list of terms printed on a complex, anatomically accurate diagram of the heart.
- Refer to the following tutorials and activities on the website:
  Tutorial 49.1 The Cardiac Cycle
  Activity 49.1 Vertebrate Circulatory System
  Activity 49.2 Structure of a Blood Vessel
  Activity 49.3 Heart Structure

Test Yourself

Knowledge and Synthesis Questions

1. Which of the following is not one of the reasons that closed circulatory systems are more efficient than open circulatory systems?
   a. Closed systems rely exclusively on simple diffusion for transport, whereas open systems rely on pumping mechanisms.
   b. Transport within closed systems is more rapid than in open systems.
   c. Blood can easily be directed to specific areas in closed systems, but not in open systems.
   d. Closed systems operate better under higher pressure than do open systems.

2. How does the pattern of blood flow change when reptiles undergo periods of intermittent breathing?
   a. Blood is shunted from the systemic circuit to the pulmonary circuit.
   b. Blood is shunted specifically to the brain.
   c. Blood is shunted from the pulmonary circuit to the systemic circuit.
   d. The pattern does not change during intermittent breathing.

3. In which of the following would you record the highest blood pressure?
   a. The ventricle supplying blood to the gills of a fish
   b. The anterior dorsal artery of an ant
   c. The pulmonary vein of a frog
   d. The ventricle supplying blood to the systemic circuit of a bird

4. Blood consists of a fluid fraction consisting of ____________ and a solid fraction consisting of ____________.
   a. plasma; water, erythrocytes, and platelets
   b. erythrocytes; leukocytes, macrophages, and platelets
   c. plasma; erythrocytes, platelets, and leukocytes
   d. leukocytes; erythrocytes and platelets

5. Red blood cells are
   a. biconcave cells containing hemoglobin.
   b. spherical cells containing hemoglobin.
   c. spherical cells capable of ameboid motion and containing hemoglobin.
   d. biconcave cells that contain platelets.

6. Blood clotting pathways cause
   a. conversion of vitamin K to prothrombin.
   b. conversion of fibrin to fibrinogen.
   c. conversion of thrombin to prothrombin.
   d. None of the above

7. Which of the following regions of the vascular bed is the actual site of gas exchange with surrounding tissue?
   a. Arteries
   b. Capillaries
   c. Lymphatic vessels
   d. Veins

8. Which is the correct sequence of parts through which cardiac action potentials pass?
   a. Purkinje fibers, AV node, SA node, bundle of His, atrial fibers
   b. AV node, atrial fibers, SA node, bundle of His, Purkinje fibers
   c. SA node, bundle of His, atrial fibers, AV node, Purkinje fibers
   d. SA node, atrial fibers, AV node, bundle of His, Purkinje fibers

9. The purpose of the AV node is to __________ and the purpose of the Purkinje fibers is to __________.
   a. create simultaneous atrial and ventricular depolarization; speed up transmission of the cardiac impulse into the ventricle
   b. delay ventricular depolarization relative to atrial depolarization; insulate the cardiac impulse from the general ventricular fibers
   c. delay ventricular depolarization relative to atrial depolarization; transmit the cardiac impulse to very small localized groups of ventricular fibers
   d. delay atrial depolarization relative to ventricular depolarization; transmit the cardiac impulse to very small localized groups of ventricular fibers

10. The atrial walls are __________ than the ventricular wall, and pressure generated in the atrial chambers is __________ than in the ventricles.
    a. thinner, higher
    b. thinner, lower
    c. thicker, higher
    d. thicker, lower

11. The left ventricle exceeds the right ventricle in
    a. the amount of blood that enters during heart contraction.
    b. the volume expelled during contraction.
    c. the pressure developed during contraction.
    d. All of the above

12. Which of the following structures of the lymphatic system acts primarily as a filter for detecting and destroying microorganisms in lymph traveling through major lymph vessels?
    a. Lymph nodes
    b. Thymus
    c. Lymph capillaries
    d. Tonsils, but not the appendix

13. The net loss of fluid from blood capillaries will increase if
    a. plasma filtration decreases.
    b. the osmotic pressure of plasma increases.
    c. blood (hydrostatic) pressure increases in the capillaries.
    d. the colloid osmotic pressure of interstitial fluid decreases.

14. Which of the following statements about the control of circulation in humans is false?
    a. Sympathetic nerve input to skeletal muscle causes the blood vessels in the muscle to dilate.
    b. Blood flow can be regulated by autonomic nerve signals emanating from cardiovascular control centers in the medulla of the brain.
    c. Carotid artery chemosensors detect low O_2 levels in the blood and promote increased blood pressure.
    d. Hormones such as angiotensin and vasopressin cause venules to constrict.

Application Questions

1. What are the important differences between the circulatory system of a fish and that of a reptile?
2. The human heart pumps blood to the pulmonary and the systemic circuit. Describe the path that blood takes as it moves through the circulatory system, starting at the left ventricle.
3. You are about to take a final exam in your freshman biology course and are very nervous. What responses will your circulatory system have to your nervousness?
4. Blood moves through the arteries and veins at a faster velocity than the blood moving through the capillaries. What is the cause and effect of blood slowing down as it moves through the capillaries?

Answers

Knowledge and Synthesis Answers

1. **a.** Both closed systems and open systems rely on pumping mechanisms to distribute fluid throughout the body.
2. **c.** During intermittent breathing, the blood is shunted from the pulmonary circuit to the systemic circuit. The resistance in the pulmonary circuit increases, resulting in the shunt.
3. **d.** Of the vertebrate groups, mammals and birds tend to have the highest blood pressures. Therefore, the ventricle supplying blood to the systemic circuit of a bird has the highest pressure.
4. **c.** The fluid portion of blood is the plasma; three components of the solid fraction are erythrocytes, platelets, and leukocytes.
5. **a.** Red blood cells are biconcave cells that contain the oxygen-binding hemoglobin.
6. **d.** Blood clotting involves none of the above. Prothrombrin is converted into thrombin and fibrinogen is converted into fibrin.
7. **b.** Gas exchange at the tissues occurs across the capillaries.
8. **d.** The cardiac action potential passes through the SA node, atrial fibers, AV node, bundle of His, and finally the Purkinje fibers.
9. **c.** The AV node delays the ventricular depolarization relative to atrial depolarization, so atrial contraction occurs before ventricular contraction. The Purkinje fibers transmit the cardiac impulse to very small localized groups of ventricular fibers.
10. **b.** The atrium has thinner walls and generates lower pressures than the ventricles.

11. **c.** The left ventricle generates a greater pressure in the blood flowing to the systemic circuit than the right ventricle with blood flowing to the pulmonary circuit.
12. **a.** The lymph nodes filter and destroy microorganisms that are traveling through the lymph system.
13. **c.** Changes in blood pressure will change the amount of filtration that occurs at the capillaries. Thus, increases in blood pressure will result in greater rates of filtration.
14. **d.** Angiotensin and vasopressin act to control constriction of the arterioles and not the venules.

Application Answers

1. Fishes have a two-chambered heart, with blood flowing from the heart of the fish to the gills and then directly to the body. Reptiles have a more complex three-chambered heart, with a subdivided ventricle that direct outflows to the pulmonary circuit and to the systemic circuit. Thus the circulatory system of fishes delivers blood to the respiratory organ and the body tissues in succession, while the reptile delivers blood to the respiratory organ and the body tissues in parallel.
2. Blood from the right ventricle exits through the pulmonary valve into the pulmonary artery and then goes to the lungs. From the lungs the blood returns to the left atrium and passes through the atrioventricular valve into the left ventricle. Blood from the left ventricle is pumped to the systemic circuit through the aortic valve and into the aorta. From the aorta, blood moves to the tissues in arteries and arterioles, passes through tissues in the capillaries, and returns to the right atrium of the heart in the venioles, veins, and the vena cava. From the right atrium blood is pumped back where we started, into the right ventricle through the atrioventricular valve.
3. Stress and nervousness elicit the fight-or-flight response. Signals from the medullary cardiovascular control center stimulate sympathetic inputs from the autonomic nervous system. In response to sympathetic inputs, the adrenal gland releases epinephrine into the blood stream. Epinephrine stimulates an increased heart rate and arterial pressure. The blood flow to the smooth muscles decreases due to constriction of blood vessels. By constriction of smooth muscles, blood is diverted away from areas such as the digestive tract to the skeletal muscle needed for fight or flight.
4. Blood is pumped from the heart and into the arteries, where it has a relatively fast velocity. Upon entering the capillaries much of the velocity is lost. The capillaries have a much higher total cross sectional area than the arteries. Blood flow velocity is inversely proportional to the cross-sectional area. The more area there is through which the blood flows, the slower it will go. This is similar to a stream that widens to become a river. Water in the upstream areas moves faster than the water in wider downstream areas. In the same way, the blood moves slowly through the capillaries and faster through the arteries and veins. The slow velocity of blood through the capillaries allows for greater exchange of materials with the tissues.

50 Animal Nutrition

Important Concepts

All animals must eat to acquire energy for survival.

- Animals are heterotrophs, meaning that they must get their energy and nutrients from the fats, carbohydrates, and proteins of other animals and plants. The energy that animals acquire from food is used to make ATP, a high-energy phosphate compound that provides energy for cellular work. The conversion of ATP during cellular work creates heat as a by-product.
- The calorie and kilocalorie (Calorie) are the measures of heat energy.
- The metabolic rate is a measure of the total energy used by an animal. Basal metabolic rate is the resting metabolic rate or the energy consumption needed for all essential physiological functions.
- Energy needed for metabolism can be obtained from food intake or from stored food. Energy is stored as carbohydrates in the liver and muscles, and as fat, which has the highest energy content.

Essential molecules must be obtained from a food source.

- Animals cannot produce acetyl groups through their own metabolism but rather must obtain them from another source (see Figure 50.4).
- Amino acids needed to make proteins are obtained by breaking down proteins in food. Some amino acids can be synthesized, but the essential amino acids can only be obtained from outside sources.
- Essential amino acids are needed for proper protein production. In humans these are isoleucine, leucine, lysine, methionine, phenylalanine, threonine, tryptophan, and valine.
- Ingested proteins are not used in their ingested form because they are too large to be absorbed as whole proteins. Their function is species specific, and they would be recognized as invaders and attacked. Instead, proteins are broken down into peptides and amino acids.
- Essential fatty acids, such as linoleic acid, must also be acquired in food and are necessary for producing membrane phospholipids.
- The nutrients needed for survival are categorized as either macronutrients or micronutrients depending on the quantity needed (see Table 50.1).
- Vitamins are used as coenzymes and are required in small amounts. Fat-soluble vitamins accumulate in the liver, while water-soluble vitamins are eliminated in urine. Vitamin D is needed to help in the uptake of calcium into the body.

Diseases can be caused by nutrient deficiency.

- Protein deficiency in humans can result in kwashiorkor which leads to swelling, distension of the abdomen, degeneration of the liver, and other problems.
- Shortages in vitamins can lead to a large number of diseases (see Table 50.2). A deficiency in thiamin or vitamin B_1 results in beriberi which causes loss of appetite and fatigue.
- Deficiencies may be due to the body's inability to absorb essential nutrients even when they are present in the diet. Anemia is caused by the inability to absorb vitamin B_{12} due to inadequate production of the intrinsic factor peptide needed for absorption.

There are a number of different ways for animals to feed.

- Two type of organisms feed on dead matter: decomposers or saprotrophs absorb nutrients from decaying organic matter, while detritivores actively feed on dead matter.
- Herbivores eat plants that are difficult to break down. They have teeth that are good for shearing, crushing, tearing, and grinding to help break down this vegetation.
- Carnivorous animals eat other animals that they capture. Their teeth are adapted for stabbing, ripping, and shredding their prey (see Figure 50.9*b*). Omnivorous animals eat both plant material and other animals and have teeth adapted for multipurposes.
- Teeth are composed of a hard outer surface (enamel), a bony layer (dentine), and a pulp cavity which contains blood vessels and nerves (see Figure 50.9a).

Digestion occurs in either a cavity or a tube in most animals.

- Food is generally digested extracellularly by enzymes that break it down. Most enzymes are produced as an inactive form, a zymogen, and are then converted to the active form at the site of use. Enzymes do not digest the gut because of a protective layer of mucus that lines it.

There are a number of classes of digestive enzymes:
- carbohydrases hydrolyze carbohydrates;
- proteases hydrolyze proteins;
- peptidases hydrolyze peptides;
- lipases hydrolyze fats;
- nucleases hydrolyze nucleic acids.

- In simple animals a single opening that functions as both a mouth and anus connects a simple gastrovascular cavity to the environment.
- The tubular gut includes the mouth where food enters and is broken up in some animals by a grinding mechanism such as teeth or mandibles. Food then goes to a chamber for digestion and storage, such as the stomach, crop, or gizzard. Food then enters the midgut or intestine where most of the nutrients are absorbed. The hindgut absorbs water and ions and stores waste until it is expelled during defecation from the anus. The interior, or lumen, of the gut can be regarded as part of the outside of the body.
- The parts of the gut that absorb nutrients have increased surface area in the form of fingerlike projections known as villi (see Figure 50.11).

The gut wall is composed of the same layers of tissue types throughout its length.

- Four layers of different cell types form the wall of the gut (see Figure 50.13).
- The first layer is a layer of mucosa that surrounds the cavity of the gut, or lumen. It secretes enzymes and hydrochloric acid and absorbs nutrients.
- The submucosa, lying under the mucosa, contains blood and lymph vessels that transport the nutrients.
- Smooth muscle surrounds the submucosa in both a circular and longitudinal layer that helps to move the contents of the digestive tract by peristalsis.
- The serosa and peritoneum make up the last two layers of the gut.

Food passes through the gut by smooth muscle contraction.

- Food enters the mouth and is swallowed, passing through the pharynx into the esophagus.
- Food is then moved along by peristalsis, which is a wave of muscle contraction which moves food through the gut (see Figure 50.14).
- The esophageal sphincter helps keep food from reentering the esophagus from the stomach, and the pyloric sphincter controls the movement of chyme into the intestines.

A number of enzymes help break down food.

- The enzyme amylase begins the breakdown of starch in the mouth.
- In the stomach, the major enzyme is pepsin, which breaks down proteins into shorter peptides. Pepsin is initially secreted by gastric glands of the stomach in the zymogen form, pepsinogen (see Figure 50.15). Low pH activates the conversion of pepsinogen to pepsin through autocatalysis.
- The secretion of hydrochloric acid by the gastric glands in the wall of the stomach maintains a low pH.
- Hydrochloric acid and pepsin can produce an ulcer under certain conditions, particularly in the presence of the bacterium *Helicobacter pylori*.
- The acidic mixture of digesting food and gastric juices is known as chyme, which slowly passes into the intestines through the pyloric sphincter.

Most chemical digestion occurs in the small intestine.

- Carbohydrate and protein digestion continue in the small intestine.
- In humans, the small intestine can be more than six meters in length, with villi and microvilli contributing to a large surface area. The small intestine is divided into the duodenum, the jejunum, and the ileum.
- Fat digestion begins in the small intestine with the help of secretions and enzymes from the liver and pancreas.
- The liver secrets bile through the hepatic duct into the gallbladder where it is stored. When lipids are present, bile is secreted by the gallbladder into the duodenum through the bile duct (see Figure 50.16). Bile molecules have hydrophobic ends which attach to the tiny fat droplets to stabilize particles of fat in the form of micelles. This increases the exposed surface area of the lipids for digestion by the fat digesting enzymes, the lipases (see Figure 50.17).
- The zymogen trypsinogen is secreted by the pancreas and is converted into active trypsin by enterokinase. The pancreas produces a large number of zymogens that are released into the duodenum through the pancreatic duct. The zymogens are converted to the active form by trypsin. The pancreas also helps maintain a slightly alkaline pH in the duodenum by secreting bicarbonate ions.

The small intestine is the site of the most nutrient absorption.

- Final digestion occurs among the microvilli of the small intestine. Proteins are broken down into amino acids by peptidases. The enzymes maltase, lactase, and surcase breakdown the disaccharides into monosaccharides. The mucosal cells secrete the peptidases and maltase, lactase, and sucrase.
- Absorption occurs via active transport for ions such as sodium, calcium, iron, amino acids, and sugars.
- Fats move through the plasma membranes as fatty acids and monoglycerides (see Figure 50.17). In the mucosal cells they are converted into triglycerides and combine with cholesterol and phospholipids to form chylomicrons which pass into the lymphatic system. Bile is not absorbed across the membrane of the duodenum, but is actively resorbed in the ileum and returned to the liver.

The large intestine is the site of water and ion absorption.

- Material entering the large intestine has had most of the nutrients removed, but still contains important water and ions. The water and ions are absorbed across the large intestine, leaving behind semisolid feces.
- Bacteria living in the large intestine produce vitamin K and biotin which is then absorbed across the large intestine.

Herbivores employ fermenting microorganisms to help break down plant matter.

- Cellulase enzymes, which break down plant cellulose, cannot be produced by most herbivores.
- Microorganisms in the digestive tract of herbivores produce the cellulase enzymes that break down cellulose into fatty acids.
- In place of a stomach, ruminants have a large four-chambered organ in which fermentation occurs. The rumen and reticulum, the first two chambers, contain many microorganisms that break down cellulose (see Figure 50.18).
- From the rumen, food passes through the omasum, where water absorption occurs, into the abomasum, or true stomach. In the abomasum, hydrochloric acid and proteases kill the microorganisms and digest them.
- Coprophagy is the reingesting of feces to increase fermentation. In coprophagous animals the fermentation vat is found in a branch off the large intestine called the cecum.

Digestion is controlled by hormones and autonomic reflexes.

- Salivation and swallowing are autonomic reflexes.
- The duodenum secretes secretin, a hormone that stimulates the pancreas to release bicarbonate ion into the small intestine to control pH (see Figure 50.19).
- Fats and proteins stimulate the mucosa of the small intestine to secrete cholecystokinin which stimulates the release of bile and pancreatic digestive enzymes.
- Gastrin is secreted into the blood and is transported to the upper regions of the stomach where it stimulates stomach secretions and movement.

Insulin and glucagon help control blood glucose levels.

- The absorptive period is the period after a meal when digestion occurs. The postabsorptive period is the period when the gut is empty and the body uses energy reserves.
- During the absorptive period, the liver takes up glucose from the blood and converts it to glycogen and fat, fat cells take up glucose, and the body preferentially uses glucose as the energy source.
- During the postabsorptive period, the liver breaks down glycogen into glucose, fatty acids are supplied to the blood by adipose tissues, and the body preferentially uses fatty acids as an energy source.
- The pancreatic hormones insulin and glucagon control metabolic fuel use (see Figure 50.21).
- Insulin is released in response to high blood glucose levels, stimulating cells to burn glucose, fat cells to use glucose to make fat, and liver cells to convert glucose to glycogen and fat.
- At postabsorptive blood glucose levels, insulin secretion is diminished and the liver and fat cells break down glycogen.
- At low levels of blood glucose, the pancreas releases glucagon which stimulates the liver to break down glycogen and produce glucose.

Lipoproteins allow fats to be transported in the circulatory system.

- Lipoproteins are particles of fat and cholesterol covered with proteins that originate in the liver.
- Very-low-density lipoproteins (VLDL) are mostly triglyceride fats. Low-density lipoproteins (LDL) contain 50 to 60 percent cholesterol. High-density lipoproteins (HDL) are thought to carry cholesterol from the tissues to the liver.

Food regulation may be controlled by leptin.

- Hunger and satiety are influenced by the hypothalamus. Both the ventromedial hypothalamus and lateral hypothalamus play a role in food intake.
- Leptin, produced by fat, acts on the hypothalamus and is involved in control of hunger and satiety.

Toxins become concentrated in animals.

- Plants and animals may contain toxic compounds. Human activities also introduce toxic compounds into the environment.
- Hydrophilic water-soluble toxins can be metabolized quickly and filtered out of the body. Hydrophobic lipid-soluble toxins are metabolized slowly and tend to accumulate in adipose tissues.
- Bioaccumulation is the concentration of toxins in animals occupying higher and higher positions in the food chain. Long-lived species at the top of the food chain run the risk of heavy body burdens of toxins that are lipid soluble.
- The cytochrome P450 enzymes are involved in detoxification of natural chemicals. Some synthetic chemicals are not broken down by P450s and therefore bioaccumulate.
- Some synthetic chemicals may also act as hormone mimics.

The Big Picture

- Animals have evolved a wide variety of mechanisms for acquiring needed energy and raw materials for metabolism. All animals are heterotrophs; they can achieve energy and nutrient input by eating plants, animals, or a combination of both.
- The digestive system is specialized for efficient digestion of the type of food it must break down. The relatively short guts of carnivores have acid and protein reducing enzymes. The longer guts of herbivores have numerous holding chambers for the laborious process of breaking down plant material.

- Food passing down the length of the gut is first broken down mechanically into small pieces and then broken down chemically into small molecules that can be easily absorbed across the gut. As nutrients and water are absorbed from plant material, the remaining material forms the feces, which are eventually eliminated.
- The process of digestion is under both neural and hormonal control. Numerous hormones control the peristaltic movements of the gut and the secretion of enzymes and other chemicals involved in digestion.

Common Problem Areas

- Sometimes students think that the study of digestive function is uninteresting or, worse, unpleasant. Yet, some of the best and clearest examples of negative feedback and homeostasis are to be found in the study of digestion.
- Sometimes students think that there is a single digestive mechanism for food and are surprised to learn of the complexity of the process. Be sure to recognize that different food components—proteins, carbohydrates, fats—are digested, absorbed, and transported by very different mechanisms.

Study Strategies

- Although the guts of higher animals are structurally complex, they can be divided functionally into a foregut, midgut, and hindgut. If you break down the structure of the gut into these sections and determine what physiological process occurs in each section, you will have divided the task into smaller pieces.
- Because different types of food are digested and absorbed by different mechanisms, learn how a particular type of food is digested and match the appropriate enzymes with that process. For example, it will be much easier to remember how proteins are digested, rather than to recount a long list of digestive enzymes and then try to pick the one that might break down protein. In short, learn the process, not the list!
- Review the following tutorial and activities on the website:
  Tutorial 50.1 Insulin and Glucose Regulation
  Activity 50.1 Mammalian Teeth
  Activity 50.2 Human Digestive System
  Activity 50.3 Mineral Elements Required by Animals
  Activity 50.4 Concept Matching: Vitamins

Test Yourself

Knowledge and Synthesis Questions

1. Certain amino acids are essential to the diet of animals because
   a. they prevent overnourishment.
   b. they are cofactors and coenzymes that are required for normal physiological function.
   c. an animal cannot directly synthesize them through the transfer of an amino group to an appropriate carbon skeleton.
   d. animals need these substances in order to make the stored fats that are used during hibernation and migration.
2. Which of the following descriptions describes digestive characteristics of a sheep?
   a. It is a saprobe: It engulfs food and performs intracellular digestion.
   b. It is an autotroph: It synthesizes organic nutrients and performs extracellular digestion.
   c. It is an herbivore: It ingests food and performs extracellular digestion
   d. It is a detritivore: It ingests food and performs intracellular digestion.
3. Protection of the walls of the stomach against the action of its own digestive juices
   a. results from the presence of an anti-enzyme chemical formed by the gastric glands.
   b. results from the nervous reactions of the lining of the stomach.
   c. is controlled by a center in the medulla of the brain.
   d. results from the neutralizing, buffering, and a coating mucus covering its inner surface.
4. Chylomicrons are produced in the
   a. mouth.
   b. stomach.
   c. lumen of the small intestines.
   d. epithelial cells of the small intestines.
5. The gallbladder
   a. produces bile.
   b. is part of the liver.
   c. stores bile which is produced in the liver.
   d. produces cholecystokinin.
6. The pancreas
   a. produces exocrine products involved in chyme digestion.
   b. is exclusively an endocrine gland which produces salivary amylase.
   c. contains villi to increase surface area.
   d. produces urobiligen (a bile pigment).
7. Hydrochloric acid
   a. is secreted by the gastric glands of the liver.
   b. is secreted by the gastric glands of the stomach.
   c. produces a low pH in the small intestine.
   d. a and c
8. Bile produced in the liver is associated with which of the following?
   a. Emulsification of fats into tiny globules in the small intestine
   b. Digestive action of pancreatic amylase
   c. Emulsification of fats into tiny globules in the stomach
   d. Digestion of proteins into amino acids
9. Most of the chemical digestion of food in humans is completed in the
   a. small intestine.
   b. appendix.
   c. ascending colon.
   d. stomach.

10. Which one of the following does not contribute to the large surface area available for nutrient absorption in the small intestines?
    a. Villi
    b. Intestinal length
    c. Microvilli
    d. Bile duct

11. Waves of muscle contractions that move the intestinal contents are
    a. caused by contraction of skeletal muscle.
    b. regulated by liver secretions.
    c. called peristalsis.
    d. voluntary.

12. What is the function of enterokinase?
    a. It converts pepsinogen to pepsin.
    b. It converts trypsinogen to trypsin.
    c. It digests proteins.
    d. It activates HCL.

13. Digestive enzymes responsible for breaking down disaccharides include
    a. pepsin, trypsin, and trypsinogen.
    b. amylase, pepsin, and lipase.
    c. sucrase, lactase, and maltase.
    d. pepsin, trypsin, and chymotrypsin.

14. Which of the following is characteristic of the large intestine?
    a. It has almost no bacterial populations.
    b. It contains chyme.
    c. It absorbs much of the water remaining in waste materials.
    d. It is the site of most of digestion.

15. The innermost layer of the digestive tract is the
    a. serosa membrane.
    b. mucosa membrane.
    c. submucosa membrane.
    d. lumen.

16. Which of the following hormones affects the rate of gluconeogenesis?
    a. Glucagon
    b. Insulin
    c. Epinephrine
    d. All of the above

Application Questions

1. You have just eaten a small pizza with lots of cheese. Because this food is high in lipids, your digestive system will have to perform specific tasks to digest and absorb the fats. What are the steps that your digestive system will take in order to digest and absorb the fats in this meal?

2. During the absorptive period your body is digesting and absorbing nutrients, while during the postabsorptive period your body is using up stored fuel. What are the main differences in fuel use and the control mechanisms during these two periods?

3. Ingested food typically contains many proteins that could be of use. However, these proteins are not used in the form they are digested but are broken down to the smaller peptides and amino acids. Why does the digestive system break down proteins rather than absorb them in their earlier form?

4. Humans have been polluting the environment for many years, dumping toxins into many of Earth's ecosystems. What effect does this have on the amount of toxins found in animals such as an eagle which is at the top of the food chain?

Answers

Knowledge and Synthesis Answers

1. **c.** Essential amino acids must be acquired through diet because an animal cannot directly synthesize all of the amino acids needed for protein production.

2 **c.** A sheep is an herbivore that ingests plant material for nutrition. Digestion occurs extracellularly in the digestive tract of the animal.

3. **d.** The stomach is protected from digestive enzymes and low pH by the neutralizing, buffering, and coating mucus secreted over its inner surface.

4. **d.** The epithelial cells of the small intestine produce chylomicrons by combining triglycerides with cholesterol and phospholipids, allowing lipids to pass into the lymphatic system.

5. **c.** The gallbladder stores bile that was created by the liver and releases it into the small intestine to aid in lipid digestion.

6. **a.** The pancreas produces exocrine products such as lipases, nucleases, amylases, and trypsin which are all involved in chyme digestion.

7. **b.** Hydrochloric acid is a strong acid secreted by gastric glands in the lining of the stomach. It lowers the pH of the stomach fluid.

8. **a.** Bile aids in the digestion of lipids in the small intestine by forming fat droplets into small fat particles.

9. **a.** The small intestines are the site for the majority of digestion in humans.

10. **d.** The length of the intestines and the presence of the microvilli and villi increase the surface area of the small intestine.

11. **c.** Food is moved through the digestive tract by wavelike contractions of smooth muscle called peristalsis.

12. **b.** Enterokinase converts the zymogen trypsinogen to trypsin.

13. **c.** The digestive enzymes sucrase, lactase, and maltase are involved in the breakdown of disaccharides into the monsaccharides glucose, galactose, and fructose.

14. **c.** The large intestine is the site of water and ion absorption. The large population of bacteria in the large intestine contributes vitamins that are useful to the host.

15. **b.** The membranes of the digestive tract are, from the inside to the outside: mucosa, submucosa, circular and longitudinal muscles, serosa.

16. **d.** Insulin reduces the rate of gluconeogenesis, and glucagon and epinephrine both increase the rate of gluconeogenesis.

Application Answers

1. The digestion of fats begins in the small intestines. Lipases are water-soluble, but lipids tend to aggregate into large droplets in an aqueous environment. Bile helps to increase the exposed surface area of the lipids by forming small micelles with the lipids. This allows lipases to break down the fats into free fatty acids and monoglycerides. The fatty acids dissolve into the plasma membrane of the intestinal epithelial cells and will be absorbed across this surface. In the mucosal cells, the free fatty acids and monoglycerides are turned into triglycerides, which are incorporated into chylomicrons. The chylomicrons can then pass out of the mucosal cells and into the lymphatic system (see Figure 50.17).

2. During the postabsorptive period, cells tend to metabolize fats, keeping the blood glucose reserves for the nervous system. Insulin levels are low during the postabsorptive period. During the absorptive period the body attempts to store the nutrients and fuel that are being absorbed. The preferred fuel source at this time is glucose, and insulin is released to help facilitate glucose uptake.

3. Proteins in digested food are not absorbed and used in their original form for a number of reasons. First, the proteins are too large to be absorbed easily across the plasma membrane of the digestive tract, while amino acids are easily moved across the wall of the gut. Second, the proteins in the food are specific for the species that was eaten and may not be compatible with the one eating it. Last, the proteins of another species would be recognized as foreign and attacked by the immune system.

4. Toxins in the environment tend to bioaccumulate in top predators such as the eagle. The lipid soluable toxins accumulate in body tissues over the life span of these top predators. Therefore, as the predator eats more contaminated food, concentrations of toxins in the body increase, resulting in increased body burdens.

51 Salt and Water Balance and Nitrogen Excretion

Important Concepts

Excretory organs control the solute concentration of animals.

- The excretory organs control the level of solutes and the volume of tissue fluid, and excrete urine. Urine composition depends on the environment surrounding the animal.
- Water movement in the body occurs either by pressure differences or differences in solute potential producing water movement by osmosis. There is no active transport of water.
- Excretory systems use filtration across membranes, active secretion, and active resorption to control water and solute balance.
- Osmolarity is the concentration of salts in a fluid.
- Osmoconformers do not actively regulate the osmolarity of their tissues; rather they allow them to come to equilibration with the environment. Marine invertebrates tend to be osmoconformers unless the concentrations in the environment are drastic.
- Osmoregulators actively regulate the osmolarity of their tissues, even as environmental osmolarity may change. The brine shrimp is an osmoregulator that is hypotonic (tissue fluid osmolarity below the environment) in high osmolarity waters and hypertonic (tissue fluid osmolarity above the environment) in low osmolarity waters (see Figure 51.1).
- Birds use a nasal salt gland to excrete salt from consumed seawater (see Figure 51.2).

Nitrogen is a potentially toxic metabolic waste.

- Ammonia (NH_4) is a highly toxic nitrogenous waste that either must be excreted or converted to the less toxic urea and uric acid (see Figure 51.3).
- Ammonotelic animals excrete ammonia to the aquatic environment, usually across the gill membranes.
- Ureotelic animals excrete nitrogenous waste as water-soluble urea.
- To conserve water, uricotelic birds and reptiles excrete nitrogenous waste as insoluble uric acid.

Reflecting their diversity, invertebrates employ a number of excretory mechanisms.

- Marine invertebrates are typically osmoconformers.
- Freshwater and terrestrial invertebrates use active mechanisms to regulate their osmolarity.
- Flatworms have a protonephridium made up of tubules and flame cells that conserve ions and excrete water to produce a dilute urine (see Figure 51.4).
- Annelid worms filter the blood across the capillaries into the coelom where the fluid enters a metanephridia in each segment through a funnel-like opening called a nephrostome. Tubules of the metanephridia actively secrete and absorb various ions and end in nephridiopores which open to the environment to excrete dilute urine (see Figure 51.5).
- Insects use Malpighian tubules as their excretory organ. Malpighian tubules actively transport uric acid and sodium and potassium ions from the tissue into the tubules, with water following passively (see Figure 51.6). The tubules lead into the hindgut where uric acid precipitates and water is reabsorbed.

The vertebrate excretory system is composed of nephrons in a kidney.

- The nephron of the vertebrate kidneys is composed of the glomerulus, renal tubules, and peritubular capillaries which are further broken down into vascular and tubule components (see Figure 51.7).
- There are two sets of capillary beds in series in a nephron. The glomerulus is composed of the first set of capillary beds with blood entering at the afferent arteriole and exiting at the efferent arteriole. From the efferent arteriole the peritubular capillaries of the second capillary bed emerge and surround the tubule component of the nephron (see Figure 51.7).
- The renal tubule begins with the Bowman's capsule which encloses the glomerulus to form the renal corpuscle. Within the Bowman's capsule, the podocyte cells wrap around the capillaries of the glomerulus. The blood is filtered by the glomerulus into the Bowman's capsule. The endothelial walls of the capillaries in the glomerulus have pores that allow water and small molecules to leave the blood. Only small molecules and water can enter the Bowman's capsule.

- Arterial blood pressure is the driving force for filtration at the glomerulus. The porous capillary beds in the glomerulus also contribute to the high filtration rate.
- The filtrate entering the tubules is similar to blood plasma, but as the fluid moves through the tubule, ions and molecules are actively resorbed and secreted. This allows for control of urine composition and dilution.

Marine and freshwater fishes employ two opposing mechanisms for controlling water and ion balance.

- Freshwater fishes live in an environment hypotonic to their own body fluids, with water tending to move into the body by osmosis. To counter this, freshwater fishes excrete large amounts of dilute water and actively take up ions.
- Marine fishes live in an environment hypertonic to their own body fluids, so water tends to be drawn out of the body by osmosis. They produce little concentrated urine, drink water, and actively secrete salts across the gills and the renal tubules.
- Cartilaginous fishes act as osmoconformers, allowing urea and tri-methyl concentrations to increase, which increases the osmolarity of their tissues. The rectal gland excretes salts in these fishes.

Terrestrial and aquatic amphibians have different mechanisms for water management.

- Amphibians living in or near fresh water have a large water influx, and in response produce large amounts of dilute urine.
- Amphibians living in terrestrial environments have skin with reduced water permeability and low metabolic rates, resulting in low water turnover. Some also use their bladders as canteens to store water for later use.

Reptiles and birds conserve water in their terrestrial habitat.

- Reptiles lay shelled eggs which allow them to live in a fully terrestrial environment. They have scaly, dry skin to decrease evaporative water loss and they excrete uric acid with little water. They do not have the ability to produce concentrated urine.
- Birds have similar mechanisms as reptiles to control and conserve water, although some birds have the ability to produce concentrated urine.

The kidneys are the main excretory organ in mammals.

- In mammals the nephron is used to help conserve water when excreting nitrogenous wastes (see Figure 51.11).
- Humans have two kidneys which filter blood and produce urine. Urine exits the kidneys through the ureter and travels to the bladder where it is stored until excreted through the urethra.
- The human kidney is shaped like a kidney bean, with a central medulla and a surrounding cortex. There are close to a million nephrons in the human kidney spanning the cortex and medulla regions.
- The glomerulus and adjoining proximal convoluted tubules are located in the cortex. The descending limb of the renal tubule runs down into the medulla and the ascending limb returns to the cortex in what is known as the loop of Henle. From the ascending limb the tubules become the distal convoluted tubule which connects to the collecting ducts and runs down the medulla.
- Blood vessels also run in a similar pattern to the tubules. An afferent arteriole carries blood to the glomerulus where it is drained into the efferent arteriole that becomes the peritubular capillaries. Capillaries run through the medulla, parallel to the loop of Henle to form a vasa recta.

Filtration occurs at the glomerulus and with resorption, at the proximal convoluted tubule.

- The typical glomerulus filters about 120 liters per day in an adult human. However, 98 percent of this fluid is resorbed into the blood, so that only 2 percent of glomerular filtrate is excreted as urine.
- The majority of the filtrate is resorbed in the proximal convoluted tubules. Sodium ions and other solutes are actively transported out of the proximal convoluted tubule and water follows passively, all to be taken up by the peritubular capillaries.

The loop of Henle sets up a concentration gradient in the medulla.

- The loop of Henle produces a concentration gradient in the medulla by acting as a countercurrent multiplier system (see Figure 51.12). The purpose of this gradient is to move water across fluid compartments by osmosis, since there is no mechanism for active transport.
- The concentration of the fluids surrounding the loop of Henle is raised by the active resorption of Cl^- and the passive flow of Na^+ into the tissue from the thick ascending limb. The thick ascending limb is impermeable to water, so only solutes leave the tubules.
- The thin descending limb is permeable to water but not sodium and chloride ions. Water moves out of the tubules into the surrounding tissues due to the higher concentration of Na^+ and Cl^- in the surrounding tissues.
- In the thin ascending limb, the fluid is more concentrated than the fluid in the surrounding tissues, so Na^+ and Cl^- move out of the tubule. Water cannot move out because of the low permeability of the thin ascending limb.
- The fluid reaching the distal convoluted tubule is less concentrated than blood plasma.

Concentration of urine occurs in the collecting duct.

- The collecting duct passes through the medulla, with its high solute concentration set up by the loop of Henle.
- Water moves out of the collecting duct and into the tissue, resulting in concentrated urine. The permeability of the collecting ducts can be adjusted hormonally to regulate the amount of water excreted.
- Longer collecting ducts result in a greater ability to concentrate urine.

Enzymes and hormones regulate glomerular filtration rate.

- Glomerular filtration rate is a function of the blood flow and pressure to the kidneys.
- The kidneys have autoregulatory mechanisms that help maintain their high filter rate (see Figure 51.14).
- In response to decreasing arteriole resistance or decreased arterial blood pressure, the afferent renal arterioles open, or dilate, to increase flow to the glomerulus. If this does not increase glomerulus filtration rate sufficiently, the kidneys release the enzyme renin into the blood. Renin converts an angiotensin precursor into active angiotensin.
- Angiotensin constricts the efferent renal arterioles to increase pressure through the glomerulus. Angiotensin also constricts peripheral blood vessels, stimulates the release of aldosterone from the kidney, and stimulates thirst.
- Aldosterone stimulates the resorption of sodium by the kidney in an effort to maintain blood volume.

Antidiuretic hormone regulates blood pressure and blood osmolarity.

- Antidiuretic hormone (ADH) is produced in the hypothalamus and released from the posterior pituitary gland (see Figure 51.15). ADH controls the permeability of the collecting ducts to water by stimulating production of water channels known as aquaporins over the long term and insertion of aquaporins into the membrane over the short term.
- Stretch receptors in the aorta and carotid arteries inhibit the release of ADH in response to increased blood pressure, causing water to leave the body and blood volume to decrease.
- Osmosensors stimulate the release of ADH in response to increased osmolarity, resulting in increased water resorption and dilution of the blood.
- ADH also causes constriction of peripheral blood vessels.
- Atrial natriuretic hormone is release by the heart in response to high blood pressure. At the kidneys, atrial natriuretic hormone decreases sodium resorption, resulting in decreased water resorption and a lowered blood volume.

The kidneys help to regulate the pH of the blood.

- Bicarbonate ion is the major buffer in the blood. The kidneys control the level of H^+ and bicarbonate ions in the blood by varying their rate of secretion of these materials into the urine (see Figure 51.16).

The Big Picture

- Different animals experience different types of stresses related to water and salt balance. Animals living in environments hypertonic to the body fluids have to overcome problems with water loss due to osmosis, while animals living in environments hypotonic to the body fluids have to overcome problems with water gain due to osmosis. Some invertebrates circumvent these problems by allowing their body fluid ion concentration to reflect that of their surroundings. Other invertebrates, and all vertebrates, retain their body fluids at more constant levels, expending energy to eliminate excess water and/or salts.
- All animals produce metabolic wastes (notably nitrogenous waste) as a by-product of metabolism. Invertebrates have evolved a variety of structures to eliminate wastes, while the vertebrates have variations on the kidney. The nephron is the "functional unit" of the vertebrate kidney. By eliminating wastes without losing valuable nutrients and salts, the kidneys cleanse the blood of metabolic waste products.
- Kidneys and other organs for urine formation work by differentially processing nitrogenous waste, ions, and water. In the kidney, large volumes of plasma are filtered from the blood into the interior of the nephron. Then, as the fluid passes through the various specialized regions of the nephron, the desirable components to be retained (most water and salts, nutrients) are pulled back into the tissues, leaving behind in the nephron the waste products in an ever more concentrated form of urine. As urine leaves the nephron, the water content is adjusted in the collecting ducts through hormonal alteration of their water permeability. If the collecting ducts are highly water-permeable, then water will be drawn by osmosis out of the collecting ducts into the region of high salt concentration created by the action of the loop of Henle.

Common Problem Areas

- Depending on whether it is living in fresh water or saltwater, fishes may drink water, excrete copious amounts of dilute urine or small amounts of more concentrated urine, or take up ions at their gills. Given that living in either of these environments creates osmoregulatory problems, think about whether the tissues are hypotonic or hypertonic relative to the surrounding water, and then predict water and ion movements on that basis.
- Understanding how the nephron functions, and in particular how the loop of Henle works as a "countercurrent multiplier system," is one of the greater challenges in this book.

Study Strategies

- It is very important to understand the concept of "tonicity" and the fact that the terms "hypertonic" and "hypotonic" always have a reference point (body fluids, typically). Thus, a liquid is not just "hypertonic," but rather is hypertonic compared with some other fluid.
- Nephron function can be learned more easily if you place yourself in the position of a single Na^+ in the blood entering the nephron. Follow your pathway—and your potential pathways—as you traverse the nephron, eventually ending up in the urine. Note especially that it may take you some time to pass through the loop of Henle as you leave the ascending limb, only to recycle back into the descending limb—a process that could be repeated many times before finally escaping the countercurrent multiplier. Repeat this imaginary journey as a water molecule.

- Review the following tutorial and activities on the website:
  Tutorial 51.1 Renal Function
  Activity 51.1 Annelid Metanephridia
  Activity 51.2 The Vertebrate Nephron
  Activity 51.3 The Human Excretory System

Test Yourself

Knowledge and Synthesis Questions

1. A marine fish lives in a hypertonic environment. What mechanisms does a marine fish use to maintain homeostasis?
   a. It drinks copious amounts of water, excretes only small amounts of water, and pumps sodium out of the body.
   b. It drinks little water, excretes large amounts of water, and pumps sodium into the body.
   c. It drinks little water, excretes only small amounts of water, and pumps sodium out of the body.
   d. None of the above

2. Which of the following statements about the excretory system of insects is *not* true?
   a. Active transport moves materials from the coelomic fluid into the Malpighian tubules.
   b. The Malpighian tubules can produce a highly concentrated urea solution, allowing insects to inhabit some of Earth's driest habitats.
   c. Resorption of salts takes place mostly in the gut.
   d. Water resorption is by osmotic movement only.

3. What pathway is taken by water and solutes as they travel through a nephron?
   a. Glomerulus, to Bowman's capsule, to proximal tubule, to loop of Henle, to distal tubule, to collecting ducts
   b. Bowman's capsule, to glomerulus, to distal tubule, to loop of Henle, to proximal tubule, to collecting ducts
   c. Glomerulus, to Bowman's capsule, to distal tubule, to loop of Henle, to proximal tubule, to collecting ducts
   d. Glomerulus, to Bowman's capsule, to proximal tubule, to collecting ducts, to distal tubule, to loop of Henle

4. Na^+ and Cl^- are actively transported out of the tubules to help set up the countercurrent multiplier. Which of the following are sites of active Na^+ and Cl^- transport in the nepheron?
   a. Proximal tubule, ascending limb of the loop of Henle
   b. Descending limb of the loop of Henle, ascending limb of the loop of Henle
   c. Ascending limb of the loop of Henle, proximal tubule
   d. Collecting duct, descending limb of the loop of Henle

5–9. Use the following diagram to complete the statements about the mammalian nephron.

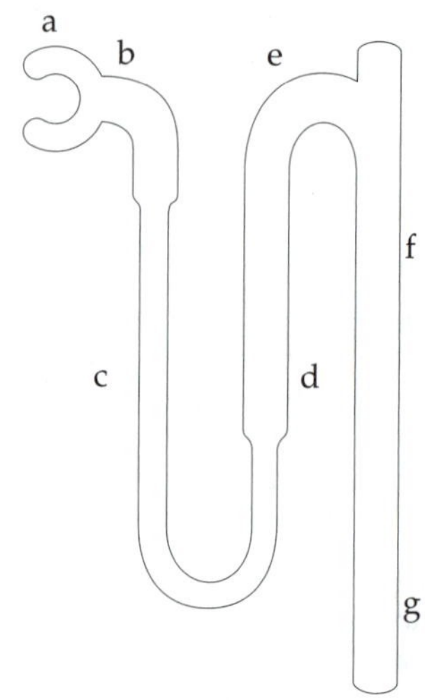

5. The composition of the filtrate would be most like plasma in the tubule next to letter _____.

6. The NaCl concentration in the extracellular fluid would be greatest in the area of letter _____.

7. The osmolarity of the filtrate next to letters _____ is similar to the osmolarity of blood plasma.

8. The urine would be most concentrated in the collecting duct next to letter _____.

9. Most of the glomerular filtrate is resorbed into peritubular capillary blood next to letter _____.

10. The sole mechanism for water resorption by the renal tubules is
    a. active transport.
    b. osmosis.
    c. cotransport with sodium ions.
    d. cotransport with bicarbonate ions.

11. A number of hormones help to regulate water and solute uptake and release in the nephron. Antidiuretic hormone (ADH) promotes _____ in response to _____.
    a. active transport of Cl^-, increased solute concentration
    b. active transport of Na^+, increased blood pressure
    c. increased permeability of the collecting duct to water, increased blood pressure
    d. decreased permeability of the collecting duct to water, increased solute concentration

12. Which of the following is not a normal constituent of the glomerular filtrate?
    a. Red blood cells
    b. Urea
    c. Sodium ion
    d. Glucose

13. If the afferent arteriole that supplies blood to the glomerulus becomes dilated,
    a. the protein concentration of the filtrate decreases.
    b. hydrostatic pressure in the glomerulus decreases.
    c. the glomerular filtration rate increases.
    d. All of the above

14. Osmoreceptors in the brain detect changes in blood ion concentration which can reflexively result in
    a. water retention by the kidneys.
    b. water loss by the kidneys.
    c. ion retention by the kidneys.
    d. All of the above

15. If the human kidneys filter 150 liters of plasma in a 24-hour period, what is the typical amount of urine produced and eliminated in that time period?
    a. 0.15 liters
    b. 1.5 liters
    c. 15 liters
    d. 30 liters

Application Questions

1. Based on your knowledge of how the nephron works, explain why mammals produce concentrated urine while reptiles cannot.
2. What would happen to the composition of excreted urine if all active transport processes in the nephron came to a stop?
3. The drug urizadole inhibits antidiuretic hormone secretion from the pituitary. How will this change the glomerular filtration rate, urine flow, and urine concentration in patients taking this medication?
4. You have just eaten a large number of very salty potato chips and the osmolarity of your blood has increased. What response will your body have in order to bring your blood osmolarity back to homeostasis?
5. Earthworms and humans are very different, yet their excretory systems share many similar characteristics. Name a few of the ways in which the excretory systems are similar in form and function.

Answers

Knowledge and Synthesis Answers

1. **a.** Marine fishes live in an environment in which salts try to enter the body and water wants to leave the body. To counter these effects marine fishes drink lots of water, excrete small amounts of water, and actively pump sodium out of the body across the gills.

2. **b.** All are true of insects except that they excrete uric acid rather than urea.

3. **a.** The route of water and solutes through the nephron is from the glomerulus, to Bowman's capsule, to proximal tubule, to loop of Henle, to distal tubule, to collecting ducts.

4. **a.** The proximal tubule is the site of active transport of Na^+ out of the tubule. Na^+ also moves out of the tubule at the ascending limb of the loop of Henle, but this is a passive transport with Cl^- being actively transported out.

5. **a.** At the Bowman's capsule the filtrate is most similar to plasma.

6. **g.** The sodium concentration is the highest in the extracellular fluid near the middle of the medulla.

7. **a., b.,** and **e.** The filtrate has a similar osmolarity to plasma in the cortex of the kidney that is in the proximal convoluted tubule, the glomerulus and Bowman's capsule, and the distal convoluted tubule.

8. **g.** The highest concentration of the filtrate in the collecting ducts will be near their end, deep in the medulla.

9. **b.** The bulk of the water and solutes that are resorbed from the filtrate occurs at the proximal convoluted tubule.

10. **b.** The sole mechanism for water resorption in the renal tubules is by osmosis.

11. **c.** Antidiuretic hormone acts on the collecting ducts by increasing the permeability to water. Antidiuretic hormone secretion is stimulated by a decrease in blood pressure.

12. **a.** Red blood cells are too large to be filtered out of the blood at the glomerulus and thus will not be found in the filtrate.

13. **c.** Changes in the afferent arteriole pressure affects glomerular filtration rate. Increases in pressure will increase filtration rate and decreases will decrease filtration rate.

14. **d.** Osmoreceptors in the brain detect changes in blood ion concentration, resulting in reflexive increases or decreases in both ion and water retention, depending on the status of blood osmolarity.

15. **b.** About 1 percent of the filtrate is excreted as urine, so about 1.5 liters of the original 150 liters would be urinated.

Application Answers

1. The loop of Henle acts as a countercurrent ion multiplier in the medulla of mammals. This allows mammalian kidneys to produce concentrated urine. Reptiles lack the countercurrent multiplier of the loop of Henle.

2. Active transport of Na^+ and Cl^- in the nephron provides the ions that set up the countercurrent multiplier, allowing for the production of concentrated urine in mammals. If active transport in the nephrons were to stop functioning, the urine produced would eventually be isotonic with blood plasma because the countercurrent multiplier would disappear.

3. Inhibition of antidiuretic hormone by urizadole will affect the permeability of the collecting ducts to water. This will result in a diuresis and increased urine flow because water will not be able to be resorbed across the collecting ducts. Blockage of ADH will have no effect on the glomerular filtration rate.

4. In response to increased blood osmolarity, osmosensors in the hypothalamus stimulate the release of ADH from the pituitary into the blood. ADH acts to increase the permeability of the collecting ducts to water so that increased amounts of water can be resorbed to bring down the blood osmolarity. The osmosensors will also stimulate thirst, causing you to increase your water intake.

5. The metanephridia of the earthworm and the nephron of humans are both composed of tubules through which filtrate flows. In the tubules of both, certain molecules are actively transported out or secreted in. Both systems produce urine that has a different osmolarity from that of the body fluid. The urine of the earthworm is hypotonic, while that of a human is usually hypertonic.

52 Animal Behavior

Important Concepts

Behaviors can be inherited or learned.

- The study of behavior is known as ethology.
- Behaviors are highly stereotypic and species-specific.
- Behaviors, such as web spinning in spiders, require no prior experience and are inherited (see Figure 52.1).
- Determination of inheritance of behaviors can be made by deprivation experiments in which all relevant experience is deprived, or by hybridization experiments in which parents with known behaviors are crossed and the offspring are studied for the behavior.
- Hybridization experiments by Konrad Lorenz on ducks, in which the mating behavior was different for similar species, showed that the hybrids performed mating behaviors that were a combination of those of the two parents (see Figure 52.2). Non hybrids were not attracted to hybrid displays.
- Triggers or releasers may be necessary for animals to exhibit genetically based behaviors. The releasers can be a simple subset of sensory information available to an animal. In herring gull chicks, a red dot on the mother's beak triggers a begging response (see Figure 52.4). The red dot on the mother's beak acts as a releaser for the genetically based begging behavior.
- Most behaviors involve the interaction between genes and learning. Spatial learning involves learning features in the environment and using them as cues. Wasps use spatial cues from the environment to help them locate their nests (see Figure 52.5).
- Behaviors with a genetic basis can be highly adaptive in species for whom the consequence is death if the behavior is dangerous. Examples of highly adaptive behaviors are predator avoidance and mating rituals.

Bird songs involve both imprinting and genes.

- Imprinting of offspring on the parents is a learned response during a critical period that helps in recognition. The critical period for imprinting is a short time during a developmental or hormonal state for the animal.
- Birds use songs in territory displays and courtship (see Figure 52.7).
- In some species, imprinting of the species-specific song is required in the nestling in order for the individual to sing the song as an adult, even though the song is never sung by the juvenile bird. Imprinting forms a memory of the song that is recalled as the bird approaches adulthood. The adults need both the initial imprinting of the song as a juvenile and the ability to match the song with auditory feedback from other adults in order to sing the correct song.
- Once the bird has perfected the song it is crystallized and will be sung in the correct form every spring.

Hormones can play an important role in controlling the behavior of animals.

- Males and females of a species frequently display different behaviors that are the result of sex steroids.
- Experiments on rats have revealed the influence of sex steroids on sexual behavior (see Figure 52.9).
- Sex steroids are necessary for the expression of sexual behavior, with testosterone affecting only males and estradiol affecting only females. Exposing male or female rates to testosterone during development results in male sexual behavior in adults of both sexes. Normal development of male sexual behavior requires the presence of testosterone, while normal development of female sexual behavior does not require estradiol.
- In male birds, testosterone levels control singing by inducing regions of the brain to grow during the breeding season (see Figure 52.10). During the nonbreeding season, the regions of the brain associated with singing are reduced in size.
- Female birds that are provided with testosterone during the spring will also develop the species-specific song. In these birds, testosterone has stimulated growth in those areas of the brain associated with singing.

There is a strong genetic component to animal behavior.

- Hybridization in crickets reveals the genetic basis for singing behavior. Crossing two species with distinct songs results in offspring song that contains features of both parental species. A female cricket's recognition of songs also has a genetic basis, with hybrid females choosing mates with the hybrid song.

- Artificial selection in dogs provides a good example of how different breeds have different genetically fixed behaviors.
- In the fruit fly, multiple and single genes have been found to influence courtship and mating behavior. The gene *per* plays a role in the wingbeat pattern of the male during courtship.
- Nest-cleaning and non-nest-cleaning strains of honeybees also provide evidence for genetic control. Hygienic bees uncap cells and remove dead larvae from the hive, making them resistant to a bacterium, while another strain does not engage in nest-cleaning behavior (see Figure 52.11). Crossing experiments have shown that two recessive genes control the cleaning behavior. In crosses involving the second generation, nonhygienic bees showed a mix of behaviors.
- Courtship in fruit flies is species-specific, requires no learning, and appears to be controlled by one gene, *fruitless* (*fru*).
- The sex-lethal (*sxl*) gene is expressed in female fruit flies and controls sexual differentiation and behavior. The *sxl* gene controls the *transformer* (*tra*) gene and its product, the *tra* protein, required for development into a female. *Tra* in turn controls the *doublesex* (*dsx*) gene involved in male differentiation and the *fruitless* (*fru*) gene involved in male courtship behavior (see Figure 52.12).

Many animal behaviors involve some form of communication between individuals of a species.

- Communication is used by animals to transmit information and reduces uncertainty between individuals of a species.
- Pheromones are chemical signals used to communicate. Pheromones can be long-lasting, like the trail-marking and territory-marking pheromones of ants or cats, or they can be short-lasting, like the sex-pheromones released by moths. The chemical properties of the pheromones dictate how long they last in the environment and how far they travel.
- Visual signals provide rapid, directional communication. One drawback to visual signals is the need for light.
- Sound has advantages over sight in that it can travel in complex environments, such as a forest, and over long distances, such as the sound produced by whales in the water. Sound also communicates directional information.
- Touch is commonly used in communication. Honeybees dance to communicate the location of food sources in the environment (see Figure 52.14). Round dances are used to communicate that food is less than 80–100 meters from the hive. A waggle dance is used to communicate distance and direction to food that is over 100 meters from the hive. Speed of the waggle indicates distance to the food source; direction of the wiggle indicates the direction of the food source.
- Some fish use electrical fields to sense their surroundings. In the glass knife fish, different individuals emit different frequencies that relate to their status in the population.

The timing of behaviors is controlled internally.

- Circadian rhythms are daily cycles in activity, sleep, foraging, and other physiological processes and behaviors that are controlled by an endogenous clock (see Figure 52.15).
- The length of one cycle in a rhythm is defined as one period. Any point in the cycle is known as a phase. Two cycles can be in phase if the rhythms match or phase-advanced or phase-delayed if they do not match.
- Circadian rhythms can be reset by environmental cues such as the light–dark cycle by the process of entrainment. In constant conditions an animal will have a natural period that is different from the 24-hour period of the day. Such an individual is said to be free-running. Shifts of no more than 20–30 minutes can occur in humans per day, explaining why you get jet-lag when crossing time zones.
- The circadian clock in many vertebrates is located in the suprachiasmatic nuclei (SCN) just above the optic chiasm (see Figure 52.16). Removal of the SCN will result in the loss of circadian rhythms. In some birds the pineal gland functions as the master clock.
- The genes that are involved in circadian rhythms are highly conserved across a wide range of organisms. In the fruit fly the *per* gene is involved in the length of the rhythms and the absence of the *tim* gene will result in the total loss of circadian rhythms (see Figure 52.17).
- The length of the day or photoperiod can trigger seasonal rhythms. An endogenous seasonal rhythm known as the circannual rhythm is found in hibernators. Seasonal rhythms are important in controlling the timing of breeding and hibernating. The changing length of available daylight may play a role in the rhythms of some animals that are know as photoperiodic.

Behaviors control navigation.

- Recognition of landmarks is used to navigate during piloting (see Figure 52.18).
- Homing is the ability to return to a specific location. Piloting is one simple method used in homing. However, many sea birds fly over the oceans where there are no environmental cues. Migrating birds are able to fly great distances seasonally and return to the same breeding grounds each season.
- Distance-and-direction navigation involves knowledge of direction and distance from the destination (see Figure 52.20). Biological rhythms can help in distance determination and the position of the sun and stars and can help in direction determination. The stars offer both moving and fixed points for navigation (see Figures 52.21 and 52.22).
- Bicoordinate navigation is the process of knowing your location and your destination, and then plotting the appropriate course.

Human behavior

- Human behavior can be modified by experiences over time.
- Learned behavior in humans is passed down from generation to generation as part of our cultures.

- Chimpanzees, like humans, also exhibit different cultures in different locations.
- Motor patterns such as facial expressions appear to have a genetic component.

The Big Picture

- Animals exhibit a wide range of species-specific behaviors that can be learned or instinctive. These behaviors may signal to members of their own species, as well as communicating with other animals that are potentially prey or predators. Instinctive, genetically based behaviors may require "triggers" or releasers to stimulate the animal.
- Hormones are important in controlling the behavior of animals. Sex steroids are frequently involved in stimulating the different behaviors of males and females of a species. Daily, seasonal, or annual rhythms are often carried out through the actions of hormones.
- Communication is an integral part of animal behavior. Communications can be chemical (involving pheromones), visual (involving displays of fins, feathers, fur, etc.), auditory (sounds created by either general or specialized structures and received by auditory sensors), and tactile. Such communications are important components of territory-marking and reproductive behavior.

Common Problem Areas

- It can sometimes be difficult to distinguish between behaviors founded in genetics (instinctive behaviors) and those that are learned during the course of an animal's lifetime. Further muddying the waters, the ability to learn new behavior is, of course, a heritable trait!
- Hormones control many behaviors, and hormone production is often a function of the time of day or the time of year. Thus, a particular hormone that stimulates (or inhibits) behaviors under one condition at one point in time may not have the same effect later on. Be sure to learn the conditions and caveats that accompany hormone actions.

Study Strategies

- As you learn about various animal behaviors, it may be useful to design "mind experiments" that you carry out in your imagination, you can distinguish between an instinctive and learned behavior. This may help you work out the nature of the behavior.
- Learn to distinguish the various behavior cycles associated with internal controls via "biological clocks" as opposed to those behaviors that are directly stimulated by an immediate event in the animal's surrounding environment.
- Review the following tutorial and activity on the website:
  Tutorial 52.1 Time-Compensated Solar Compass
  Activity 52.1 Honeybee Dance Communication

Test Yourself

Knowledge and Synthesis Questions

1. A newborn male rat is castrated and then allowed to mature. When he reaches adulthood, he is given an injection of male sex steroids and exposed to a normal adult female rat exhibiting lordosis. Which of the following will occur?
   a. The male will mount the female and attempt to copulate.
   b. The male will show lordosis.
   c. The male will attempt to copulate and show lordosis.
   d. The male will show no sexual behavior at all toward the female.

2. Select the following example that is least likely to be a fixed action pattern.
   a. Egg laying in a turtle
   b. Courtship display
   c. Predator's prey-search behavior
   d. Web spinning in a spider

3. Which of the following statements about hormonal influences on bird song is *false*?
   a. Male songbirds castrated as nestlings do not sing proper songs as adults.
   b. Female songbirds can be induced to sing by treatment with testosterone.
   c. Females learn their species song, but do not normally express it under natural conditions.
   d. Testosterone is not necessary for singing in males once they have learned their song.

4. Studies of which of the following organism types have led most directly toward connecting a specific gene with a specific behavior?
   a. Honeybees
   b. Dabbling ducks
   c. *Drosophila* fruit flies
   d. Domestic animals

5. Which of the following is not a valid conclusion resulting from studies done on hygienic behavior in honeybees?
   a. The inheritance of this behavior can be explained in a simple Mendelian manner.
   b. The uncapping and larval-removal components of this behavior seemed to be determined by a single gene.
   c. Since all the F_1 bees resulting from a cross between hygienic and nonhygienic strains were nonhygienic, the trait must be determined by recessive alleles.
   d. A backcross of the F_1 to the nonhygienic strain should yield one-quarter hygienic to three-quarters nonhygienic individuals.

6. Which of the following statements about the action of releasers in herring gull chicks is *false*?
   a. The red dot on the mother's beak acts as a releaser.
   b. The shape of the head is not important in eliciting a response from the chick.
   c. This behavior is learned rather than inherited genetically.
   d. Begging is the behavior triggered by the releaser.

7. Which of the following statements about genetically determined behaviors is *false*?
   a. Inherited behaviors that involve mating and courtship are highly adaptive.
   b. Hybridization experiments mate closely related species to determine the behavior of the offspring.
   c. The web spinning in spiders is a natural deprivation experiment.
   d. Most behaviors have only a genetic component.

8. Animals exhibit daily rhythms in their behavior and physiology. Which of the following is *false* about circadian rhythms?
   a. Animals that are active at night do not display circadian rhythms.
   b. The circadian clock can be entrained by environmental cues.
   c. The circadian clock is located in the suprachiasmatic nuclei in mammals.
   d. In fruit flies the loss of the *tim* gene will result in the loss of circadian rhythms.

9. Which of the following is an example of piloting?
   a. Silkworms following a trail of hormones
   b. Marine migration over the featureless ocean
   c. Bees learning the directions to a food source from a dance
   d. Gray whale migration along the pacific coast of America

10. There are a number of different navigational methods used by animals. Which definition of a navigational method listed below is correct?
    a. Distance-and-direction navigation requires knowledge of latitude and longitude.
    b. Bicoordinate navigation may involve using the circadian clock to determine direction in relation to the sun.
    c. The stars offer two sources of information about direction, fixed points and moving constellations.
    d. Piloting involves the magnetic mineral magnetite.

11. Which one of the following animals would not be expected to have an endogenous circannual rhythm?
    a. A ground squirrel that hibernates
    b. An Arctic hare
    c. A cave salamander
    d. A bird that summers in North American and migrates to South America for the winter

12. In studies using fruit flies it has been found that the _____ gene is involved in the development of male courtship behavior, and the ______ gene is involved in anatomical differentiation of males.
    a. *sex-lethal, transformer*
    b. *sex-lethal, doublesex*
    c. *fruitless, doublesex*
    d. *fruitless, transformer*

Application Questions

1. You go to a nature center that has a honeybee hive that you can see into. As you look at all the busy bees you notice that one bee is doing a dance. What is the purpose of this dance and what may it be telling the other bees?

2. A starling has been trained to feed at food bins at the eastern end of a circular cage from which it can see the sky. There are food bins at the N, NE, E, SE, S, SW, W, and NW ends of the cage. Knowing this, answer the following questions.
   a. If the cage is covered and a fixed light source is presented at the east of the cage at the time of sunrise, where will the animal search for food?
   b. If a mirror is used to shift the apparent position of the sun at noon from the South to the Northwest, where will the animal search for food?
   c. The bird has been placed in a light-controlled environment for three weeks and phase-delayed by six hours. If the bird is returned to the cage under natural lighting conditions at sunset, where will the bird search for food?

Answers

Knowledge and Synthesis Answers

1. **d.** The development of normal male sexual behavior requires the presence of testosterone. In castrated new born males, the testosterone is lacking. As an adult this animal will not show a sexual behavior towards a female.
2. **c.** Fixed action behaviors are acquired by inheritance and are not learned. The behavior of a predator may be a learned response rather than one that is inherited.
3. **d.** In songbirds, the new born males must hear the species song it must become imprinted. In the adult, the presence of testosterone is needed to increase the size of regions in the brain associated with singing during the breeding season. Absence of testosterone will result in the inability to perform the correct song.
4. **c.** The fruit fly has been used in many genetic studies on behavior.
5. **b.** In hygienic bees, the uncapping and larval removal appears to be controlled by two separate genes.
6. **c.** The herring gull chicks have inherited the begging response that occurs when they observe a red dot.
7. **d.** Behaviors tend to involve a complex interaction between learning and genetics.
8. **a.** All animals exposed to a daily cycle display circadian rhythms regardless of when they are active.
9. **d.** Piloting is a means of navigation using visual cues and landmarks. Following the coast during migration is considered piloting.
10. **c.** Many animals use the stars for navigation. The stars present two types of information. Some stars are fixed in the sky, providing a fixed point, while others provide moving constellations.
11. **b.** Hibernators, cave-dwellers, and equatorial migrators lack the needed daily cues at some stage during the year, so they most likely have endogenous circannual rhythms. Thus the Arctic hare would be the only animal that could use circadian cues.

12. **c.** The *fruitless* gene regulates development of courtship behavior and the *doublesex* gene regulates male anatomical development.

Application Answers

1. The bees dance to inform the other members of the hive about the distance and location of a food source. If food is less than 80—100 meters from the hive, the bees will perform a round dance. When the food is farther from the hive, the bees will a waggle dance. Only the waggle dance contains information about direction.

   a. The starling will eat from the northern bin at noon. Normally at noon the sun is in the south and the bird would eat from the eastern bin that is 90° to the left of the sun. With the light in the east, the bird would eat in the North, which is 90° to the left of the light.

   b. The bird will go to the bin that is 90° from the sun at noon. This will have the bird feeding from the southwestern food bin.

   c. At sunset the eastern bin with food would normally be 180° from the setting sun in the west. After the bird has been phase-delayed six hours it thinks that the setting sun is the noon sun. At noon the bird usually eats from the eastern bin, which is located 90° to the left of the sun Therefore, the bird will eat from the southern bin.

STUDY GUIDE

53 Behavioral Ecology

Important Concepts

Ecology is the study of the interactions of organisms with one another and with their nonliving environment.

- Ecology is a complex field of inquiry dealing with levels of organization ranging from the interactions of individual organisms with the environment to the structure of ecosystems.
- The environment of an organism includes both abiotic (chemical and physical) and biotic factors. Organisms both influence and are influenced by their environments.

Behavioral ecology is the study of how animals decide when and where to carry out their activities; how to respond to predators, competitors, and other members of their own species; and how to select the resources they need.

- Animals do not make conscious "decisions," but their behavioral choices do influence their chances of survival and reproduction and thus are molded by natural selection.
- Natural selection molds behavior in accordance with costs and benefits. There are three aspects to the cost of behaving:
  1. *Energetic cost*—the amount of energy the animal expends during the behavior.
  2. *Risk cost*—the amount of risk the behavior entails for the animal.
  3. *Opportunity cost*—the benefits the animal forfeits by not engaging in other behaviors instead.

Where an animal chooses to live and how it uses that environment strongly influence its survival and reproductive success.

- The habitat of an animal is the environment in which it normally lives. In choosing their habitat, animals use cues that are good predictors of general conditions suitable for future survival and reproduction.
- Foraging theory is an attempt to understand how animals select what kind of food to eat and where and when to search for it. These questions are investigated by optimality modeling, which assumes that natural selection has molded the behavior of animals so that they make the best choices available to them. To decide whether a model is realistic, scientists test its predictions by experiments that measure the actual foraging behavior of animals.

The most important choice of associates an animal makes is mate selection.

- Because the quantity of sperm produced by males is typically far greater than the number of eggs produced by females, males and females approach courtship very differently. Males of most species can increase their reproductive success by mating with many females. In contrast, the reproductive success of females depends primarily on the quality of genes she receives from her mate, the resources he controls, and the amount of care he provides for their offspring. As a result, males usually initiate courtship and frequently fight for access to a female. Females seldom fight over males and often reject courting males.
- Sexual selection favors traits that benefit their bearers in the competition for mates or resources or during courtship. This type of selection often leads to exaggerated traits in one sex or the other.
- A male may attract females by courtship behavior that signals his good health and genetic quality. The behavior also may show that he is a good provider of parental care. In some species, he may attract females by possessing resources such as a territory that contains food or nesting sites.
- Males have evolved behaviors that increase the probability that it will be their sperm (not another male's) that fertilize a female's eggs.
- By their ability to recognize signals at which males cannot cheat, females have favored the evolution of "reliable" signals of male quality.
- DNA fingerprinting methods have revealed that social fathers often are not genetic fathers.

Social behavior evolves when individuals that cooperate with others of the same species have higher rates of survival and reproductive success than those achieved by solitary individuals.

- Social systems are best understood not by asking how they benefit the species as a whole, but by asking how the individuals that join together benefit by the association.
- Social systems are dynamic: Individuals constantly communicate with one another and adjust their relationships.
- Group living may benefit both predator and prey species. For predators, it may improve hunting efficiency or

increase the size of the prey that can be captured. For prey, it may provide increased protection against predators.

- Living in a group imposes costs as well as benefits. Individuals in groups may compete for food, interfere with one another's foraging, injure one another's offspring, inhibit one another's reproduction, or transmit diseases to their associates.

Social acts include altruistic acts, selfish acts, cooperative acts, and spiteful acts.

- Altruistic acts benefit another individual at a cost to the performer. Selfish acts benefit the performer but impose a cost on another individual. Cooperative acts benefit both the performer and the recipient. Spiteful acts inflict costs on both the performer and the recipient.
- The terms used for these acts do not imply conscious motivation on the part of the animal. If a genetic basis for an act exists, the genes governing the act will increase in frequency if performing it increases the fitness of the performer.
- Altruistic acts evolve most easily through kin selection, by which an individual can increase the representation of its own genes in the population by helping its relatives. Together, individual fitness and kin selection determine the inclusive fitness of an individual. An altruistic behavior pattern can evolve if it increases the inclusive fitness of the altruist—that is, if the benefit of increasing the reproductive success of relatives is greater than the cost of decreasing the altruist's own reproductive success.
- Eusocial species are those whose social groups include sterile individuals. The evolution of eusociality in the Hymenoptera (ants, bees, and wasps) has been facilitated by their sex determination system, in which males are haploid and females are diploid, with the result that sisters are genetically more similar to one another than to their own offspring. In eusocial species in which both sexes are diploid (termites, naked mole-rats), inbreeding and high predation rates may have favored the evolution of eusociality.
- Altruistic behavior among unrelated individuals can evolve through reciprocal altruism, in which helpers are in turn recipients of beneficial acts by individuals they have helped.

Because behavior leaves few traces in the fossil record, the evolution of social systems must be inferred primarily from current patterns of social organization.

- The origins of all animal societies lie in the association of parents with their offspring.
- Vertebrate groups show major differences in the amount and kind of parental care given by males and females. In mammals, females nurse the young and usually provide other parental care; in most bird species, both sexes feed the young; in fish, parental care (consisting primarily of guarding eggs and young) is given frequently by males.
- The most widespread social system is the family, an association of one or more adults and their dependent offspring. More complex systems have probably evolved via the extended family route.
- The type of social system a species evolves is related to the environment in which it lives. For example, striking differences in social organization may be observed between species that inhabit forests and related species that live in grasslands or savannas.

The Big Picture

- Behavioral ecologists attempt to understand how the decisions animals make are related to their survival and reproductive success. In this effort, they not only draw on concepts from evolution, ecology, and animal behavior, but also make important contributions to those fields.
- The discovery, based on DNA fingerprinting, that the nestlings of many bird species that form pair bonds may have different fathers is an example of how recent advances in molecular biology have benefited other areas of biological research.
- Since, according to standard Darwinian theory, only traits that contribute to individual fitness are favored by natural selection, explaining the evolution of altruistic behavior has presented a major challenge to evolutionary biologists. The concepts of kin selection, inclusive fitness, and reciprocal altruism have proved to be very fruitful in understanding altruism in animals and have also been applied to many aspects of human social behavior (though not without controversy).

Study Strategies

- In studying any area of biology, you should try to develop a sense of the kinds of questions that scientists in the field are trying to answer. Behavioral ecologists are especially interested in the adaptive value of behavior (i.e., how it contributes to fitness). As you know from reading this text, biologists in many other fields focus primarily on understanding the underlying mechanisms (e.g., physiological) that produce the properties (including behavior) of living organisms. These two approaches are complementary and both are essential for a full understanding of life.
- Review the following tutorial and activity on the website:
  Tutorial 53.1 Foraging Behavior
  Activity 53.1 Concept Matching

Test Yourself

Knowledge and Synthesis Questions

1. Which of the following is not usually included within the domain of ecology?
   a. Interactions between conspecifics
   b. Modifications of the environment by organisms
   c. Evolution of different social organizations
   d. Modification of the environment by physical processes

2. Which of the concepts listed below best describes the following situation: Birds spend some of their time scanning the horizon for predators. While scanning, they cannot be foraging for food.
   a. Cooperative hunting
   b. Energetic cost

c. Risk cost
d. Opportunity cost

3. Two species of mice live in the same geographical region, but species 1 prefers open fields, whereas species 2 lives in forests. When presented in an experiment with simulated "fields" and "forests," individuals of each species born in a laboratory preferred the environment in which they normally lived. This experiment illustrates the concept of
a. habitat selection.
b. optimal foraging strategy.
c. territoriality.
d. a and b

4. Which of the following animals would likely have a territory that is used exclusively for a nest site?
a. The song sparrow
b. The Florida scrub jay
c. The red-winged blackbird
d. A colonial marine bird

5. You perform a series of experiments in which the density of prey species 1 is varied while the density of prey species 2 is kept constant. At each prey species 1 density, you determine how many prey of species 1 and 2 the predator takes. The two curves (*a* and *b*) plotted in the following graph show the outcome of this study.

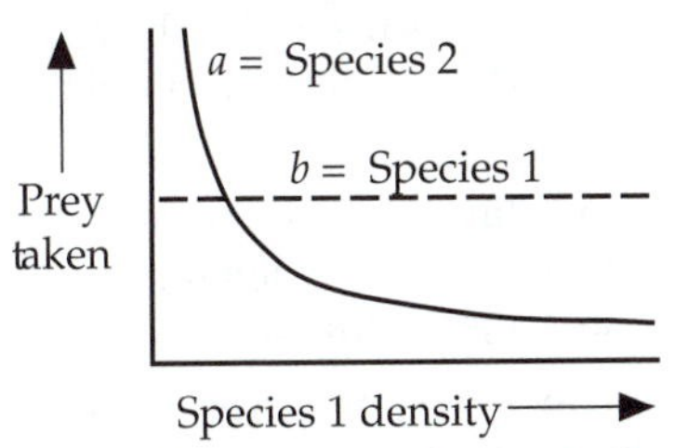

Select the statement that correctly interprets these curves.
a. Curve *a* indicates that species 2 is the preferred prey.
b. Curve *b* indicates that species 1 is the preferred prey.
c. Curve *b* indicates that species 1 is not the preferred prey.
d. Neither of these curves provides insight into the prey preference of the predator.

6. You observe an example of an apparently altruistic act by an animal that seems to reduce its near-term likelihood of reproductive success. What would be the *least* plausible explanation for its behavior?
a. The act aids the reproductive success of individuals sharing a high proportion of genes with the altruistic individual.
b. The act is only apparently altruistic; over the long term, the behavior actually contributes to individual fitness.
c. The behavior is based on the expectation that performing the act increases the probability of being repaid in the future by other members of the social group.
d. The act is advantageous because it helps the species to survive and reproduce, even if the altruist himself does not.

7. Some birds give a species-specific vocalization called an "alarm call" when they see a predator, although this call may direct the predator toward them. Other members of their species respond to these calls by taking cover. This would be an example of altruistic behavior that is beneficial to the calling bird if
a. the bird giving the vocalization survives the attack.
b. the inclusive fitness of the bird giving the vocalization is increased.
c. all birds survive the attack.
d. the birds benefiting are offspring of the bird giving the vocalization.

8. A dominant individual displaces a subordinate individual from food that the subordinate has discovered. This is an example of a(n)
a. selfish act.
b. cooperative act.
c. altruistic act.
d. spiteful act.

9. For a behavioral act to be considered an example of reciprocal altruism,
a. both individuals must potentially benefit from the act at some time.
b. the act must have a genetic basis.
c. the individuals involved must be closely related.
d. a and b

10. Which of the characteristics listed below is (are) shared by all eusocial species?
a. A sex determination system in which males are haploid and females are diploid
b. Queens that mate with a single male
c. Presence of sterile classes
d. All of the above

11. Which of the following statements describing the social system of the Florida scrub jay is *false?*
a. It is a eusocial organization.
b. Most helpers at the nest are offspring from the previous breeding season that remained with their parents.
c. The social system probably evolved because offspring have little chance of establishing new territories on their own.
d. Only the parent birds breed.

12. Of the following factors, select the one that was *not* mentioned as being important in the evolution of social organization in African hoofed animals.
a. Metabolic needs
b. Predation
c. Dispersal of breeding territories
d. Dispersal of food resources

13. Which of the following choices would *not* be a factor that can influence the social organization typical of a species?
a. Presence of a dominance hierarchy
b. Availability of nest sites
c. Sex determination mechanism
d. Asymmetry in reproductive investment

Application Questions

1. Describe the experiments done with bluegill sunfish preying on water fleas and the implications of these studies for foraging theory.
2. Describe what is meant by sexual selection and how the sexual selection of tail length in male widowbirds was studied.
3. For each of the two y-axes in the following graph, draw a labeled curve that correctly summarizes observations made on goshawks attacking wood pigeons, as described in the textbook.

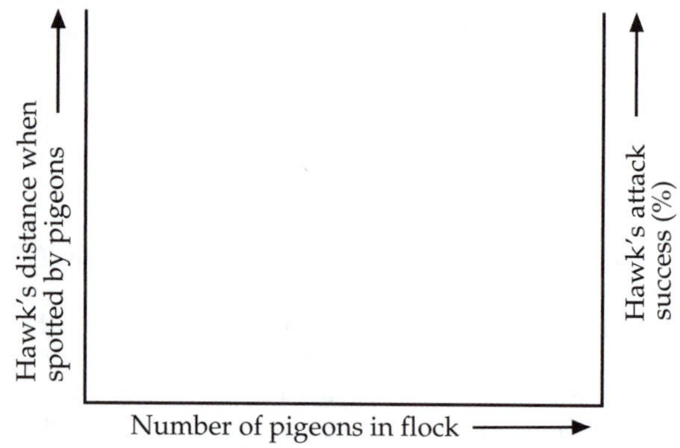

4. Based on the unusual sex determination system in Hymenoptera, describe W. D. Hamilton's hypothesis for the evolution of eusociality.
5. Describe important environmental variables that seem to have guided the evolution of social organization in species of African weaverbirds.

Answers

Knowledge and Synthesis Answers

1. **d.** All except the modification of the environment by physical factors are normally included within the domain of ecology.
2. **d.** The forfeited benefits of behaviors that could not be performed as a result of performing a different behavior, like scanning, comprise the opportunity cost of the performed behavior.
3. **a.** The environment in which a species normally lives is its habitat.
4. **d.** A territory that is used exclusively for nesting and rearing young is characteristic of many marine birds. All of the other species have territories that are also used for foraging.
5. **b.** Since the number of species 2 taken by the predator increases markedly at low species 1 densities, you can conclude that the predator includes species 2 in its diet only when insufficient species 1 individuals are available. The number of species 1 taken is independent of species 1 density.
6. **d.** For altruistic behavior to evolve by natural selection, it must increase the fitness of the individual performing the behavior, either through kin selection or reciprocal altruism. Any explanation of an altruistic act based on its purported benefit to the species as a whole violates this principle.
7. **b.** Altruistic behavior is beneficial to the performer when the improvement in the reproductive success of relatives exceeds the reduced reproductive success of the individual performing the act. This would be true if the behavior improves the inclusive fitness (individual fitness + kin selection) of the performer.
8. **a.** A selfish act benefits the performer at the expense of the recipient of the act.
9. **d.** Reciprocal altruism involves cooperative behavior between unrelated individuals; the behavior is reciprocated so all benefit, and the behavior has a genetic basis.
10. **c.** By definition, eusocial species live in social groups with sterile castes. The sex determination mechanism in which males are haploid and females diploid is found in the Hymenoptera (ants, bees, and wasps) but not in termites and naked mole rats. Within the Hymenoptera, there are species in which queens mate with many males.
11. **a.** Though only the parent birds breed on a Florida scrub jay territory, the helper offspring do not constitute a sterile caste (as would be required for eusociality). They may later become breeders themselves by taking over their parents' territory or establishing one of their own.
12. **c.** The distribution of breeding territories was not mentioned as an important factor in the evolution of social organization in African hoofed animals.
13. **a.** A dominance hierarchy is a type of social organization, not a determining factor.

Application Answers

1. In studies in which bluegill sunfish were provided with small, medium, and large waterfleas at three different prey densities, the fish took equal proportions of the three prey sizes at low densities, but mostly large waterfleas at high prey densities. This agrees with predictions from foraging theory on how a predator should behave to maximize its energy intake.
2. Sexual selection is the spread of a trait that improves the reproductive success of the individual, as determined by mating preferences of individuals of the opposite sex. By artificially lengthening or shortening the tails of male widowbirds, behavioral ecologists were able to show that increased tail length attracted more females but did not confer an advantage in interactions with other males.
3. Your curves should show a positive relationship between the hawk's distance when spotted and pigeon flock size and a negative relationship between the

hawk's attack success and pigeon flock size (see below).

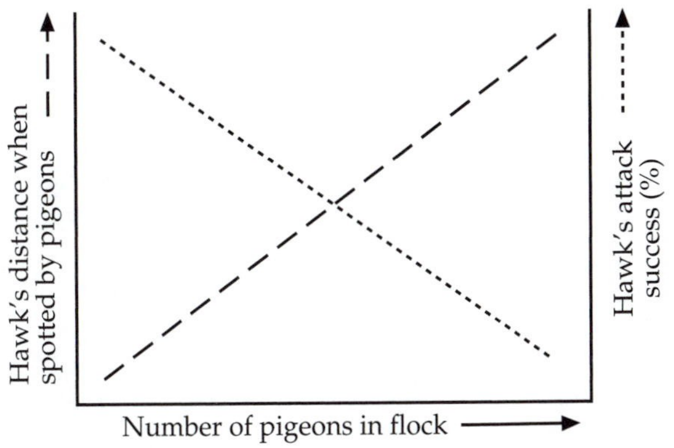

4. All species of the Hymenoptera (ants, bees, and wasps) have a sex determination system in which males are haploid and females are diploid; thus females share 75 percent of their genes with their sisters but only 50 percent of their genes with any offspring they could produce. Hamilton's hypothesis predicts that in species whose sexes are determined in this way, females can increase their inclusive fitness by foregoing reproduction and helping to raise sisters. This hypothesis cannot explain the evolution of eusociality in species without this mode of sex determination or in the many species of hymenopterans in which the queen mates with many males.

5. Forest-dwelling African weaverbird species are monogamous, eat insects, build inconspicuous nests, and have males and females that are similar in appearance. Savanna-dwelling species are polygynous, eat seeds, build elaborate nests in colonies in *Acacia* trees, and have males that are brightly colored. The major environmental variable causing these differences seems to be that nest sites are widely dispersed in forests, but are highly clumped in savannas.

54 Population Ecology

Important Concepts

A population consists of all the individuals of a species within a given area.

- Population structure includes such features as the population's size, density, spacing pattern, and age distribution.
- Population density is usually measured by determining the number of individuals of a species per unit of area (or volume), though for plants and some animals, total biomass is a more useful measure than the number of individuals. Density is commonly estimated by sampling a representative area or by marking and recapturing individuals.
- A modular organism consists of many genetically identical units (modules). For a population ecologist, the number of the modules in a population may be more important than the number of genetically distinct individuals.
- Individuals in a population may be clumped, evenly spaced, or randomly scattered.
- The age distribution of a population describes the proportions of individuals in all age categories, from newborn to postreproductive adults. It reveals much about the recent history of births and deaths in the population. In populations of long-lived species such as humans, the timing of births and deaths may influence age distributions for many years.

Demography is the study of the changes in the size and structure of populations.

- To study the dynamics (changes over time) of a population, demographers measure its rates of birth, death, immigration, and emigration.
- A life table for a population is constructed by determining for a cohort (a group of individuals born at the same time) the number still alive at specific times and the number of offspring they produced during each time interval. Life tables are used to predict future growth rates of populations.
- Survivorship curves show the number of individuals in a cohort still alive at different times over the life span. Graphs of survivorship in real populations often resemble one of three types.
  1. Most individuals survive for the entire potential life span and die at about the same age (curve I).
  2. Survivorship is the same throughout the life span (curve II).
  3. Survivorship of the young is very low, but is high for most of the remainder of the life span (curve III).

All populations have the potential to grow exponentially, but because of environmental limitations, no population can maintain exponential growth for very long.

- In exponential growth, the per capita rate of increase remains constant per unit of time while the number of individuals added to the population accelerates. This occurs because, as the population grows, more and more individuals are alive to reproduce.
- Exponential growth can be expressed mathematically as
  $$\Delta N/\Delta t = (b - d)\,N$$
  where $\Delta N/\Delta t$ is the rate of change in the population size (ΔN = change in number of individuals; Δt = change in time), b and d are the per capita birth and death rates, respectively, and N is the population size.
- The difference between the per capita birth rate (b) and the per capita death rate (d) is the per capita rate of increase (r). Where there are no limits on population growth, r has its highest value, designated r_{max}, the intrinsic rate of increase. Therefore, the rate of growth of a population under optimal conditions is
  $$\Delta N/\Delta t = r_{max}\,N.$$
- The carrying capacity (K) is the maximum population size that the environment can support. It is determined by resource availability, as well as by disease, predators, and, in some cases, social interactions. Because growth tends to slow as the carrying capacity is approached, an S-shaped growth curve is characteristic of many populations growing in environments with limited resources. This type of growth can be expressed mathematically by the logistic growth equation:
  $$\Delta N/\Delta t = r\,((K - N)/K)\,N.$$
  Population growth stops when $N = K$ because then $(K - N)/K = 0$.
- The overall densities of many populations are determined by "births" (colonizations) and "deaths" (extinctions) of local subpopulations, among which some exchange of individuals occurs. In the rescue effect, immigrants may prevent declining subpopulations from becoming extinct.

Factors affecting the growth of populations are categorized as density-dependent or density-independent.

- Density-dependent environmental factors—such as resource depletion, predation, and disease—cause per capita birth or death rates of a population to change in response to changes in population density. As a result, population growth slows down as density increases. A population whose dynamics are influenced primarily by density-dependent factors is said to be regulated.
- Factors (such as adverse weather) that change per capita birth or death rates of a population irrespective of its density are said to be density-independent. A population whose dynamics are controlled by density-independent factors is unregulated.
- The density of a population is determined by the combined effects of all density-dependent and density-independent factors affecting it.
- Disturbances are short-term physical or biological events that change the environment and hence affect population densities. They typically decrease the carrying capacity for some species while increasing it for others. The frequency of some disturbances (fires, for example) is influenced by populations themselves.
- Dispersal—movement to another habitat—is a common way by which organisms cope with environmental change. Migration is the regular seasonal movement of an animal population from one place to another.

Life histories influence population growth.

- Life histories include several common features. For at least part of their lives, all organisms grow by gathering and assimilating energy and nutrients. In many species individuals change form during their lives: for example, by undergoing metamorphosis from larva to adult, or by forming a resting stage such as a spore. The life history of all organisms also includes at least one dispersal stage, in which individuals change their location.
- Life histories inevitably represent trade-offs between number and size of offspring, between number of offspring and amount of parental care, between survival and reproduction, and between reproduction and growth.
- Reproductive value is the average number of offspring that remain to be born to individuals of a particular age. It rises to a peak when an individual first begins to reproduce and declines to zero after reproduction ceases.
- Senescence—an increased probability of dying with increasing age—has evolved because natural selection cannot act on alleles, no matter how detrimental, that first produce their phenotypic effects after reproductive value has dropped to zero.

Humans use the principles of population dynamics to manage and control populations.

- To maximize the number of individuals that can be harvested from a population, the population should be held far enough below carrying capacity to have high birth and growth rates.
- Life history traits, particularly reproductive capacity, determine how heavily a population can be exploited.
- The populations of undesirable species are most effectively controlled by lowering the carrying capacity of the environment for those species. For non-native pest species, the introduction of predators and parasites from the pest's original habitat is sometimes an effective control measure.
- To preserve a rare species, the most important step usually is to provide it with suitable habitat.
- Over the past 10,000 years, Earth's carrying capacity for humans has been increased several times by technological advances: the agricultural revolution, the tapping of fossil fuels, and the development of modern medicine and hygiene. Earth's human population grew explosively in the twentieth century and now exceeds 6 billion people.

The Big Picture

- An understanding of concepts of population ecology such as exponential growth and carrying capacity is fundamental to your ability to think intelligently about issues such as human population growth, the preservation of biodiversity, and the control of undesirable species.
- Darwin's realization that all populations have the inherent capacity for exponential growth was crucial to the development of his theory of natural selection.

Common Problem Areas

- Students frequently have problems interpreting graphs depicting concepts of population ecology. Look carefully at how the axes of a graph are labeled. Ask yourself questions such as: Are both scales arithmetic? Is the fate of a cohort of the population being traced, or is the growth pattern of the entire population being described?

Study Strategies

- Redrawing graphs from the text is a good method of reinforcing your understanding of the concepts being presented.
- Take advantage of laboratory activities involving computer simulations of population growth or other aspects of population ecology.
- Review the following tutorials and activities on the website:
  Tutorial 54.1 Applet: Exponential Growth
  Tutorial 54.2 Applet: Logistic Growth
  Activity 54.1 Logistic Growth
  Activity 54.2 Population Regulation

Test Yourself

Knowledge and Synthesis Questions

1. From an ecologist's point of view, population structure does *not* include the
   a. distribution of genotypes within a population.
   b. population density.
   c. spacing of population members.
   d. biomass of the population.

2. A random distribution of population members is usually caused by
   a. settling near one's birthplace.
   b. territoriality.
   c. competition for a uniformly distributed resource.
   d. the interaction of several factors that affect survival.

3. In the following graph, which of the expressions (*a* through *e*) of the exponential growth equation should be *increased* in order for curve 1 to become more like curve 2?

$$\frac{\Delta N}{\Delta t} = (b - d)\,N$$

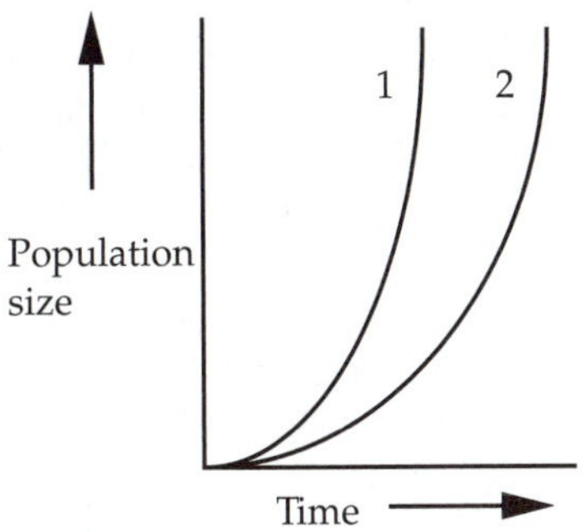

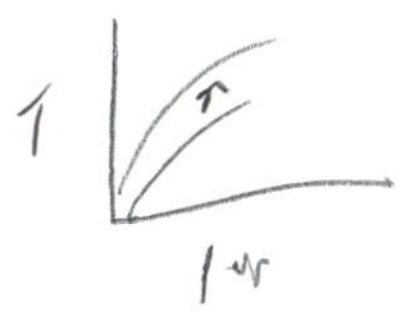

   a. *N*
   b. *d*
   c. *b*
   d. $(b - d)$

4. Species A and B have intrinsic rates of increase (r_{max}) of A = 0.25 and B = 0.50. In reference to the graph shown in question 3, the population growth curve for A should be more like curve _______.

5. In the logistic population growth curve shown below, the rate of growth is greatest at which point?

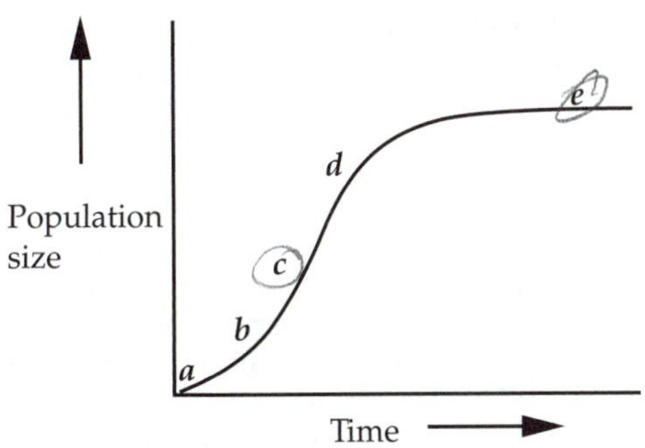

6. At which point in the graph shown above would there be zero population growth ($\Delta N/\Delta t = 0$)?

7. Based on the following life table, during what time interval is survivorship greatest?

Age (years)	Number Alive
0	800
1	770
2	550
3	125
4	75
5	0

   a. 0–1 years
   b. 1–2 years
   c. 2–3 years
   d. 3–4 years
   e. 4–5 years

8. Based on the graph shown below, which of the following events is *least likely* to be true of this population?

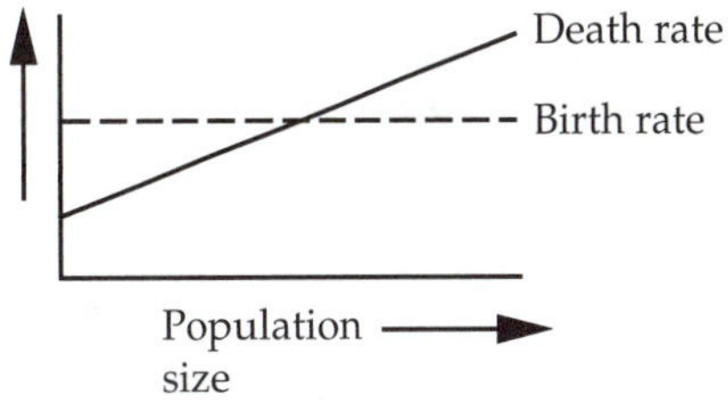

   a. Parasites spread between population members.
   b. Increased competition for food causes some individuals to delay reproduction.
   c. The number of predators in the area varies with population density.
   d. Territorial disputes led to injury and death of some males.

9. In the graph shown in question 8, the population size where the curves for the birth and death rates intersect is
   a. an estimate of the carrying capacity.
   b. an estimate of the intrinsic rate of increase.
   c. the point at which density-dependent regulation begins.
   d. the point at which density-independent regulation begins.

10. Which of the following would probably *not* be true of a population whose dynamics are primarily influenced by density-independent factors?
    a. The population's growth pattern is similar to the logistic growth curve.
    b. The birth rate of the population is dependent on the nutritional status of its adult females.
    c. The most important source of mortality in the population is unfavorable weather conditions.
    d. a and b

11. For a population that is stable in size, the following age distribution indicates that

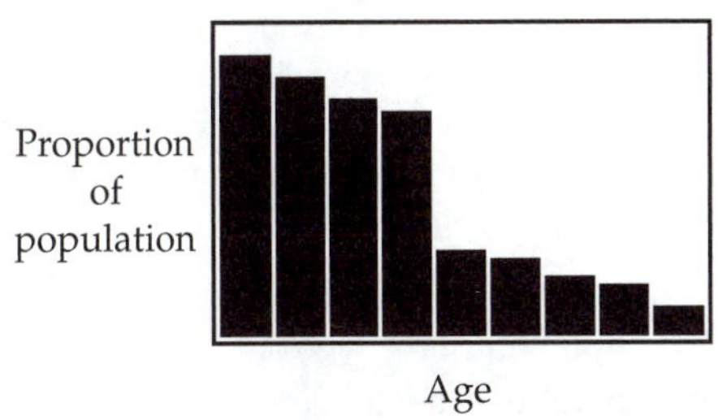

    a. the population's birth and death rates are both high.
    b. the population's birth and death rates are both low.
    c. the population's birth rate is low but its death rate is high.
    d. the population's birth rate is high but its death rate is low.

12. Select all of the following that are *not* adaptations to seasonal changes in habitat quality.
    a. Migration
    b. Expansion of the species range
    c. Production of seeds
    d. Hibernation

13. The *best* way to decrease the population size of a pest species is to
    a. poison it.
    b. introduce additional predators.
    c. decrease the carrying capacity of the habitat for the species.
    d. add competitors.

14. The number of seeds that a tree produces increases, but the resources available to the tree remain the same. Put a "+", "–", or "0" before each of the following variables to express the expected effect of this change on the variable.

    ______ Weight of each seed

    ______ Width of the tree's growth rings

15. In which of the following life-history stages are annual plants and Pacific salmon most similar?
    a. Growth
    b. Dispersal
    c. Reproduction
    d. Senescence

16. For a species that breeds only during one season and lives in a disturbed environment where new habitat is constantly being created, it would be best to produce
    a. many small offspring only once.
    b. a few small offspring only once.
    c. many small offspring more than once.
    d. many large offspring more than once.

Application Questions

1. You collect a sample of 100 bats that are hibernating in a cave. Each bat is marked and released. Two weeks later you return and collect another sample of 100 bats, 20 of whom are marked. How many bats are estimated to be in the cave
2. How does the division of some populations into discrete subpopulations relate to the process known as the rescue effect?
3. Draw the three major types of survivorship curves, taking care to label both axes of the graph accurately with respect to their scales. For each curve, name a species whose survivorship resembles it.
4. Discuss the three life history trade-offs presented in the textbook.
5. Discuss the implications that the concept called "reproductive value" has for hopes of greatly increasing human life expectancy.
6. Describe the history of human population growth over the past 12,000 years. What factors have raised Earth's carrying capacity for our species?

Answers

Knowledge and Synthesis Answers

1. **a.** Population structure includes numbers, spacing, and age distribution of population numbers, but not the distribution of genotypes. Biomass is often the best measure of density for some organisms like plants.
2. **d.** A uniform distribution normally results from competition for a uniformly distributed resource or from territoriality. Settling near one's birthplace can cause a clumped distribution. The interaction of several factors may lead to a random distribution.
3. **b.** Both curves show exponential growth, but the rate of exponential growth for curve 2 is less than the rate for curve 1. Since the rate of growth is determined by the expression ($b - d$), increasing d, the death rate, would cause curve 1 to be more like curve 2.
4. **Curve 2**. Since $r_{max} = (b - d)$, you would expect the growth curve for species A to correspond to the curve with the smaller rate of exponential growth.
5. **c.** The curve showing the relationship between population size and time is steepest at point *c*, so the growth rate would be greatest at that point.
6. **e.** The growth rate at *e* would be zero. This is the point called the carrying capacity, the point at which the birth and death rates are equal.
7. **a.** The decrease in consecutive age classes is least between 0 and 1 year, so the survivorship is greatest during that interval.
8. **b.** The graph indicates that the birth rate is independent of population size, so event *b* would not be true. The graph shows that the death rate is density-dependent
9. **a.** The carrying capacity is the equilibrium population size when the birth rate equals the death rate. The birth rate equals the death rate at the intersection of the two curves.
10. **d.** The logistic growth curve describes density-dependent growth because the rate of increase in the size of the population steadily decreases as the carrying capacity is approached. The nutritional status of the females in the population would depend on the availability of food and hence would be density-dependent. Unfavorable weather, on the other hand, generally occurs without any predictable relationship to the size of a population and thus is a density-independent factor.
11. **a.** When a population's birth rate and death rate are both high, the age distribution is dominated by young individuals. A similar age distribution would occur in a population with a high birth rate and a low death rate, but such a population would not be stable in size.
12. **b.** Expansion of a species' range is a long-term population change and not a response to seasonal changes in habitat quality.
13. **c.** The best way to control a pest species is to reduce the carrying capacity of the habitat for the species.

14. If the resources available to the plant are fixed, increasing the number of seeds produced would result in a decrease in both variables.

15. **c.** Both Pacific salmon and annual plants reproduce and then die, so they are most similar in the reproductive stage of their life histories.

16. **a.** Since new, unoccupied habitat is constantly being created in disturbed areas, a good strategy is to produce large numbers of smaller offspring.

Application Answers

1. The marked individuals were 20 percent of the second sample. If you assume that the original 100 marked bats redistributed themselves randomly within the whole population, then 100 is 20 percent of the total population. So the total population is $100/0.2 = 500$ individuals.

2. The rescue effect occurs if a subpopulation that has undergone a large decline is saved from extinction by the immigration of individuals from other subpopulations.

3. See Figure 54.4 in the text. Notice that the vertical axis of a survivorship curve has a logarithmic, rather than an arithmetic, scale. (This explains why curve II, in which survivorship is the same for all equal time intervals during the life span, is a straight line.)

4. Given fixed resources, there is an inverse relationship between the number of offspring produced and the amount of energy allocated to each. Also, there is an inverse relationship between the number of offspring produced and the amount of care each can receive. A final trade-off is that energy expended in reproduction is not available for growth and maintenance of the parent, so reproduction usually reduces the probability of parental survival.

5. Reproductive value is the number of offspring remaining to be born to an individual of a particular age. Reproductive value decreases with age and is zero for postreproductive individuals. Natural selection cannot act on traits that are expressed after reproduction is complete. Thus, many of the diseases afflicting postreproductive segments of populations in developed countries (cancer, cardiovascular disease, etc.) are beyond the reach of natural selection.

6. See Figure 54.16 in the text.

STUDY GUIDE

55 Community Ecology

Important Concepts

An ecological community consists of all the organisms that live together in a particular area.

- Within a community, the interactions of organisms may be
  1. mutually harmful (competition);
  2. beneficial to one organism and harmful to the other (predator–prey, parasite–host interactions);
  3. mutually beneficial (mutualism);
  4. beneficial to one organism while leaving the other unaffected (commensalism); or
  5. harmful to one organism while leaving the other unaffected (amensalism).
- These categories of species relationships are a useful guide, but not all interactions fit into them neatly.
- A resource is anything directly used by an organism that can potentially lead to the growth of the population and whose availability is reduced when it is used.
- The ecological niche of a species is the set of environmental conditions under which it can persist. The fundamental niche of a species is the set of physical conditions in which it can persist in the absence of other species in the environment. The fundamental niche is often larger than the realized niche, which takes into account the effects (usually negative) of biotic interactions.
- Limiting resources are those whose supply is less than the demand made upon them by organisms. Which resources are limiting differs among environments.

Competition occurs when two or more individuals use the same resources and those resources are insufficient to meet their demands.

- Intraspecific competition occurs among members of the same species, whereas interspecific competition involves individuals of different species. Either type may result in reduced rates of growth or reproduction, exclusion from favorable habitats, or death of some individual competitors.
- Competitive exclusion occurs when, as a result of interspecific competition, an entire species is kept out of habitats where it cannot compete successfully. Extinction or range restriction may result from competitive exclusion.
- Plants are good subjects for competition studies because the resources for which they compete are easily manipulated.
- Competition often reduces the abundance of competing species rather than eliminating them. Two forms that competition may take among animals are
  1. exploitative competition, in which one competitor reduces the food supply of another;
  2. interference competition, in which one competitor defends a space containing resources against another competitor.
- Species that use similar resources commonly coexist in nature because nature is spatially and temporally complex, many resources typically are available, and other factors often keep populations below carrying capacity so that they do not compete strongly.

Predator–prey and parasite–host relationships differ in important ways.

- Predators are typically larger than, and live outside the bodies of, their prey. They generally kill prey individuals when they eat them. Parasites are smaller than their hosts, live in or on them (often without killing them), and may live for many generations within a single host individual.
- Microparasites—such as viruses, bacteria, and protists—are much smaller than their host and typically reproduce repeatedly within their host's body. The host population generally has three classes: susceptible, infected, and recovered and immune. For a microparasite population to survive in a host population, on average at least one susceptible host individual must become infected before each infected host dies.
- Predator–prey interactions often result in oscillations in the densities of both populations, just as microparasites often cause fluctuations in populations of their hosts. However, factors in addition to predation (such as variation in food supply) may also be causes of population cycles of prey species.
- In heterogeneous environments, predators may eliminate their prey in some places but not in others.

In response to predation, prey species have evolved a rich variety of defensive adaptations, including toxic hairs, spines, noxious chemicals, camouflage, and mimicry.

- Mimicry is an evolved resemblance of a prey species to some inedible or unpalatable item. In Batesian mimicry, a palatable species mimics an unpalatable or noxious one. In

Müllerian mimicry, two or more unpalatable species converge in appearance over evolutionary time.

- Most plants have evolved secondary defensive chemicals for protection. Acute toxins interfere with herbivore metabolism or defend plants against viruses and bacteria. Digestibility-reducing compounds such as tannins make plant tissues difficult to digest by herbivores.

Amensalism, commensalism, and mutualism are common and important types of species interactions.

- In amensalism, one participant is harmed by the relationship but the other is unaffected. Commensalism benefits one partner but has no effect on the other.
- Mutualisms are interactions that benefit both participants. Some of the most ecologically important mutualisms are between members of different kingdoms or domains (between plants and prokaryotes, between fungi and protists, and between animals and protists). Animals have mutualistic relationships both with other animals and with plants, such as pollination and seed dispersal.

Coevolution is the process by which two species mutually influence one another's evolution.

- In diffuse coevolution, species, traits are influenced by interactions with a wide variety of predators, parasites, prey, or mutualists.
- In species-specific coevolution, much rarer than diffuse coevolution, two particular species strongly affect one another's evolution.

Some species have major influences on community composition.

- Through their interactions with other species, the organisms of a community may affect the number of species that live there (species richness). Organisms can also influence the physical environment and distribution of resources in a community.
- Keystone species have effects on ecological communities that are greater than would be expected on the basis of their abundance, but abundant species also have major influences on community structure.
- Vascular plants, mammals that change vegetation structure, and predators on dominant competitors often have major influences on ecological communities.

Ecological succession is the sequence of change in the species composition of a community.

- Patterns and causes of ecological succession are varied, but species that first colonize a site often alter the conditions under which later-arriving species grow.
- Primary succession begins with the establishment of organisms on newly available sites that previously had no organisms.
- Secondary succession begins when organisms reestablish themselves on disturbed sites where some organisms survived the disturbance. Secondary succession is also exemplified by the sequence of organisms that decompose all or part of the dead body of an organism.

Interactions between species often have indirect effects on many other species populations.

- To understand and quantify indirect effects, investigators need to observe and manipulate several species.
- An example of an indirect effect is the role of predation by mice in preventing gypsy moth populations from recovering quickly after they defoliate oak trees, thereby allowing the trees to recover.

The Big Picture

- In Chapter 54 ("Population Ecology"), you learned that density-dependent factors frequently regulate population size. The present chapter describes in greater depth how these factors—competition, predation, parasitic disease—actually operate.
- The discussion in this chapter of the dynamics of microparasite–host interactions is relevant to human disease epidemics, including the spread of the AIDS virus, HIV.
- As you learned in previous chapters, some mutualistic relationships have extraordinary evolutionary or ecological significance. Examples are the role of endosymbiosis in the evolution of eukaryotic cells, and the association of nitrogen-fixing bacteria with their host plants.
- In urging a cautious approach to human manipulation of the natural environment, ecologists like to say that "You can never do just one thing." The descriptions in this chapter of the effects of keystone species and of other complex community interactions illustrate the meaning of this saying.
- As mentioned in Chapter 54, if we wish to preserve a rare species, usually the most important step is to protect its habitat. This often requires an understanding of (and manipulation of) ecological succession in order to prevent changes in habitat unfavorable to the rare species.

Study Strategies

- Much of this chapter consists of descriptions of experiments on competition, predator–prey interactions, and so forth. As you study these descriptions, focus on what question the experiment was designed to answer, and how the results support the conclusions drawn from the experiment.
- Review the following tutorial and activity on the website:
  Tutorial 55.1 Applet: Predator-Prey
  Activity 55.1 Ecological Interactions

Test Yourself

Knowledge and Synthesis Questions

1. A bird eats the fruit of a plant species. The seeds are not digested and germinate in the bird's excrement at some distance from the parent plant. This is an example of
   a. predation.
   b. competition.
   c. commensalism.
   d. mutualism.

2. Certain birds follow swarms of foraging army ants and prey upon the insects that the ants flush. The relationship between these birds and the ants is an example of
   a. competition.
   b. commensalism.
   c. mutualism.
   d. a and b

3. According to the definition of resource developed in the text, which of the following would *not* be a resource?
   a. Oxygen for a terrestrial animal
   b. Sunlight for a plant species
   c. Oxygen for an aquatic insect
   d. Nesting boxes for a bluebird

4. As shown below, species A and B originally had nonoverlapping ranges. Species A expanded its range and now overlaps with species B in area 2. The two species still live separately in areas 1 and 3.

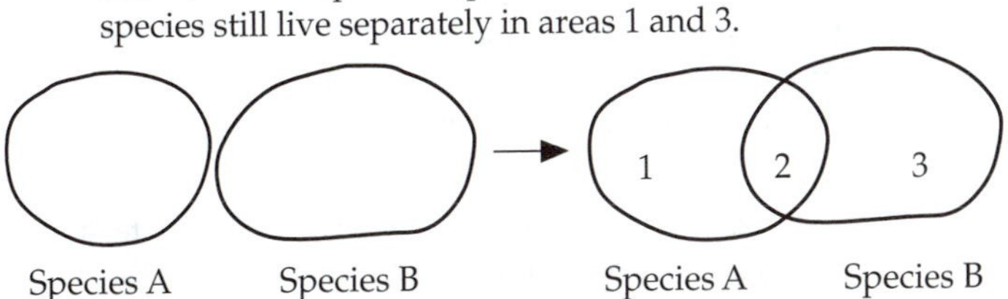

   If the two species compete for common, limiting resources, which of the following statements is not a likely outcome of the range expansion of species A?
   a. The range of conditions under which species B can exist will be greater in area 3 than in area 2.
   b. The ecological niche of species B in area 2 would be the same as it had been before the range expansion.
   c. The ecological niche of species A in area 1 would be the same as it had been before the range expansion.
   d. The range of conditions under which species A can exist will be greater in area 1 than in area 2.

5. In laboratory experiments, two species of the protist *Paramecium* were grown alone and in the presence of the other species. The following graphs show growth of species 1 (left) and species 2 (right), both alone and when in mixed culture.

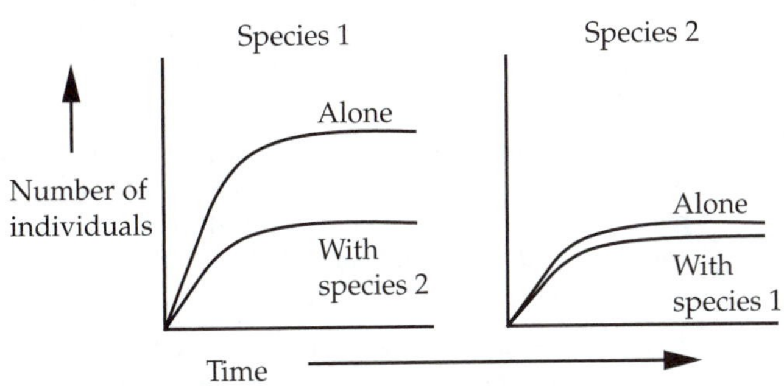

   Interpretation of these graphs shows that
   a. competitive exclusion occurred in these experiments.
   b. both species are affected by interspecific competition but species 1 is affected less.
   c. both species are affected by interspecific competition but species 2 is affected less.
   d. both species are affected equally by interspecific competition.

6. Which vertical line in the following graph represents the time at which the predator population (gray) is *increasing* and the prey population (black) is *decreasing*?

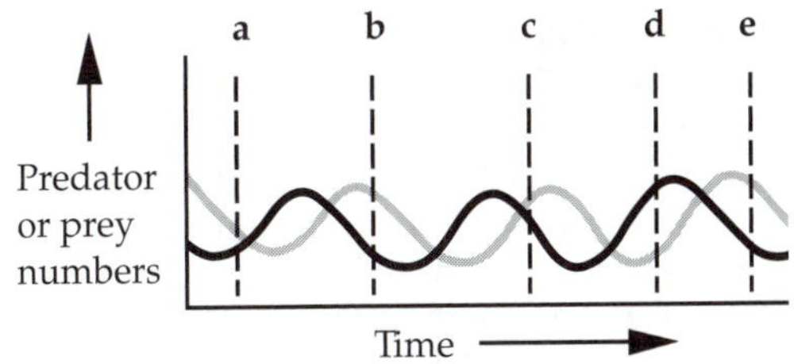

7. You are studying three butterfly species (1, 2, 3) that have very similar, bright coloration patterns. You collect the following data on the abundance of these species in two different regions, A and B. Based on your data shown below, which of the following conclusions is *most probable*?

Region	**Numbers collected for species:**		
	1	**2**	**3**
A	90	20	120
B	190	35	155

   a. Species 1 is the model, species 2 and 3 are Batesian mimics.
   b. Species 1 is a Batesian mimic, species 2 and 3 are Müllerian mimics.
   c. Species 1 and 3 are Müllerian mimics, species 2 is a Batesian mimic.
   d. All three species are Müllerian mimics.

8. If a species of plant that is trampled by an animal eventually evolves sharp spines that prevent trampling, we can say that its association with the animal has changed from
   a. amensalism to competition.
   b. amensalism to commensalism.
   c. commensalism to competition.
   d. commensalism to mutualism.

9. Which of the following is *not* an influence that plants normally have on biological communities?
   a. Moderating the climate
   b. Determining which animal species will be present
   c. Increasing CO_2 levels
   d. Reducing moisture reaching the soil

10. Which of the following is most characteristic of a keystone species?
   a. Very abundant
   b. Removal having a great effect on community structure
   c. Herbivore
   d. Uniform distribution

11. If the following list of stages of ecological succession occurring on glacial moraines in Alaska were put in correct chronological sequence, which would be *third*?
   a. Arrival of willows and alders
   b. Arrival of lichens
   c. Arrival of conifers
   d. Increase in soil nitrogen content

12. Which of the following would represent primary succession?
    a. The development of an aquatic community in a newly excavated farm pond, eventually followed by the filling in of the pond with sediment and its conversion to a forest
    b. The gradual establishment of a forest following a fire that destroyed the previous community
    c. The invasion of weeds followed by shrubs and trees on an abandoned farm field
    d. a and b

Application Questions

1. Describe how you would assess the degree of interspecific competition that exists between two rodent species living in the same habitat and how you would interpret the results.
2. Every year, during the "flu season," one or more strains of influenza virus infect many millions of people around the world, only to die out and be replaced by new strains the following year. How are the principles governing the dynamics of microparasite–host interactions relevant to this phenomenon?
3. Describe how plants defend themselves against herbivores.
4. Discuss the costs and benefits for both the moth and the plant in the yucca–yucca moth coevolution described in the textbook.
5. Describe how the interactions among gypsy moths, mice, oak trees, and predators of mice influence the frequency of gypsy moth outbreaks in the forests of eastern North America.

Answers

Knowledge and Synthesis Answers

1. **d.** Both species have benefited; the bird dispersed and provided fertilizer for the plant's seed, the plant provided food for the bird.
2. **d.** Both ants and birds are competing for the same food supply; the birds also benefit from the ants' activity in flushing the insects, but the ants might not be affected.
3. **a.** A resource must be needed by an organism, but also must be present in limited supply (use of a resource depletes it). Atmospheric oxygen is so abundant that terrestrial animals never reduce its concentration.
4. **b.** If the species compete for limited resources, the ecological niches of *both* species will be more constrained in areas where the ranges overlap and will differ from their niches before the range expansion.
5. **c.** Both species grow less well in the presence of the other; the effect is greater for species 1.
6. **c.** The key to answering this question is realizing that prey cycles lead predator cycles, so the black line indicates the prey and the gray line the predator.
7. **c.** Although it is true that all three could be Müllerian mimics, the observation that species 2 is much less abundant than the other two species suggests that it may be a Batesian mimic
8. **a.** Initially the plant is harmed but the animal is unaffected (amensalism). After the evolution of sharp spines, the two species are competitors for space.
9. **c.** Most plants decrease CO_2 levels by incorporating it within their biomass.
10. **b.** By definition, removal of a keystone species has a larger effect on the community than would be expected based on its abundance.
11. **d.** Of the stages listed, the arrival of lichens would be first, followed by the arrival of willows and alders, increase in soil nitrogen content (the correct answer), and finally arrival of conifers.
12. **a.** Succession that begins on a newly available site, like a freshly excavated pond, is primary. Succession that occurs after disturbances such as fire or the conversion of a natural community to farmland is considered secondary.

Application Answers

1. A removal experiment could be done in which one or the other species is removed in two similar areas. One then looks for changes in the abundance of the remaining species. The species with the largest increase in population density was most affected by competition with the removed species.
2. When a new strain of the influenza virus arises via mutation, the human host population contains a very high proportion of people susceptible to infection. As the infection spreads, the proportion of the population that is susceptible falls and the proportion of recovered and immune individuals rises. Eventually a point is reached at which infectious individuals fail, on average, to transmit the virus to at least one other person, and the epidemic dies out.
3. Plants have both physical and chemical defenses against herbivores. Physical defenses include tough leaves, hairs, or spines. Many plants also produce secondary compounds that are either acutely toxic to herbivores or make plant tissues difficult to digest.
4. The moth has given up the opportunity to feed from other species of plants and the yucca has given up the opportunity to use other pollinators. Both benefit from being able to specialize at exploiting the other as an exclusive resource.
5. See Figure 55.22 in the text.

STUDY GUIDE

56 Ecosystems

Important Concepts

An ecosystem is made up of all the species living in a particular area, together with the physical environment with which they interact.

- Ecosystems can be studied on many spatial scales, from local units to the entire globe. On the global scale, Earth is a single ecosystem.
- The dynamics of ecosystems are the result of the capture, transformation, and transfer of energy and materials by myriad individual organisms.

Climates exert a powerful influence on the distribution, abundance, and evolution of species.

- Solar energy varies with latitude. Regions near the poles receive less energy per unit of ground area than regions near the equator because of the lower angle of the sun. At high latitudes there is also more variation over the course of a year in both day length and the angle of arriving solar energy than at latitudes closer to the equator.
- Rising air expands and cools, releasing moisture, whereas descending air warms and dries and takes up moisture. As a result, the movement of air over mountains often results in a rain shadow—a dry area on the leeward side of the range.
- Unequal heating of the atmosphere at low and high latitudes produces vertical and latitudinal movements of air masses. These movements cause very moist climates to occur at the equator and at 60° north and south latitudes (where air rises), and arid climates at about 30° north and south latitudes and near the poles (where air descends).
- The spinning of Earth on its axis causes air masses moving latitudinally to be deflected to the right in the Northern Hemisphere and to the left in the Southern Hemisphere. Thus winds blowing toward the equator at low latitudes veer to become the northeast and southeast trade winds, whereas winds blowing away from the equator at midlatitudes are deflected to become the prevailing westerlies.
- Because Earth's axis is tilted, the amount of solar energy that reaches a given region varies seasonally as Earth orbits the sun. In the tropics, this variation causes seasonal shifts in the position of the intertropical convergence zone, where trade winds converge and air rises. As a result, the tropics experience rainy and dry seasons.
- Ocean currents are driven primarily by prevailing winds but are deflected by continents. Oceans play an important role in world climates both because their waters move long distances and because water has a high specific heat.
- Wind circulation patterns and the properties of ocean water cause the west coasts of continents at middle latitudes to have maritime climates, with relatively small differences between winter and summer temperatures. In contrast, the interiors of large continents at high latitudes have continental climates, characterized by very cold winters and hot summers.

Energy flows through ecosystems.

- Energy flow through ecosystems is governed by the second law of thermodynamics, which states that when energy is converted from one form to another, some of it becomes unavailable to do work.
- With very few exceptions, energy flow in ecosystems originates with photosynthesis, which captures about 5 percent of the solar energy that arrives on Earth.
- Gross primary productivity is defined as the *rate* at which plants assimilate energy through photosynthesis; gross primary production is the *amount* of energy assimilated by plants over the course of a year.
- Net primary production is the portion of assimilated energy left over after subtracting the energy plants use to power their own metabolism. All other organisms in an ecosystem directly or indirectly derive their energy from net primary production, which takes the form of new plant tissues.
- The major limits on primary production in terrestrial ecosystems are low temperatures and lack of moisture. In aquatic systems, production is limited by light, nutrients, and temperature.
- Energy flows through ecosystems when organisms eat one another. Organisms are grouped into trophic levels according to the number of steps through which energy passes to get to them. Energy passes, in sequence, from primary producers (photosynthesizers), to herbivores, to primary carnivores, to secondary carnivores, and so on. Detritivores (decomposers) feed on dead bodies and waste products of other organisms. Omnivores obtain their food from more than one trophic level.

- A food chain is a sequence of feeding linkages beginning with a primary producer. Food chains in a community are usually interconnected in a food web, which can be diagrammed.
- Most of the energy captured at any trophic level is used for respiration and ultimately dissipated as heat. Only the net production (growth plus reproduction) of organisms at a trophic level is available to organisms at the next level. The efficiency of energy-transfer through a food web depends on (1) the fraction of net production at one trophic level that is consumed by organisms at the next level, and (2) how those organisms divide the ingested energy between production and respiration.
- A pyramid of energy shows how energy (production) decreases in moving from lower to higher trophic levels. A pyramid of biomass shows the mass of organisms existing at different trophic levels at a given moment of time. For the same ecosystem, these pyramids usually have similar shapes. Major exceptions are aquatic communities, in which a small mass of rapidly dividing unicellular photosynthesizers can support a much larger biomass of herbivores.
- Detritivores are essential for the continued productivity of ecosystems because they release nutrients from dead bodies that can be taken up again by plants.
- A few deep-sea and cave ecosystems are powered by chemosynthesis rather than photosynthesis.
- Through agriculture, humans manipulate ecosystem productivity in order to increase the yield of products useful to us. In modern agriculture, the energy required to do this is provided by fossil fuels.
- Integrated pest management (IPM) relies on a variety of approaches—chemical, cultural, and biological—to control crop pests while minimizing the problems of pollution and genetic resistance associated with heavy use of chemical pesticides.

Chemical elements cycle repeatedly through organisms and the physical environment.

- The main compartments of the global ecosystem are the oceans, fresh waters, atmosphere, and land, among which materials are constantly being exchanged.
- Elements entering the ocean gradually sink to the seafloor, where they may remain for many millions of years. Although vertical mixing of ocean waters is retarded in most regions by the thermocline (a depth at which temperatures drop abruptly), it occurs freely in zones of upwelling near the coasts of continents. The nutrient-rich, cool, bottom water brought to the surface stimulates high primary productivity in these zones.
- Most mineral nutrients that enter fresh water are released by the weathering of rocks and are carried to lakes and rivers via groundwater or by surface flow.
- In temperate lakes, turnovers that occur in the spring and fall when the water temperature is uniformly 4°C allow nutrients and oxygen to become evenly distributed in the water column. The bottom waters of permanently stratified tropical lakes lack oxygen.
- The lowest layer of the atmosphere is the troposphere, where almost all water vapor is located and where most global air circulation takes place. Above the troposphere is the stratosphere, where incoming ultraviolet radiation is absorbed by ozone (O_3).
- The most abundant gases in the atmosphere are nitrogen (about 78 percent) and oxygen (about 21 percent). Although carbon dioxide constitutes only 0.03 percent of the atmosphere, it is the source of the carbon used by terrestrial photosynthetic organisms. In the greenhouse effect, atmospheric carbon dioxide, water vapor, and ozone trap outgoing infrared (heat) radiation emitted by Earth and thus raise its surface temperature.
- About one-fourth of Earth's surface is covered by land, most of which is cloaked by a layer of soil. The type of soil that forms in an area and the elements it contains depend on the underlying rock, as well as on climate, topography, the organisms living there, and the length of time that soil-forming processes have been acting.

The pattern of movement of a chemical element through organisms and compartments of the physical environment is called its biogeochemical cycle.

- The hydrological cycle is driven by evaporation, most of it from the ocean surfaces.
- Nearly all the carbon in organisms comes from carbon dioxide in the atmosphere or dissolved bicarbonate (HCO_3^-) in water. Carbon is incorporated into organic molecules by photosynthesis; respiration and decomposition break down these organic molecules and return carbon to the atmosphere and water.
- Fossil fuels exist because in the remote past large quantities of organic carbon were removed from the carbon cycle by burial of organisms in sediments lacking oxygen. The ever-increasing rate of burning of fossil fuels over the past 150 years has elevated the carbon dioxide concentration of the atmosphere.
- Much evidence indicates that the rising concentration of atmospheric carbon dioxide is a cause of global warming, which is probably already underway. Shifting climatic zones and rising sea levels are likely consequences of global warming. In addition, ecologists are investigating the direct effects of elevated carbon dioxide concentrations on ecosystems.
- Atmospheric nitrogen (N_2) is converted into biologically useful forms by a few species of bacteria and cyanobacteria that can carry out nitrogen fixation, which converts elemental nitrogen into ammonia. Thus nitrogen is often in limited supply in ecosystems. Other chemical conversions in the nitrogen cycle are also performed by unicellular organisms.
- Through the burning of fossil fuels, humans have altered the nitrogen and sulfur cycles. Industrial smokestacks and automobiles produce emissions that form sulfuric and nitric acid in the atmosphere. The resulting acid precipitation affects all industrialized countries and has been shown to have harmful effects on lake ecosystems and on terrestrial plants.

- The phosphorus cycle differs from the cycles of carbon, nitrogen, and sulfur in that it lacks a gaseous phase. On land, most phosphorus becomes available through the weathering of phosphate-containing rocks.
- Phosphorus is often a limiting nutrient in soils and lakes. Thus adding phosphorus (from detergents and fertilizers) to lakes is a major cause of eutrophication (enrichment). Algal blooms and the depletion of oxygen in affected lakes are among the undesirable consequences of this process.
- Human activities are increasing the cycling of other elements such as lead and creating cycles of synthetic chemicals. These changes can cause serious environmental problems.

The Big Picture

- The second law of thermodynamics (first mentioned in Chapter 6) is one of the major unifying principles of biology as well as of other scientific fields. It provides the most fundamental explanation for the pattern of energy flow through ecosystems.
- The productivity of ecosystems, both terrestrial and aquatic, is relevant to the problem of providing food for the exploding human population (discussed in Chapter 54). Also relevant are the environmental effects of agricultural practices designed to increase ecosystem productivity and to channel it to meet human needs.
- Knowledge of how materials move through biogeochemical cycles is crucial for understanding and predicting the effects of human alterations of these cycles. Global warming and eutrophication of lakes are two important examples.
- The important roles of nitrogen-fixers and other microorganisms in the nitrogen cycle illustrate the sometimes overlooked ways in which rather obscure members of the biotic community may be essential for the healthy functioning of ecosystems.

Common Problem Areas

- Students sometimes have difficulty understanding the difference between pyramids of energy and pyramids of biomass. Pyramids of energy can *never* be inverted because there must always be a lower rate of energy flow through any given trophic level than through the trophic level below it. Pyramids of biomass *can* be inverted because they do not represent energy flow rates, but rather the amount of living matter present in the various trophic levels of the community at a given moment in time.

Study Strategies

- In studying the material on biogeochemical cycles, focus on these basic questions: Where are the abiotic reserves of the element located? How does the element leave the reserve and enter living organisms? Why is the element needed by living organisms? How does the element return to its abiotic reserve?
- Review the following tutorials and activities on the website:
  Tutorial 56.1 Rain Shadow
  Tutorial 56.2 Biogeochemical Cycles
  Activity 56.1 Concept Matching
  Activity 56.2 Ecosystem Energy Flow

Test Yourself

Knowledge and Synthesis Questions

1. In comparing an acre of land in Colombia to an acre of land in Michigan, which of the following would *not* differ?
   a. The angle of the sun reaching the ground in the month of July
   b. The solar energy flux in the month of July
   c. The annual solar energy flux
   d. The total hours of daylight per year

2. The following diagram shows a mountain with a sea breeze blowing as indicated by the arrow. Circle the letter for the area with air that would be *both* relatively warm and dry.

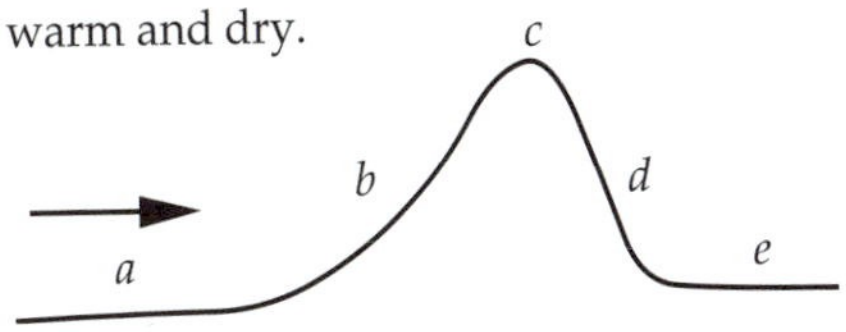

3. In what area of the above diagram in question 2 is a process occurring that is similar to the process that occurs in the intertropical convergence zone?

4. If Earth did not spin on its axis, from what direction would the northeast trade winds blow?
   a. Northeast
   b. South
   c. North
   d. East

5. Ocean circulation patterns are influenced by all of the following *except*
   a. circulation of Earth's atmosphere.
   b. deflection by land masses.
   c. upwelling of deeper water.
   d. rotation of Earth on its axis.

6. Differences in the specific heat of water and land have most to do with creating
   a. the intertropical convergence zone.
   b. maritime climates.
   c. continental climates.
   d. oceanic upwelling.

7. Which of the following could *not* be considered an ecosystem?
   a. A small pond
   b. All the fish in a coral reef
   c. Earth
   d. A pile of dung in a pasture

8. A plant in the dark uses 0.02 ml of O_2 per minute. The same plant in sunlight releases 0.14 ml of O_2 per minute. A correct estimate of its rate of *gross* primary production is
   a. 0.02 ml of O_2 per minute.
   b. 0.12 ml of O_2 per minute.
   c. 0.14 ml of O_2 per minute.
   d. 0.16 ml of O_2 per minute.

9. Examine the following food web. Organism 9 is a(n)

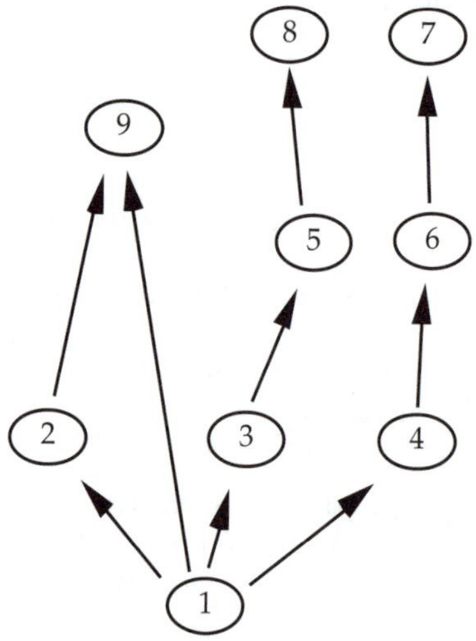

   a. herbivore.
   b. primary carnivore.
   c. secondary carnivore.
   d. omnivore.

10. The food web shown in question 9 has _____ trophic levels.

11. Which of the following organism types would probably rank the *lowest* in terms of their expected efficiency of energy transfer between trophic levels?
   a. Herbivorous mammal
   b. Carnivorous mammal
   c. Invertebrate herbivore
   d. Invertebrate carnivore

12. Which of the following would be *least* likely to be a method employed in integrated pest management?
   a. Mixed plantings of crop plants
   b. Development of pest-resistant strains of crop plants
   c. Aerial spraying of insecticides
   d. Use of chemical attractants for controlling pest species

13. Which of the following features would be characteristic of oceans but *not* of freshwater ecosystems?
   a. Receives material from land mostly via groundwater
   b. Seasonal mixing of materials
   c. Elements buried in bottom sediments for long periods of time
   d. Bottom waters frequently lacking oxygen

14. Which of the following features would be characteristic of the stratosphere but *not* of the troposphere?
   a. Most ozone resides here.
   b. Most water vapor resides here.
   c. Most of the mass of the atmosphere lies here.
   d. Circulation of this layer influences ocean currents.

15. Which of the following compartments of the global ecosystem would be characterized by very slow movement of materials within the compartment?
   a. Oceans
   b. Fresh waters
   c. Atmosphere
   d. Land

16. In which of the following compartments of the global ecosystem would circulation of materials be affected by Earth's revolution around the sun?
   a. Oceans
   b. Fresh waters
   c. Atmosphere
   d. All of the above

17. In the following diagram, the temperature of a freshwater lake is plotted against depth.

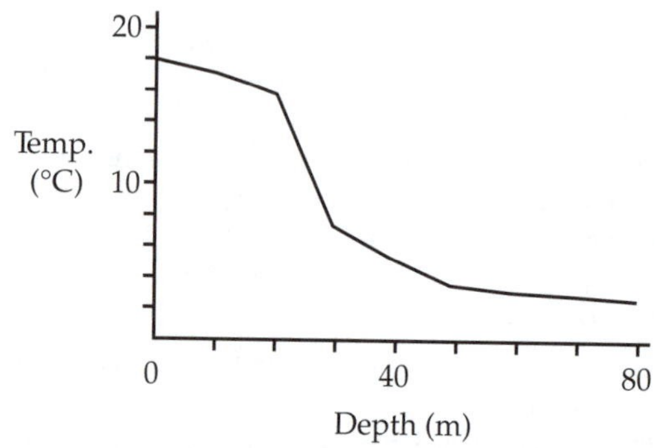

This temperature profile would be most characteristic of the lake during
   a. winter.
   b. spring.
   c. summer.
   d. fall.

18. Based on data in the graph from question 17, the thermocline for the lake is located between
   a. 0 and 20 meters.
   b. 20 and 30 meters.
   c. 30 and 50 meters.
   d. 50 and 80 meters.

19. Which of the following statements about biogeochemical cycles is *not* true?
   a. Most elements remain longest in the living portion of their cycle.
   b. Gaseous elements cycle more quickly than elements without a gaseous phase.
   c. You may have some atoms in your body that were once part of a dinosaur.
   d. Biogeochemical cycles all include both organismal and nonliving components.

20. Next to each of the following features, place a *c*, *n*, *p*, or *s* if it is characteristic of the biogeochemical cycles of *carbon, nitrogen, phosphorus,* or *sulfur*. (Note: some features may apply to more than one cycle.)
   _____ Major reservoir is atmospheric
   _____ Major reservoir is in sedimentary rocks
   _____ Often in short supply in ecosystems
   _____ Fossil fuel reserve is part of this cycle
   _____ Involved in cloud formation
   _____ Lacks a gaseous phase
   _____ Includes a form that is a greenhouse gas
   _____ Most fluxes involve organisms

21. Which of the following statements is *not* a major concern about our altering of the carbon cycle?
    a. The Greenland and Antarctic ice caps are expected to melt if global warming continues.
    b. High-sulfur fuels are used by power plants because they are less expensive than low-sulfur fuels.
    c. The increase in atmospheric CO_2 exceeds the ability of the oceans to absorb the increase.
    d. CO_2 is a gas that traps infrared radiation.

22. Which of the following statements concerning acid precipitation is *false*?
    a. Acids that enter the atmosphere primarily affect ecosystems located less than 100 kilometers from the source of the pollution.
    b. The effects of acidification of lakes appear to be quickly reversible if their pH can be returned to normal.
    c. While regulation of emission sources has raised the pH of precipitation in parts of the eastern United States, precipitation in many parts of the West is becoming more acidic.
    d. Sulfuric acid and nitric acid from smokestack and automobile emissions are the major causes of acid precipitation.

23. The process by which a lake ecosystem is altered by eutrophication involves several stages, each causing the next. Which of the following stages would occur *second* in the chain of causation?
    a. Algal blooms occur.
    b. Oxygen levels drop in deeper water.
    c. Phosphorus input from sewage and agricultural runoff increases.
    d. Respiratory demand from decomposers increases.

Application Questions

1. If the dry biomass of the primary producers of an ecosystem is 1000 g per square meter and the efficiency of transfer between trophic levels is 20 percent, what dry biomass weight would you expect at the *secondary carnivore* level?

2. Where do biomass pyramids like the one shown below occur and what are the conditions that create them?

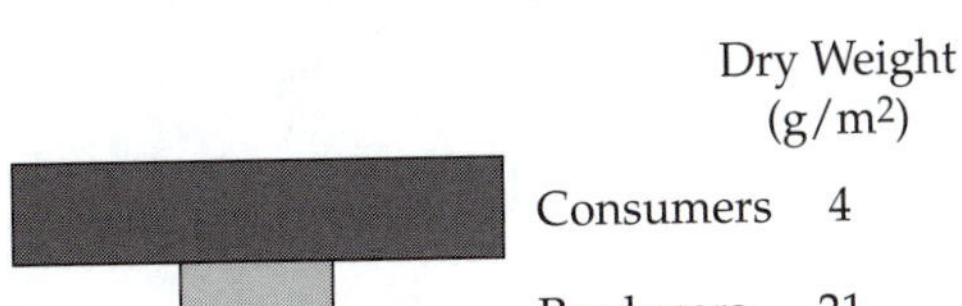

3. Why would you expect a small lake to be able to recover more quickly from eutrophication than a large lake such as Lake Erie?

Answers

Knowledge and Synthesis Answers

1. **d.** All areas on Earth receive equal hours of daylight per year. The seasonal distribution of those hours, the solar energy input, and the sun angle do vary latitudinally.

2. **e.** The air would have lost most of its moisture while rising on the windward side of the mountain, and in descending to area *e* it would have warmed again.

3. **b.** In the equatorial convergence zone, warm air rises and loses much of its moisture. These same events occur when moist air rises over a mountain.

4. **c.** If Earth did not spin on its axis, air flowing south toward the equator would not be deflected to the right. Therefore, the northeast rate winds would blow directly from the north instead of from the northeast.

5. **c.** The upwelling of deeper water, especially on the western side of continents, is an effect of the pattern of ocean current circulation and not a cause of the pattern.

6. **b.** Because of the high specific heat of water, a large body of water moderates the climate of adjacent land masses, creating a maritime climate.

7. **b.** An ecosystem includes the environment *and* the biological community living there. A coral reef ecosystem would include much more than just the fish.

8. **d.** The amount of production used in maintenance and biosynthesis is added to net primary production to determine gross primary production. Photosynthetic release of O_2 in the light is an estimate of net production; O_2 use in the dark is an estimate of maintenance and biosynthesis costs. You would add 0.02 ml to 0.14 ml to get an estimate of gross primary production of 0.16 ml per minute.

9. **d.** Since organism 9 eats from both the primary producer level (1) and the herbivore level (2), it would be an omnivore.

10. 4 trophic levels. The levels are primary producer (1), herbivore (2, 3, 4), primary carnivore (5, 6), and secondary carnivore (7, 8). The omnivore (9) occupies either the herbivore or primary carnivore level depending on whether it is feeding on a primary producer or on a herbivore.

11. **a.** Because they spend so much energy maintaining constant high body temperatures, mammals (and birds) have very low production efficiencies. Herbivores are less efficient than carnivores because plant tissues take more energy to digest than animal tissues do.

12. **c.** Indiscriminate aerial spraying that kills both useful and harmful insects alike is not normally employed in integrated pest management.

13. **c.** Most elements remain in bottom sediments in the ocean for millions of years, until elevated above sea level by movements of Earth's crust.

14. **a.** The ozone that shields the surface of Earth from most incoming ultraviolet radiation is located in the stratosphere.

15. **d.** Unlike their behavior in air and water, elements on land move slowly, and usually only short distances.

16. **d.** The revolution of Earth around the sun (combined with the tilt of Earth's axis) is the cause of the seasons. Circulation patterns of materials in the oceans, fresh

water, and the atmosphere are all affected by seasonal changes in temperature, precipitation, and prevailing wind direction.

17. **c.** The steep thermocline evident in this temperature profile would typically be established only by mid- to late summer.

18. **b.** The range of depths over which the temperature changes most abruptly is 20 to 30 meters.

19. **a.** Most elements cycle through organisms more quickly than they cycle through the nonliving world.

20. **n** Major reservoir is atmospheric
c, p Major reservoir is in sedimentary rocks
n, p Often in short supply in ecosystems
c Fossil fuel reserve is part of this cycle
s Involved in cloud formation
p Lacks a gaseous phase
c Includes a form that is a greenhouse gas
n Most fluxes involve organisms

21. **b.** The use of high-sulfur coal has more to do with acid precipitation than it does with global warming.

22. **a.** Acid precipitation is a regional, not a local, environmental problem. Acids may travel hundreds of kilometers before they settle to Earth in precipitation or as dry particles.

23. **a.** The correct sequence is:
1. Phosphorus input from sewage and agricultural runoff increases.
2. Algal blooms occur.
3. Respiratory demand from decomposers increases.
4. Oxygen levels drop in deeper water.

Application Answers

1. The answer is 8 grams. The flow of energy is from primary producers, to herbivores, to primary carnivores, to secondary carnivores. If each of these three energy transfers is only 20 percent efficient, then the biomass of secondary carnivores should be:
$1000 \text{ g} \times 0.2 = 200 \text{ g}$
$200 \text{ g} \times 0.2 = 40 \text{ g}$
$40 \text{ g} \times 0.2 = 8 \text{ g}$

2. In most aquatic ecosystems, the dominant primary producers are unicellular algae. Their populations multiply so rapidly and are cropped so efficiently by slower growing herbivores that they frequently support a larger herbivore mass than their own. Note that the diagram depicts an inverted pyramid of biomass. A pyramid of energy can *never* be inverted.

3. The rate at which a lake can recover from eutrophication depends on the rate of replacement of its water. A small lake is likely to have a quicker replacement time for its water than a very large body of water like Lake Erie.

STUDY GUIDE

57 Biogeography

Important Concepts

The goal of biogeography is to explain patterns in the distribution of life on Earth. One part of this goal is to understand why particular species are found where they are.

- If a species occupies an area, either it evolved there, or it evolved elsewhere and dispersed to that area.
- If a species is not found in an area, either it evolved elsewhere and never dispersed to that area, or it was once present in that area but no longer lives there.
- Continental drift has influenced the evolution and mixing of species throughout the history of life on Earth.
- Ecological biogeographers study how current ecological interactions influence where species are found today.

Past events influence today's patterns of distribution.

- Historical biogeographers attempt to determine the influence of past events on present patterns of species distributions on the basis of the evidence provided by "molecular clocks," fossils, and the distributions of living species.
- The technique of transforming taxonomic phylogenies into "area phylogenies" involves substituting the taxa's geographic distributions for their names. Distribution patterns thus identified may suggest routes of dispersal of species or point to the splitting of biotas due to the appearance of barriers to dispersal.
- A split distribution of a species can be accounted for in two ways.
  1. A barrier may appear that splits a species' distribution (a vicariant event).
  2. A species may cross an existing barrier and establish a new population (dispersal).
- Taxa found in only one place are said to be endemic to that location. Endemic taxa may be very old ones in the process of becoming extinct, or very young taxa that have recently evolved in a restricted area. The longer an area has been isolated, the more endemic taxa it is likely to have.
- In reconstructing the biogeographic histories of taxa, biogeographers use evidence provided by distribution maps, phylogenies, and knowledge of ancient climates and ancient geography. In interpreting this information, they apply the parsimony principle—that is, they prefer the simplest hypothesis consistent with the evidence.
- On the basis of the taxonomic similarities among the organisms living in them, animal biogeographers divide Earth into six biogeographic regions: the Nearctic, Neotropical, Ethiopian, Palearctic, Oriental, and Australasian regions. Plant biogeographers recognize two additional regions: southern South America and the Cape Region of South Africa.

Ecological biogeographers test theories that explain the numbers of species in different communities, the ways in which species disperse, and the effectiveness of barriers to movement.

- According to the island biogeographic model, equilibrium species richness (the number of species living in an area) on islands is determined by the rates of arrival of new species and extinction of species already present.
- The island biogeographic model predicts that the equilibrium number of species should increase with island size and decrease with distance from the mainland. Biogeographers have confirmed the predictions of the model both by counting existing species on real islands and by measuring arthropod recolonization rates following fumigation of very small islands.
- Species richness varies latitudinally, with more species living in tropical than in high-latitude regions.

A terrestrial biome is a major ecosystem type that differs from other types in the structure of its dominant vegetation.

- The distribution of biomes on Earth is strongly influenced by annual patterns of temperature and precipitation.
- Tundra is found at high latitudes in the Arctic and in high mountains. It is a treeless biome dominated by short perennial plants. In the Arctic, permanently frozen soil called permafrost underlies tundra vegetation.
- Boreal forests are located equatorward from, or at lower elevations on mountains than, tundra. The short summers favor evergreen trees, which are the dominant vegetation.
- Temperate deciduous forests are dominated by deciduous trees, which produce leaves that photosynthesize rapidly

during the warm, moist summers and lose their leaves during the cold winters.

- Temperate grasslands occur in areas that are relatively dry for much of the year. Grasses dominate this biome because they are well adapted to grazing and fire.
- Cold deserts are found in dry regions at middle to high latitudes. Seasonal changes in temperature are great. They are dominated by a few species of low-growing shrubs.
- Hot deserts, characterized by very warm and dry conditions year-round, are found in two belts around the 30°N and 30°S latitudes. Succulent plants that store large quantities of water in their stems are conspicuous.
- The chaparral biome, found on the west side of continents at midlatitudes, is dominated by evergreen shrubs and low trees that are adapted to survive periodic fires. Winters are cool and wet; summers are hot and dry.
- Thorn forests and savannas are found in semiarid climates on the equatorial side of hot deserts. Savanna, a grassland punctuated by scattered trees, is maintained by grazing, browsing, and burning. In their absence it reverts to dense thorn forest dominated by small, spiny shrubs and trees.
- Tropical deciduous forests are dominated by trees that lose their leaves during the long, hot, dry season. Because their soils are less leached of nutrients than the soils of wetter areas, most of these forests have been cleared for grazing cattle and growing crops.
- Tropical evergreen forests, found in equatorial regions where annual rainfall exceeds 250 cm annually, have the greatest species richness and the highest productivity of all terrestrial communities. Nevertheless, their soils are poor and usually cannot support long-term agriculture. Epiphytes—plants that grow on other plants, deriving their nutrients and water from the atmosphere—thrive in tropical mountain forests where clouds form regularly.

Aquatic ecosystems occupy three-fourths of Earth's surface.

- Because the oceans form one interconnected water mass, there are no obvious barriers to dispersal in the oceans. For freshwater organisms that cannot survive out of water, terrestrial habitats are a barrier to dispersal.
- The oceans are divided into zones based on sharp horizontal and vertical environmental gradients. The ocean floor constitutes the benthic zone, which coincides with the abyssal zone below the limit of light penetration. The column of water above the ocean floor constitutes the pelagic zone, The coastal zone affected by wave action is called the littoral zone.
- Though dispersal in the oceans occurs easily, living successfully in a particular zone requires that organisms be adapted to the light, hydrostatic pressure, nutrient, and temperature conditions characteristic of that zone. Thus organisms from one zone almost always survive poorly if they attempt to live in another.
- Boundaries between regions of the pelagic zone typically occur where the temperature of surface water changes abruptly as a result of ocean currents.
- Species that live in shallow water disperse with difficulty across wide deep-water barriers.
- Recent vicariant events, notably the formation of the Isthmus of Panama, influence current distributions of marine species.

The distributions of plants, animals, and continents have had a strong influence on human history.

- Eurasia happened to have a larger number of species of plants and animals suitable for domestication than Africa or the Americas.
- Eurasian peoples acquired immunity to many human diseases as a result of their association with domestic animals.When Europeans colonized the New World, they brought with them diseases that devastated indigenous peoples.
- Dispersal of people and their domesticated plants and animals was relatively easy in Eurasia because most mountain ranges are oriented in an east-west direction.

The Big Picture

- As you will learn in Chapter 58, the island biogeographic model has been applied to the problem of saving endangered species from extinction.
- Much evidence supporting the theory of continental drift is based on the distributions of organisms on Earth. By the same token, the acceptance by geologists and biologists of the reality of continental drift revolutionized the field of historical biogeography.
- The distributions of particular plant and animal species that have been domesticated in various regions of the world have had major consequences for human history and culture. So have the distributions of many species of microorganisms and animals that cause or carry disease.

Common Problem Areas

- The graphical representations of the concepts of the island biogeographic model are troublesome for some students. Try redrawing the graphs for yourself, paying particular attention to how the axes of the graphs are labeled.

Study Strategies

- In attempting to master the information concerning the many biomes described in this chapter, focus on the graphs that summarize the essential points about each one.
- You also should study Figure 57.9 to get a sense of the geographical location and extent of each biome. A useful exercise to help you learn the relative locations of biomes is to imagine a journey (e.g., from equatorial South America to Alaska, or from Maine to California), naming the biomes that you would pass through.
- Review the following tutorial and activity on the website:
  Tutorial 57.1 Biomes
  Activity 57.1 Biogeographical Regions

Test Yourself

Knowledge and Synthesis Questions

1. Which of the following types of information would be *least* used by a historical biogeographer?
   a. Experiments on species richness
   b. Fossils showing past distributions of organisms
   c. Application of the principle of parsimony
   d. Present distributions of organisms

2. The following diagram shows the present locations of four continents (W, X, Y, Z) and the number of endemic beetle families found on each. Three were part of a supercontinent in the past, the other was always separate.

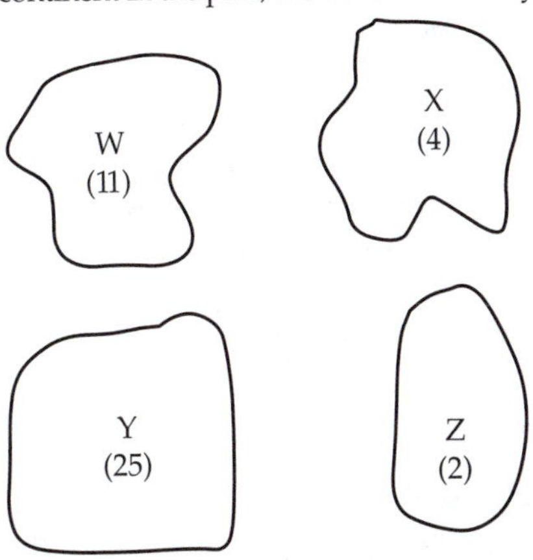

   Based on this information, choose the correct conclusions.
   a. W was always separate, Y separated from the supercontinent first.
   b. X was always separate, Y separated from the supercontinent first.
   c. Y was always separate, W separated from the supercontinent first.
   d. Z was always separate, X separated from the supercontinent first.

3. The following diagram shows three islands. Islands A and B were connected in the past, C was always separate. A species of land snail is found on all three islands.

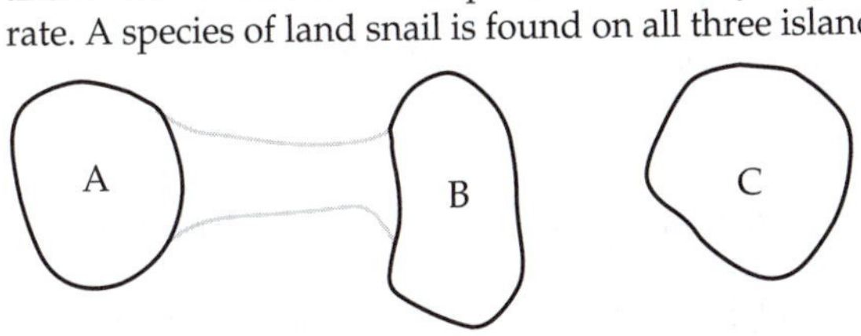

   Which of the following does not correctly describe the distribution of this snail relative to the three islands?
   a. Vicariant distribution relative to A and B
   b. Dispersal distribution relative to A and C
   c. Dispersal distribution relative to B and C
   d. Vicariant distribution relative to A and C

4. If the species of snail mentioned in question 3 is found *only* on the three islands shown in the above diagram, which statement would be correct concerning the endemism of this species?
   a. Endemic relative to A and B
   b. Endemic relative to C
   c. Endemic relative to A, B, and C
   d. All of the above

5. Which of the following biogeographical regions represents the largest area?
   a. Nearctic
   b. Palearctic
   c. Neotropical
   d. Oriental

6. Circle an area on the following graph that would most apply to the spatial and temporal scale of processes studied by ecological biogeographers.

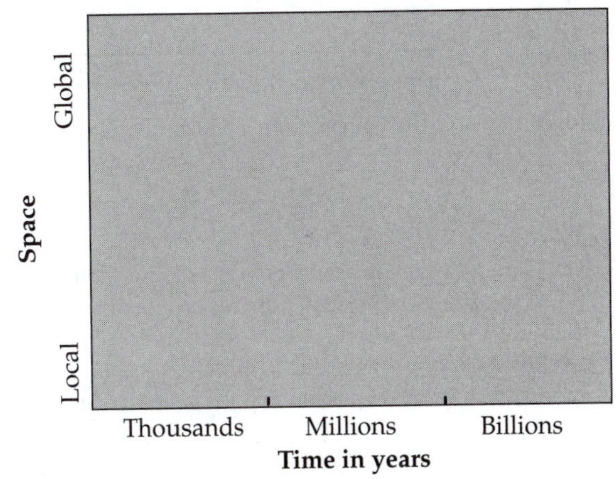

7. According to the island biogeographic model, which of the statements about the following graph showing the effect of species number on arrival and extinction rates is *not* true?

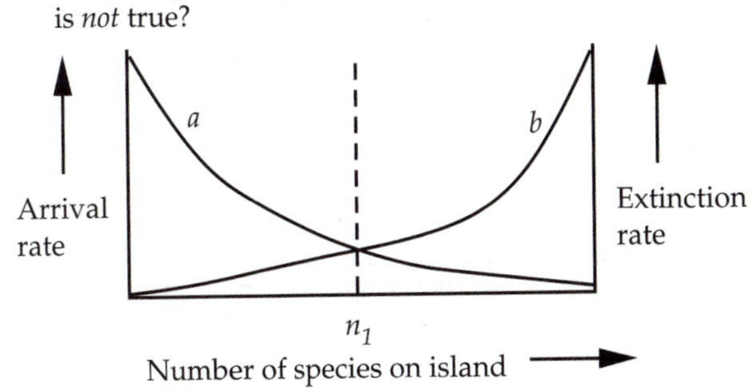

   a. Curve *a* is the arrival rate curve.
   b. Curve *b* is the extinction rate curve.
   c. The arrival rate equals the extinction rate at n_1.
   d. The extinction rate is zero at n_1.

8. The species numbers of many islands of different sizes are plotted against their distance from the mainland. Data points for islands of similar size were connected to form the four curves shown in this figure.

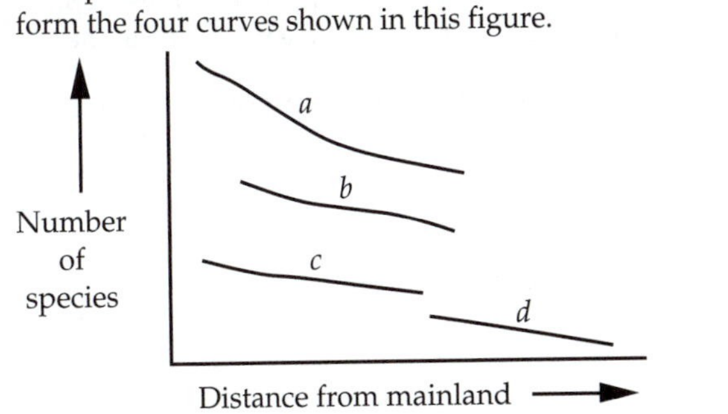

   Circle the letter of the curve corresponding to the group of islands with the smallest size.

9. A small volcanic island was destroyed by an eruption. The following graph shows recolonization by a number of different organism types.

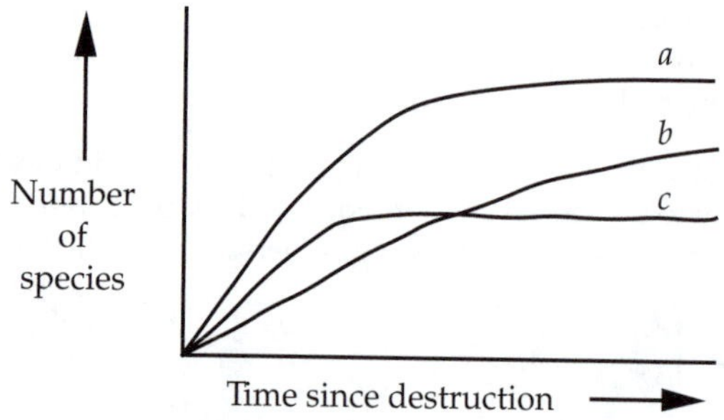

Identify the correct curve for each organism type.
_____ Birds
_____ Plants
_____ Insects

10. The island biogeographic model makes predictions about the effects of species number on the rate of extinction. In the following figure, the solid curve shows this relationship for a large island. Which of the curves shows the expected relationship for a small island?

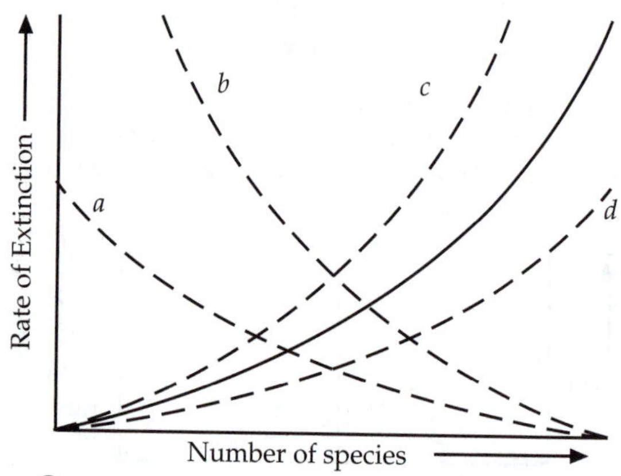

a. Curve *a*
b. Curve *b*
c. Curve *c*
d. Curve *d*

11. Which of the following locations would you expect to have the greatest number of species?
a. Mountainous forest area on a mainland in the temperate zone
b. Mountainous forest area on a mainland in the tropics
c. Lowland forest area on a mainland in the tropics
d. Mountainous forest area on a peninsula in the tropics

12. The biome that is maintained by browsers, grazers, or fire is the
a. tundra.
b. savanna.
c. cold desert.
d. tropical deciduous forest.

13. Match the letters of the following biomes with the descriptions that follow.
a. Tundra
b. Boreal forest
c. Hot desert
d. Chaparral
e. Tropical deciduous forest

_____ Mostly coniferous, wind-pollinated and wind-dispersed tree species
_____ Leaves lost during dry season; agriculturally desirable land
_____ Cool winters, hot dry summers; maritime climate
_____ Succulent plants prominent; found at 30°N and 30°S latitudes
_____ Distribution altitudinally or latitudinally determined; permafrost present

14. The following climograph shows yearly variation in rainfall and temperature for four biomes. Select the correct curve for each of the following biomes.

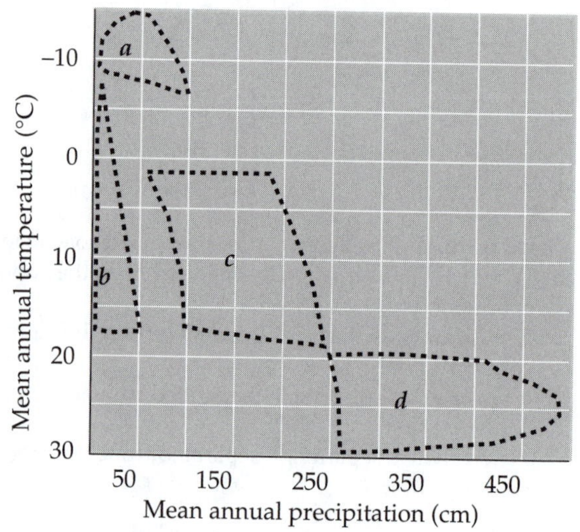

_____ Tundra
_____ Tropical evergreen forest
_____ Temperate deciduous forest
_____ Cold desert

15. The tropical evergreen forest biome has relatively constant _____________, but _____________ varies seasonally; the temperate deciduous forest biome has relatively constant _____________, but _____________ varies seasonally.

16. Which of the following comparisons of freshwater and marine habitats is *not* true?
a. About 10 percent of all aquatic species live in fresh water.
b. There are many more insect species in freshwater habitats than in marine habitats.
c. The global volume of marine habitats is much greater than that of freshwater habitats.
d. Freshwater species richness is less than marine species richness in proportion to the relative extent of the two habitats.

17. Next to each of the following features, place *p*, *l*, *b*, or *a* if it is characteristic of the *pelagic*, *littoral*, *benthic*, or *abyssal* zones of marine communities. (Note: some features may apply to more than one zone.)
_____ The ocean floor below the penetration of light
_____ Coastal zone affected by wave action
_____ Zone in which organisms must have adaptations to avoid sinking
_____ Divided into a lighted and dark region
_____ Globally, the zone occupying the smallest total area

18. Which of the following is most likely to be true of a family of freshwater fish with species distributed on several continents?
 a. They are a recently derived group.
 b. They have an ancient lineage that was widely distributed in Laurasia or Gondwana.
 c. Their distribution was not affected by continental drift.
 d. Their distribution can be explained by movements of glaciers.

19. Which of the following statements about the biogeography of marine biomes is *false*?
 a. Vertical gradients of light and temperature divide the ocean into different zones.
 b. In interpreting the biogeography of the marine biome, oceanic currents are more important than physical barriers.
 c. Organisms adapted for life in nutrient-poor waters usually flourish in nutrient-rich areas.
 d. Continental drift does not explain the distribution of many marine organisms.

Application Questions

1. Based on an examination of the following phylogenetic tree and information on the distribution of the taxa members on three continents, discuss what can be inferred about the histories of the three continents.

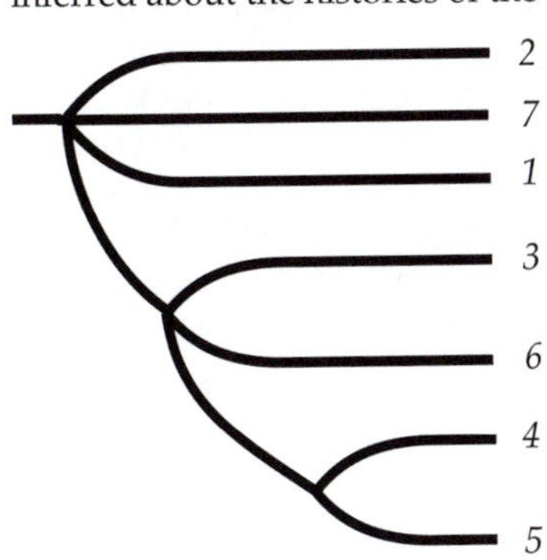

	Continent		
Species	A	B	C
1	✔	✔	✔
2	✔	✔	✔
3	✔	✔	
4		✔	
5		✔	
6	✔	✔	
7	✔	✔	✔

2. How does a knowledge of continental drift help explain the past and present distribution of the family of plants called proteads (family Proteaceae)?

3. Many species of large mammals were domesticated in Eurasia, but none in Africa, and only one in the Americas. Why? What were the consequences of this biogeographical pattern?

Answers

Knowledge and Synthesis Answers

1. **a.** Experimental results to explain the distribution and abundance of present-day species would be used more typically by an ecological biogeographer.

2. **c.** Because the number of endemic taxa in an area is related to the period of time that the area has been separated from other areas, we would expect that continent Y had always been separate and continent W had been separated first from the other two.

3. **d.** A vicariant distribution requires that the species once had a continuous distribution in what are now separate areas.

4. **c.** The snail is endemic to the group of islands, but not to any single one or pair of them, since it is also found on the remaining islands or island.

5. **b.** See Figure 57.4 in the text.

6. Ecological biogeography is mostly concerned with a local spatial scale and events occurring within the past few thousand years.

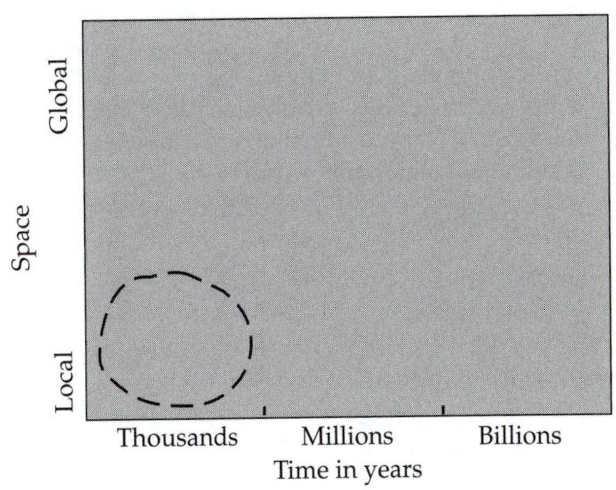

7. **d.** The extinction rate is nonzero and equals the arrival rate at n_1, which is therefore the equilibrium species number on the island.

8. **d.** Each curve corresponds to a group of similar-sized islands whose species number is plotted against their distance from the mainland. For each curve, species number decreases with distance from the mainland and the curve that is lowest relative to the vertical axis would be the group of smallest islands.

9. More mobile groups like insects and birds should reach an equilibrium species number on the island more quickly than the plants, and we would expect more insect species than bird species.

   c Birds
   b Plants
   a Insects

10. **c.** With less space available, populations of the different species would be smaller. This would subject them to higher extinction rates than would be expected on larger islands.

11. **b.** Species richness increases toward the equator. It is also higher in mountainous regions than in lowland areas, and is higher on mainland areas than on peninsulas.

12. **b.** If savanna vegetation is not grazed, browsed, or burned, it typically reverts to dense thorn forest.

13. **b.** Mostly coniferous, wind-pollinated and wind-dispersed tree species
   e. Leaves lost during dry season; agriculturally desirable land
   d. Cool winters, hot dry summers; maritime climate
   c. Succulent plants prominent; found at 30°N and 30°S latitudes
   a. Distribution altitudinally or latitudinally determined; permafrost present

14. **a.** Tundra
   d. Tropical evergreen forest
   c. Temperate deciduous forest
   b. Cold desert

15. The tropical evergreen forest biome has relatively constant **temperature**, but **rainfall** varies seasonally; the temperate deciduous forest biome has relatively constant **rainfall** (or precipitation, since snow may occur in winter) but **temperature** varies seasonally.

16. **d.** Even though freshwater habitats occupy a very small global area, they have about 10 percent of all aquatic species, which by proportion gives them a very large species richness.

17. *a* The ocean floor below the penetration of light
   l Coastal zone affected by wave action
   p Organisms in this zone must have adaptations to avoid sinking
   p, b Divided into a lighted and a dark region
   l Globally, the zone occupying the smallest total area

18. **b.** The most likely explanation for families of freshwater fish with representatives on several continents is that their lineages originated prior to the breakup of Laurasia or Gondwana.

19. **c.** Usually organisms adapted for one zone do not do well in a different zone.

Application Answers

1. Species *2*, *7*, and *1* represent the clade that first separated from the others. That each species in this clade is found on all these continents suggests that the three continents were once joined. The next clade includes species *3* and *6*, which are found on continents *A* and *B*. This suggests that the original supercontinent split to form two continents. One became continent *C*. Species *3* and *6* evolved on the other continent, which then split to form continents *A* and *B*. Species *4* and *5*, in the final clade, evolved on continent *B* after the split.

2. Proteads are found in South America, Africa, and Australia, but the African species are members of a specialized endemic subfamily. The distribution of the family suggests that they were broadly distributed in Gondwana before that continent began to break up. After separation, protead lineages on the three continents evolved major differences from one another.

3. Many large mammal species that happened to be found in Eurasia had several social characteristics that made them amenable to domestication: they lived in herds with male dominance hierarchies and were nonterritorial. All the large mammals of Africa lacked one or more of these traits. Domestication of these mammals provided several advantages to Eurasian cultures, the most obvious of which was increased efficiency of agricultural production. Others were the availability of horses for use in warfare and immunity to many diseases (such as smallpox and measles) that originated in domesticated species. When Europeans colonized the New World, these diseases devastated the indigenous peoples.

STUDY GUIDE

58 Conservation Biology

Important Concepts

Conservation biology is the scientific study of how to preserve the diversity of life.

- The science of conservation biology draws on concepts and knowledge from population genetics, evolution, ecology, biogeography, wildlife management, economics, and sociology.

Conservation biologists use several methods to estimate the rates of extinction.

- Estimates of current rates of extinction worldwide are based primarily on species-area relationships and rates of tropical deforestation.
- Rates of extinction have been much higher on islands than on the mainland during the past 400 years, but extinction rates on continents are also rising fast.
- To estimate the risk of extinction of a particular population, biologists study the interactions between its genetic variation, morphology, physiology, and behavior and its environment, both physical and biological.

There are many reasons to care about species extinctions.

- Many individual species are important because they supply useful products such as medicinal drugs. Extinction may eliminate many species before their potential uses can be investigated.
- Extinctions deprive us of opportunities to study the structure and functioning of communities.
- Extinctions lessen the aesthetic pleasure many people derive from interacting with other organisms. Extinctions caused by human activities also raise ethical issues.
- Ecosystem processes provide many services that can be replaced only by expensive and continuing human effort. Examples of such services are the maintenance of fertile soils, detoxification and recycling of waste products, control of agricultural pests, and pollination.

There are many causes of endangerment and extinction of species.

- Rare species—for example, those that live in a rare habitat or that occupy a high trophic level—are the most vulnerable to extinction, but common species can also become extinct.
- Habitat destruction is the most important cause of species extinction today. It is caused primarily by agriculture and forestry, which convert diverse natural communities into communities dominated by one or a few species, while diverting for human use about one-third of all net terrestrial primary production.
- As habitats are fragmented into small patches, populations of species that require large areas cannot be maintained.
- The proportion of a habitat patch subject to detrimental edge effects increases as patch size decreases. For example, species from surrounding habitats may invade the edges of the patch to compete with or prey upon the patch inhabitants.
- The persistence of species in small patches may be improved if the patches are connected by corridors of suitable habitat through which individuals can disperse.
- Exotic pests, predators, and competitors introduced by humans have driven many species to extinction.
- Overhunting was, until recently, the most important cause of extinction. Overexploitation for commercial reasons continues to threaten some species, such as elephants, orchids, and parrots.
- Loss of mutualists threatens some species. For example, the disappearance of pollinating species may cause some plant species to become extinct.
- Global warming may result in the extinction of species because new climates will develop and existing climatic zones will shift their location or disappear.

Appropriate management of endangered species can prevent their extinction.

- Designing successful recovery plans for endangered species requires an accurate diagnosis of the reasons they are endangered. Recovery plans often involve human manipulation of the species' biotic environments.
- For a small fraction of endangered species, captive propagation can help to prevent extinction by maintaining the species during critical periods and by providing individuals for reintroduction into the wild. The costs of captive propagation are modest compared to the costs of many other human activities.

Preserving a wide array of ecological communities and habitats is vital for species preservation.

- Parks, sanctuaries, and reserves maintain species and ecosystems relatively free of human disturbance. High-priority areas for parks and reserves are regions that have unusually high species richness or high numbers of endemic species.
- In most countries, new parks must be created in already settled areas. In addition to parks and reserves, lands that are now exploited for economic purposes must also play an important role in conservation.
- Ecoregions—areas that have a relatively uniform climate and a biota dominated by a group of widely distributed species—are used as the basis for large-scale conservation planning.
- Restoration of degraded ecosystems is an important component of conservation biology but is often difficult to accomplish.

The preservation of biodiversity depends ultimately on human value systems.

- Preserving genes, species, and habitats provides many current and future economic benefits, but these are difficult to measure precisely. If we exterminate a species, we have irreversibly destroyed a resource of unknown value.
- If species are valued only for their economic utility, increased losses of species are inevitable. Only when we value biological diversity and ecosystem functioning as the heritage of all humankind will we begin to reduce the rates of ecosystem destruction and species extinction.

The Big Picture

- Conservation biology provides a final example of the importance of the synthetic approach to science. In this case, knowledge and concepts derived from different areas of biology, as well as the social sciences, are being applied to the solution of the problem of species and ecosystem preservation
- Preserving biological diversity is considered by many informed individuals to be one of the most important challenges facing humanity in the twenty-first century.

Study Strategies

- The discussions of particular endangered species in this chapter are interesting and important in their own right, but in studying this material you should focus more on the general principles than on the specifics of the illustrative examples.
- Review the following tutorial and activity on the website:
  Tutorial 58.1 Habitat Fragmentation
  Activity 58.1 Concept Matching

Test Yourself

Knowledge and Synthesis Questions

1. The concepts of conservation biology come mainly from all of the following fields *except*
   a. ecology.
   b. evolutionary biology.
   c. population genetics.
   d. immunology.

2. The fynbos shrub community in South Africa
   a. is threatened by introduced species of taller, faster-growing plants.
   b. is a fire-adapted community.
   c. helps to maintain a regional supply of high-quality water.
   d. All of the above

3. Which of the following is *not* currently a major cause of the global reduction in biodiversity?
   a. Overexploitation
   b. Global warming
   c. Habitat destruction
   d. Introduction of foreign predators and disease

4. Based on species-area relationships, ecologists predict that
   a. 50 percent of Earth's species may become extinct in the next 50 years.
   b. a 90 percent loss of habitat will result in loss of about nine percent of the species living there.
   c. the area required by most species will need to be reduced.
   d. extinction is inevitable.

5. In the United States, the groups of organisms with the highest proportion of endangered or extinct species live in
   a. grasslands.
   b. the deciduous forest biome.
   c. freshwater habitats.
   d. deserts.

6. In the following graph, select the curve (*a*, *b*, or *c*) that *correctly* shows the expected relationship between habitat patch area and the proportion influenced by edge effects.

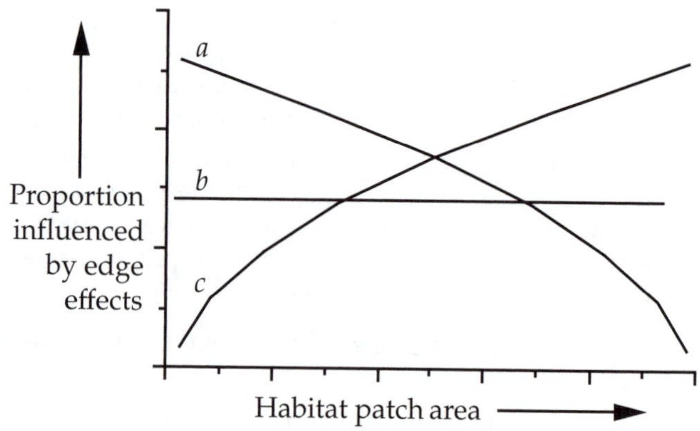

7. Why do conservation biologists believe that global warming may lead to extensive decimation of species?
   a. Since little change in plant community composition has occurred in the past, we cannot expect present communities to adapt to climate change.
   b. The magnitude of climate change will be much greater than past periods of climatic change.
   c. Many plant species may not be able to shift their ranges at the same pace as the northern movement of temperature zones.
   d. All of the above

8. Which of the following is endangered because of the loss of pollinating agents?
   a. Furbish's lousewort
   b. Madagascar rosy periwinkle
   c. American chestnut
   d. Hawaiian *Lobelia*

9. Two endangered species are Kirtland's warbler and the California sea otter. What do the recovery plans for these species have in common?
   a. The populations of both species were studied carefully to determine the reasons for their endangerment.
   b. The recovery plans for both species entail manipulation of their habitats.
   c. Captive propagation is an important part of the recovery plans of both species.
   d. All of the above

10. Which of the following statements about the successful reintroduction of the peregrine falcon is *not* true?
    a. The peregrine's habitat became more suitable after use of DDT was restricted.
    b. DDT adversely affected the peregrine's reproductive physiology.
    c. Captive propagation was an important part of the peregrine reintroduction effort.
    d. Reintroducing peregrines into habitats where they were not at the top of the food chain was critical to the success of the program.

11. Which of the following characteristics would cause a country to be considered a high-priority region for conservation efforts?
    a. A high degree of endemism
    b. Having low species richness
    c Having little natural habitat remaining
    d. All of the above

Application Questions

1. Aside from the aesthetic benefits of biodiversity, list three economic reasons people should care about species extinctions.
2. What are corridors, and how have experiments demonstrated that they help species persist in patchy environments?
3. Why are exotic species frequently a threat to the biological diversity of the region where they are introduced?
4. What are ecoregions? What is their significance for the preservation of biodiversity?
5. Why is cattle raising being encouraged as part of the effort to restore the tropical deciduous forest in Guanacaste National Park in Costa Rica?

Answers

Knowledge and Synthesis Answers

1. **d.** Although immunology is sometimes a useful tool in assessing the amount of genetic variation that exists in a population, its importance is less than the other fields listed.
2. **d.** See Figure 58.5 in the text.
3. **b.** Global warming is predicted to be a major cause of extinction in the future, but is not yet having an impact on populations.
4. **a.** Based on estimates of current and future disappearance of tropical forests, perhaps 50 percent of Earth's species will become extinct in the next 50 years. This loss is not inevitable if we reduce the rate at which these forests are converted to pasture and cropland.
5. **c.** Habitat destruction and pollution have caused extinction or endangerment of a very high proportion of aquatic freshwater species.
6. a. Since the edge of a habitat patch equals the perimeter of the patch, it is proportionally greater for smaller habitat patches. Therefore, the relationship between edge effect and habitat patch is an inverse relationship, as shown by curve *a*.
7. **c.** Although the expected magnitude of the climate change due to global warming may be similar to past climate changes, the rate of warming will be greater. This may make it impossible for many species to extend their ranges at the same rate as the northward movement of the temperature zones.
8. **d.** The declining population of the iiwi (*Vestiaria coccinea*), a Hawaiian honeycreeper, threatens the *Lobelia* plant, which has no other pollinator.
9. **a.** Protection of these two species involves very different measures. However, the recovery plans for both are based on a diagnosis of the reasons why they are endangered.
10. **d.** Peregrines were reintroduced into the same habitats from which they had disappeared due to the effect of DDT on eggshell thickness. Reintroduction took place after restricted use of DDT had made those habitats safe for peregrines.
11. **a.** A country with a high proportion of endemic species should receive high conservation priority because if the endemic species are lost there, they generally become completely extinct.

Application Answers

1. A list of economic reasons people should care about species extinctions would include the importance of natural products, ecosystems services, and nature tourism.

2. Corridors are relatively thin strips of habitat of a particular type that connect larger patches of the same type of habitat. Their importance in permitting individuals to disperse from one patch to another was demonstrated in the experiment described in Figure 58.11 in the text.
3. Native species that live in a particular community have evolved to cope successfully with the particular predators, competitors, and diseases that are part of its environment. Introduced species represent a change in the environment for which native species may not have been prepared by natural selection.
4. Ecoregions are large areas with fairly uniform climates and biotas dominated by groups of widely distributed species. They are useful to scientists as they formulate large-scale, long-term plans for the preservation of biological diversity. For example, scientists can identify sites within an ecoregion that are particularly important to protect for reasons such as their high species richness or their critical role as corridors for the movement of wildlife.
5. Domestic livestock help to keep the grasses cropped down so they cannot serve as fuel for extensive fires that damage adjacent forests. Cattle are also good seed-dispersers for some tree species. Cattle grazing will be discontinued after woody succession has reached a point at which fire is no longer a danger.